21st International Conference for Students and Young Scientists: Modern Techniques and Technologies (MTT'2015)

IOP Conference Series: Materials Science and Engineering
Volume 93

Tomsk, Russia
5 - 9 October 2015

ISBN: 978-1-5108-1464-6
ISSN: 1757-8981

Printed from e-media with permission by:

Curran Associates, Inc.
57 Morehouse Lane
Red Hook, NY 12571

CURRAN ASSOCIATES INC. proceedings.com

Some format issues inherent in the e-media version may also appear in this print version.

Printed by Curran Associates, Inc. (2016)

Institute of Physics
Dirac House, Temple Back
Bristol BS1 6BE UK

Phone: 44 1 17 929 7481
Fax: 44 1 17 920 0979

techtracking@iop.org

Additional copies of this publication are available from:

Curran Associates, Inc.
57 Morehouse Lane
Red Hook, NY 12571 USA
Phone: 845-758-0400
Fax: 845-758-2634
Email: curran@proceedings.com
Web: www.proceedings.com

Table of contents

Volume 93

21th International Conference for Students and Young Scientists: Modern Technique and Technologies (MTT'2015)
5–9 October 2015, Tomsk, Russia

Accepted papers received: 7 September 2015
Published online: 13 October 2015

Preface

Papers

SESSION 1: POWER INDUSTRY

SESSION 2: ELECTRONIC ENGINEERING

SESSION 3: MATERIALS AND TECHNOLOGIES NEW GENERATIONS

SESSION 4: CONTROL SCIENCE AND SYSTEM ENGINEERING

SESSION 5: PHYSICAL METHODS IN SCIENCE AND ENGINEERING

SESSION 6: DESIGN AND ARTISTIC ASPECTS OF ENGINEERING

SESSION 7: SOCIAL AND HUMANITARIAN ASPECTS OF ENGINEERING

IOP Conf. Series: Materials Science and Engineering **93** (2015) 011001 doi:10.1088/1757-899X/93/1/011001

Involving young researchers in the scientific process, and allowing them to gain scientific experience, are important issues for scientific development. The International Conference for Students and Young Scientists “Modern Technique and Technologies” is one of a number of scientific events, held at National Research Tomsk Polytechnic University aimed at training and forming the scientific elite.

During previous years the conference established itself as a serious scientific event at an international level, attracting members which annually number about 400 students and young scientists from Russia and near and far abroad.

An important indicator of this scientific event is the large number of scientific areas covered, such as power engineering, heat power engineering, electronic devices for monitoring and diagnostics, instrumentation, materials and technologies of new generations, methods of research and diagnostics of materials, automatic control and system engineering, physical methods science and engineering, design and artistic aspects of engineering, social and humanitarian aspects of engineering. The main issues, which are discussed at the conference by young researchers, are connected with analysis of contemporary problems, application of new techniques and technologies, and consideration of their relationship.

Over the years, the conference committee has gained a lot of experience in organizing scientific meetings. There are all the necessary conditions: the staff of organizers includes employees of Tomsk Polytechnic University; the auditoriums are equipped with modern demonstration and office equipment; leading scientists are TPU professors; the status of the Tomsk Polytechnic University as a leading research university in Russia also plays an important role. All this allows collaboration between leading scientists from all around the world, who are annually invited to give lectures at the conference.

The editorial board expresses gratitude to the Administration of Tomsk Polytechnic University (TPU Rector, Professor P.S. Chubik and Vice Rector for Research and Innovation, Professor A.N. Dyachenko) for financial support of the conference. Also, we heartily thank both chairmen of the conference sections and the organizing committee’s members for the great, effective, creative work in organizing and developing the conference as well as a significant contribution to the safeguarding and replenishment of the intellectual potential of Russia.

Organizing Committee

IOP Conf. Series: Materials Science and Engineering **93** (2015) 011001 doi:10.1088/1757-899X/93/1/011001

S. Panin	3rd Section Chairman, TPU, Tomsk, Russia
V. Vereschagin	
A. Malyshenko	4th Section Chairman, Professor, TPU, Tomsk, Russia
V. Rudnitskiy	
A. Potylitsyn	5th Section Chairman, Professor, TPU, Tomsk, Russia
M. Kukhta	6th Section Chairman, TPU, Tomsk, Russia
A. Dulzon	7th Section Chairman, TPU, Tomsk, Russia
E. Galanina	

Peer review statement

All papers published in this volume of *IOP Conference Series: Materials Science and Engineering* have been peer reviewed through processes administered by the proceedings Editors. Reviews were conducted by expert referees to the professional and scientific standards expected of a proceedings journal published by IOP Publishing.

Application of Hybrid Real-Time Power System Simulator for Designing and Researching of Relay Protection and Automation

Yu S Borovikov[1] **A O Sulaymanov**[1] **and M V Andreev**[1]
[1] *634050, Tomsk, Russia*, National Research Tomsk Polytechnic University

E-mail: andreevmv@tpu.ru

Abstract. Development, research and operation of smart grids (SG) with active-adaptive networks (AAS) are actual tasks for today. Planned integration of high-speed FACTS devices greatly complicates complex dynamic properties of power systems. As a result the operating conditions of equipment of power systems are significantly changing. Such situation creates the new actual problem of development and research of relay protection and automation (RPA) which will be able to adequately operate in the SGs and adapt to its regimes. Effectiveness of solution of the problem depends on using tools – different simulators of electric power systems. Analysis of the most famous and widely exploited simulators led to the conclusion about the impossibility of using them for solution of the mentioned problem. In Tomsk Polytechnic University developed the prototype of hybrid multiprocessor software and hardware system – Hybrid Real-Time Power System Simulator (HRTSim). Because of its unique features this simulator can be used for solution of mentioned tasks. This article introduces the concept of development and research of relay protection and automation with usage of HRTSim.

1. Introduction

Currently, task of development, research and exploitation of smart grids (SG) is one of the highest priorities task in power engineering. The development of SG associated with the solution of sophisticated problems focused on radical increasing of controllability and observability of electrical networks and power systems by usage of FACTS (Flexible Alternating Current Transmission Systems).

Planned large-scale integration of high-speed FACTS devices which are used for voltage regulation, power flows and etc. in quasi-steady-state and transient normal and abnormal operating regimes is greatly complicates are dynamic properties of power systems. As result the operating conditions of equipment of power systems are significantly changing.

Such situation creates the new actual problem of development and research of relay protection and automation (RPA) which will be able to adequately operate in the SGs and adapt to its regimes.

Effectiveness of solution of the designated problem depends on using tools. The development and research of RPA requires a large amount of information about the regimes and processes in power systems. Because of it simulators are main tools for such task. Alternative for mathematical simulation in this case is absent because of commonly understood reasons.

[1] M V Andreev, andreevmv@tpu.ru .

The "ideal" simulation system must meet the following requirements:

- ensure satisfactory simulation's accuracy;
- usage of high-adequate three-phase mathematical models of all primary and auxiliary equipment of SG;
- lack of decomposition of regimes and processes in primary and auxiliary equipment of SG;
- lack of limit of simulation time-interval;
- flexibility of simulated power system scheme;
- availability of integration of RPAs algorithms;
- possibility of physical RPA connection for closed-loop testing;
- possibility of simulation of big (real dimension) power systems.

The complex of these requirements allows removing from the list of "suitable" tools all software simulators. Among hardware and software simulators the most famous and widely exploited are digital simulators – RTDS [1, 2] and eMEGASim [3, 4]. But these devices cannot provide the key requirement – guarantee satisfactory simulation's accuracy, because of usage of numerical integration methods [5, 6]. Moreover, due to the presence of restrictive conditions [5-8]: solutions of the differential equation must satisfy the Lipchitz's condition, the length of the interval solutions is limited by Dahlquist's theorem and others; in most cases we have compromise: to limit the order of differential equations system or the interval of integration, or the total simulation time. The increasing of dimension of the simulated SG necessarily leads to simplification of the power equipment models. Because of it the adequacy and accuracy of the simulation's results are reduced. The simplification (equivalenting) of simulated power system significantly reduces the adequacy of testing and researching of RPA.

Analysis of numerous works devoted to the numerical simulation of power systems and the numerical integration methods shows that achieved results generally remains within the existing theory of discretization methods for ordinary differential equations. Any prospects of fundamental positive changes and deformations of these bounds are absent.

In Tomsk Polytechnic University developed the prototype of hybrid multiprocessor software and hardware system – Hybrid Real-Time Power System Simulator (HRTSim) (Figure 1) [9-13]. Inherent in HRTSim principles allows eliminating methodological error of solving of the elements and SG models, regardless of the differential order, stiffness and time-interval. Therefore, the accuracy of the solution is guaranteed and is determined only by instrumental error of simulator's hardware part. This error is can be minimized by usage of integral precision components.

2. Application of HRTSim for designing and researching of relay protection and automation

However, current version of this simulator not allows implementing the requirement of flexibility of simulated SG scheme. Also there is problem with requirement # 6: user cannot integrate any algorithms or models in HRTSim without source code of its software. These shortcomings, however, are not insurmountable.

In case of elimination of designated shortcomings HRTSim can serve as the universal platform for RPA developing (Figure 2), including wide-area protection systems (WAPS), as well as for its testing and researching, including closed-loop testing (Figure 3).

The presence of steps 1 and 2 (Figure 2) do not require, according to the authors, any comments.

The step 3 involves the initial verification of developed RPA's algorithms. Usage of HRTSim is possible here, however, this stage of development, usually accompanied by a large number of tests and adjustments, which requires frequent usage of the SG model. In the presence of a plurality of parallel developments and only one HRTSim obvious difficulties may arise. And although the simulator allows multi-user mode, this requires concerted actions. If such problem is absent, the apparent barriers for HRTSim using on the stage of pilot studies are absent too. But for this designated shortcomings must be eliminated.

The step 4 allows checking of the developed and debugged RPA's algorithms in operating conditions close to reality for elimination of any errors which was not resolved on step 3.

The step 6 is the last stage which involves usage of the HRTSim. Physical currents and voltages generated in the simulator during a simulation process are sent to the analog outputs of HRTSim and from there, by means of external amplifiers, can be transmitted to the analog inputs of tested RPA.

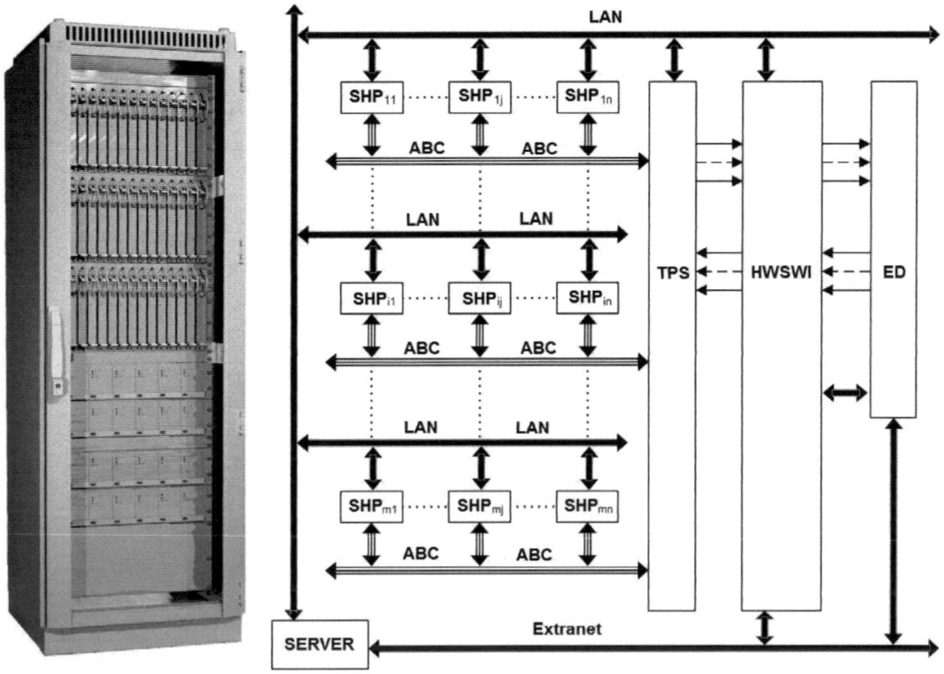

Figure 1. HRTSim external view and structural scheme: SHP - specialized hybrid processors; LAN – local area network; TPS – three-phase switch; HWSWI – hardware/software interface; ED – external device; ABC – three-phase input/output of SHP.

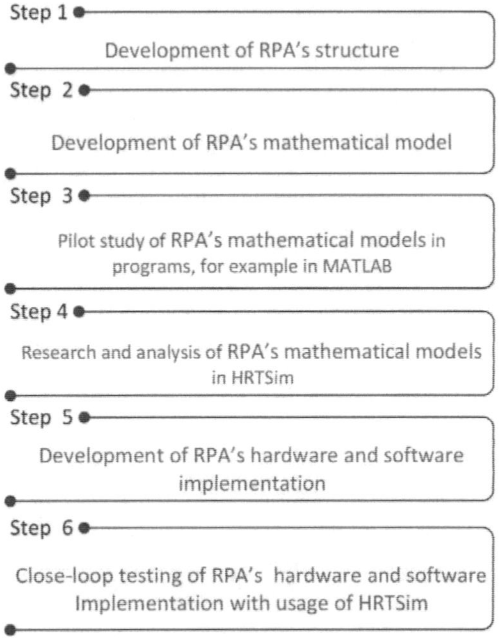

Figure 2. The proposed scheme of RPA's development with usage of HRTSim.

I want to emphasize once again that the efficiency of research on the steps 4 and 6 will be maximized by using of real-dimension SGs model and accordingly the minimum amount of simplifications and restrictions.

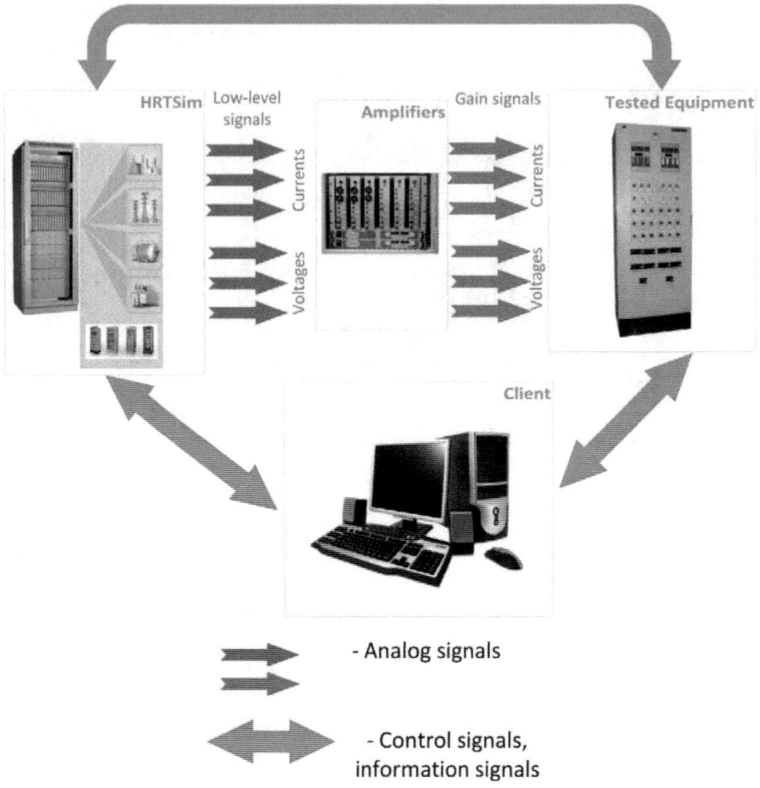

Figure 3. Close-loop testing scheme.

It should be noted that the HRTSim allows integrating mathematical models of RPA devices, adequately and reliably reproducing processes in their specific implementations and measuring current and voltage transformers – detailed models [14]. Usage of such models increasing the effectiveness of testing of RPA devices, including the close-loop testing. It is possible because of a detailed analysis of the key functional units of RPA, as well as tracking the reaction of a real device and the model to changes in SG and reliably identifies the causes of inconsistencies in the case of its occurrence.

3. Conclusion

In conclusion, I want to emphasize that the "foundation" laid in HRTSim has a unique potential and after additional work the simulator may be used as the universal platform for research and development of any RPA devices, including wide-area protection systems (WAPS).

Acknowledgment
This work was supported by mega-grant "Hybrid simulation and control of smart grids".

References

[1] Qi L, Woodruff S and Steurer M 2007 Study of power loss of small time-step VSC model in RTDS *Proc. Int. Conf. Power Engineering Society General Meeting (Tampa)* (Washington: IEEE) pp 1–7.
[2] Haibo Z and Dandan G 2012 The research and implementation of experimental simulation

platform based on RTDS and EMS (OPEN-3000) *Proc. Int. Conf. Innovative Smart Grid Technologies - Asia (ISGT Asia) (Tianjin)* (Beijing: IEEE) pp 1–4.
[3] Paquin J-N, Belanger J, Snider L A, Pirolli C and Li W 2009 Energy Conversion Congress and Exposition *Proc. Int. Conf. ECCE 2009 (San Jose)* (Washington: IEEE) pp 3194–202.
[4] Jalili-Marandi V, Robert E, Lapointe V and Belanger J 2012 A real-time transient stability simulation tool for large-scale power systems *Proc. Int. Conf. Power and Energy Society General Meeting (San Diego)* (Washington: IEEE) pp 1–7.
[5] Hall J and Watt M 1976 Modern numerical methods for ordinary differential equations (Oxford: Clarendon Press).
[6] Butcher J C 2009 Numerical methods for ordinary dfferential equations: early days in the birth of numerical analysis (New Jersey: World Scientific).
[7] Gusev A S, Gurin S V, Svetchkaryov S V and Plodisty I L 2004 Hybrid methods as a way out of electric power system modeling problem *Proc. Int. Symp. KORUS 2004 (Tomsk)* vol 1 (Tomsk: Tomsk Polytechnic University) pp 215–19.
[8] Gusev A S, Svechkarev S V and Plodisty I L 2005 Basic aspects of modeling problem for electrical power systems, perspectives and methods of their solution *Proc. Int. Symp. KORUS 2005(Novosibirsk)* (Novosibirsk: Novosibirsk State Technical University) pp 322–26.
[9] Borovikov Yu and Sulaymanov A 2013 Information and control system for use in the real time multiprocessor simulation of power equipment *Russian Electrical Engineering* **84** 290–95.
[10] Borovikov Yu S, Gusev A S, Sulaymanov A O and Ufa R A 2014 Synthesis of a hybrid model of the VSC FACTS devices and HVDC technologies *Proc. Int. Conf. Modern Techniques and Technologies (MTT 2014) (Tomsk)* (London: Institute of Physics Publishing) pp 1–6.
[11] Borovikov Yu S, Sulaymanov A O, Gusev A S and Andreev M V 2014 Simulation of automatic exciting regulators of synchronous generators in hybrid real-time power system simulator *Proc. Int. Conf. Systems and Informatics (ICSAI 2014) (Shanghai)* (Beijing: IEEE) pp. 153–58.
[12] Borovikov Yu S, Prokhorov A Vand Andreev M V 2012 Application of Hybrid Real Time Simulator for solution of Smart Grid tasks on the example of Elgaugol energy cluster pilot project *Proc. Int. Conf. IFOST 2012 (Tomsk)* vol 2 (Tomsk: Tomsk Polytechnic University) pp 604–08.
[13] Borovikov Yu S, Gusev A S, Sulaymanov A O and Ufa R A 2014 Hybrid real-time simulator of power system for advanced simulation of the FACTS and HVDC system based on Voltage Source Converter *Proc. Int. Conf. Systems and Informatics (ICSAI 2014) (Shanghai)* (Beijing: IEEE) pp 148–52.
[14] Ruban N Yu, Borovikov Yu S and Sulaymanov A O 2014 Simulation of transmission lines phase-comparison protection for optimization its settings *Proc. Int. Conf. IFOST 2014 (Cox's Bazar)* (Bangalore: IEEE) pp 264–67.

Thermal analysis of IRT-T reactor fuel elements

A Naymushin, Yu Chertkov, I Lebedev and M Anikin
National Research Tomsk Polytechnic University, TPU, Tomsk, Russia

E-mail: agn@tpu.ru

Abstract. The article describes the method and results of thermo-physical calculations of IRT-T reactor core. Heat fluxes, temperatures of cladding, fuel meat and coolant were calculated for height of core, azimuth directions of FA and each fuel elements in FA. Average calculated values of uniformity factor of energy release distribution for height of fuel assemblies were shown in this research. Onset nucleate boiling temperature and ONB-ratio were calculated. Shows that temperature regimes of fuel elements at rated power of the reactor keep within limit values and meet safety requirements. Obtained results could be applied at feasibility study of increasing thermal power of IRT-T research reactor to 12 MW level.

1. Introduction

IRT-T research reactor uses IRT-3M type fuel assemblies in the reactor core [1]. The technical report of IRT-3M safety provision gives the values of the following parameters:

- inlet coolant temperature — 45 °C;
- outlet pressure — 133 kPa;
- pressure drop — 37 kPa.

The values of nonuniformity of energy release at ~220 kW/m^2 average heat flux density can reach ~2 times higher than the maximum values due to sufficiently high heterogeneity of materials arrangement in the reactor core. Therefore, in order to establish correspondence of thermal parameters of fuel elements to passport values it is needed to determine values of nonuniformity of energy release over fuel assemblies and reactor core. For this purpose, it was used MCU-PTR software designed for precision neutron-physical calculations [2]. Distributions of energy release were obtained as a result of three-dimensional calculations, also, the most stressed fuel assemblies and most stressed areas in them (directions) were selected.

2. The method of calculation of thermal-physical parameters

Thermal analysis was carried out for steady state of reactor core at rated power (6 MW) in accordance with specially designed method [3].

The distributions of energy release over calculation cells, height and azimuth of all fuel assemblies of reactor core were obtained as a result of neutron-physical calculation. 8-tube and 6-tube fuel assemblies with the most stressed parameters were used for the calculation. These distributions were input data to carry out thermal analysis.

Coolant temperature distribution and temperature distribution of fuel element's wall over the height of the most stressed fuel assemblies were calculated within approximation of one-dimensional cylindrical channel with coaxial fuel elements. Inlet coolant temperature, pressure and pressure drop in reactor core were set constant [4].

The results of calculation over height areas of the most stressed azimuth and height direction of fuel assemblies determined:

- coolant temperature between fuel elements;
- fuel element's wall temperature;
- temperature of cladding-core border of fuel elements;
- maximum temperature of fuel element's core;
- surface boiling point;
- surface boiling safety margin;
- heat fluxes from fuel element's surface;
- coolant heating over each ring area between fuel elements.

3. Calculation models for fuel assemblies

Thermal regime of fuel assemblies was calculated by using ASTRA software [5], designed for calculation of fuel assemblies with tubular coaxial fuel elements. The software makes it possible to calculate distribution of heat fluxes from fuel element's surface, distribution of coolant temperatures in the gaps between fuel elements, distribution of outer surface temperatures of fuel elements, distribution of temperatures of a contact between cladding and core of the fuel elements, as well as distributions of maximum temperatures of fuel elements' cores over the height of reactor core. The software also allows carrying out calculation for direct flow as well as for Field scheme flow of the coolant.

Distributions of temperature over the volume of cylindrical models for the most stressed fuel assemblies in (r,z)-geometry were obtained by using ASTRA software. The influence of azimuth unevenness of energy release in these fuel assemblies was estimated by using distributions of energy release of the most stressed directions in the model.

The calculation of three-dimensional fuel assembly was carried out by using two-dimensional ASTRA software with the following conditions:

- Perimeters of cross-section, cladding thickness and core thickness should be preserved since heat flux is moving across fuel elements' walls. This condition cannot be fulfilled for all the gaps and fuel elements. However, since in the most stressed conditions generally works the outer fuel element, its perimeters of cross-section were preserved along with cladding thickness, core thickness and thickness of the surrounding gaps with the coolant.
- Thicknesses of areas were preserved for the rest fuel elements and gaps.
- Azimuth unevenness of energy release distribution was taken into account as follows: radial distributions of energy release over eight azimuths with 45° step (see Figure 2, 3) was obtained; averaged radial distribution of energy release was used to calculate coolant heating; the distribution with maximum irregularity factor was used for calculation of the most stressed parameters [6].

4. Calculation results

During the reactor operation, the fuel burns up and distributions of energy release over reactor height and over cross section of reactor core are evened. Therefore, maximum values of heat flux densities from fuel elements' surface, maximum values of fuel and coolant temperature should be observed in "fresh" fuel assembly [7].

Averaged calculation values of irregularity factors of energy release distribution over the fuel assembly height are:

- for eight-tube — 1,28;
- for six-tube — 1,27.

Average value of specific energy release in fuel element's core for the most stressed eight-tube fuel assembly is 1.093 kW/cm^3 (for six-tube fuel assembly – 0.974 kW/cm^3).

The boiling point (T_b) of coolant is 114.5 °C at the outlet coolant pressure of 1.33 bar. Maximum value of coolant temperature does not exceed 63.4 °C when the coolant of eight-tube fuel assembly is heated up to average 9.4 °C. Maximum values of fuel temperature (82.5 °C), cladding temperature (81.0 °C) and maximum values of heat flux density from fuel elements to coolant — 487 kW/m^2 (on the surface of the outer fuel element) are achieved in the outer fuel element, which has the highest specific power in fuel assembly. This value of heat flux density is much lower than designed value, which is 800 kW/m^2.

Figure 1 shows the ONB ratio for the outer fuel element of eight-tube fuel assembly at inlet coolant temperature 45 °C.

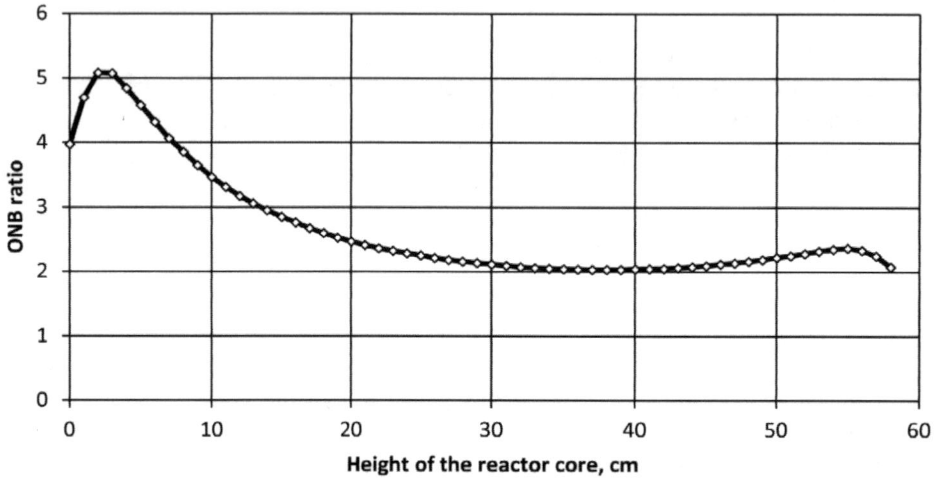

Figure 1. ONB ratio for the outer fuel element of the most stressed eight-tube fuel assembly during the first cycle at inlet coolant temperature 45 °C.

Unlike energy release distribution over "fresh" reactor core, in the real cartograms of fuel loading for IRT-T reactor the amount of "fresh" fuel assemblies generally does not exceed 1-2 pcs. In the irradiated fuel assemblies the burnup lies in the range 5–60 %, thereby, there is not only reduced power in these assemblies but also significantly reduced irregularity factors of energy release. Therefore, in the real cartograms of loading maximum values of heat flux density and fuel temperature are found in "fresh" fuel assemblies with zero burnup.

As an example the cycle was considered which has irradiated fuel assemblies as well as 2 "fresh" ones. Figure 2 shows burnup distribution over reactor cells and designed energy release distribution obtained by means of MCU-PTR software.

Average ^{235}U burnup over reactor cells (% of initial loading)			
10.07	53.31	54.53	11.20
0.00	60.09	54.03	60.22
51.90			56.98
53.16			56.08
0.00	60.93	55.63	58.90
10.37	54.19	54.36	11.01

Distribution of the density of number of fission over reactor core cells (normalized to the average value), rel. u.			
1.21	0.89	0.86	1.21
1.26	0.80	0.90	0.74
1.08			1.08
1.08			1.08
1.31	0.81	0.87	0.74
1.24	0.90	0.86	1.20

Figure 2. Burnup and energy release (normalized to the average value) distributions over reactor core cells at the beginning of the third cycle.

As in the case of "fresh" reactor core, the maximum values of relative power are ~1.3 and found in "fresh" fuel assemblies with zero burnup, which, in this case, are six-tube assemblies. Azimuth distribution of energy release for the most stressed six-tube fuel assembly is shown in Figure 3.

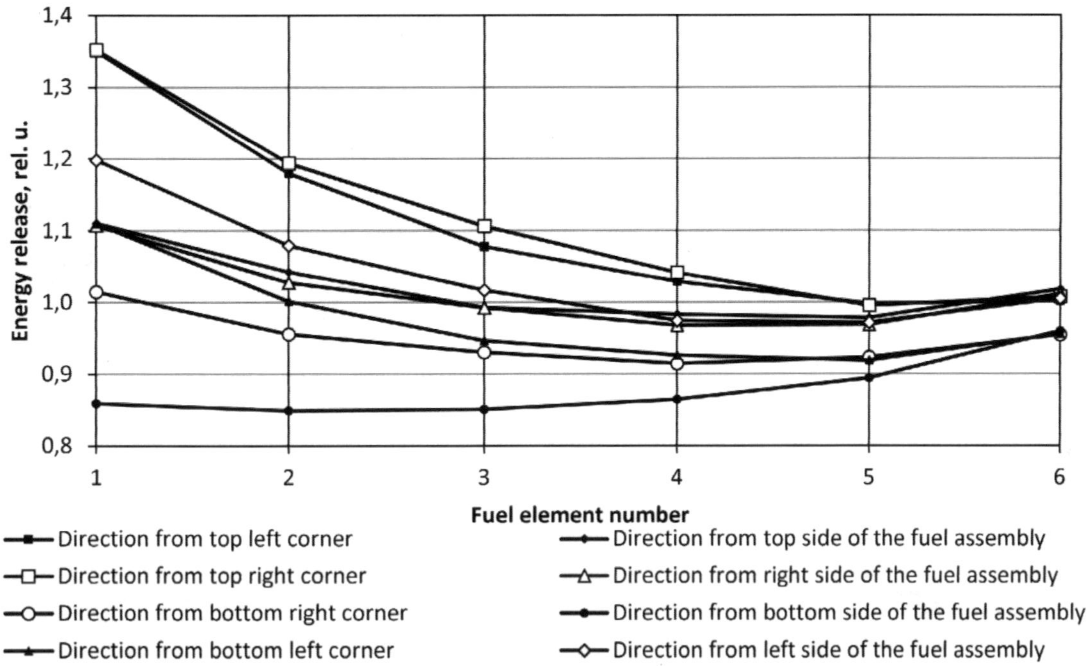

Figure 3. Radial distributions of energy release for six-tube fuel assembly over different directions (normalized to average value over fuel assembly, count from inner to outer fuel element).

Figure 4 shows distribution of ONB ratio for the outer fuel element of six-tube fuel assembly at inlet coolant temperature 45 °C.

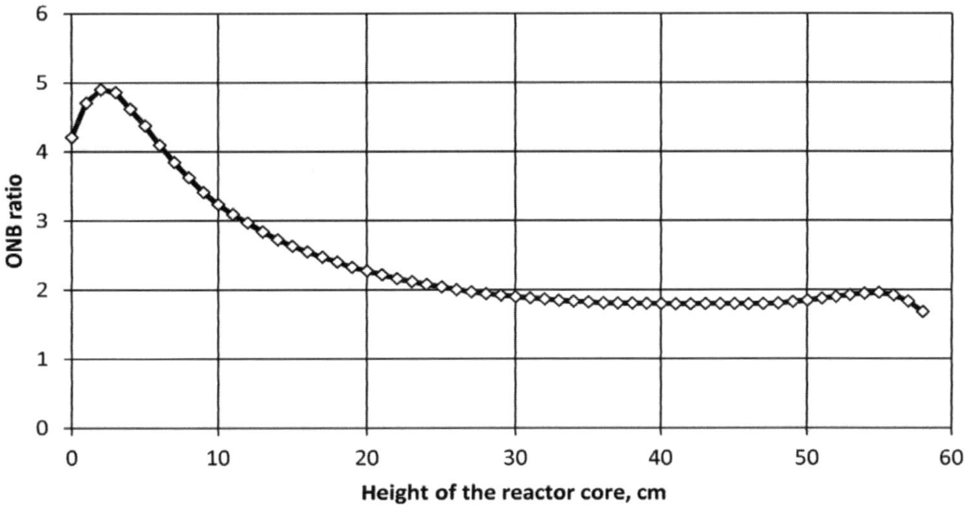

Figure 4. ONB ratio at inlet coolant temperature 45 °C for the outer fuel element of the most stressed six-tube fuel assembly during the 3rd cycle.

The maximum value of specific energy releases in fuel element's core for six-tube fuel assembly is 2.69 kW/cm^3. At inlet coolant temperature 45 °C the maximum value of coolant temperature in given azimuth direction does not exceed 67.2 °C. The outer fuel element has the highest specific power in fuel assembly, therefore, maximum values of fuel temperature (87.4 °C), cladding temperature (85.6 °C) and maximum values of heat flux density from fuel elements to coolant — 538.3 kW/m^2 are achieved in this outer fuel element

Coolant ONB ratio is 1.8.

5. Conclusion

This paper describes the method of determination of thermal-physical and energy parameters for IRT-T reactor. The calculation of thermal-physical regimes of fuel elements was carried out during different cartograms of fuel loading. It was shown that at IRT-T reactor rated power (6 MW) temperature regimes of fuel elements keep within the limit values and meet safety requirements of reactor operation. Maximum values of fuel elements' surface temperature and fuel elements' cores at 6 MW reactor power does not exceed 86 °C. Maximum coolant heating in the gaps between fuel elements is 15 °C and maximum heat flux density from fuel element's surface does not exceed 540 kW/m^2.

One of the reactor safety criteria is that in case of not exceeding the designed value of relative power of the fuel assembly (which is 1.3 in "fresh" fuel assembly) heat flux density in its most stressed area does not exceed 540 kW/m^2 and ONB ratio is not lower than 1.8.

Acknowledgment

This work was performed on the unique scientific IRT-T equipment and financially supported by Government represented by the Ministry of Education and Science of the Russian Federation (RFMEFI59114X0001).

References

[1] Varlachev V A, Glukhov G G and Skuridin V S 2011 *Research Nuclear Reactor IRT-T*, Tomsk Polytechnic University (Tomsk:TPU Press)

[2] Alekseev N I and Gomin E A, 2011 MCU-PTR Program for High-precision Calculations of Pool and Tank Type Research Reactors, (Atomic En. Vol 109/3) ed N N. Ponomarev-Stepnoi (Moscov: Atomnaya Energiya Press) pp 149–158

[3] Naymushin A G, Chertkov Yu B, Lebedev I I and Anikin M N, 2013 Method and Results of Thermo-physical Calculations of IRT-T Fuel Elements (Proc. of Institution of Higher Education, Physics vol 11/2), (Tomsk:TSU Press) pp 360–366

[4] Nagy, Mohamed E., and Amgad M. Shokr. Hot channel factors evaluation for thermalhydraulic analysis of MTR reactors. Nukleonika 49.2 (2004): 61-67.

[5] Taliev A V, 2006 Modernized ASTRA code for thermal calculation of research reactor FAs consisting from tubular coaxial fuel elements (Preprint IAE-6405/5, Moscow)

[6] Zagar, Tomaz, M. Ravnik, and A. Peršic. "Analysis of TRIGA reactor thermal power calibration method." International Conference Nuclear Energy in Central Europe. 1999. pp. 91-98

[7] Naymushin A G, Chertkov Yu B, Lebedev I I and Elpashev S N, 2012 Feasibility Study of Energy Realise Distributions in IRT-T Reactor for Different Control Rods Positions (Proc. of Institution of Higher Education, Physics vol 11/2), (Tomsk:TSU Press) pp 353–359

Numerical Simulation of Physical and Chemical Processes in Fluidized Bed

D A Baturin[1] and A V Gil[1]
Steam generating machinery engineering department, Tomsk Polytechnic University, Lenin Avenue, 30, 634050, Tomsk, Russia

E-mail: gutair@sibmail.com

Abstract. The paper presents a numerical simulation of the furnace with a circulating fluidized bed. Numerical study carried out for the bottom of the combustion chamber with the varying heights of volume filling. The results contours of particulate matter concentration and of velocities, as well as a graphical representation of changes in the concentration of particles on the bed height are shown. Simulation performed in Eulerian - Eulerian representation on a 2D model.

1. Introduction

Currently, much attention is paid to technology for to achieving low harmful emissions during effective combustion of low-quality solid fuels. Boilers with circulating fluidized bed (CFB), widespread in the energy sector where fuel is supplied from both the burners and of returning ash chute, became in the bottom of the furnace there is a fluidized bed where a constant particle removal is organized [1].

High-quality design of large stakes of a circulating fluidized bed depends on the analysis of dynamic processes in the furnace. Experimental studies of combustion processes are limited to local areas. Time-consuming for testing and incomplete information about the conditions of functioning the real object as opposed to the numerical simulation of the interest in which is constantly increasing.

Formation of the structure, particle distribution, the speed of solids and other physical and chemical processes are of scientific interest because insufficiently studied objects in the major energy [2, 3]. At present, special attention is paid to two and three-dimensional modeling using Euler-Lagrangian and Eulerian-Eulerian approach.

The object of investigation is selected CFB boiler height of 37.74m and a depth of 4m at the bottom of the furnace and 8m in the main. The boiler is equipped with 8 burners and 2 rows of air nozzles 4 in a row.

Side wall bottom portions form ramps for the purpose of narrowing the furnace and increasing the thermal stress in the fluidized bed. The walls of the furnace in the zone of high thermal stresses are lined with an abrasive material. The upper part of the furnace is divided into 8 flues equipped with inertial separators. Some of the fuel and the ash is returned to the lower part of the furnace after eight burners arranged oppositely four on each side wall. The combustion chamber is equipped with 48 air jets arranged oppositely on the side walls of the furnace in three tiers (figure 1).

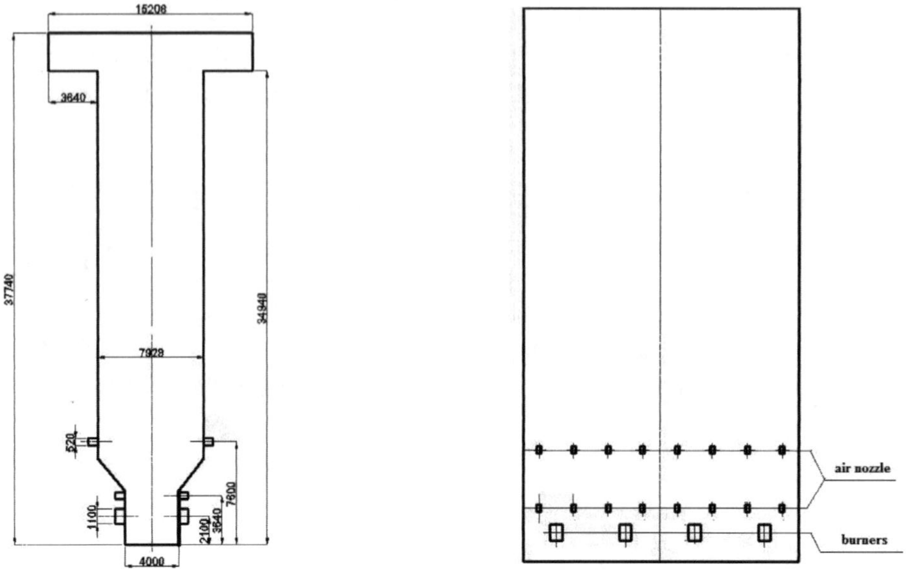

(a) (b)

Figure 1. Sketch of the combustion chamber: (a) – a cross-section; (b) – side view.

2. Mathematical model

Mathematical modeling of combustion devices is one of the most important ways of obtaining information about the aerodynamics, the local and the total heat transfer. Obtained information it is necessary for the design and evaluation of new thermal energy units [4, 5].

Simulation of the gas phase was carried out by Eulerian-Eulerian. The continuity equation for gas phase is

$$\frac{\partial(\varepsilon_g \rho_g)}{\partial t} + \frac{\partial}{\partial x_j}\left(\varepsilon_g \rho_g u_{gj}\right) = 0 \tag{1}$$

Here is ε_g – bed voidage; ρ_g – density of gas phase (kg/m^3); t – tracer sampling time (s); u_{gj} – gas velocity (m/s); x_j – direction coordinate.

The continuity equation for particulate phase, is written as

$$\frac{\partial(\varepsilon_s \rho_p)}{\partial t} + \frac{\partial}{\partial x_j}\left(\varepsilon_s \rho_p u_{pj}\right) = 0 \tag{2}$$

Here is ε_s – solid fraction; ρ_p – particle density (kg/m^3; u_{pj} – particle velocity (m/s).

The conservation of momentum for the gas phase is described by

$$\frac{\partial}{\partial t}\left(\varepsilon_g \rho_g u_{gi}\right) + \frac{\partial}{\partial x_j}\left(\varepsilon_s \rho_g u_{gj} u_{gi}\right) = -\varepsilon \frac{\partial p}{\partial x_i} + \frac{\partial \tau_{g,ij}}{\partial x_j} - \beta\left(u_{gi} - u_{pi}\right) + \rho_g \varepsilon_g g_i \tag{3}$$

Here is τ – stress tensor (Pa); β – inter-phase momentum exchange coefficient (kg/(m3 s)).

The stress tensor of gas phase is given by

$$\tau_{g,ij} = \mu_g \left(\frac{\partial u_{gj}}{\partial x_i} + \frac{\partial u_{gi}}{\partial x_j} \right) \tag{4}$$

Here is μ – viscosity (Pa s).

Table 1 shows the boundary conditions necessary for the mathematical modeling.

Table 1. The boundary conditions.

Fuel mass flow Bg, kg / s	1.89
Fuel type	anthracite
Excess air ratio	1.2
The diameter of the coal particles, mm	2.5
Excess air ratio	1.2
Air temperature, °C:	
Burners inlet	460
Air nozzles inlet	300
Air velocity, м/s:	
Burners inlet	1.5
Air nozzles inlet	7.7

The study was carried out using the program ANSYS Fluent 12.1.4. For calculation simplication productivity and the two-dimensional computational grid was used (figure 2).

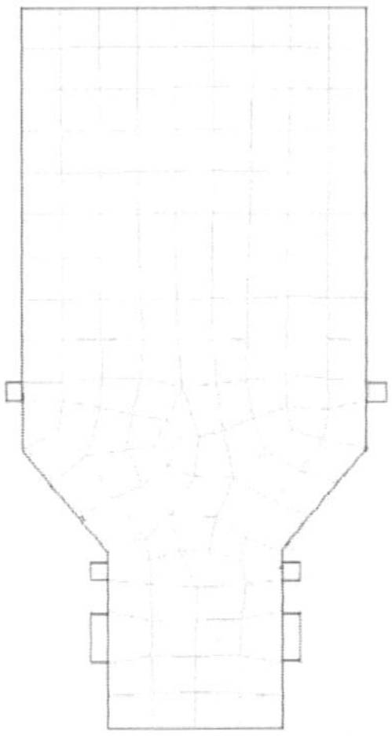

Figure 2. Estimated Net furnace.

3. Results of the study

Figures 3 and 4 show the concentration of fuel particles, fractions of volatilizes and concentration change with the height of the combustion chamber for variants of fuel loading at 1.2m, 2.4m, 4m.

Figures 3,a and b show that when loading the fuel to a height of 1.2 m of the solid phase rate reaches the average value of 0.8m/s. Concentration of the fractions of fuel reaches a maximum value of 0.47. Comparing Figures 3,c and d to Figure 3, it is possible to see that there is output of fuel fractions from a lower portion of the combustion chamber. The average solids velocity is 0.85 m/s.

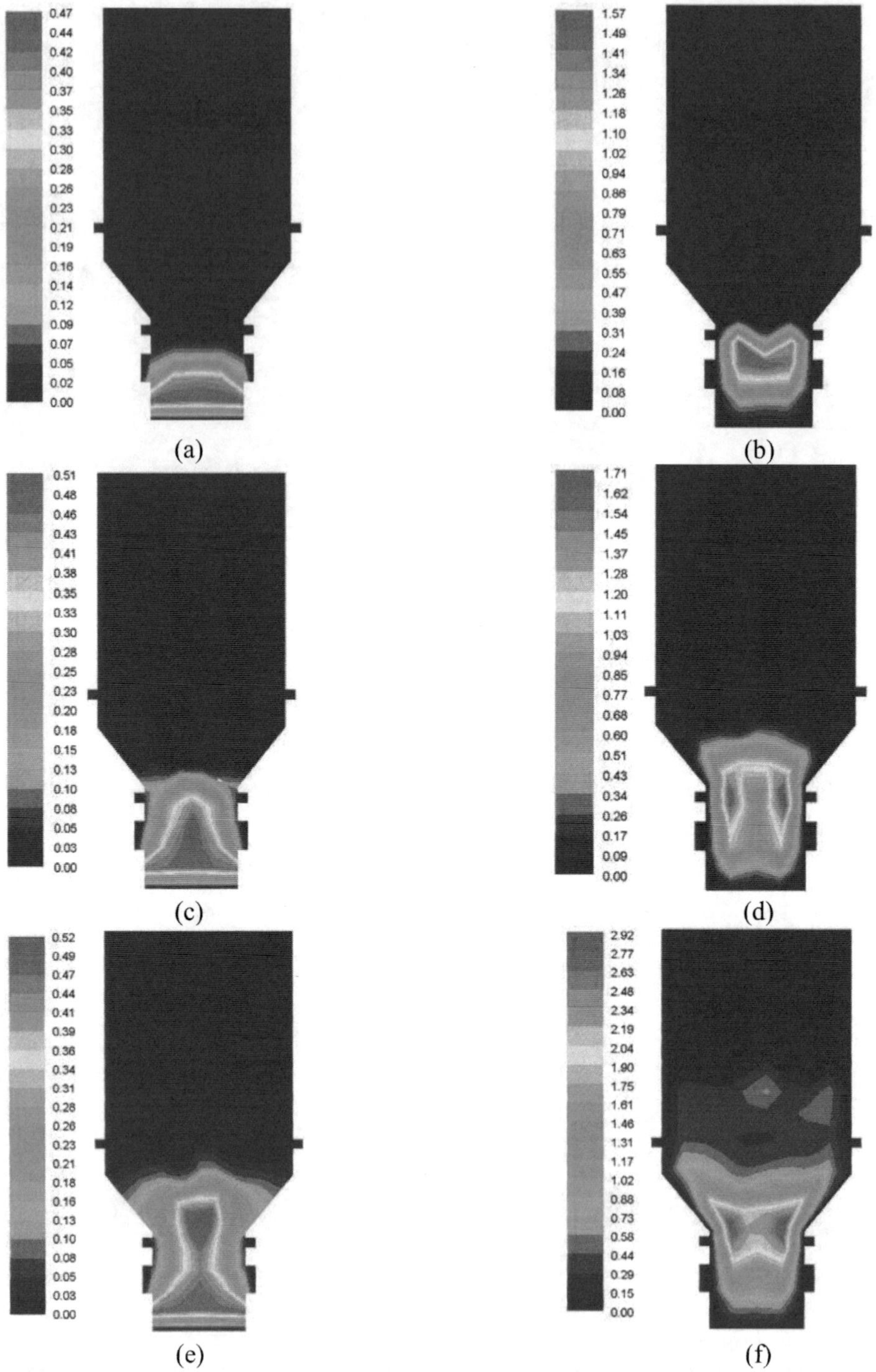

Figure 3. Results of numerical simulation: (a) – the concentration of the fractions of fuel in a fluidized bed when loading fuel at 1.2m; (b) – solids velocity when loading fuel at 1.2m; (c) – the concentration of the fractions of fuel in a fluidized bed when loading fuel at 2.4m; (d) – solids velocity when loading fuel at 2.4m; (e) – the concentration of the fractions of fuel in a fluidized bed when loading fuel at 4m; (f) – solids velocity when loading fuel at 4m.

The most rational pattern can be observed in figures 3,e and f. The highest concentration is focused on the center of the combustion chamber and exposed to air, both below and on the sides of the air nozzles, climbs up the combustion chamber. The maximum speed of the solid phase is observed in the vicinity of the first row of air nozzles and reaches 2.92m/s.

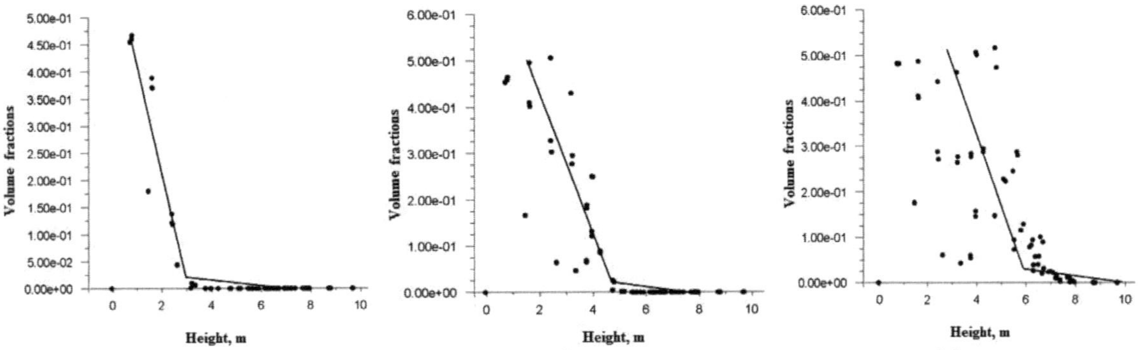

(a) (b) (c)

Figure 4. Change in concentration depending on the altitude: (a) – loading fuel at 1.2m; (b) – loading fuel at 2.4m; (c) – loading fuel at 4m.

According to figure 4 it can be concluded that the more fuel loading, the higher the height distributed fuel fractions. These visual results of a series of numerical modeling parameters investigated were prepared using Eulerian.

4. Conclusion

The most rational pattern is observed when loading the fuel to a height in the range of 2.4 m - 4 m as when loading fuel at a lower height is difficult to draw any conclusions on the dynamic processes. This range allows that race gas dynamic processes in the furnace of a fluidized bed boiler. Theredemolitionfuelfractionsclosertothecenterofthebottomofthecombustionchamberanditssubsequentdistributionadjustment. Speed also remain with in the range limits.

This approach is the study of gas-dynamic processes can rationally exploit the boiler units with circulating fluidized bed.

References

[1] Ryabov G A, Folomeev O M, Sankin D A, Khaneev K V, Bondarenko I G and Mel'nikov D A 2010 *Power Technology and Engineering.* **44** 2–11

[2] Gil A V and Starchenko A V 2012 *Thermophysics and Aeromechanics.* **19** 503-519

[3] Gil A V, Zavorin A S, Starchenko A V and Obukhov S V 2011 *Power Technology and Engineering.* **45** 42-49

[4] Armstrong L M, Luo K H and Gu S 2010 *Chemical Engineering Journal.* **160** 239–248

[5] Wojciech P A, Gabriel W, Marcin K, Paweł K, Adam K and Ryszard A B 2014 *Particuology.* **16** 29–40

Study of melting of a pure gallium under influence of magnetic field in a square cavity with a local heat source

N S Bondareva[1] and M A Sheremet[2]
Tomsk State University, 36, Lenin ave., Tomsk, 634050, Russia

E-mail: [1]r0dniki@mail.ru, [2]sheremet@math.tsu.ru

Abstract. Numerical analysis of the unsteady natural convection with phase transitions inside the pure galluim under the influence of a uniform magnetic field with a heat source of constant temperature has been conducted in the presence of magnetic field. The vertical walls of the cavity are kept at low constant temperatures whereas the top and bottom horizontal walls are adiabatic with the exception of the heat source of high constant temperature. The mathematical model formulated in dimensionless stream function, vorticity and temperature variables is solved using the implicit finite difference schemes of the second order. The governing parameters are the Rayleigh and Hartmann numbers, and the dimensionless time. The effects of these parameters on the streamlines and isotherms are analyzed.

1. Introduction

Numerous studies of magnetic natural convection with solid-liquid phase change occupy one of the central stages of the heat and mass transfer problems. The interaction of magnetic field with natural convection in the melting process can have a very strong effect on the resulting flow structure and performance of the material. Because of the high electrical conductivity in metals, magnetic induction significantly suppresses natural convection and consequently smoothes the temperature differences. During solidification using imposed magnetic field produced semi-conductors with suitable quality.

Melting and solidification coupled with natural convection under the influence of magnetic field is a topic of many experimental and numerical studies. Wittig and Nikrityuk [1] presented flow visualization of two- and three-dimensional pure gallium melting. The authors compared the obtained results of 2D and 3D simulations and noted an essential influence of the third coordinate. Bouabdallah and Bessaih [2] performed numerical simulations of solidification with magnetic field in cubical enclosure. Authors showed that the liquid phase is suppressed by the Lorentz force and in this connection the phase front becomes regular.

Present numerical study is devoted to natural convection with melting under the influence of the magnetic field in square cavity with a local heat source. The heat source of a constant temperature is located on the bottom wall of the considered enclosure and has a square shape. Vertical walls of the cavity are cooled while the horizontal walls are adiabatic. Mathematical simulation of unsteady melting process is carried out on the basis of dimensionless variables such as stream function, vorticity and temperature. The special smoothing function is introduced to simplify the solving process. Governing equations are solved using the finite difference method.

2. Mathematical model and numerical method

Natural convection with phase transition and uniform magnetic field in a square cavity with an internal heat source is considered numerically. Schematic illustration of physical model is shown in Figure 1. Square heat source is located at the centre of the bottom wall and it has a constant temperature that is grater than the temperature of melting. Vertical walls of the cavity are cooled and have constant temperature that is less than the melting temperature. Horizontal walls are supposed to be adiabatic. Initially material has the solid state and its temperature is equal to the fusion temperature. For the description of this problem the following assumptions have been introduced in the model: the Boussinesq approximation is valid, the melt is Newtonian and incompressible, the flow is laminar, physical properties of material is not depend on the temperature.

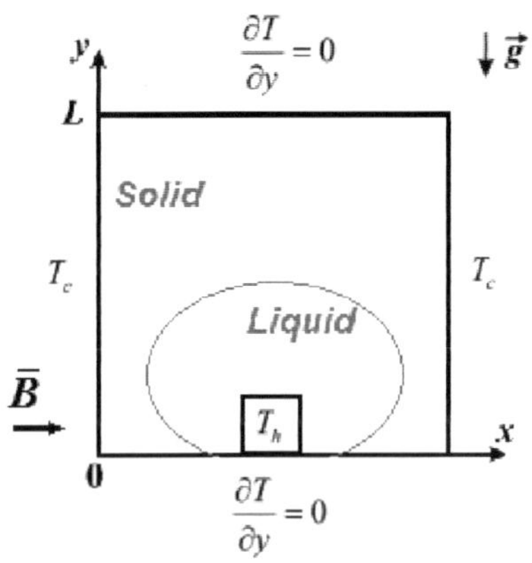

Figure 1. Schematic illustration of the cavity.

The governing equations in this statement can be written using the following conservation laws:
– conservation of mass:

$$\frac{\partial u}{\partial x}+\frac{\partial v}{\partial y}=0, \tag{1}$$

– conservation of momentum in liquid phase:

$$\frac{\partial u}{\partial t}+u\frac{\partial u}{\partial x}+v\frac{\partial u}{\partial y}=-\frac{1}{\rho}\frac{\partial p}{\partial x}+\nu\nabla^2 u+\left(\vec{j}\times\vec{B}\right)_x, \tag{2}$$

$$\frac{\partial v}{\partial t}+u\frac{\partial v}{\partial x}+v\frac{\partial v}{\partial y}=-\frac{1}{\rho}\frac{\partial p}{\partial y}+\nu\nabla^2 v+\left(\vec{j}\times\vec{B}\right)_y+g\beta\left(T-T_F\right), \tag{3}$$

– conservation of energy:

$$\frac{\partial h}{\partial t}+u\frac{\partial h}{\partial x}+v\frac{\partial h}{\partial y}=k\left(\frac{\partial^2 T}{\partial x^2}+\frac{\partial^2 T}{\partial y^2}\right), \tag{4}$$

where

$$h = \begin{cases} \rho_s c_s T, & T < T_F, \\ \rho_s c_s T_F + \rho_l L_F + \rho_l c_l (T - T_F), & T \geq T_F. \end{cases} \quad (5)$$

Here *x*, *y* are the Cartesian coordinates, respectively; *t* is the time; *g* is the gravity acceleration; ν is the viscosity; β is the thermal expansion coefficient of liquid; ρ is the density; *u*, *v* are the velocity components in *x*- and *y*-directions, respectively; *p* is the pressure; $\vec{j}$ is the current density; $\vec{B}$ is the magnetic intensity; *T* is the temperature; T_h is the heat source temperature; T_F is the melting temperature; *h* is the enthalpy, *k* is the thermal conductivity; L_F is the latent heat; c_s and c_l are the specific heats of solid and liquid phases; ρ_s and ρ_l are the densities of solid and liquid phases, respectively.

To remove discontinuous from the energy equation (4) at the phase front the following smoothing function is introduced:

$$\varphi = \begin{cases} 0, & T < T_F - \eta \\ \dfrac{T - (T_F - \eta)}{2\eta}, & T_F - \eta \leq T \leq T_F + \eta \\ 1, & T > T_F + \eta \end{cases} \quad (6)$$

New variables such as the stream function ψ and vorticity ω have been used to simplify solving of equations:

$$u = \frac{\partial \psi}{\partial y}, \quad v = -\frac{\partial \psi}{\partial x}, \quad \omega = \frac{\partial v}{\partial x} - \frac{\partial u}{\partial y}. \quad (7)$$

To reduce the governing equations to non-dimensional form we used the following variables:

$$X = x/L,\ Y = y/L,\ U = u/\sqrt{g\beta\Delta TL},\ V = v/\sqrt{g\beta\Delta TL},\ \tau = t\sqrt{g\beta\Delta T/L},$$
$$\Theta = (T - T_F)/\Delta T,\ \Psi = \psi/\sqrt{g\beta\Delta TL^3},\ \Omega = \omega L/\sqrt{g\beta\Delta TL}$$

where *L* is the cavity size; $\Delta T = T_h - T_F$; *X*, *Y* are the dimensionless coordinates corresponding to *x* and *y* coordinates, respectively; *U*, *V* are the dimensionless velocities corresponding to dimensional velocities *u*, *v*, respectively; τ is the dimensionless time, Θ is the dimensionless temperature, Ψ is the dimensionless stream function; Ω is the dimensionless vorticity.

Reduced governing equations (1)–(4) can be rewritten as follows:

$$\frac{\partial^2 \Psi}{\partial X^2} + \frac{\partial^2 \Psi}{\partial Y^2} = -\Omega, \quad (8)$$

$$\frac{\partial \Omega}{\partial \tau} + U\frac{\partial \Omega}{\partial X} + V\frac{\partial \Omega}{\partial Y} = \sqrt{\frac{\mathrm{Pr}}{\mathrm{Ra}}}\left(\frac{\partial^2 \Omega}{\partial X^2} + \frac{\partial^2 \Omega}{\partial Y^2}\right) + Ha^2\sqrt{\frac{\mathrm{Pr}}{\mathrm{Ra}}} \times \left[\left(\frac{\partial U}{\partial X} - \frac{\partial V}{\partial Y}\right)\sin(\varphi)\cos(\varphi) - \right.$$
$$\left. - \frac{\partial V}{\partial X}\cos^2(\varphi) + \frac{\partial U}{\partial Y}\sin^2(\varphi)\right] + \frac{\partial \Theta}{\partial X} \quad (9)$$

$$\beta(\varphi)\left[\frac{\partial \Theta}{\partial \tau} + U\frac{\partial \Theta}{\partial X} + V\frac{\partial \Theta}{\partial Y}\right] + \mathrm{Ste}\left[\frac{\partial \varphi}{\partial \tau} + U\frac{\partial \varphi}{\partial X} + V\frac{\partial \varphi}{\partial Y}\right] = \frac{\xi(\varphi)}{\sqrt{\mathrm{Ra}\cdot\mathrm{Pr}}}\left(\frac{\partial^2 \Theta}{\partial X^2} + \frac{\partial^2 \Theta}{\partial Y^2}\right) \quad (10)$$

where $\beta(\varphi)=\frac{\rho_s c_s}{\rho_l c_l}+\varphi\left(1-\frac{\rho_s c_s}{\rho_l c_l}\right)$, $\xi(\varphi)=\frac{k_s}{k_l}+\varphi\left(1-\frac{k_s}{k_l}\right)$.

One can see that magnetic natural convection with melting is governed by six dimensionless parameters such as the Rayleigh, Prandtl, Hartmann and Stefan numbers, ratio of volumetric heat capacities and thermal conductivity ratio.

The following boundary conditions for the stream function are set on the solid walls and the phase front $\Psi=0$, $\partial\Psi/\partial\overline{n}=0$. Vorticity values on the walls are found from the Poisson equation (8). Stream function and vorticity inside the solid phase are equal to zero.

The governing equations (8)–(10) with corresponding initial and boundary conditions are solved using the finite difference method [3, 4]. Validation of the created algorithm is based on the benchmark problem proposed by Gau and Viskanta [5]. It should be noted, that obtained solid-liquid interphases are in good agreement with experimental data of Gau and Viskanta [5].

3. Results and discussion

Physical parameters of pure gallium used in this investigation are: $\rho_l=6093\,\text{kg/m}^3$, $\rho_s=6095\,\text{kg/m}^3$, $\beta=1.2\cdot10^{-4}\,\text{K}^{-1}$, $k=32\ \text{W/(m}\cdot\text{K)}$, $T_f=29.78\ °\text{C}$, $L_f=8.016\cdot10^4\ \text{J/kg}$, $c=381.5\ \text{J/kg}$, $\mu=1.81\cdot10^{-3}\ \text{kg/(m}\cdot\text{s)}$. Numerical studies have been carried out at the following values of the governing parameters: $7.17\cdot10^5\le Ra\le3.59\cdot10^6$, $0\le Ha\le100$, $2.1\le Ste\le10.51$, $\rho_s c_s/(\rho_l c_l)=1$ and $k_s/k_l=1$. Prandtl number is set 0.0216. Dimensionless parameters selected correspond to properties of the pure gallium. Size of the enclosure is L = 5 cm. Dimensionless temperature of the heat source is $\Theta_h=1$ and on the vertical walls we have $\Theta_h=-0.3$.

The effect of the magnetic field on heat transfer is illustrated in Figure 2. This figure shows isotherms for Rayleigh number Ra = $3.59\cdot10^6$ at several times. It can be seen that strong magnetic field suppresses natural convection and as a result of the velocity decreases the melting process slows down and the phase front becomes smooth and symmetric. During the melting process without magnetic field few convection cells are formed inside the melt zone. The multi-cellular mode is a time-dependent because the temperature difference leads to more complicated flow patterns. With imposed magnetic field the inhibition of fluid occurs, that is presented in Figure 3. Also this Figure 3 shows the comparison of streamlines for two cases.

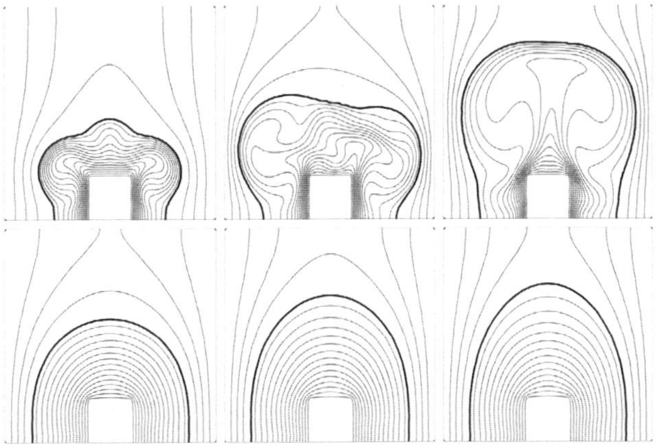

Figure 2. Isotherms for Ha = 0 (top row) and Ha = 100 (bottom row) for the time moments t = 12, 30 and 60 seconds (from left to right).

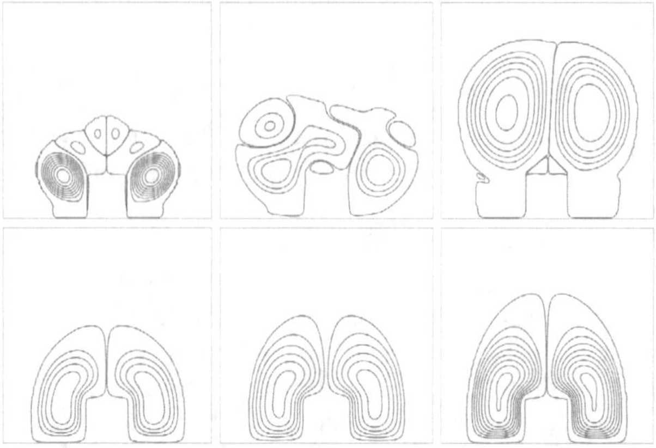

Figure 3. Streamlines at Ha = 0 (top row) and Ha = 100 (bottom row) for the time moments t = 12, 30 and 60 seconds (from left to right).

4. Conclusions

Natural convection of pure gallium under the influence of a uniform magnetic field with phase change inside a square cavity with a local heat source has been studied numerically. Governing equations has been formulated using the special enthalpy formulation with the dimensionless stream function and vorticity variables. The local characteristics of melting process have been obtained in wide ranges of the Rayleigh and Hartmann numbers. The effect of the magnetic field on the fluid flow and heat transfer has been analyzed. It should be noted that increasing of magnetic intensity leads to an attenuation of convective flow and slowing of melting.

Acknowledgments

This work was supported by the Russian Foundation for Basic Research, Grant No. 14-08-31137 mol_a and by the Grants Council (under the President of the Russian Federation), Grant No. MD-6942.2015.8.

References

[1] Wittig K and Nikrityuk P A 2011 *Materials Sci. Engineering: IOP Conf. Series* **27** 012054.
[2] Bouabdallah S and Bessaïh R 2012 *Int. J. Heat Fluid Flow* **37** 154–166.
[3] Bondareva N S, Volokitin O G, Morozova O O and Sheremet M A 2013 *Thermophysics and Aeromechanics* **20** 621-629.
[4] Sheremet M A 2010 *J. Engineering Thermophysics* **19** 119–127.
[5] Gau C and Viskanta R 1986 *J. Heat Transfer* **108** 174-181.

Applying the technology of hydrodynamic cavitation treatment of high-viscosity oils to increase the efficiency of transportation

A E Brand[1], S V Vershinina[1], A A Vengerov[1] and N A Mostovaya[1]
Tyumen, 625000, Russia, Tyumen State Oil and Gas University

E-mail: Allbert-@mail.ru

Abstract.. The article investigates the possibility of applying hydrodynamic cavitation treatment to reduce oil viscosity in Russian pipeline transportation system and increase its performance. The result of laboratory tests and suggestions on technology application are given

1. Introduction

Modern oil industry of the Russian Federation is characterized by the growth of viscous (further VO) and high-viscosity (further HVO) oils in a total structure of hydrocarbon reserves. Developing fields with prevalence of VO and HVO is a perspective direction from the point of view of their high quality and special rheological characteristics. According to data of 2014, the HVO fraction from the total makes up almost 30 % or 7.5 billion ton, and over the last years, since 1992, this fraction has been constantly growing [1].

The conducted analysis of economic efficiency of VO and HVO extraction and transportation is characterized by their unprofitability or small profitability. Large Russian companies, such as OJSC "NK "Rosneft" and OJSC "AK "Transneft", annually invest sufficient financial resources in perfecting the transportation process and modifying VO and HVO rheological characteristics. The most common ways of treating raw materials in order to achieve large transportation rates are:

- Thermal heating
- Adding dopants and diluents
- Applying ectromagnetic radiation

Thermal heating is the most common and effective, but at the same time the most expensive method of treatment, performed by means of oil heaters. According to JSC "Transneft – Siberia", along the entire route of the oil main Zapolyarye – Purpe there will be constructed eight points of oil heating to treat northern VO under low temperature conditions [1]. Currently, for oil preheating tubular sectionalized furnaces (PTB-10/TSF-10) are used. Their basic characteristics are presented in Table 1.

Fuel used for operation of such furnaces is gas distillate. Taking into account that the cost of gas for large industrial enterprises of Yamal-Nenets autonomous region and Tyumen region is on average 2848 rbl./m^3, approximate losses on oil heating per 1°C account for 280 000 roubles a year at pumping volume of 1 million ton [2] . Over the year, the expenditure on heating will account for 252 million roubles.

Unfortunately, thermal oil heating is dependent on economic conditions in the country. If we assume that on average the price for 1000м^3 of gas rises annually by 15 % for the needs of industrial enterprises, and the minimum gas consumption at heating points (based on the throughput capacity of the oil main "Zapolyarye – Purpe" of 30 million ton) will comprise 2.8 million m^3 a year, application of more power-intensive and energy resource effective technologies of transportation is of vital importance[3].

Table 1. Passport data of tubular sectionalized furnaces (PTB10/TSF-10)

Parameter	**Value**
Output by product	900 m^3/hour
Fuel power	4.2 GJ/hour
Operating pressure	To 6.4 MPa
Oil temperature at the output	Not above 70 °C
Consumable fuel	Gas distillate
Number of burners	4 pieces KC-400
Performance efficiency, %	71
Heat losses at operation, %	8

2. Research

According to foreign scientists, such as Hemmit F. and Dejli J., from the American Society of Mechanical Engineers, the most perspective is the hydrodynamic cavitation treatment characterized by its performance, profitability and possibility to use internal reserves of the material to change rheological characteristics of oil (plastic viscosity, pour point, etc.). Within the framework of realization of the grant given by OJSC "AK "Transneft" the authors have investigated the cavitation phenomenon and the possibility of combining cavitation with pour point depressants to improve the efficiency, and the hydrodynamic equipment has been designed.

The cavitation phenomenon arises at the moment when the fluid pressure and the fluid saturated vapor pressure become close. At close values the fluid emanates plentiful bubbles (caverns) filled up with dissolved gas[4]. The collapse of the bubble is characterized by sharp jumps of temperature and pressure, and propagation of the energy wave facilitates breaking down of nearby carbonaceous chains and molecular compounds. The quantity of bubbles can vary from 10^4 to 10^6.

Figure 1 shows simulation of the hydrodynamic cavitation process due to pipeline section change by means of ANSYS/FLUENT software. The obtained data and dependences have been taken into consideration when conducting laboratory tests.

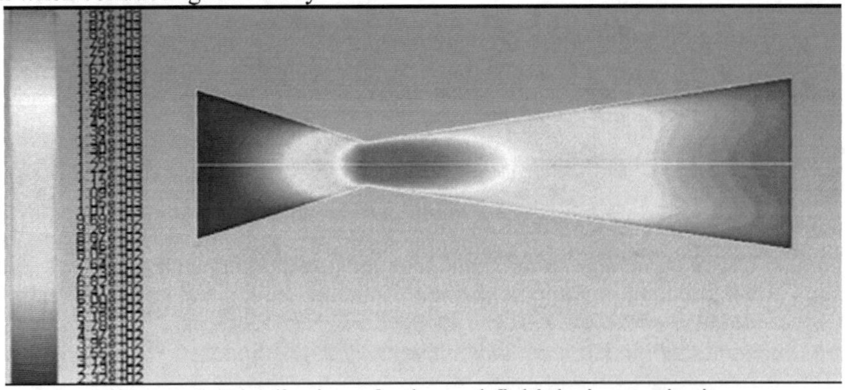

Figure 1. Distribution of a thermal field during cavitation
Zone 1 – a preheat zone, zone 2 – a zone of cold boiling, zone 3 – flow and thermal field stabilization.

Simulation has demonstrated that when high-viscosity oil passes through a narrowed section there is a local rise in temperature on the working chamber – diffuser boundary, which is a precondition for economic profit.

Figure 2 shows a laboratory unit; cavitation equipment is in the bottom left-hand corner and Figure 3. Tests involved such equipment as the centrifugal pump and cylindrical tubes of 3 sm^2 in diameter. Pour point depressant DPN-1R (produced in Russia) was tested; about 200 liters of high-viscosity Usinsk oil $\rho = 945$ kg/m^3 and $\nu = 150$ mPa·s have been treated.

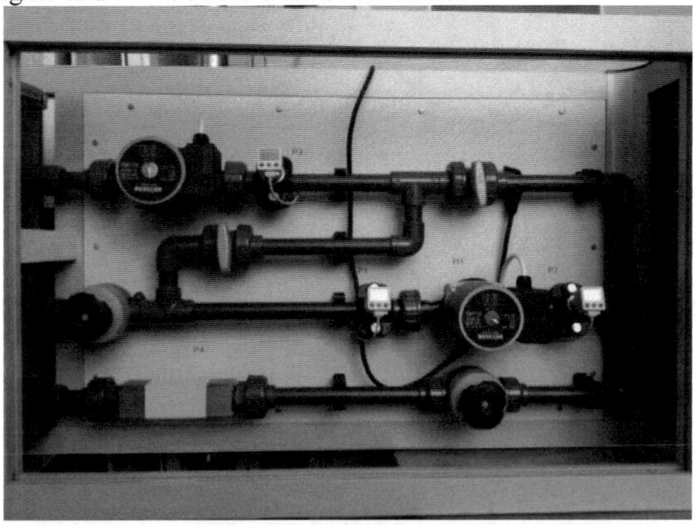

Figure 2. Laboratory unit.

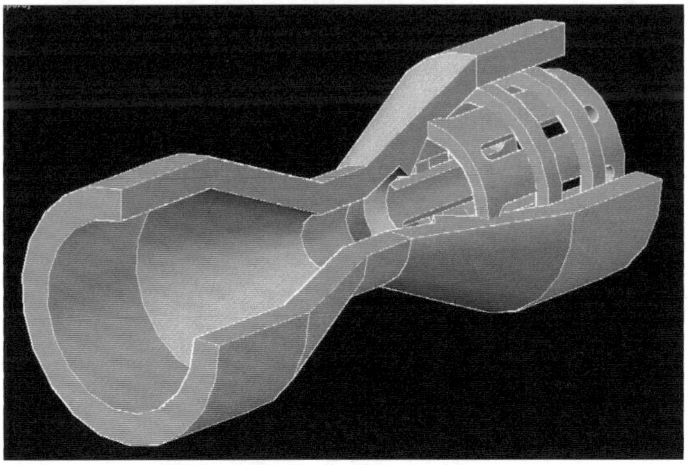

Figure 3. Cavitation equipment

The minimum oil flow rate in the module required for cavitation to emerge is determined by the formula (1):

$$V_{kav} = \frac{64 \cdot P_n}{g \cdot P_{kr}} \left(\frac{29 \cdot \rho \cdot g}{P_n} \right)^{1.9}, \quad (1)$$

P_{kr} – critical cavitation pressure (Pa)

P_n – saturated vapor pressure (Pa)

Required pump head for cavitation to emerge:

$$N_{To} = \frac{v^2_{To}}{2g} + h_M , \tag{2}$$

where h_M– total head loss in the module, m;

Total head loss in the module is made up of head losses in the confuser, diffuser and cylindrical part of the module and is calculated by the formula:

$$h_M = \left(\frac{\lambda}{8\sin\frac{\alpha}{2}}\left(1-\frac{1}{n^2}\right) + \lambda\frac{L}{d} + \frac{\lambda}{8\sin\frac{\beta}{2}}\left(1-\frac{1}{n^2}\right) + \sin\beta\left(1-\frac{1}{n^2_1}\right) \right) \cdot \frac{v^2_{TO}}{2g} , \tag{3}$$

where β –expansion angle of a diffuser, degree; n_1 - expansion ratio of a diffuser, α – expansion angle of a confuser, degree; n – contraction ratio.

During laboratory tests it has been revealed that the least pressure loss is observed in confusers with the expansion angle of up to 40° and the contraction ratio of 1.2 – 3.0 and in diffusers with the expansion angle of less than 50°.

The final form of the formula of head looks as follows:

$$N_{To} = \frac{v^2_{To}}{2g} \cdot \left(\xi_{To} + \xi_{TS} + \xi_D + 1\right), \tag{4}$$

The conducted studies have demonstrated that complex oil treatment is a perspective direction Table 2 and Figure 4.

Table 2. Comparative analysis of efficiency of oil viscosity reduction methods.

Characteristics	DPN-1R	Cavitation treatment	Complex treatment
Depression T, °C	10–14	5–7	17–20
Viscosity reduction, %	42-57	32-40	66-70
Dynamic stress reduction	2–3	4	6-7
Static stress reduction	4-5	3-4	5-7
Time to restore rheological 1 characteristics, days	5	4	7

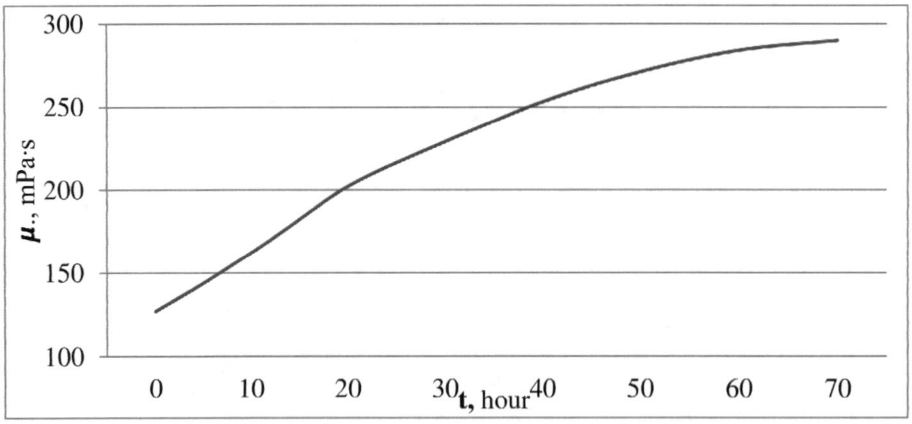

Figure 4. Storage Age-Dependence of Dynamic Viscosity.

IOP Conf. Series: Materials Science and Engineering **93** (2015) 012005 doi:10.1088/1757-899X/93/1/012005

Due to local pressure changes and, thereof, emergence of conditions for cavitation and breaking of fluid continuity, energy is released. The released energy is equivalent to oil temperature changes at an average density of 945 kg/m^3 and viscosity of 150 mPa·s by 2-3 °C. Both computer and real models have demonstrated a 94.6 % convergence. Because initial oil heating temperature can be changed, financial expenditures [5, 6] also change

3. Conclusion

The analysis and calculations carried out by the authors demonstrate that application of cavitation treatment to reduce economic load is substantiated and leads to economic benefits [7]. The expediency and efficiency of applying complex oil treatment consists in the following:

- Increase of initial heating temperature of a high-viscosity oil by 2-3°C due to cavitation treatment;
- Reduction of oil viscosity due to a complex method that in turn will lead to reduction of the power and fuel consumption by heaters;
- Improvement in rheological characteristics of oil due to breaking down of carbonaceous connections that will positively affect further petroleum transportation and refining, allowing for increase in the yield of light cuts during atmospheric rectification.

The applied technology can be used on field or technological pipeline systems with diameters up to 300 mm and viscosity values of transported oil 150-200 mPa·s

References

[1] Shiryaev A M and Zholobov V V 2015 *J. Science and Technology* **24** 60-68
[2] Ilyin E G and Ivantsova S G 2014 *J. Crude oil pipeline* **6** 67-70
[3] Vyboyschik M A and Ioffe AV 2015 *J. JSC Rosneft* **36** 45-52
[4] Vengerov A A and Brand A E 2014 Oil cavitation treatment to prevent formation of paraffin deposits *J. Natur. and intellect. resource. of Siberia.* **15** 36-37
[5] Ponomareva T G and Brand A E 2014 Manifestation of super-abnormality and viscosity on flow curves of high-setting oils and their mixtures *J. Natur. and intellect. resource. of Siberia.* **15** 42-43.
[6] Tarasenko M A and Brand A E 2014 Reducing oil viscosity by a method of hydrodynamic *J. Natur. and intellect. resource. of Siberia.* **15** 44-45.
[7] Stashkovskaya N V and Brand A E 2013 Pipelines of innovation and prosperity *J. Prob. of funct. of transp. sys.* 25-29.

IOP Conf. Series: Materials Science and Engineering **93** (2015) 012006 doi:10.1088/1757-899X/93/1/012006

Experimental research of solid waste drying in the process of thermal processing

V V Bukhmirov[1], O B Kolibaba[2] and R N Gabitov[3]
[1] Ivanovo State Power Engineering University, Ivanovo, Rabfakovskaya st.34, apt. 315, 153003, Russia
[2] Ivanovo State Power Engineering University, Ivanovo, Rabfakovskaya st.34, apt. 325, 153003, Russia
[3] Ivanovo State Power Engineering University, Ivanovo, Stroitelnaya st.4, apt. 205, 153003, Russia

E - mail: tevp@tvp.ispu.ru.

Abstract. The convective drying process of municipal solid waste layer as a polydispersed multicomponent porous structure is studied. On the base of the experimental data criterial equations for calculating heat transfer and mass transfer processes in the layer, depending on the humidity of the material, the speed of the drying agent and the layer height are obtained. These solutions are used in the thermal design of reactors for the thermal processing of multicomponent organic waste.

1. Introduction

Thermal methods, which include pyrolysis and gasification are widespread waste technologies. Thermal processing of raw materials, makes it possible to obtain not only the combustible gas, but also to solve the environmental problems associated with urban pollution and land reclamation.

The process of thermal processing of solid waste is most often carried out in shaft furnaces (thermal reactors) [1,2]. Municipal solid waste is loaded into the upper part of the reactor and consistently held drying, pyrolysis and gasification zones.

The drying zone is carried out heating and drying of solid waste layer by heat exchange with a stream of hectares of the area downstream from the pyrolysis, and allocated in conjunction with the moisture of light-volatile compounds is supplied to the gasification of solid carbonaceous residues (SCR).

In the pyrolysis zone the combustible components of MSW pyrolyzed to form the pyrolysis gas, a liquid product comprised of polyaromatic hydrocarbons, and SCR. Solid carbonaceous residue, gradually falling to the bottom of the reactor, feds to the gasification zone. There also oxidant gas (steam mixture) is directed. The resulting SCR gasification fuel gas in a mixture with the pyrolysis gas is a fuel quality suitable for use.

Solid waste can be attributed to particulate material, which in the presence of moisture internal require high energy consumption for the drying process[3]. In order to select a rational regime of the reactor it is necessary to determine the temperature field of solid waste layer, drying time raw material and energy costs in the process. Knowledge of these parameters is possible to calculate the speed of charging, the height of the drying zone of the thermal reactor and, therefore, to design it properly.

2. Problem statement

For the convective drying conjugate problem of heat and mass transfer, which is characterized by the following is solved:

- drying agent gives heat to the dried up material and takes the moisture evaporates;
- drying material receives the heat of the drying agent and gives it moisture.

The drying agent is used as the transport means.

In drying theory, the method of the analysis of external heat and mass transfer [4], based on the joint consideration of the equations of motion and continuity of viscous incompressible flow:

$$\frac{\partial \vec{\upsilon}}{\partial \tau} + \vec{\upsilon} \cdot grad\vec{\upsilon} = \vec{g} - \frac{1}{\rho} gradP + \nu\nabla^2 \vec{\upsilon} \quad (1)$$

$$\frac{\partial \rho}{\partial \tau} + div\rho\vec{\upsilon} = 0 \quad (2)$$

convection-diffusion transport of moisture in a moving drying agent:

$$\frac{\partial C}{\partial \tau} + \vec{\upsilon} \cdot gradC = D\nabla^2 C \quad (3)$$

and the equation describing the temperature field in the flow of coolant is used:

$$\frac{\partial t}{\partial \tau} + \vec{\upsilon} \cdot gradt = a\nabla^2 t \quad (4)$$

υ - speed of the drying agent, m/s; g - gravitational acceleration, m/s^2;ρ - the density of the drying agent, kg/m^3;P - pressure, Pa; ν - kinematic viscosity coefficient, m^2/s; C - moisture concentration, m^3 / m^3; D - molecular diffusion coefficient, m^2/s; t – temperature, °C; a - coolant thermal diffusivity, m/ s^2.

The difficulties of external heat and mass transfer analysis by the theoretical methods result in the need of a similarity theory with the using the experimental data.

3. Experiment

We carried out experimental studies of the convective drying process of wet solid waste layer in the four temperature conditions during gas flow rates: 0.1, 0.5, 1, 1.5, 2 m / s. The first mode of drying is carried out at a constant temperature of the drying agent, 107 ° C. The second, third and fourth - at temperatures of 127 ° C, 147 ° C, 167 ° C.

Figure 1 shows the experimental setup for study of the heat and mass transfer process between the solid waste layer and a drying agent (air). The experimental setup consisted of a pipe section (1) for retaining a metal mesh MSW layer (working zone), and the electric heater (2) for heating the drying medium (air). Supply air is a fan (4) located in front of heater. The air flow rate and its temperature is controlled by means of the control unit (3) with a screen for displaying information. The temperature and the humidity was recorded by sensors RHP-2S11 (5) readout on the screen of the control unit. The air flow rate was measured by an anemometer Testo 416 (6). During the experiment, the temperature of the sample was measured 5 thermocouples TCAestablished in the layer of solid waste. Analog-to-digital converter of the signal OWEN MVA8 transmited a signal to a personal computer. Drying of the sample continued to establish its constant weight.

Before experiment samples of municipal solid waste average morphological composition were placed in a water bath for 20 hours. With the help of centrifuge the excess moisture in samples is removed. Then the MSW sample is placed in the installation working zone. To determine the moisture content of MSW sample weighed portion was recovered. It was weighed on electronic scales laboratory (VL E134) 4th grade accuracy, placed in an oven heated to a temperature of 107 ° C, and kept up-achievements of constant weight.

During the drying process the change in mass of the material was captured (Fig.2). Then the calculation of humidity of raw materials and the rate of change of moisture content is carry out.

4. Results and Discussion

Figure 2 shows experimental data of temperature change MSW layer during its drying, and change in humidity sample at four temperature conditions of the drying agent, at a speed of 0.1 m/s.

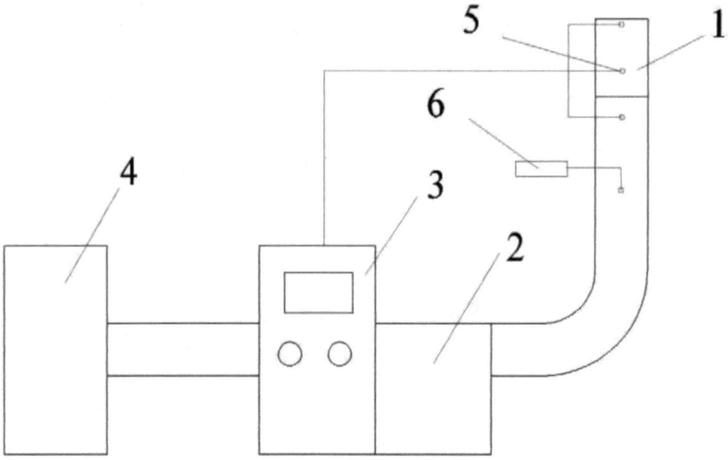

Figure 1. Experimental installation for the investigation of solid waste drying processes:
1 – Operating zone; 2 – Electric heater; 3 – Control block;
4 – Fan; 5 – The temperature sensors; 6 – Anemometr.

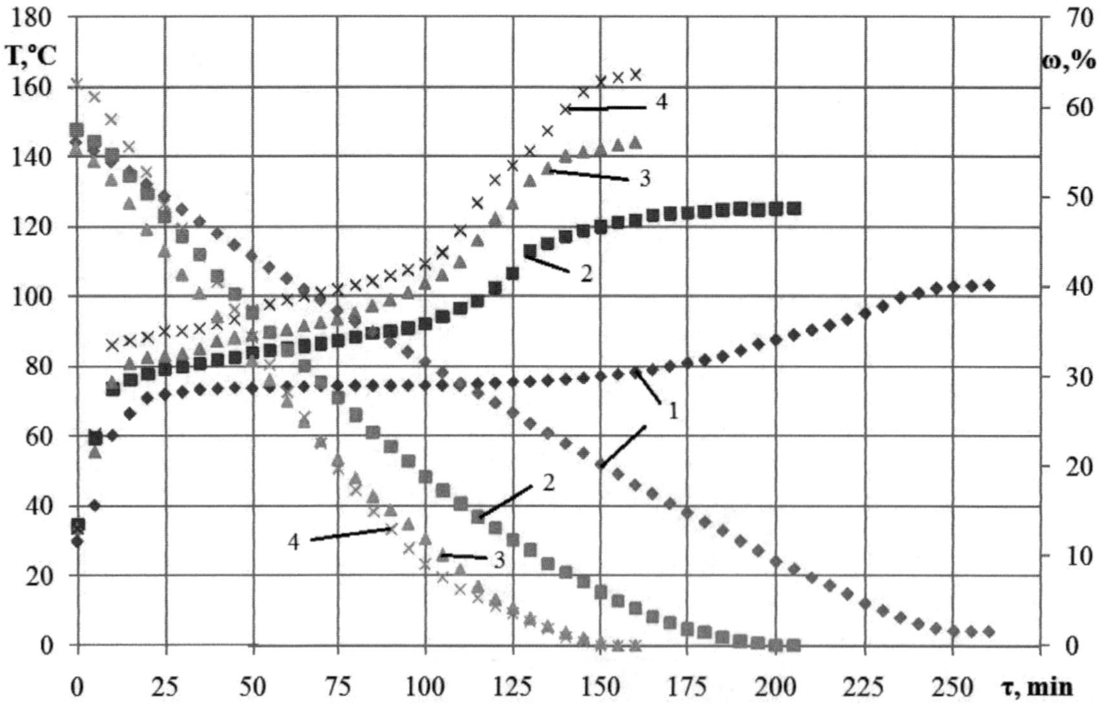

Figure 2. The temperature and humidity of the sample MSW layer in the drying process at four temperature regimes of the drying agent
(1 – 107 ° C, 2 – 127 ° C, 3 – 147 ° C 4– 167 ° C) at a speed of 0.1 m/s.

The experimental data obtained in the process of convective drying of wet MSW layer at temperature regimes of drying agent: 107 ° C, 127 ° C, 147 ° C, 167 ° C and at gas flow rates: 0.1, 0.5, 1, 1.5, 2 m/s, it was approximated as a criteria equation:

$$Nu = A\left(\frac{d}{H}\right)^{k} Re^{n} E^{m} \tag{7}$$

where H – the height of the layer, m; d– equivalent to the size; E – simplex, taking into account the change in moisture content (MSW), i.e, the impact of mass transfer on heat transfer.

$$E = \frac{1-\omega}{\omega} \tag{8}$$

where ω– integrated relative humidity of the material.

Calculation of mass transfer can be reduced to the calculation of heat transfer by using the ratio [5]:

$$Nu = Sh \cdot K_{c} \tag{9}$$

where Sh – Sherwood criterion; K_c– correction factor for the diffusion flux.

$$K_{c} = \frac{1}{1-p^{cp}} \tag{10}$$

where p^{cp}– mean log vapor pressure in the boundary layer.

After processing of the experimental data criterial equations for determination the coefficients of heat transfer and mass transfer were obtained.

Criteria equation for laminar:

$$Nu = 1,24\left(\frac{d}{H}\right)^{0.3} Re^{0.33} E^{0.084}, \tag{11}$$

$$Sh = 0,911\left(\frac{d}{H}\right)^{-0.2} Re^{0,33} E^{0.084} \tag{12}$$

Criteria equation for turbulent regime:

$$Nu = 0,107\left(\frac{d}{H}\right)^{0.38} Re^{0.82} E^{0.12} \tag{13}$$

$$Sh = 0,066\left(\frac{d}{H}\right)^{-0.49} Re^{0,82} E^{0.12} \tag{14}$$

In the case of heat exchange for defining the size the equivalent diameter of the piece found by direct measurement of samples with the Gaussian distribution is adopted. In the case of mass transfer - the length of the evaporation surface, i.e. the minimum length of the pore canal in the direction of movement of the drying agent. For defining the temperature the average temperature of the drying agent is accepted.

Figure 3 and 4 shows the average heat transfer coefficients of the drying agent to the material and the mass transfer in the drying MSW layer at different speeds of the drying agent. Analysis of the graphs shows that with a decrease in the relative humidity of the material intensity of the heat and mass transfer is reduced.

5. Conclusion.

An experimental study of the organic raw materials convective drying process, modeling layer of municipal solid waste, was fulfilled. As the result criterial equations for calculating heat transfer and mass transfer processes in the porous layer, depending on the humidity of the material, the speed of the drying agent, the layer height were obtained.

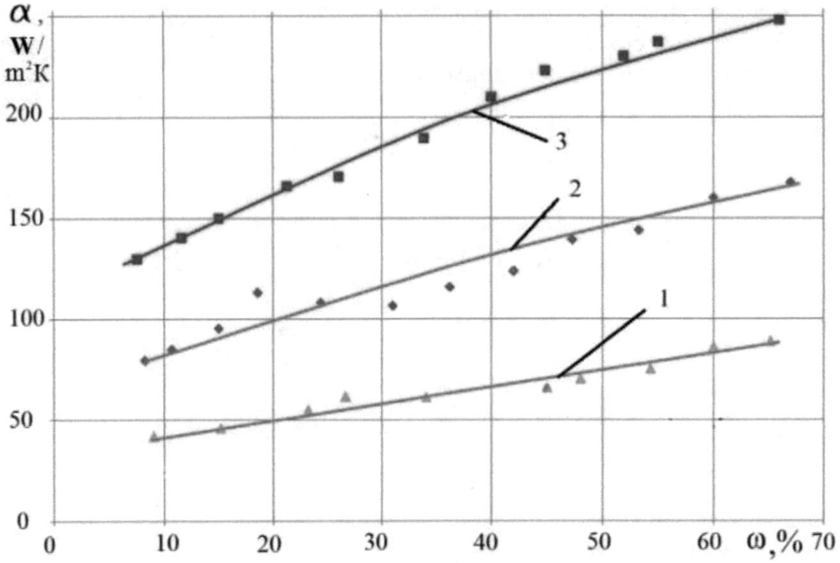

Figure 3. The dependence of the heat-transfer coefficient from the humidity during the drying process at a constant temperature of 107 ° C and the air velocity: 1 - 1 m / s 2 - 1.5 m / s, 3 - 2 m / s.

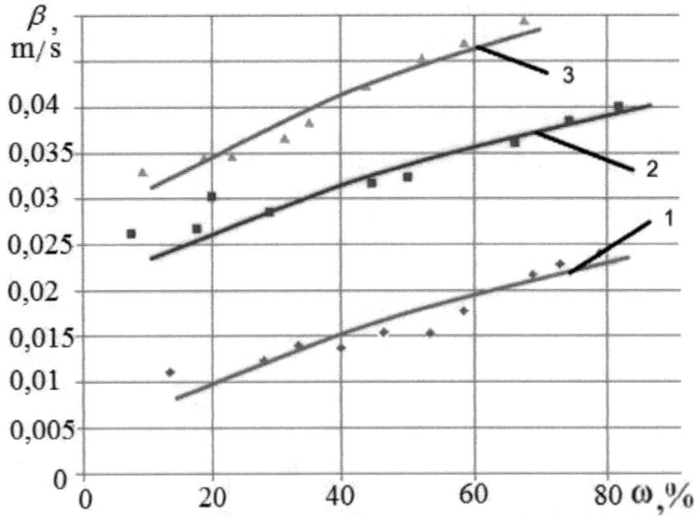

Figure 4. The dependence of the mass-transfer coefficient on the relative humidity during the drying process at a constant temperature of 107 ° C and the air velocity: 1 - 1 m / s 2 - 1.5 m / s, 3 - 2 m / s.

References

[1] Dolinin D A, Gabitov R N, Semin E S, Gorinov O I, Kolibaba O B, Gorinov P O and Samyshina O V, 2014 *J. Bulleten №23* (The Russian Federation patent №2525558)

[2] Dolinin D A, Gorinov P O, Semin E S, Gorinov O I, Kolibaba O B, Gabitov R N, and Samyshina O V, 2014 *J. Bulleten №36* (The Russian Federation patent №2536896)

[3] Mitrofanov A V 2015 *J. Vestnik ISPU* **2** 67-70

[4] Sokolsky A I 2005 Heat treatment of disperse materials in devices with vortex two-phase flows *Ivanovo: ISABU* p 320

[5] Lykov M V 1970 Drying in the chemical industry *Moscow: Chemistry* p 430

Numerical modelling and analysing of conjugate radiation-convective heat transfer of fin-tube radiator of spacecraft

A Delcov, A Hodenkov and D Zhuikov
Reshetnev Siberian State Aerospace University, Krasnoyarsk, Russia

E-mail: delkov-mx01@mail.ru

Abstract. This paper covered the problem of assessing the effectiveness of the section of the fin-tube radiator of space thermal control system. The task of calculating the conjugate radiation-convective heat transfer is presented. The results of numerical simulation are described.

1. Introduction

Currently there is the problem of calculating the conjugate radiation and convective heat transfer characteristic for a wide range of technical problems (the nozzles of rocket engines, industrial furnaces, chemical reactors, etc.). For this heat exchange contact the current environment and the heat-conducting surfaces as well as the presence of significant temperature gradient between the surface and the environment are the defining conditions.

The mathematical description of this problem causes some difficulties: the need to localize the heat transfer equations for the two areas – liquid and solid – followed by a joint decision. In this paper we consider the heat radiation panel on the spacecraft designed to remove heat energy into space. In space systems such panels are an important part of the thermal power system (for example, solar power plant or thermal control system) [1,2].

Radiation panel is one of the important elements of the thermal control system. In terms of space radiation discharge is advantageous way to remove heat from the spacecraft [3]. Consider the section of the radiator (Figure 1), which is designed as a pipe with a coolant being connected to the radiation plate.

On the one side of the section contacts the coolant flow is correspondingly the heat transfer from the coolant in the wall; the other side is that a panel of radiation with which the heat flow is discharged into the open space. Sections are grouped in panel radiators.

The problem of calculating the conjugate heat transfer in this case is divided into two sub-tasks: the temperature distribution in the solid (radiator) and the temperature distribution in the liquid coolant. These two problems are solved together with the general boundary condition – convective heat transfer through the wall of the pipe.

2. Governing equations

This problem is described by the equations of heat and mass transfer [4,5]. The major difference of this problem is that we have two domains – solid domain and fluid domain.

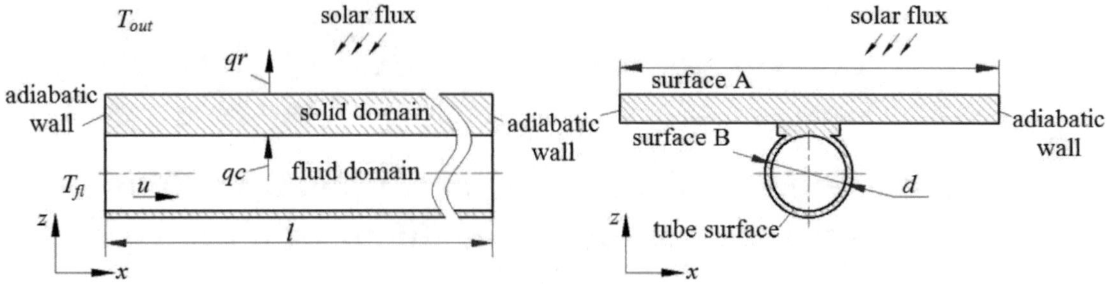

Figure 1. Geometry definition and boundary conditions.

In the solid domain we have heat equation:

$$\rho c_v \frac{\partial T}{\partial t} = \nabla \cdot (k \nabla T) \tag{1}$$

The boundary conditions for the present fin geometry (fig. 1) are defined as:

for convection surface: $$k \frac{\partial T}{\partial n}\bigg|_{tube\_surf} = q_c = \alpha (T_{fl} - T)$$

for radiation panel: $$k \frac{\partial T}{\partial n}\bigg|_{surf\_A} = q_r = \varepsilon \sigma \left(T^4 - T_{out}^4 \right) + q_{solar}$$

for adiabatic walls: $$k \frac{\partial T}{\partial n}\bigg| = 0$$

In this equations ρ – density of solid domain, c_v – heat capacity, T – temperature, t – time, k – thermal conductivity, α – heat transfer coefficient, n – normal to surface, ε – emissivity, σ – the Stefan–Boltzmann constant, T_{fl} – temperature of incident flow in convection heat transfer, T_{out} – temperature of surrounding area in the radiation heat transfer, q_{solar} - solar flux.

In the fluid domain we have three equations:

$$\frac{\partial \rho}{\partial t} + \nabla \cdot (\rho u) = 0 \quad \text{(continuity equation)} \tag{2}$$

$$\rho \left(\frac{\partial u}{\partial t} + u \cdot \nabla u \right) = \rho g + \nabla p \quad \text{(cauchy momentum equation)} \tag{3}$$

$$\rho \frac{dh}{\mathrm{dt}} = \nabla \cdot (k \nabla T) + \rho \frac{dq}{\mathrm{dt}} \quad \text{(energy equation)} \tag{4}$$

The boundary conditions for the present fin geometry (fig. 1) are defined as:

at x=0: $$T = T_{fl}, \quad v = v_{fl}$$

$$k \frac{\partial T}{\partial n}\bigg|_{tube\_surf} = q_c = \alpha (T - T_{solid})$$

Empirical correlation for the Nusselt number is used to the determine heat transfer coefficient α:

$$Nu = 0.33 \mathrm{Re}^{0.5} \mathrm{Pr}^{0.43} \quad \text{(laminar)} \tag{5}$$

$$Nu = 0.021\,\mathrm{Re}^{0.8}\,\mathrm{Pr}^{0.43}\ \text{(turbulent)} \qquad (6)$$

where $Nu = \alpha d k^{-1}$

In this equations ρ – density of fluid domain, t – time, u – velocity, T – temperature, k – thermal conductivity, h – enthalpy, q – internal heat source, p – pressure, Re – Reynolds number, Pr – Prandtl number.

3. Methods of numerical modeling

The equations for calculation of processes in liquid and solid areas recorded for the three-dimensional case [6]. Depending on the purpose of the simulation is possible to simplify the equations for one-dimensional and two-dimensional cases. To study the problem of radiation-convective heat transfer in the framework of this work were used three methods:

- An algorithm developed by the authors (1-d on two domains);
- Calculation using a package Syrthes (3-D on solid and 1-d of fluid);
- Calculation using a combination of packet Syrthes and Code_Saturne (3-D on the two domains).

For a quick evaluation of the parameters of heat exchange in the first approximation, the authors have developed a 1-D model of the process based on the finite volume method. Area calculation model is the contact liquid and a solid in which the temperature change occurs only in length. The calculation is carried out in two phases: the first (predictor) detects the temperature of the liquid and solid region, taking into account the thermal equilibrium of only two contact-ties and volumes. At the second stage of the calculation work together to solve two problems for liquid and solid (each sub-task the assumption of fixed boundary conditions). The second stage uses a stationary iterative method for minimizing the residual temperature areas. This algorithm is suitable only for preliminary parameter estimation coupled heat (e.g., in the case of multi-parameter optimization), and is ineffective for the qualitative evaluation of complex cases of heat exchange with the presence of complex geometry and boundary conditions.

To calculate the nature of the temperature distribution in the solid area was used package Syrthes [7]. This package allows modeling of heat transfer in solids with different boundary conditions, including the conditions of radiative heat transfer. The interaction of solid and liquid regions is conducted through convective heat transfer boundary condition recalculation of the flow temperature with floor length.

For complex cases, the dual radiant and convective heat exchange with the 3-D representation of the liquid and solid areas requires the use of CAE-packages. For reasons of availability, the authors used open packages Syrthes (solid domain) and Code_Saturne (fluid domain) [7], united to work together with a script language pyton. The main solver is Code_Saturne. It uses temperature distribution regions obtained in the previous step for each subsequent iteration.

4. Test Case

As part of this work as a test problem the problem was solved with the following boundary conditions:

- Working medium - iso-octane;
- Material of the radiator – aluminum;
- A pipe with a diameter of 12 mm (inner) and 13 mm (external) and a length of 0.5 m;
- Radiation pad size 0.5 * 0.1 m, thickness 2mm pad;
- Emissivity for surface A is 1, for the surface B is 0 (multi-layer insulation - MLI);
- Solar flux is 0.

For the numerical solution using CAE-packet computational domain was divided into subregions of a solid and a liquid. The calculation results are presented in Fig. 2-3. The temperature of the coolant at the outlet shows the average over the cross section.

IOP Conf. Series: Materials Science and Engineering **93** (2015) 012007 doi:10.1088/1757-899X/93/1/012007

Figure 2. The nature of the temperature distribution on the radiating surface at a flow rate of 0.05 ms^{-1}.

As it is seen from the transverse profile of temperature distribution, temperature divergence in the central part and on the boundary is 1.5 K. The magnitude of discrepancies depends on the conductivity of the radiator. Obviously, the smaller the gap means better performance of the radiator. To obtain a uniform longitudinal profile a number of sections are combined with common collector. The longitudinal profile of the temperature has the nature of the spline, which is explained by the finite thermal conductivity of the material of the battery. The temperature decreases along the length of the coolant flow as a result of the backlog.

Meet the challenges of the three methods showed good consistency. The accuracy of the first method in many ways determined by the errors in the reduction of geometry and boundary conditions to the 1-d mean (especially the complex profile of the pipe, which is replaced by an equivalent rectangle). It should be noted that the third method causes considerable time for calculation and is not very suitable for solving multiparameter optimization. The second method was used as a basis for further research (Syrthes-only).

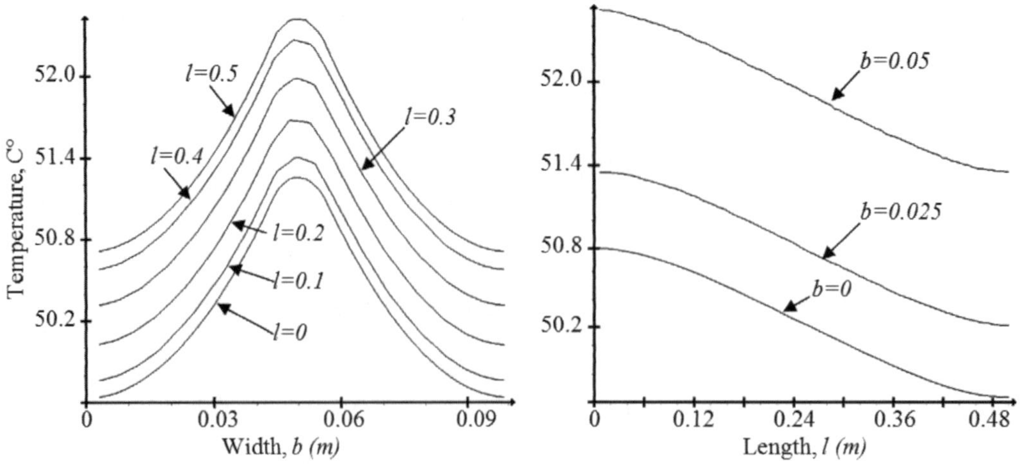

Figure 3. Radiating surface temperature profiles across and along fin-tube radiator.

5. Results and discussions

The impact of speed and coolant temperature on the performance of the section was estimated. The thermodynamic properties from the database REFPROP were used for calculating the heat transfer coefficient of isooctane. Results of numerical modeling are shown in tables 1,2.

To evaluate the effectiveness we using [6]:

$$\eta = \frac{Q_f}{Q_{f,ideal}} = \frac{\iint \varepsilon\sigma T^4 df}{f\,\varepsilon\sigma T^4}$$

Table 1. Results of the calculation for the temperature of +60 C°.

heat transfer agent			radiation surface		
velocity (ms^{-1})	heat transfer coefficient ($W(m^2K)^{-1}$)	outlet temperature (C°)	max temperature (C°)	min temperature (C°)	fin performance (Wt)
0,02	157,42	50,48	49,65	41,31	29,05
0,05	248,91	55,85	53,48	49,63	31,29
0,1	352,01	57,91	55,02	53,01	32,42
0,2	397,55	58,92	55,56	54,56	32,83
0,5	827,45	59,55	57,81	57,35	33,86
1	1440,68	59,77	58,71	58,49	34,28

Table 2. Results of calculation for the temperature of +30 C°.

heat transfer agent			radiation surface		
velocity (ms^{-1})	heat transfer coefficient ($W(m^2K)^{-1}$)	outlet temperature (C°)	max temperature (C°)	min temperature (C°)	fin performance (Wt)
0,02	160,21	23,49	22,78	16,96	20,81
0,05	253,32	27,23	25,28	22,72	22,04
0,1	358,25	28,58	25,65	25,27	22,63
0,2	372,02	29,29	29,72	26,05	22,77
0,5	774,30	29,78	28,37	28,14	23,34
1	1348,17	29,85	29,06	28,91	23,58

The result of numerical studies has shown that increasing the coolant flow rate calculation and the efficiency of the radiator (see Fig. 4). This is due to two factors: the increase in the coefficient of heat transfer by increasing the flow rate and a decrease in the temperature difference in the coolant inlet and outlet. The absolute value of the efficiency of the section also increases with increasing temperature of the coolant. The efficiency for the temperature + 30 C° is more than for the temperature + 60 C°. It is connected with the fact that the heat flow at + 30 C° is less than correspondingly smaller temperature difference on the surfaces that directly affects the efficiency.

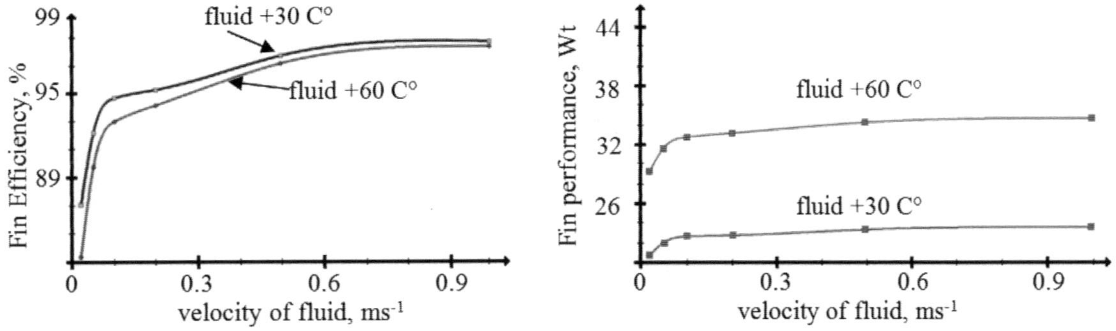

Figure 4. Comparison of the effect of various factors on the performance of the radiator.

IOP Conf. Series: Materials Science and Engineering **93** (2015) 012007 doi:10.1088/1757-899X/93/1/012007

6. Conclusion

It should also be noted that the increase in the flow of coolant in the thermal control system increases the size and weight of the coolant pumps. In general, the results showed the possibility of the numerical solution of coupled heat transfer area for the radiator. For solving the optimization problem for several parameters (the area of the panel, flow geometry) it must be simplified the consideration of the problem (for example, using 1-D algorithm described above).

References

[1] Massardo A F, Tagliafico L A, Fossa M and Agazzani A 1997 *J. Propul. Power.* **13(4)** 560–4
[2] Siamidis J 2006 Heat rejection concepts for lunar fission surface power applications *4th Int. Energy Conversion Engineering Conference and Exhibit (IECEC)* (San Diego:AIAA)
[3] Delcov A V, Hodenkov A A and Zhuikov D A 2014 *12th Int. Conf. on Actual Problems of Electronics Instrument Engineering (APEIE)* (Novosibirsk: IEEE)
[4] Modest M 2003 *Radiative heat transfer* (New York: Academic Press)
[5] Minkowycz W J, Sparrow E M and Murthy J Y 2000 *Handbook of numerical heat transfer* (Hoboken: John Wiley and Sons)
[6] Kamrava M H and Bazdidi-Tehrani F 2011 *Applied Mechanics and Materials* **110-116** 23–8
[7] Peniguel C, Rupp I, Rolfo S and Hermouet D 2014 Conjugate heat transfer study of a wire spacer SFR fuel assembly thanks to the thermal code SYRTHES and the CFD code Code_Saturne *Joint International Conference on Supercomputing in Nuclear Applications + Monte Carlo* ed D Caruge et al (Paris: EDP Sciences)

IOP Conf. Series: Materials Science and Engineering **93** (2015) 012008 doi:10.1088/1757-899X/93/1/012008

Effect of buoyancy force on turbulent modes of complex heat transfer in an air-filled square cavity

I V Miroshnichenko and M A Sheremet
Tomsk State University, Tomsk, Russia

E-mail: miroshnichenko@land.ru, michael-sher@yandex.ru

Abstract. Turbulent natural convection with the interaction of surface radiation in square enclosure has been numerically studied. The governing equations are solved contemporaneously by finite difference method to obtain the velocity, temperature and heat flux distributions. Turbulence has been modeled using the standard k–ε model. The change of convective and radiative Nusselt numbers with respect to time has been described. The analysis is carried out over a wide range of Rayleigh number from 10^8 to 10^{10}. The effect of this key parameter on temperature and velocity distributions, convective and radiative Nusselt numbers has been investigated. The results clearly demonstrate a significant effect of buoyancy ratio on unsteady turbulent heat transfer.

1. Introduction

Heat and mass transfer induced by natural convection has been well studied for the last two decades due to its importance in many areas such as building insulations, solar energy collectors, nuclear reactor design and so on. Radiative heat transfer plays a significant role in heat transfer in enclosures. In the literature, many researches have been done in the area of numerical analysis of natural convection with the interaction of surface radiation [1-3].

Lari et al. [2] studied the effect of radiative heat transfer on natural convection heat transfer in a square cavity under normal room conditions. The flow characteristics of the cavity were analyzed at a broad range of Rayleigh numbers from 10^2 to 10^6. They showed that the radiation plays a significant role on temperature distribution and flow pattern in the cavity. Shati et al. [3] investigated the effects of natural convection with and without the interaction of surface radiation in square and rectangular enclosures. It was shown that the ratio between Rayleigh number and Nusselt number follows a similar trend.

The goal of this paper is to investigate the effect of Rayleigh number on unsteady turbulent modes of complex heat transfer in an enclosure.

2. Physical and mathematical models

The geometry of the problem is shown in Figure 1. This is a two-dimensional fluid flow in a square enclosure. Two vertical walls ($x=0, x=L+2h$) are kept at constant temperatures T_h and T_c respectively.

The external surface of the bottom wall ($y=0$) and top wall ($y=H+2h$) were considered to be adiabatic. The internal surfaces of the walls are suggested to be gray diffusive emitters and reflectors of radiation. Turbulence is modeled using the standard k–ε model [4].

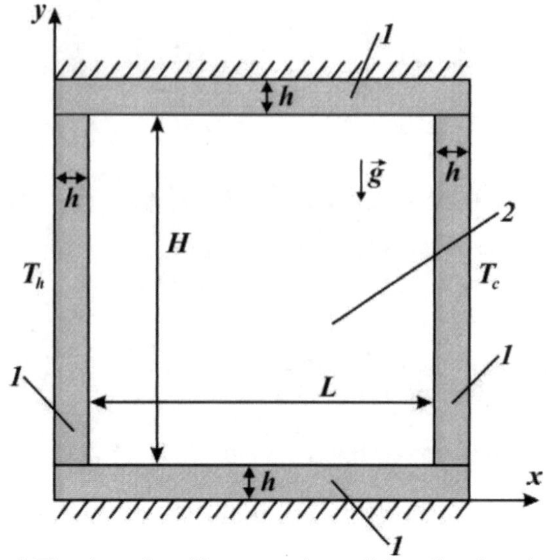

Figure 1. The domain of interest: *1* – solid walls, *2* – air cavity.

The continuity, Navier–Stokes and energy equations are given below:
inside the air cavity (*2* in Figure 1):

$$\frac{\partial u}{\partial x}+\frac{\partial v}{\partial y}=0 \tag{1}$$

$$\frac{\partial u}{\partial t}+u\frac{\partial u}{\partial x}+v\frac{\partial u}{\partial y}=-\frac{1}{\rho}\frac{\partial p}{\partial x}+2\frac{\partial}{\partial x}\left[(\nu+\nu_t)\frac{\partial u}{\partial x}\right]+\frac{\partial}{\partial y}\left[(\nu+\nu_t)\left(\frac{\partial u}{\partial y}+\frac{\partial v}{\partial x}\right)\right] \tag{2}$$

$$\frac{\partial v}{\partial t}+u\frac{\partial v}{\partial x}+v\frac{\partial v}{\partial y}=-\frac{1}{\rho}\frac{\partial p}{\partial y}+\frac{\partial}{\partial x}\left[(\nu+\nu_t)\left(\frac{\partial u}{\partial y}+\frac{\partial v}{\partial x}\right)\right]+2\frac{\partial}{\partial y}\left[(\nu+\nu_t)\frac{\partial v}{\partial y}\right]+g\beta\Delta T \tag{3}$$

$$\frac{\partial T}{\partial t}+u\frac{\partial T}{\partial x}+v\frac{\partial T}{\partial y}=\frac{\partial}{\partial x}\left[\left(\alpha_2+\frac{\nu_t}{\sigma_T}\right)\frac{\partial T}{\partial x}\right]+\frac{\partial}{\partial y}\left[\left(\alpha_2+\frac{\nu_t}{\sigma_T}\right)\frac{\partial T}{\partial y}\right] \tag{4}$$

$$\frac{\partial k}{\partial t}+u\frac{\partial k}{\partial x}+v\frac{\partial k}{\partial y}=\frac{\partial}{\partial x}\left[\left(\nu+\frac{\nu_t}{\sigma_k}\right)\frac{\partial k}{\partial x}\right]+\frac{\partial}{\partial y}\left[\left(\nu+\frac{\nu_t}{\sigma_k}\right)\frac{\partial k}{\partial y}\right]+P_k+G_k-\varepsilon \tag{5}$$

$$\frac{\partial \varepsilon}{\partial t}+u\frac{\partial \varepsilon}{\partial x}+v\frac{\partial \varepsilon}{\partial y}=\frac{\partial}{\partial x}\left[\left(\nu+\frac{\nu_t}{\sigma_\varepsilon}\right)\frac{\partial \varepsilon}{\partial x}\right]+\frac{\partial}{\partial y}\left[\left(\nu+\frac{\nu_t}{\sigma_\varepsilon}\right)\frac{\partial \varepsilon}{\partial y}\right]+ +\left(c_{1\varepsilon}f_1\left(P_k+c_{3\varepsilon}G_k\right)-c_{2\varepsilon}f_2\varepsilon\right)\frac{\varepsilon}{k} \tag{6}$$

in the solid walls (*1* in Figure 1):

$$\frac{\partial T}{\partial t} = \alpha_{1,2}\left(\frac{\partial^2 T}{\partial x^2} + \frac{\partial^2 T}{\partial y^2}\right) \tag{7}$$

Where

$$P_k = \nu_t\left(2\left(\frac{\partial u}{\partial x}\right)^2 + 2\left(\frac{\partial v}{\partial y}\right)^2 + \left(\frac{\partial u}{\partial y} + \frac{\partial v}{\partial x}\right)^2\right),\ G_k = -\frac{\nu_t}{\sigma_T} g\beta\frac{\partial T}{\partial y},\ \nu_t = c_\mu f_\mu \frac{k^2}{\varepsilon}$$

For $k-\varepsilon$ turbulence model, the constants used are [4]:

$$c_\mu = 0.09,\ c_{1\varepsilon} = 1.44,\ c_{2\varepsilon} = 1.92,\ c_{3\varepsilon} = 0.8, \sigma_k = 1.0, f_\mu = f_1 = f_2 = 1.0, \sigma_T = 1.0,\ \sigma_\varepsilon = 1.3.$$

The initial and boundary conditions for the formulated governing equations (1) – (7) have been described in [5, 6]. The mathematical model has been formulated in terms of the dimensionless stream function, vorticity and temperature [4-6].

The following algebraic transformation is used for thickening of the computational grid:

$$\xi = \frac{1}{2}\left\{1 + \mathrm{tg}\left[\frac{\pi}{2}(2x-1)\gamma\right] \Big/ \mathrm{tg}\left[\frac{\pi}{2}\gamma\right]\right\},\quad \eta = \frac{1}{2}\left\{1 + \mathrm{tg}\left[\frac{\pi}{2}(2y-1)\gamma\right] \Big/ \mathrm{tg}\left[\frac{\pi}{2}\gamma\right]\right\}.$$

The dimensionless net radiative heat flux Q_{rad} is determined as follows [6]:

$$Q_{rad,k} = R_k - \sum_{i=1}^{N} F_{k-i} R_i, \tag{8}$$

$$R_k = (1-\varepsilon'_k)\sum_{i=1}^{N} F_{k-i} R_i + \tilde{\varepsilon}_k (1-\zeta)^4\left(\Theta_k + 0.5\frac{1+\zeta}{1-\zeta}\right)^4, \tag{9}$$

Where ν – kinematic viscosity, ν_t - turbulent viscosity, $\alpha_{1,2}$ - thermal diffusivity ratio, α_2 - air thermal diffusivity, α_1 - thermal diffusivity of the wall material, β - coefficient of volumetric thermal expansion, ζ - temperature parameter, Θ - dimensionless temperature, R_k – radiosity of the *k*th element of an enclosure, F_{k-i} – view factor from kth element to the *i*th element of an enclosure, $Q_{rad,k}$ – dimensionless net radiative heat flux of the *k*th element of an enclosure, γ - compaction parameter.

The governing equations (1)–(7) are solved by the finite difference method. The solution procedure is described in [4]. The developed computational code has been verified for several benchmark problems [4-6].

3. Results and discussion

Numerical analysis has been conducted in a wide range of the Rayleigh number from 10^8 to 10^{10}. Figure 2 shows the graphic dependences of the average convective and radiative Nusselt numbers at the left solid-fluid interface on the dimensionless time and the Rayleigh number.

It was shown that at dimensionless time $\tau = 900$ the convective Nusselt number increases up to 2.1 times at changing of Ra from 10^8 up to 10^{10}. It is worth noting that $\overline{\mathrm{Nu}}_{con}(\tau) < \overline{\mathrm{Nu}}_{rad}(\tau)$ at surface emissivity $\tilde{\varepsilon}$ = 0.6 for a wide range of Rayleigh number. An increase in this parameter leads to an intensification of fluid flow inside the cavity.

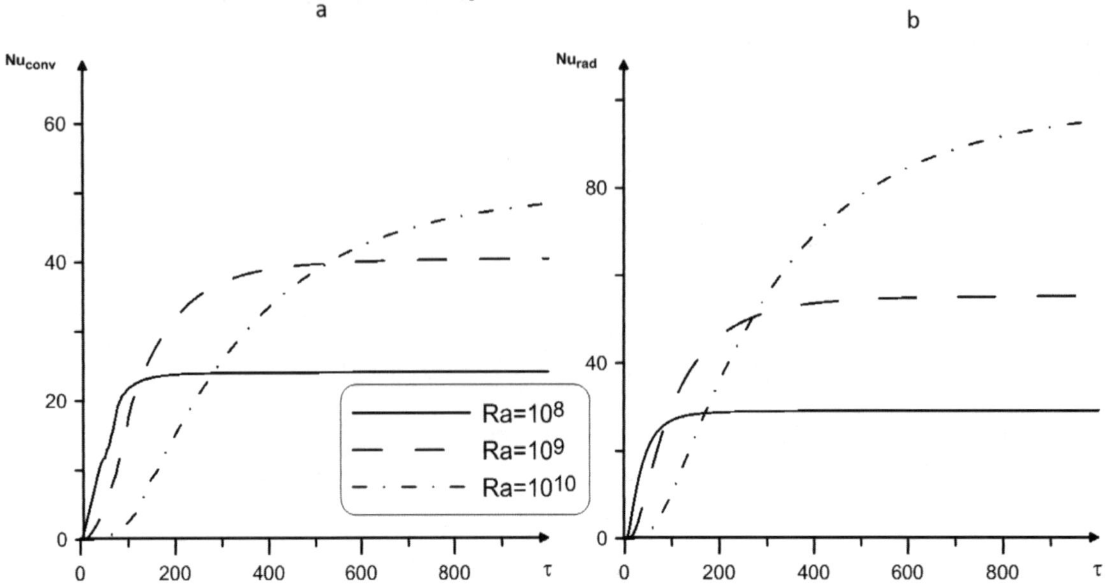

Figure 2. Variation of the average convective (a) and radiative (b) Nusselt numbers at the left vertical wall versus the dimensionless time and the Rayleigh number at surface emissivity $\tilde{\varepsilon}$ = 0.6 and thermal conductivity ratio $k_{1,2}$ = 100.

4. Conclusion

The distributions of streamlines and isotherms describing the main features of the flow pattern and heat transfer have been obtained. The influence of the buoyancy force on both velocity and temperature fields and the convective and radiative Nusselt numbers at solid–fluid interface has been analyzed. It has been found that thermal boundary layers are formed close to the vertical solid-fluid interfaces for high values of the Rayleigh number. It is therefore hoped that this type of investigation would lead to an improved knowledge in the area of numerical analysis heat transfer characteristics.

Acknowledgments

This work was supported by the Grants Council (under the President of the Russian Federation), Grant No. MD-6942.2015.8 and by the Russian Foundation for Basic Research, Grant No. 14-08-31137 mol_a.

References

[1] Vivek V , Sharma A K and Balaji C 2012 *Int. J. Therm. Sci.* **60** 70-84
[2] Lari L, Baneshi M, Gandjalikhan N S A, Komiya A and Maruyama S 2011 *Int. J. Heat Mass Tran.* **54** 5087-5099
[3] Shati A K A, Blakey S G and Beck S B M 2012 *J. Eng. Sci. Technol.* **7** 257-279
[4] Miroshnichenko I and Sheremet M 2015 *EPJ Web of Conf.* **82** 01057
[5] Martyushev S G, Miroshnichenko I V and Sheremet M A 2014 *J. Engineering Physics and Thermophysics* **87** pp 124-134
[6] Martyushev S G and Sheremet M A 2014 *Int. J. Therm. Sci.* **76** pp 51-67

The dynamics of thermal regime changes of a local working zone in conditions of its heating by gas infrared radiators

A Nee
Tomsk Polytechnic University, Tomsk, Russia

E-mail: nee_alexander@mail.ru

Abstract. Mathematical modeling of unsteady heat transfer in a closed rectangular area with a local heat supply object in a conjugate formulation in working conditions of radiation source of energy is passed. Fields of temperatures and stream functions, illustrating the influence of a local typical object on thermal regime are received. The effect of Grashof number on dimensionless heat transfer coefficient - Nusselt number is investigated. The influence of nonconducted heat supply object on heat transfer rate in solution domain is showed.

1.Introduction

For modeling of temperature fields of local heat supply objects located in large industrial premises and heated by gas infrared radiators (GIR) was developed an approach [1], which is based on natural convection model in air-filled cavity with solid enclosing walls of finite thickness [2]. Heat sink to the enclosure structures and accumulation of heat in them was allowed at problem formulation [1], but was assumed that the radiation coming from GIR evenly distributed only on the lower horizontal boundary of the heating region. In addition, specific working area (heat supply object) was not considered as an obstruction for the movement from the lower boundary of the heated air in formulation problem [1].

It should be noted that the results of experimental researches analogically considered in [1] of conjugate convective – conductive heat transfer processes still haven't been published. Also the results of theoretical studies of fundamental mechanisms of heat conduction, convection and radiation processes jointly progressing with the work of gas infrared radiators haven't been published as well.

The aim of this study is the numerical simulation of unsteady heating process of a typical manufacturing object in the gas cavity of a closed rectangular area in conjugate formulation.

2.Problem formulation

The boundary value problem of convective – conductive heat transfer in a closed rectangular area consisting of six rectangular subdomains (Fig. 1) was solved. Heat insulation conditions were adopted at its external borders. At internal borders "air – enclosure structures" are the fourth type boundary conditions.

Assumptions that thermal properties of air and enclosure structures don't depend on temperature are introduced. Flow regime is laminar. The air is considered as a Newtonian fluid, incompressible, satisfies the Boussinesq approximation and absolutely transparent for thermal radiation. Total radiant flux coming from the GIR can be represented as the sum of heat fluxes values of which were determined by zonal method as shown in [3]. Investigated heat transfer process is described by the unsteady Navier - Stokes equations and the energy for the air and the heat conduction equation for enclosure structures within the adopted model. Dimensionless equations of Navier - Stokes and energy in variables "vorticity - stream function - temperature " are as follows [1, 2]:

$$\frac{\partial \Omega}{\partial \tau}+U\frac{\partial \Omega}{\partial X}+V\frac{\partial \Omega}{\partial Y}=\frac{1}{\sqrt{\mathrm{Gr}}}\cdot\nabla^2\Omega+\frac{1}{2}\frac{\partial \Theta_1}{\partial X}, \quad (1)$$

$$\nabla^2\Psi=-2\cdot\Omega, \quad (2)$$

$$\frac{\partial \Theta_1}{\partial \tau}+U\frac{\partial \Theta_1}{\partial X}+V\frac{\partial \Theta_1}{\partial Y}=\frac{1}{\mathrm{Pr}\sqrt{\mathrm{Gr}}}\cdot\nabla^2\Theta_1, \quad (3)$$

$$\frac{\partial \Theta_2}{\partial Fo_2}=\nabla^2\Theta_2, \quad (4)$$

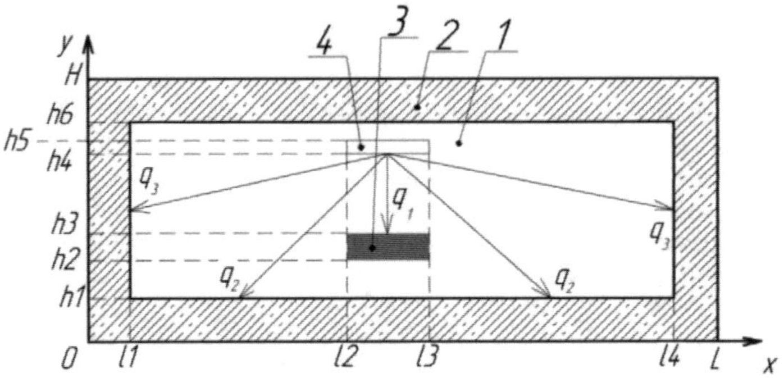

Figure 1. Solution domain: 1) air; 2) enclosure structures; 3) heat supply object; 4) gas infrared radiator (symbolic notation).

The initial conditions for the system of equations (1) – (5) are as follows:

$$\Psi(X,Y,0)=0,\ \Omega(X,Y,0)=0,\ U(X,Y,0)=0,\ V(X,Y,0)=0,$$
$$\Theta_1(X,Y,0)=\Theta_2(X,Y,0)=0.$$

The boundary conditions at the outer boundaries of the solution domain are as follows:

$$X=0,\ X=1,\ 0<Y\le 1:\ \frac{\partial \Theta_2(X,Y,\tau)}{\partial X}=0,$$
$$Y=0,\ Y=1,\ 0<X\le 1:\ \frac{\partial \Theta_2(X,Y,\tau)}{\partial Y}=0.$$

at internal interfaces “heat supply object – air”, “solid wall – air”, parallel to the axis OX:

$$\Psi=0,\ \frac{\partial \Psi}{\partial Y}=0,\ \begin{cases}\Theta_i=\Theta_j,\\ \dfrac{\partial \Theta_i}{\partial Y}=\dfrac{\lambda_j}{\lambda_i}\cdot\dfrac{\partial \Theta_i}{\partial Y}+Ki_{q_k},\end{cases}\ \text{где}\ \left|\begin{array}{l} i=\overline{1,3}\\ j=\overline{1,3}\\ k=1,2.\end{array}\right.$$

at internal interfaces “heat supply object – air", "solid wall - air", parallel to the axis OY:

$$\Psi=0,\ \frac{\partial \Psi}{\partial X}=0,\ \begin{cases}\Theta_i=\Theta_j,\\ \dfrac{\partial \Theta_i}{\partial X}=\dfrac{\lambda_j}{\lambda_i}\cdot\dfrac{\partial \Theta_i}{\partial X}+Ki_{q_3},\end{cases}\ \text{где}\ \left|\begin{array}{l} i=\overline{1,3}\\ j=\overline{1,3}\end{array}\right.$$

where Fo – Fourier number; Gr – Grashof number; Ki –, Kirpichev number; Pr – Prandtl number; X,Y – dimensionless Cartesian coordinates; U, V – dimensionless velocities along X,Y directions correspondingly; $\nabla -$ dimensionless Laplace operator. Indexes: 1,2,3 - design elements; for Ki: - heat flow values. Indexes: 1, 2, 3 – element number; for Ki: q_1, q_2, q_3 – heat flux values.

Equations (1)–(4) with the corresponding initial and boundary conditions were solved by applying the finite difference method, as in [4]. Locally one scheme of A.A. Samarskiy was used for approximation of equations (1), (3) – (4) [5]. Approximation of Poisson's equation was done by the scheme of variable directions [5]. Woods condition [5] was used to determine the boundary conditions for the vortex velocity. One - dimensional difference analogues were solved by the sweep method [5]. To evaluate the reliability of results of the computational modeling, the conservatism of the difference scheme was checked analogously to [6].

3.The results of the numerical simulation

Numerical investigation was performed for the following values of dimensionless criteria: $Gr=10^7$, $Pr=0{,}71$, $Ki_1=42$, $Ki_2=25$, $Ki_3=5$. The results of solving boundary value problem for three possible variants of the heated air flow are shown in Figures 2, 3 and 4.

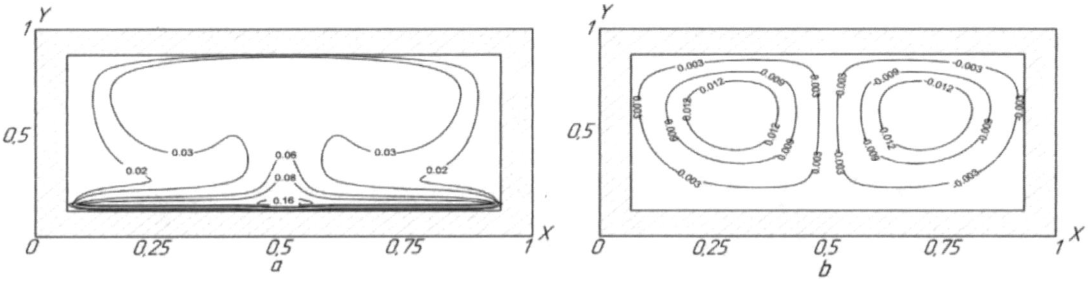

Figure 2. Fields of temperature (a) and the stream function (b) (τ = 3600) in the absence of an heat supply object in the gas cavity.

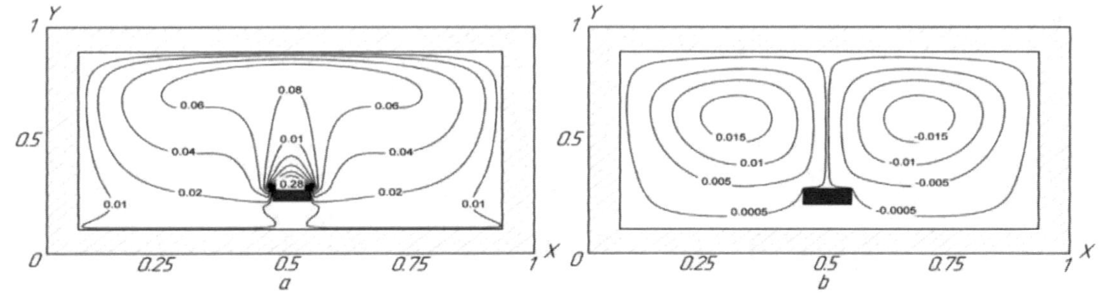

Figure 3. Fields of temperature (a) and the stream function (b) (= 3600) in the presence of an nonconducted heat supply object in the gas cavity (symmetric variant).

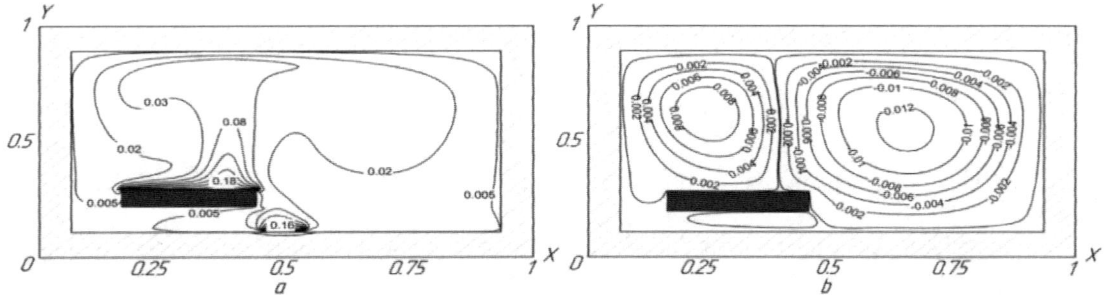

Figure 4. Fields of temperature (a) and the stream function (b) (= 3600) in the presence of an nonconducted heat supply object in the gas cavity (asymmetric variant).

Comparing the results of numerical simulations, it can be concluded that the nonconducted object in the gas cavity significantly effects the character of heat transfer in this solution domain. If the heat supply object is located under GIR (Fig. 3) and coincides with it in size, a significant proportion of the energy comes to the border in y= h3, l2 <x<l3, what is well illustrated by the position of the isotherms. Heated air flows around the object and rises in the result of natural convection then cools down due to heat sink to the enclosure structures and descends along the vertical walls. As the result a symmetrical circulation flow is formed relatively to the section X = 0.5, which clearly demonstrates the field of the stream function (Fig. 3 b). Movement along the stream lines with the sign "-" is oriented clockwise, with the "+" counterclockwise.

For the purpose of analyzing the possible asymmetric flow, the variant with offset of nonconducted object larger lateral dimension to the left is considered (Fig. 4). In this case, an intense air heating occurs at the border = h1, l2 <x <l3. In the section X = 0.39 of the solution domain above the heat supply object an ascending heated air is formed. Besides the two main vortices additional one (under the object) is formed.

By analogy with [7, 8] the influence of the Grashof number at the average Nusselt number was analyzed. At the interface "enclosure structures - air» (y = h1, l1 <x <l4) average dimensionless heat transfer coefficient is defined [8] as:

$$\overline{Nu} = \frac{1}{L}\int_{0,12}^{0,88}\left(\frac{\partial\Theta}{\partial Y}\right)_{Y=0,12} dX.$$

Dependences $\overline{Nu}$ from Gr are presented in Figure 5.

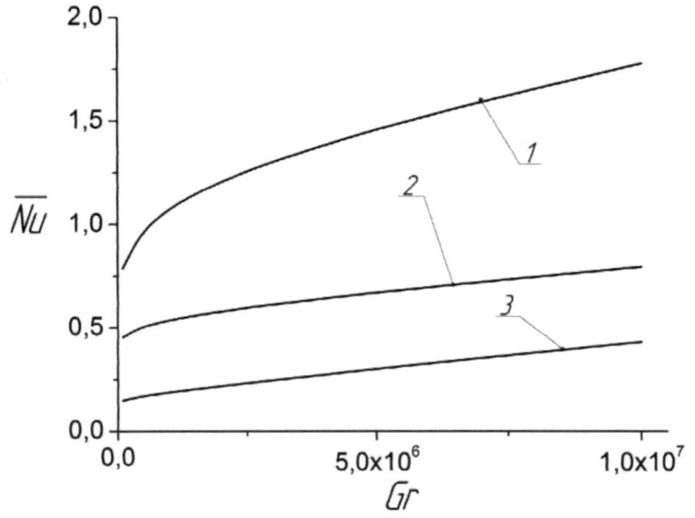

Figure 5. Dependences from Gr (= 3600): 1 – heat supply object is absent; 2 – nonconducted object in the gas cavity (asymmetric variant); 3 – nonconducted object in the gas cavity (symmetric variant).

Figure 5 shows that is increased with rising Gr. Also nonconducted object in the gas cavity (and its location) significantly effects the intensity of convective heat transfer. If the heat supply object is absent, the average dimensionless heat transfer coefficient in section Y = 0,12 of the solution domain 4.5 times greater compared with the case of its presence. It can be concluded that with placement of the GIR the location of the workplace in order to optimize heat transfer conditions must be allowed.

4.Conclusion

Numerical simulation results show that the presence of the heat supply objects in the gas cavity significantly changes the temperature field and the intensity of the air movement in the working area of industrial premises.

Acknowledgments

Scientific supervisor: Head of Thermal Theory and Engineering Department of the Energy Institute of the National Research Tomsk Polytechnic University, Doctor of Physical and Mathematical Sciences Kuznetsov G.V.

This work is the part of the Research Works of the State publisher "Nauka" №13.1339.2014/K (code of the target-oriented scientific-technological program 2.1410.2014)

References

[1] Kuznetsov G.V., Maksimov V.I., Nagornova T.A., Kurilenko N.I and Mamontov G.Ya 2013 *J. of Eng. Phys. and Thermoph.* **86** 519 – 524.
[2] Kuznetsov G.V. and Sheremet M.A. 2009 *T. and A.* **16** 119 – 128.
[3] Bukhmirov V.V, Krupennikov S.A. and Solnyshkova Yu.S. 2010 *Vestn. Ivanovsk. Gos. Energ. Univ.* **4** 23–25.
[4] Kuznetsov G.V. and Sheremet M.A. 2008 *Rus. Microel.* J. **3** 131 – 138.
[5] Paskonov V.M., Polezhaev V.I. and Chudov L.A. 1984 *Numerical Simulation of the Heat and Mass Transfer Processes* (Nauka)
[6] Kuznetsov G.V. and Strizhak P.A. 2009 *J. of Eng. Thermoph.* **18** 162 – 167.
[7] Kuznetsov G.V. and Sheremet M.A 2006 *Fl. Dyn.* **41** 881 – 890.
[8] Nee A.E and Nagornova T.A. 2014 *IOP Conf. S.: M. Sci. and Eng.* **66** 1 – 5.

Dynamic contact angle and three-phase contact line of water drop on copper surface

E G Orlova, D V Feoktistov and K A Batishcheva
Thermal Theory and Engineering Department, Power Engineering Institute, National Research Tomsk Polytechnic University, 30, Lenin Avenue, 634050, Tomsk, Russia

E-mail: enja1991@rambler.ru

Abstract. Nowadays there is a lack of experimental data describing the physical process of drop spreading on a solid metal surface for developing wetting and spreading theory. The experimental data obtained by using the high speed video-recording will allow to identify unknown previously spreading modes as well as the change of the dynamic contact angle and the three-phase contact line. The purpose of the work is to determine the effect of the drop growth rate and the copper substrate surface roughness on the dynamic contact angle and the three-phase contact line speed at distilled water drop spreading. Shadow and Schlieren methods are used to obtain experimental data. Three drop spreading modes on the rough surfaces were identified. Time dependences of the dynamic contact angle and contact line speed were obtained. Experimental results can be used for assessing the validity of the developed mathematical models of wetting and spreading processes in the field of micro- and nano-electronics, ink jet printing, thin-film coatings, spray cooling, and optoelectronics.

1. Introduction

The surface microstructure and conditions of drop formation (growth rate) are well known [1-6] to influence the hydrophobic and hydrophilic properties of materials. The theory of wetting and spreading processes is not developed at a level providing an opportunity of heat exchangers' construction with a patterned surface that intensifies heat transfer. One of the factors suppressing the development of scientific achievements in this sphere is an insufficient number of experimental results describing the spreading process taking into account the droplet formation conditions. Experimental data of the dynamic contact angle (DCA) and the three-phase contact line speed can be used to assess the validity of the developed mathematical models of wetting and spreading processes.

The purpose of the presented work is to obtain experimentally dependences on DCA and the three-phase contact line speed from the drop growth rate and the surface roughness of copper substrates.

2. Research technique

Research was conducted by using the experimental setup (figure 1) consisting of the equipment for shadow and Schlieren methods implementation [7, 8].

In the shadow optical method light source 3 with lens (placed in the widest part of the light source case) was used to produce a beam of plane-parallel light illuminating the drop on the substrate. The distance between the lens and the object under investigation must be greater or equal to the focal length ($h \geq F$).

In Schlieren method incoherent light source 4, ground glass 10 and coding filter 11 were used for a light flux with a stepped decrease of intensity on the space. A light beam from source 4 passed through collimating lens 5, which transformed it into a plane-parallel one. Then it was reflected from beam splitter 9, fell on to the substrate, passed to lens 8 and projected on the sensor of the high-speed video camera 7. The transparent shield with an opening 6 is set for reducing the influence of external light sources.

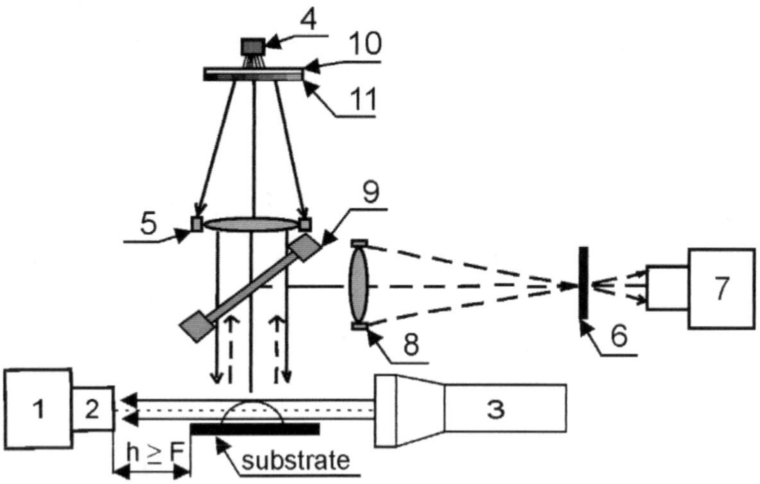

Figure 1. A schematic view of the experimental setup: 1 – photographic camera; 2 – macro lens; 3 – light source of Shadow method; 4 – light source of Schlieren method; 5 – collimating lens; 6 – transparent shield with an opening; 7 – high-speed video camera; 8 – lens; 9 – beam splitter; 10 – ground glass; 11 – coding filter.

Video recording of spreading drops on the surface was carried out simultaneously in two coordinate directions. The equipment for Schlieren method implementation was used to control the drop symmetry. If the drop lost symmetry, the experiment was repeated.

Three copper substrates (54 mm in diameter and 4 mm thick) with a through opening (1 mm in diameter) located on the center were used. The surface microstructure of two substrates was formed by bombardment with Al_2O_3 10 and 100 μm sized particles. The third substrate was made of the flexible copper.

According to results of the preliminary experiment, the values of influencing factors were defined (table 1).

Table 1. The main influencing factors.

Parameter	Value
Drop volume (ml)	0.3
Drop growth rate (is controlled by syringe pump, Figure 5) (ml/s)	0.005; 0.02; 0.04; 0.08
Substrate material	Copper
Roughness parameter of surface – Ra (μm)	Sample № 1 – flexible copper with Ra 0.591; Sample № 2 – copper with Ra 5.190; Sample № 3 – copper with Ra 6.210;
Wetting liquid	nondeaerated distilled water

Surfaces of substrates were investigated on the profilometer "Micro Measure 3D station" and microscope TM-3000. The parameter of roughness (arithmetic average roughness Ra) and microstructure were defined. Surfaces microstructure and substrates view are presented in figures 2-4.

(a) (b)

Figure 2. Sample No 1 – flexible copper Ra 0.591 [μm]: (a) - surface microstructure (Magnification: x3000); (b) - substrate view.

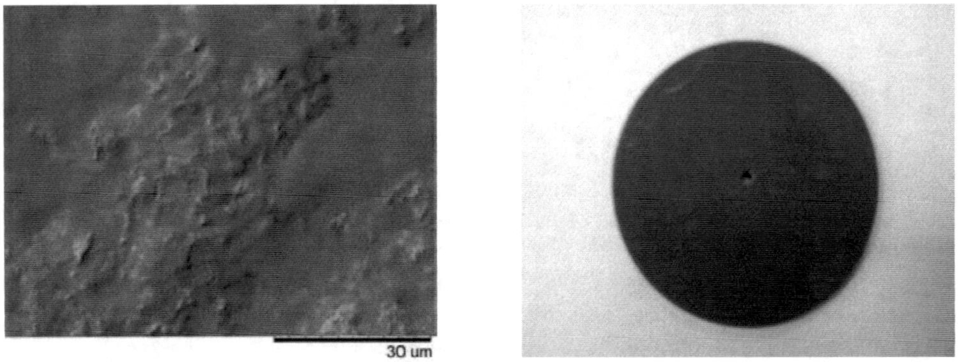

(a) (b)

Figure 3. Sample No 2 – copper Ra 5.190 [μm]: (a) - surface microstructure (Magnification: x2000); (b) - substrate view.

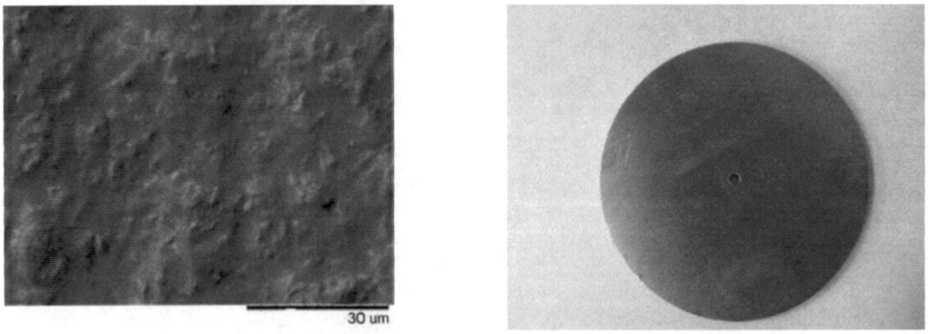

(a) (b)

Figure 4. Sample No 3 – copper Ra 6.210 [μm]: (a) - surface microstructure (Magnification: x2000); (b) - substrate view.

The roughness of the sample No 1 (flexible copper) is formed by longitudinally arranged grooves.

The roughness of the samples No 2 and No 3 (copper substrates) is formed by chaotically arranged asperities and cavities.

The drop was formed on the surface by the syringe pump (Cole-Parmer Touch Screen) (Figure 5). The nondeaerated distilled water was squeezed on the surface through the channel placed in the substrate. This bottom-up methodology of droplet formation in comparison with the known by syringe dispenser [9, 10] facilitates a precise control of droplet formation and size, as well as allows to reduce the error at maintaining the initial volume.

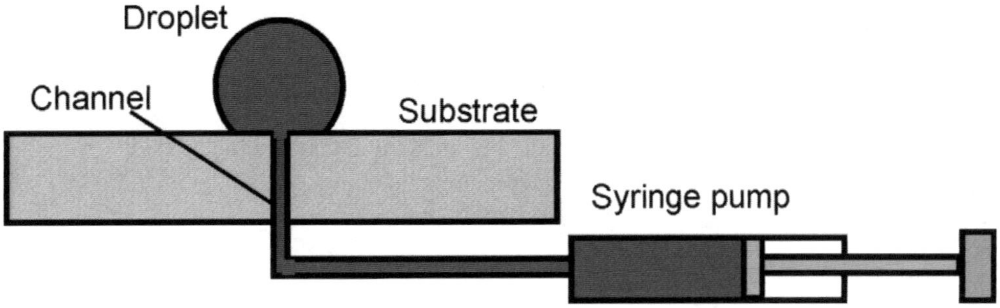

Figure 5. The bottom-up methodology of droplet formation on the surface.

The drop profile was obtained by the Shadow method. The equipment for implementation of this method includes the high-speed video camera (FastVideo-500M) and the light source (MI-150 Edmund). Geometrical parameters (contact diameter, height and contact angle of the drop) were defined at processing images in Drop Shape Analysis (DSA) software (KRUSS Company).

According to the obtained experimental data, the contact line speed was calculated. This parameter describes the rate of spreading drops on a surface and is determined from the following equation:

$$Vk = \frac{C \cdot (d_3 - d_1)}{2 \cdot (n_3 - n_1)}, \tag{1}$$

where C is a frame capture speed, frame/s; d_1, d_3 are droplet diameters in the previous and the next moment, mm; n_1, n_3 are numbers of the frame at previous and the next moment in time, frame.

3. Results and discussion

Time dependences on DCA and the three-phase contact line speed at the drop growth rate from 0.005 to 0.08 ml/s are shown in figure 6.

Three drop spreading modes on the rough copper surfaces were conditionally subdivided. The first mode is characterized by an abrupt increase in the three-phase contact line speed. The advancing DCA and speed decrease monotonously in the second mode. The equilibrium contact angle at the constant wetted area is formed in the third mode.

Duration of identified modes in experiments from all time of spreading equaled: the first – 1-2%, the second – 39-50%, the third – 48-60%.

According to the data obtained by the authors [11, 12], the spreading process after collision of a drop with a solid occurs simultaneously in the two modes. The first mode is the movement of liquid within the grooves of a roughened surface. The second is spreading liquid over the grooves of a surface. At high velocities of collision the first mode dominates. At low velocities the movement over the grooves prevails. It is obvious, that these modes are implemented at drop spreading under the gravity force without collision with a substrate. However, when applied the syringe pump at drop placing (bottom-up method), at high drop growth rate (above 0.010 ml/s) the movement of liquid over the grooves is dominant. In this case cavities of microstructure are filled with gas, and the contact of liquid with a solid substrate represents a heterogeneous interphase "liquid-solid-gas". So that, the liquid drop moves over "air cushion" [13]. At low drop growth rate (0,005ml/s) the dominant mode is the motion of liquid within the grooves, and the interphase is "liquid-solid".

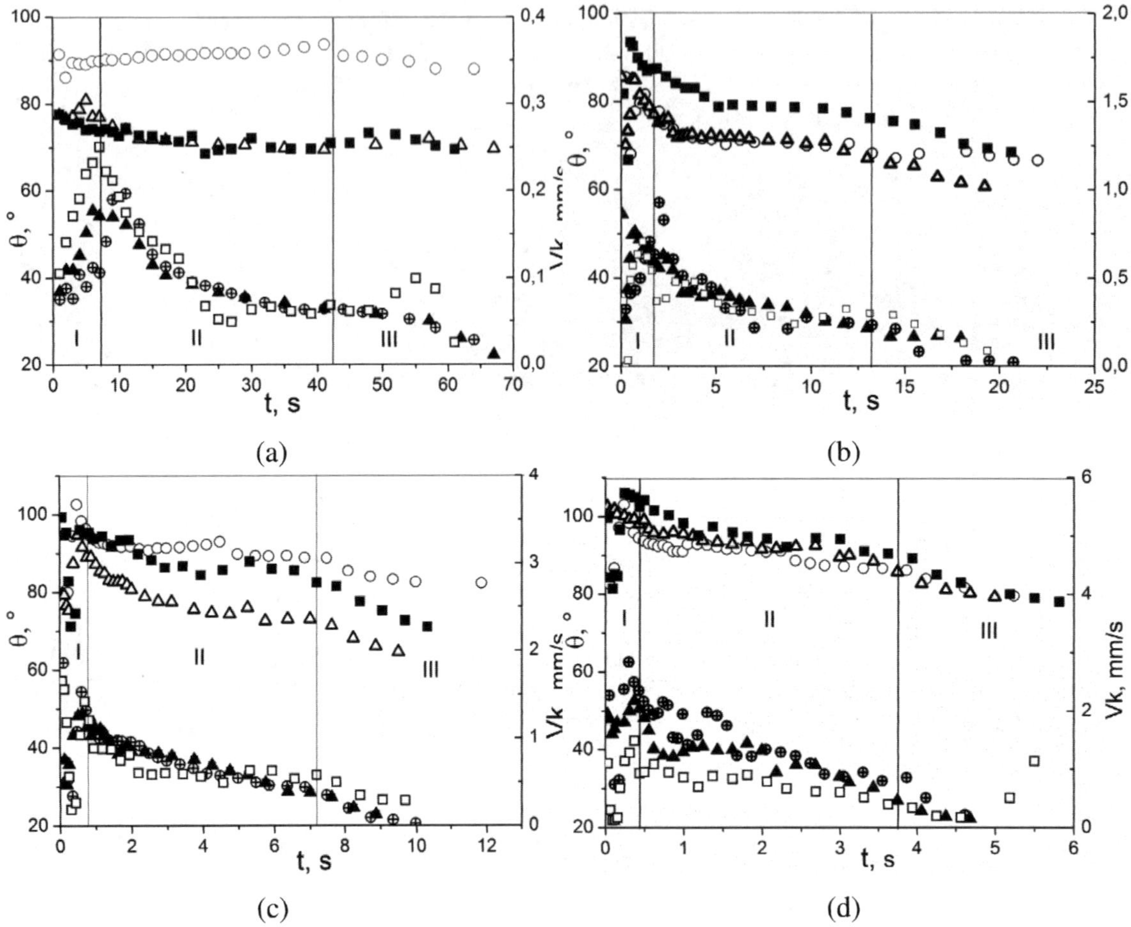

Figure 6. Time dependences on DCA and the three-phase contact line speed. The drop growth rate, ml/s: a – 0.005; b – 0.02; c – 0.04; d – 0.08. The advancing DCA: ○ – sample No 1; △ – sample No 2; ■ – sample No 3. The three-phase contact line speed: ⊕ – sample No 1; ▲ – sample No 2; □ – sample No 3. Spreading modes are shown by Roman numbers: I, II, III.

During the second mode slight increase in DCA was observed at drop spreading on the surface of sample No1 (flexible copper) with the drop growth rate 0.005 ml/s. According to the "model of parallel grooves" proposed in [14-16], at a parallel arrangement of contact line relative to the grooves on the surface, a great number of pinning is possible. The increase in DCA on the surface with parallel grooves can be explained as follows: by increasing roughness, firstly, natural decrease of θ_d is found, but when cavities becomes deep enough, there is a complex structure with lower energy barriers, the value of θ_d increases.

It was found that the increase in the drop growth rate at spreading on the surface with Ra=5.190 μm (Sample No 2, Figure 3) led to increasing the maximum value of DCA of 30%; on the surface with Ra=6.210 μm (Sample No 3, Figure 4) of 37%. DCA increasing on the surface with Ra=0.591 μm (Sample No 1, Figure 2) equaled no more than 8%.

According to the analysis of the experimental results, it was established that non-wetting is replaced by wetting at drop spreading on the copper substrates. The advancing DCA during the contact line movement decreases and after a while (depending on the drop growth rate and the surface microstructure) the contact angle becomes smaller than 90 °.

At high drop growth rate (0.8ml/s) change of the contact line speed occurs non- monotonically. Since at the movement on surface the drop loses an equilibrium state and oscillates.

Random error is calculated based on the experimental results and equals to 3.9% for contact angle and to 1.87% for the contact line speed. Three experiments at fixed factors were performed. Systematic error was measured by syringe pump error Cole Parmer ± 0.355%.

Acknowledgments

Authors thank to prof. Geniy V. Kuznetsov for his help with analyzing the results.

The work was held within the research state assignment “Science” №13.1339.2014/K (Code of Federal Target Scientific and Technical Program 2.1410.2014).

References

[1] Gatapova E Y, Graur I A, Sharipov F and Kabov O A 2015 *Int. J. Heat Mass Transfer* **83** 235
[2] Janardan N and Panchagnula M V 2014 *Colloids Surf.* A **456** 238
[3] Nakoryakov V E, Misyura S Y and Elistratov S L 2011 *J Eng Thermophys Rus* **20** 1
[4] Nakoryakov V E, Misyura S Y and Elistratov S L 2012 *Thermal Science* **16** 997
[5] Gatapova E Ya, Semenov A A, Zaitsev D V and Kabov O A 2014 *Colloids Surf.* A **441** 776
[6] Nakoryakov V E, Misyura S Y, Elistratov S L and Dekhtyar R A 2014 *J Eng Thermophys Rus* **23** 257
[7] Orlova E G, Kuznetsov G V and Feoktistov D V 2015 *EPJ* Web of Conferences **82** 1
[8] Orlova E G and Kuznetsov G V 2014 MATEC Web of Conferences **19** 01006
[9] Pierce E, Carmona F and Amirfazli A 2008 *Colloids Surf.* A **323** 73
[10] Brutin D, Zhu Z, Rahli O, Xie J, Liu Q and Tadrist L 2009 *Microgravity Sci. Technol.* **21** 67
[11] Stapelbroek B B J, Jansen H P, Snoeijer J H and Eddi A 2014 *Soft matter* **10** 2641
[12] Sivakumar D, Katagiri K, Sato T and Nishiyama H 2005 *Phys. fluids* **17** 1
[13] Blake T D and Ruschak K J 1997 *Wetting: Static and Dynamic Contact Lines*. In: Liquid Film Coating ed S F Kistler, P M Schweizer (London: Chapman Hall) pp 63-97.
[14] Johnson R, Dettre R 1964 Contact angle hysteresis. In: *Contact Angle, Wettability, and Adhesion* ed F M Fowkers (Washington, D.C.: American Chemical Society) pp 112-144
[15] Cox R G 1983 *J. Fluid Mech* **131** 1
[16] Mason S G 1978 *Wetting, spreading and adhesion* ed J F Padday (New York: Academic Press) p 321

Experimental Investigation of the Contact Angle at Wetting the Non-ferrous Metals

D V Feoktistov, E G Orlova and K O Ponomarev
Thermal Theory and Engineering Department, Tomsk Polytechnic University, Lenin Avenue, 30, 634050, Tomsk, Russia

E-mail: kotyaponomarev@gmail.com

Abstract. Experimental dependences on the effect of the drop volume from the contact angle under the conditions of the static three-phase contact line formation during wetting the non-ferrous metals (aluminium, magnalium, copper and brass) are presented in the work. The surface of the substrates was investigated by modern equipment (profilometer "Micro Measure 3D station" and microscope TM-3000). The drop was placed on the surface by the precision electronic single-channel pipette (Thermo scientific). Shadow method was used to obtain the drop profile; symmetry of the drop was controlled by Schlieren method. The comparison of the methods used to determine the contact angle on the image of the drop profile was executed. It was established that in spite of influencing the friction and gravity forces the structure of metal surfaces affects greatly the value of the contact angle.

1. Introduction
Over the last twenty years heat and mass transfer processes in mini and microsystems [1-5] have attracted attention of many researchers.

It is caused by the intensive development of microelectronics and medicine, as well as the miniaturization of various devices and control of technological processes (aerospace industry, automobile manufacturing, and transportation). There is an intensive development of the micro- and nanosized heat exchange systems. It is found [6] that such systems are much more efficient than their analogues with characteristic dimensions of units and tens of millimeters. For example, in recent years thermosiphons and flat heat pipes with the transverse sizes of 2 millimeters or less are used for microelectronic equipment cooling. However researches in this area are conducted primarily as a search of functional construction versions (it concerns to coolants, materials, regime parameters, sizes, configuration of elements). Such search does not include analyzing the physics of the basic processes occurring in such equipment. Materials for manufacturing the mini-sized heat exchange devices are non-ferrous metals (brass, magnalium, copper, aluminium and others). Studying the physics of the wetting process is possible with using the high-speed video recording and subsequent analysis of frames in order to determine the contact angle.

The purpose of the present work is to determine experimentally the effect of the drop volume on the contact angle under the conditions of the static three-phase contact line formation during wetting the non-ferrous metals.

2. Research technique
Substrates made of aluminium, magnalium, copper and brass were used in experimental research.

Surfaces of substrates were investigated on the profilometer "Micro Measure 3D station" and microscope TM-3000. The parameter of roughness (arithmetic average roughness) and microstructure were defined for substrates under investigation (figures 1-4). The roughness of these substrates is formed by longitudinally arranged grooves.

Figure 1. Microstructure of aluminium substrate (Ra=0.374 μm).

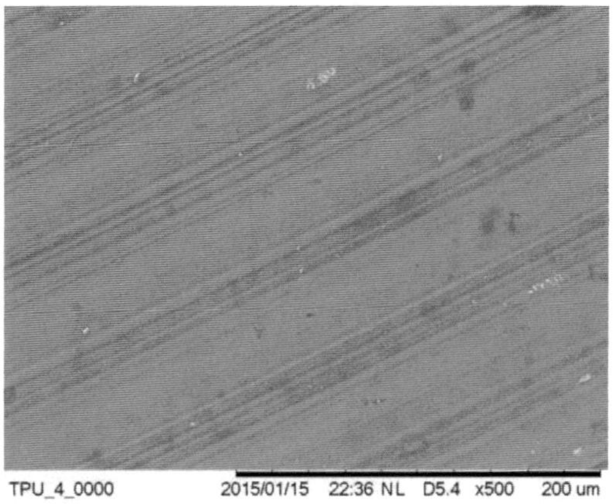

Figure 2. Microstructure of magnalium substrate (Ra=0.51 μm).

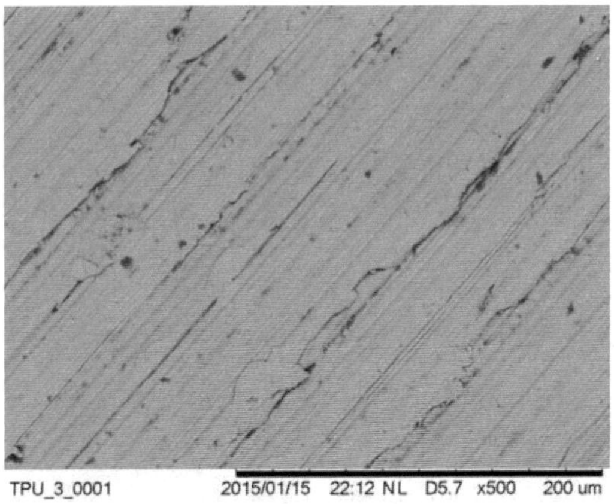

Figure 3. Microstructure of copper substrate (Ra=0.361 μm).

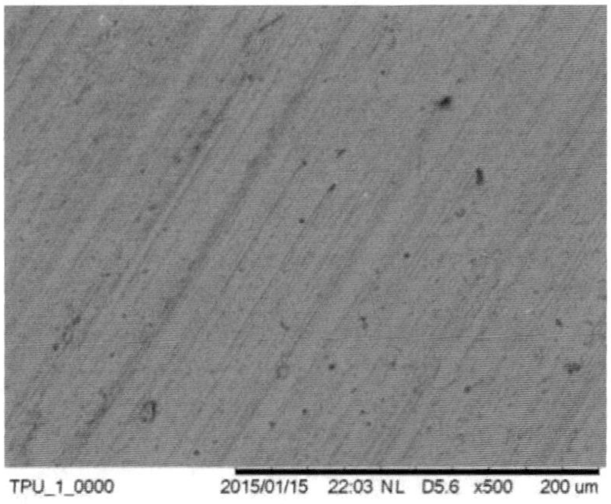

Figure 4. Microstructure of brass substrate (Ra=0.939 μm).

The research was conducted by using the experimental setup (figure 5) consisting of the equipment for Shadow and Schlieren methods implementation [7, 8].

Drop with predetermined volume was placed on the surface by the single-channel Thermo Scientific Finnpipette Novus Electronic pipette. Shadow method was used for obtaining the drop profile. The photographic camera 1 (NIKON D7100) with micro lens (Nikon 105mm f/2.8G IF-ED AF-S) and the light source 2 (MI-150 Edmund) were used for implementation of this method. The contact angle was obtained during image processing in software Drop Shape Analyses (DSA) by two methods (tangential, Young-Laplace). Schlieren method was used to control the drop symmetry.

In shadow optical method the light source 2 with a lens (placed in the widest part of the light source case) are used to produce a beam of plane-parallel light illuminating the drop on the substrate. The distance between the lens and the object under investigation must be greater than or equal to the focal length ($h \geq F$).

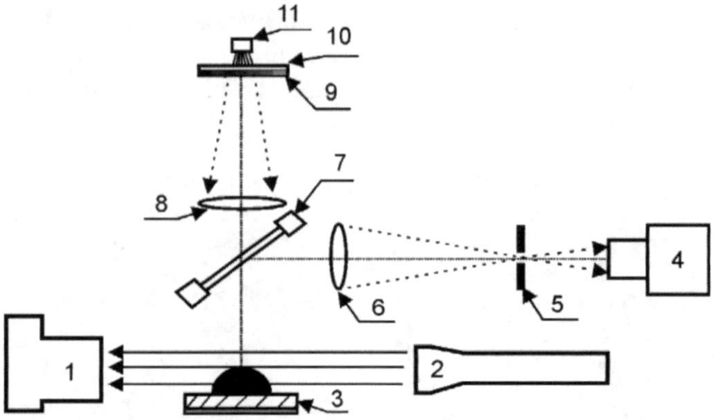

Figure 5. A schematic view of the experimental setup: 1 – photographic camera; 2 – light source of Shadow method; 3 – substrate; 4 – high-speed video camera; 5 – transparent shield with an opening; 6 – lens; 7 – beam splitter; 8 – collimating lens; 9 – coding filter; 10 – ground glass; 11 – light source of Schlieren method.

In Schlieren method the source of incoherent light 11, the ground glass 10 and the coding filter 9 are used for the light flux with a stepped decrease of intensity in space. A beam of light from the source 11 passed through the collimating lens 8, which transformed it into a plane-parallel. Then it was reflected from the beam splitter 7, fell to the substrate and passed to the lens 6 and projected on the sensor of the high-speed video camera 4. The transparent shield with an opening 5 is set for reducing the influence of external light sources .

3. Results and discussion

According to the results of experiments, we obtained the values of contact angle (θ) depending on the drop volumes (V). The experimental data were processed by the least squares method. The maximum value (θ_{max}) and the corresponding drop volume (V) were determined according to the first derivative of the second order polynomial dependencies. The results are presented in table 1.

Table 1. Results of calculation of the contact angle and the drop volume.

Substrate	Young-Laplace method		Tangential method	
	θ_{max} (°)	V (μl)	θ_{max} (°)	V (μl)
Aluminium	86.2	67.4	104.2	78.0
Magnalium	85.5	41.7	85.1	43.4
Copper	85.0	34.3	84.0	35.5
Brass	101.3	35.7	97.9	35.8

The contact angle is plotted over the drop volume for different substrates: brass, magnalium, aluminium, and copper (figure 6).

The maximum divergence of θ_{max} at obtaining by Young-Laplace and tangential methods was 17% for the aluminium substrate. For magnalium, copper, and brass this value is not more than 4%. Probably, it is due to the most highly reflective surface of the aluminium substrate as compared to other non-ferrous metals, used in the experiments.

According to the classical theory of wetting and spreading, it is known [9] that at increasing the drop volume the contact angle can both increase and decrease. This uncertainty can be explained the following way: at increasing the drop volume the friction force and the weight of the drop impact on the contact angle. Increase or decrease in the contact angles happens depending on the ratio of these forces and volumes.

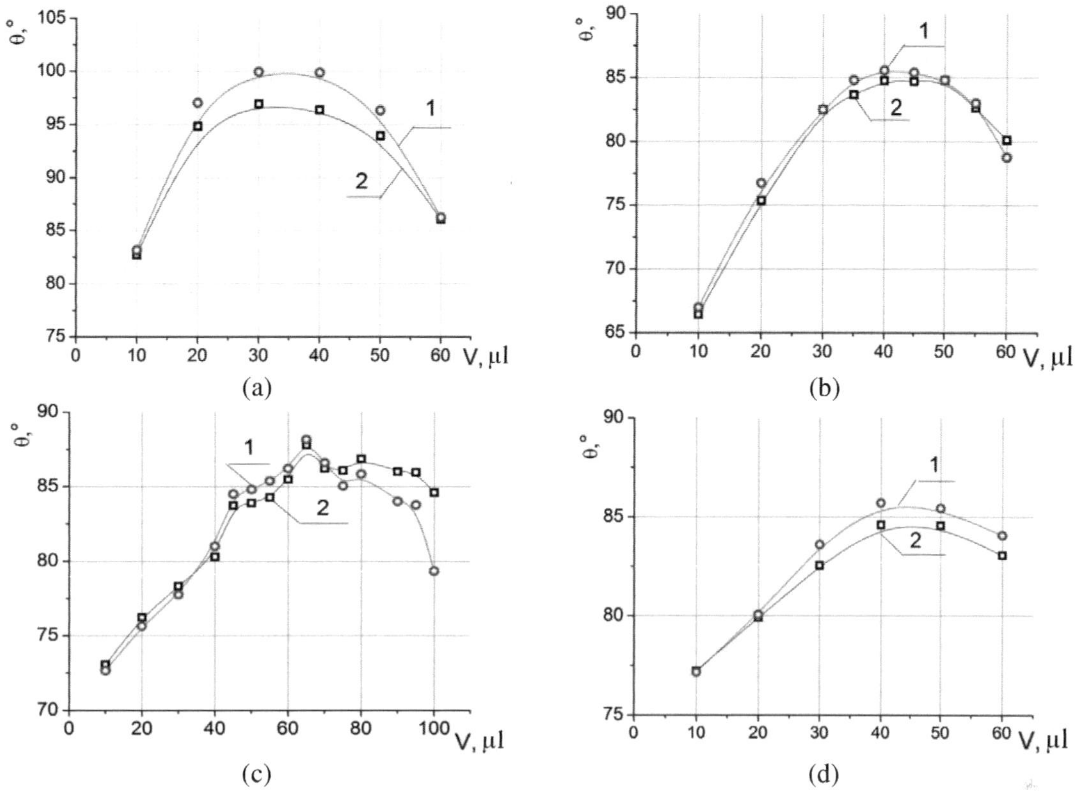

Figure 6. Dependences on the contact angle from the drop volume. Material of the substrate: (a) – brass; (b) – magnalium; (c) – aluminium; (d) – copper. The method of processing the experimental data: 1 - tangential 2 - Young-Laplace.

According to the results of the experiments, it was found that, except the listed forces, material of the substrate affects significantly. The maximum value of θ for each substrate is achieved at different drop volumes (table 1).

It is known [9] that the surface tension at the interface of "liquid-solid" system depends on the nature of the contacting medium and the electric potential.

The dielectric capacitance of the surface of non-ferrous metal is affected by composition and structure. If the surface has a looser structure, so polar molecules can be implemented into such structure (for example, molecules of water or alcohol). It leads to increasing the dielectric capacitance of the surface. In this case, the adhesion between a solid surface and wetting liquid increases. According to the analysis of surface structures of non-ferrous metals used in this experiment (figures 1-4), it was established that the brass and magnalium have a more compact structure. In this case, the implementation in the structure of these metals for the polar liquid molecules is difficult, that affects the large contact angle for the same volume of liquid for the magnalium and aluminium, brass and copper substrates.

4. Conclusion

The influence of the drop volume on the value of the contact angle under the conditions of the static three-phase contact line formation during wetting the non-ferrous metals was defined. A comparison of methods for image processing, allowing to identify the contact angle was executed. The maximum divergence of θ_{max} at obtaining by Young-Laplace and tangential methods was 17% for the aluminium substrate. For magnalium, copper and brass this value is not more than 4%. Probably it is due to the most highly reflective surface of the aluminium substrate as compared to other non-ferrous metals, used in the experiments. It was established that, except the friction and gravity forces, the structure of non-ferrous metals influences greatly on the value of the contact angle. The polar molecules (e.g. water or alcohol) can be implemented into the more loose structures of metals. This increases the dielectric capacitance of the surface. As a result, the adhesion between a solid surface and wetting liquid increases. In it turns, it leads to increasing the hydrophobic properties of the surface.

Acknowledgments

Authors thanks to prof. Geniy V. Kuznetsov for his help with analysing the results.

The work was realized within the research state assignment “Science” №13.1339.2014/K (Code of Federal Target Scientific and Technical Program 2.1410.2014).

References

[1] Misyura S Ya 2014 *Int. Journal Heat Mass Transfer* **71** 197
[2] Nakoryakov V E, Misyura S Ya and Elistratov S L 2012 *Thermal Science* **16** 997
[3] Rioboo R, Marengo M and Tropea C 2002 *Exp Fluids* **33** 112
[4] Gunjal P R, Ranade V V and Chaudharim R V 2005 *AlChE J* **51** 59
[5] Roques-Carmes T, Mathieu V and Gigante A 2010 *J. Colloid Interface Sci.* **344** 180
[6] Misyura S Ya, Nakoryakov V E and Elistratov S L 2013 *J Eng Thermophys Rus.* **22** 1
[7] Orlova E G, Kuznetsov G V and Feoktistov D V 2015 *EPJ Web of Conferences* **82** 01053
[8] Orlova E G and Kuznetsov G V 2014 *MATEC Web of Conferences* **19** 01006
[9] Summ B D and Goryunov Yu V 1976 *The Physochemical Principles of Wetting and Spreading* [in Russian] (Moscow: Khimiya)

IOP Conf. Series: Materials Science and Engineering **93** (2015) 012012 doi:10.1088/1757-899X/93/1/012012

Real-time comprehensive simulation of electric power systems for the task of overvoltages value determination

N Yu Ruban[1], **A S Gusev**[1] **and V A Sulaymanova**[1]
[1]634050, Tomsk, Russia, National Research Tomsk Polytechnic University

E-mail: rubanny@tpu.ru

Abstract. For a correct determining of the overvoltage limiters characteristics, as well as the optimal location of its installation is required to use reliable information about the primary electrical quantities, especially voltage level, in different operation modes of electric power systems (EPS). This problem is highly complex, due to the danger of the experiments in real EPS. Methods and tools for calculation and researches of quasi-stationary and switching overvoltages which are applied to solve this problem not always allow to obtain satisfactory results. The developed tools of comprehensive real-time simulation of electric power systems, providing adequate reproduction of processes associated with overvoltages in EPS, are presented in the article. Fragments of the researches of switching overvoltage in the electrical network of the real power system are given.

1. Introduction

Reliability and efficiency of design, research and operation of the electric power systems (EPS) depend on the completeness and accuracy of the information about working conditions of the equipment of EPS in normal, emergency and post-emergency modes. In particular, such issues are regulated by guidelines and recommendations for the use of surge arresters in EPS. However, the validity of the choice of types and locations of their installation in accordance with these instructions depends on the adequacy of the information about the possible overvoltage value and duration, the most dangerous of which are switching overvoltage. In view of the known specificity and complexity of modern power systems to obtain this kind of information is possible only by mathematical modeling of EPS. To ensure sufficient completeness and accuracy of the mentioned information mathematical models for all significant power system equipment must be three-phase and comprehensive, continuously and adequately reproduce the real range of operating modes and processes. In this way a model of any real power system, taking into account the allowable partial network reduction contains a very stiff, non-linear high dimensionality system of differential equations. According to the theory of discretization techniques for ordinary differential equations, such systems of equations are poor conditioned on the restrictive conditions of applicability of the numerical integration methods and cannot be solved satisfactorily. The only way to improve the conditioning is to reduce the stiffness of the differential order, nonlinearity and in accordance with Dahlquist's Theorem interval solutions restriction [1]. Necessity of these simplifications and limitations inherent to all software and software-hardware purely numerical EPS simulators and prevents adequate modeling of a single continuous spectrum of quasi-steady-state and transient

[1] N Yu Ruban, rubanny@tpu.ru

IOP Conf. Series: Materials Science and Engineering **93** (2015) 012012 doi:10.1088/1757-899X/93/1/012012

processes, including switching surge in EPS with different sorts of normal, emergency and post-emergency conditions of their operation.

2. Hybrid power system simulation

The alternative to the currently used numerical power system simulation [2-5] can be a comprehensive approach, which is hybrid EPS simulation [6-9]. The problem of adequate simulation of large dynamical systems, including EPS, in recent years is discussed in many countries: United States, Europe, China and others. Since 1998 till the present time the annual International Conference on Hybrid Systems: Computation and Control (HSCC) is held.

2.1. Hybrid real-time power system simulator

The concept of continuous, adequate, real-time modeling with an unlimited range of quasi-steady-state and transient processes, including the surge in equipment of EPS, in the case of normal, emergency and post-emergency conditions of their operation was developed to solve problems mentioned above [10, 11]. To implement this concept Hybrid real-time power system simulator (HRTSim) was created [12] and adapted for the researching of all kinds of overvoltage and using this sufficiently complete and accurate information to select the types and locations of surge arresters, including, according to the previously named instructions and recommendations.

2.2. Features of the HRTSim

1. To achieve the necessary completeness and accuracy of the simulation elements of EPS are reproduced by a synthesized for this purpose mathematical models that exclude decomposition of processes in the equipment and the EPS as a whole.

2. The implicit integration method for the continuous, stiff, non-linear systems of differential equations of high dimensionality is used.

3. Formation of simulated three-phase EPS circuit's nodes is at the physical layer by converting the continuous mathematical variable of phase currents of simulated EPS elements to the corresponding physical model currents and voltages. This allows: the almost complete absence of restrictions on the dimension of the simulated power system; adequate representation of the spectrum of all possible three-phase series and shunt switching, including a phase switching; connection of physical models of various equipment and other.

4. Using a digital-to-analog and analog-to-digital conversion of information allows to realize all kinds of automated and automatic control of modeling parameters and settings and obtaining information about operating of relay protection and automation models. Block diagram of the HRTSim is shown in figure 1.

5. Implementation of mentioned above is carried out by developed for this specialized processors (SP), standardized for all kinds of elements EPS and universal for each of them. Each SP includes corresponding physical and mathematical models of power equipment EPS specialized coprocessors (CP) and hybrid microprocessor unit (MPU), consisting of a central processing unit (CPU) and functionally oriented peripheral processors. The three-phase model physical inputs / outputs of all SP equipped with digital-controlled series-shunt switches, which are implemented in the SP. This makes it possible to carry out all kinds of three-phase and per phase switching, including controlled moment of operating.

6. Implementation of mentioned above are based on the latest achievements of IT technologies for potentially necessary range of information and control capabilities for real-time simulation of EPS and effective solutions identified and other problems of design, research and service. For this functional three-phase input / output of A, B, C phases (ABC in figure 1) of the SP are interconnected according to the topology of the simulated power system using a three-phase switch units (TSU), and their CPU of microprocessor units combined through network switches (NS) by local computer network (LCN) with Server, which uses developed for this purpose specialized software.

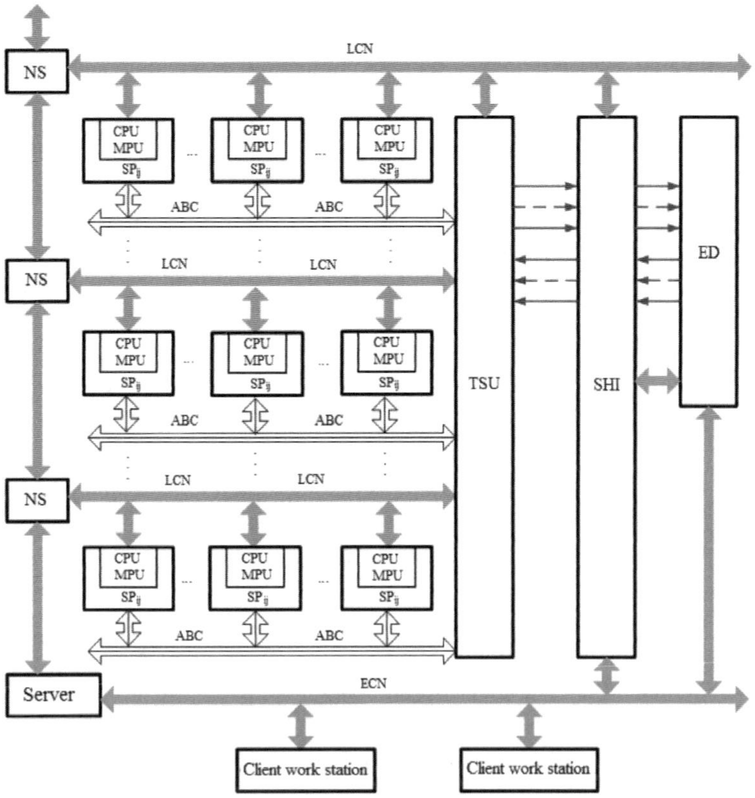

Figure 1. Block diagram of the HRTSim.

7. Professional-oriented software of Client work station is created to use HRTSim. There is no limit to set this software on the server or on users' computers in the external computer networks (ECN), communicating with the server via the Ethernet.

8. For the interaction with real equipment (FACTS control stations, relay protection and automation, SCADA systems, a variety of information management system and other external devices (ED)) are developed appropriate software and hardware interfaces (SHI) that have matching amplifiers for physical interaction with ED and information linkages to the server and SP for the local and external computer networks.

9. For reliable and effective solving of design, research and operation of EPS tasks operational and technical specifications of HRTSim are provided by using of the latest achievements in integrated microelectronics, microprocessor technology and IT.

Developed tools provide continuous and methodically accurate solution in a real time and on an unlimited range with guaranteed acceptable instrumental inaccuracy of stiff, non-linear, high dimensionality systems of differential equations. This allows one to reproduce the spectrum of steady state and transient processes, surges of the real EPS, which in turn makes it possible to produce reliable and valid selection of types and locations of surge arresters installation.

In addition, these features and capabilities of developed simulation tools also provide a reliable solution to the urgent problem of simulated power system model verification, including the specific power equipment, relay protection and emergency automation, various automatic control systems and SCADA systems. Different abilities of HRTSim application with examples of its realization is given in [13-15].

3. Research results

IOP Conf. Series: Materials Science and Engineering **93** (2015) 012012 doi:10.1088/1757-899X/93/1/012012

HRTSim reproduces processes in EPS, including overvoltages by continuous and methodically accurate solution of the same comprehensive model of this EPS. Thus, any quasi-steady mode is the current result of the same solution at a frequency close to 50 Hz, and can be reliably verified by the data of operational information complex server of the simulated power system, especially by dispatching measurements. Consequently in the frequency band defined by the gain frequency and phase-frequency characteristics and by microelectronic components used in the submitted tools, also will reproduce significantly all of the processes of this spectrum. Considered verification and identification thoroughly and repeatedly carried out at implementing specific practical researches with HRTSim using for Tyumen, Tomsk EPS and for united power grid of the Far East. As an example, figure 2 shows switching overvoltage oscillograms in the 500 kV network of Tyumen EPS.

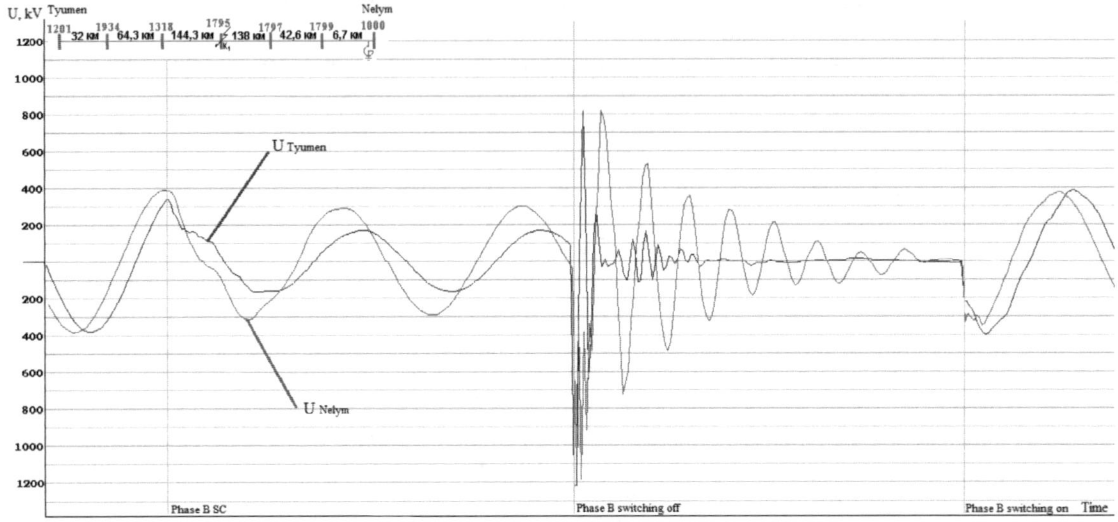

Figure 2. Phase B short-circuit experiment (node 1795) on the 500 kV overhead transmission line Tyumen - Nelym of Tyumen EPS.

These oscillograms display the voltage changes on the buses of Tyumen and Nelym substations of Tyumen EPS over time. In the upper left corner of the figure 2 is presented fragment of the investigated power system scheme, where intermediate nodes and the length of intervals between them are indicated, as well as a fault location is shown here. The experiment as an example of the 500 kV overhead transmission line Tyumen - Nelym of Tyumen EPS includes the following sequence of actions: phase B short circuit (node 1795), switching off the faulted phase and its subsequent switching on.

Switching overvoltages occur after any commutation (planned or emergency) and due to the overcharge of capacities and network parameters changing in the transition process of the system from one state to another. Duration of switching overvoltages can range from a few milliseconds to a few seconds. Switching overvoltages, as well as lightning surges, in most cases higher than acceptable levels of surges for all insulation kinds and therefore should be limited, especially in extra-high voltage electrical installations, which include the 500 kV network. However, the danger of switching overvoltages is determined not only by their large amplitude but also by duration, which is an average of two orders longer than the duration of lightning impulse. Therefore, the switching overvoltage energy is much higher than the energy of lightning surge. As a result the conditions of the equipment operation during the switching overvoltages primarily non-linear resistances of overvoltage limiters is significantly harder because much more power is dissipated in them. The basis for the solution of issues related to the switching overvoltages is initial information about them. It is obvious that the best way of such information obtaining is using of switching overvoltages oscillograms. A set of oscillograms for various power system nodes is a complete map of overvoltages for the region. As a result of our researches the overvoltage map for 500 kV network of Tyumen EPS was made. This map has helped greatly to solve the task of designing and setting up of protection and automation devices for this region including substantiated selection of types and locations of overvoltage limiters installation, especially nonlinear. Part of the research results was verified according to the dispatching department.

4. Conclusion

Confirmed by results of experimental researches the properties and capabilities of developed comprehensive EPS simulation tools allow to reliably verify and adequately reproduce the entire spectrum of steady-state and transient processes in EPS. This also applies to overvoltage, especially switching overvoltage that forms the most dangerous combination of conditions of the equipment operation according to the values and duration. The use of this sufficiently complete and reliable information about the processes in EPS for all possible normal, emergency and post-emergency conditions of its work provides a reliable and effective solution to most of the important tasks of design, research and service.

Acknowledgments

This work was supported by mega-grant "Hybrid simulation and control of smart grids".

References

[1] Hall J and Watt M 1976 Modern numerical methods for ordinary differential equations (Oxford: Clarendon Press)

[2] Feng G, Herrera L, Murawski R, Inoa E, Chih-Lun W, Yi H, Ekici E, Jin W and Beauchamp P 2011 Real time simulation for the study on smart grid *Energy Conversion Congress and Exposition (ECCE)* (Phoenix: IEEE) pp 1013–18

[3] Li W and Xiao X 2010 Electromagnetic and Electromechanical Transient Hybrid Real-time Simulation Technology Based on RTDS used in Subsynchronous Resonance Research *International Conference on Power System Technology (POWERCON)* (Hangzhou: IEEE) pp 1–6

[4] Mahseredjian J, Dinavahi V and Martinez J A 2009 Simulation Tools for Electromagnetic Transients in Power Systems: Overview and Challenges *IEEE Trans. Power Delivery*. **24** 1657–69

[5] Paquin J-N, Belanger J, Snider L A, Pirolli C and Li W 2009 Monte-Carlo study on a large-scale power system model in real-time using eMEGAsim *Energy Conversion Congress and Exposition* (San Jose: IEEE) pp 3194– 3202

[6] Aström K J, Elmqvist H and Mattsson S E 1998 Evolution of Continuoustime Modelling and Simulation *Proc. The 12th European Simulation Multiconference* (Manchester: ESM) pp 1–10

[7] Bond J A, Dorsey J F, Brooke M, Magill J and Tabler J 1994 Fast Solution of Nonlinear Equations Using Parallel Analog Hardware *Analog Integrated Circuits and Signals Processing*. **6** 135–156

[8] Carullo S P, Olaleye M, Nwankpa C O 2004 VSLI Based Analog Power System Emulator for Fast Contingency Analysis *Proc. Hawaii International System Sciences Conf.(Hawaii: IEEE)* pp 1–8

[9] Zhang Y, Gole A M, Wu W, Zhang B, Sun H 2013 Development and Analysis of Applicability of a Hybrid Transient Simulation Platform Combining TSA and EMT Elements *IEEE Transaction on power system*. **5(1)** 357–366

[10] Prokhorov A, Borovikov Yu and Gusev A 2012 Real Time Hybrid Simulation of Electrical Power Systems: Concept, Tools, Field Experience and Smart Grid Challenges *International Journal of Smart Grid and Clean Energy*. pp 67–68

[11] Borovikov Yu and Sulaymanov A 2013 Information and control system for use in the real time multiprocessor simulation of power equipment *Russian Electrical Engineering* **84** 290–295

[12] Borovikov Yu S, Prokhorov A V and Andreev M V 2012 Application of Hybrid Real Time Simulator for solution of Smart Grid tasks on the example of Elgaugol energy cluster pilot project *Proc. Int. Conf. IFOST 2012 (Tomsk)* vol 2 (Tomsk: Tomsk Polytechnic University) pp 604–608

[13] Borovikov Yu S, Sulaymanov A O, Gusev A S and Andreev M V 2014 Simulation of automatic exciting regulators of synchronous generators in hybrid real-time power system simulator *Proc. Int. Conf. Systems and Informatics (ICSAI 2014) (Shanghai)* (Beijing: IEEE) pp 153–158

[14] Ufa R A, Sulaymanova V A and Gusev A S 2014 Hard- and software of real time simulation tools of Electric Power System for adequate modeling power semiconductors in voltage source convertor based HVDC and FACTS *MATEC Web of Conferences, The 2nd International Youth Forum "Smart Grids"* Volume 19

[15] Ruban N Yu, Borovikov Yu S and Sulaymanov A O 2014 Simulation of transmission lines phase-comparison protection for optimization its settings *Proc. Int. Conf. IFOST 2014 (Cox's Bazar)* (Bangalore: IEEE) pp 264–267.

IOP Conf. Series: Materials Science and Engineering **93** (2015) 012013 doi:10.1088/1757-899X/93/1/012013

Method of determining the optimal settings of automatic excitation regulators of synchronous machines in EPS

A Ilina[1] and S Svechkarev [1]
[1] National Research Tomsk Polytechnic University
30 Lenin av., Tomsk, 634050, Russia

E-mail: alina_ilina25@mail.ru

Abstract. The stability of the electric power system can be improved by forming of the correct settings of automatic exciting regulators. Currently, there is no unified methodology of automatic exciting regulators, so analysis of their impact is still an urgent task. The article describes the approach to solving above-mentioned problem, which combines several methods. Research based on Hybrid Real Time Simulator of EPS developed in Tomsk Polytechnic University.

1. Introduction

Normal and stable operation of electric power systems (EPS) is largerly determined by quality of tuning parameters of automatic voltage regulators (AVR) of synchronous machines. The increasing of damping time of transition (emergency) processes in EPS and the emergence of synchronous oscillations, which can lead to an asynchronous operation, are the result of improper adjustment of AVRs. [1]. Since all the processes in the power system are interrelated, and the adjustment of regulators for a generator (generator groups) have an impact on processes in other parts of the power system. Therefore, changing (improving) the adjustment parameters for the AVR of the generator can observe degradation of the regime in other parts of the power system, if before they were tuned optimally. Therefore the problem of an complex approach of the optimal determination of AVRs settings of generator units is still relevant.

2. Materials and methods

Existing approach of optimal settings determination of the AVRs the includes several methods, each of which has some disadvantages:

1. Instructive materials and recommendations. Recommendations and guidelines are developed by the task of adjusting parameters of AVR based on large practical and theoretical experience, which usually includes only ranges of variation these parameters, but doesn't regulate their exact values [2].
2. Determination of optimal parameters by field tests in EPS. This approach is often used by staff of small power plants. However the unjustifiably high risk of accidents and the lack of long-term positive results limit the application of such an approach [3].
3. Testing AVR on simplified network models [4-6]. This is the most modern approach of functional testing and determination of optimal values of parameters AVR. It allows to perform complex inspection the whole station excitation and functioning of protection systems. The equivalenting and using of simplify EPS in particular one generator (bus of infinite capacity) and

communications link that leads to wrong settings in channels of voltage derivatives, excitation current and frequency are the main disadvantage of this approach.

4. Mathematical modeling. The simulation software should calculate the dynamic modes of EPS for reliable determination of the influents of adjustment AVR parameters to oscillatory stability of EPS. However, limitations of the EPS model size and the time step simulation, which should take into account the guaranteed simulation of the electromagnetic processes in mathematical models of AVR (and equipment of EPS in general), don't provide the results in a reasonable time. Another limitation of numerical methods is the accumulation of errors resulted the loss of simulation data [7, 8].

The article describes the approach to solving above-mentioned problem, which combines all the above methods and based on Hybrid Real Time Simulator of EPS developed in Tomsk Polytechnic University [9]. This simulation approach allows to reproduce the three-phase dynamic processes of modeled EPS without errors of numerical methods, as well as to connect to it with various external devices via analog and digital interfaces, such as excitation station. Thus the modeling of processes of EPS is made in real-time on mathematical models, the station of excitation may be connected to Hybrid Real Time Simulator of EPS or reproduced on a built-in model [10], determination of optimal parameters AVRs produced by algorithms automated scripts, and the initial parameters are set on AVR guidance material.

It is necessary to perform only two requirements for the simulated EPS to ensure accurate reproduction of all the processes in the EPS: the similarity of structure and parameters (design and adjusting). This is related to the fact that the mathematical models used in the Hybrid Real Time Simulator of EPS contains fairly complete and detailed description of all power facilities.

It is important to set the disturbance and control of the selected criteria to perform optimization. As the disturbance using pulse frequency change of tires infinite capacity (system, G-1, figure 1). Optimality criteria are the following factors: the lack of asynchronous operation and minimizing decay time of transition for most indicative parameters of synchronous machines, such as:

- Frequency (ω, Hz)
- Voltage of excitation winding.

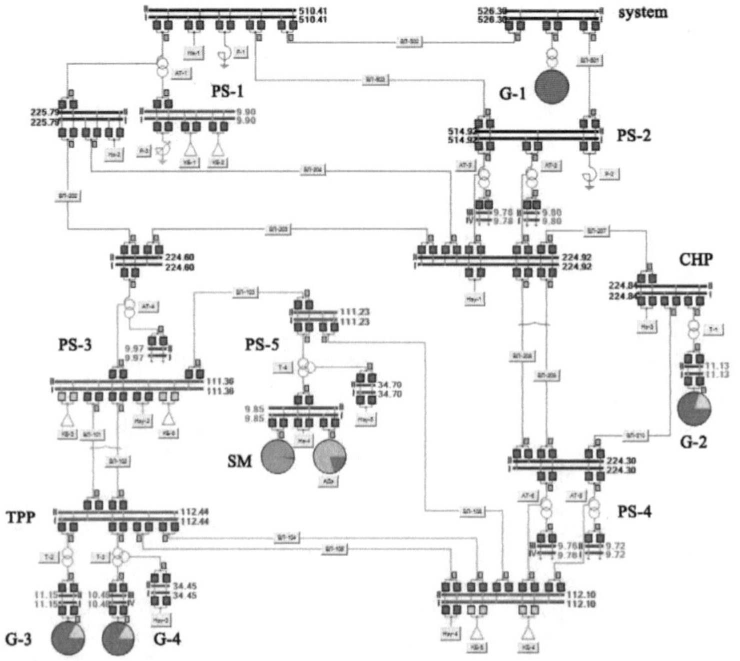

Figure 1. Scheme the simulated power system.

The scheme of simulated EPS is shown in Figure 1.

Synchronous oscillation occurs in the simulated power system with total for the entire power system perturbations in the form of a pulse frequency changes of infinite capacity bus, which lead to the asynchronous operation of synchronous motor (Figure 2).

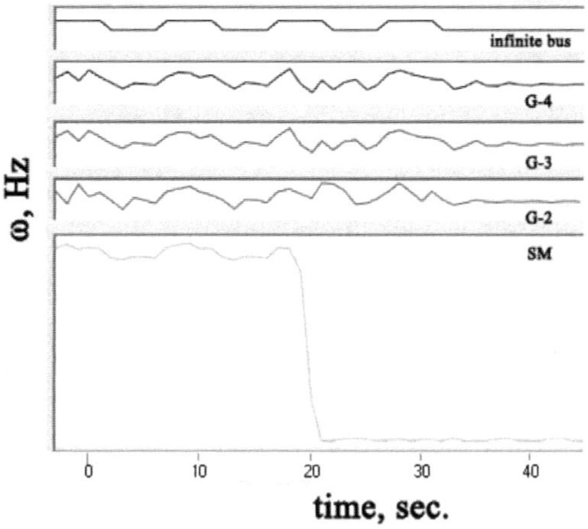

Figure 2. Frequency characteristics.

Determining the optimal settings AVRs produced by a successive approximation algorithm for each value of the adjustment parameters AVR of all synchronous machines, at which measured decay time of electromechanical transients in EPS. Minimizing the decay time is the criterion of determining the optimal parameter values AVRs. In Figure 3 shows the excitation voltage Uf and frequency ω. Figure 3 shows that Uf on G-2 is set longer than other values, so the decay time is determined for this parameter.

3. Results

Implementation of the method is shown by the example of the synchronous machine (SM) in the substation 5. Figures and graphics transient dependencies are shown below.

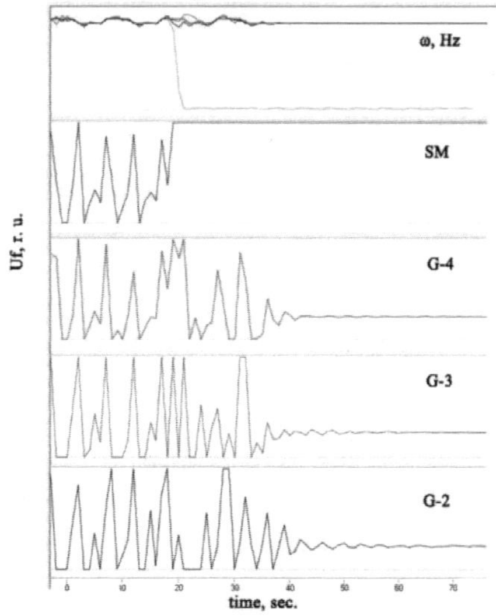

Figure 3. The dependence of the excitation voltage from the oscillation frequency.

In Figure 4 shows the effect of parameter kΔU changes in the excitation voltage Uf in the studied electrical machines. Minimum damping time shaded gray that corresponds value kΔU = 5.

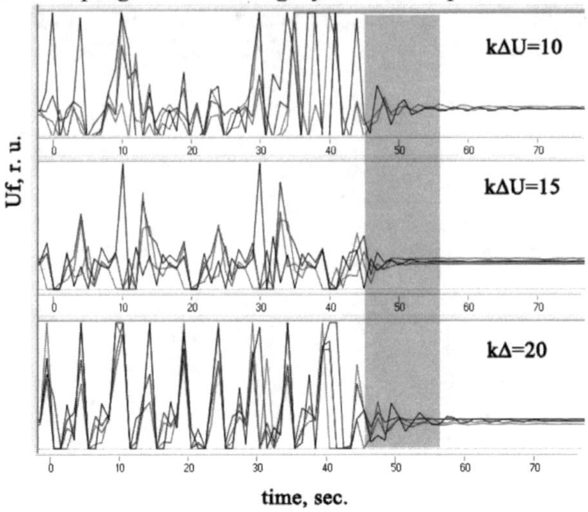

Figure 4. Dependence Uf from the kΔU.

Figure 5 shows a graph of the damping time of the setting kΔU on SM. Black dotted lines show the values at which occurs asynchronous operation of the synchronous motor. The decay time is equal to 100 seconds conditionally.

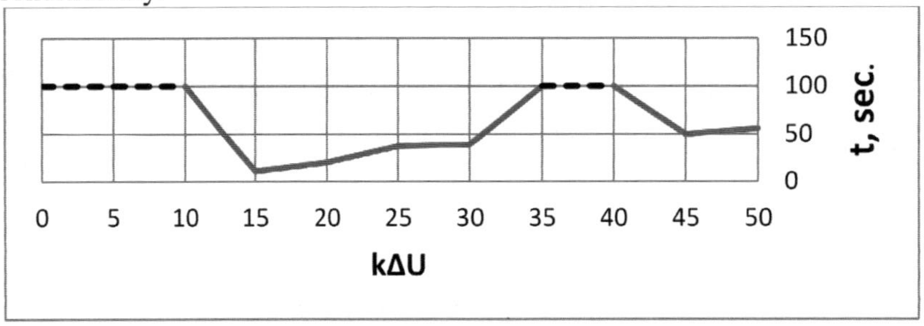

Figure 5. The dependence of damping time from the kΔU.

This experience is done sequentially for each of the synchronous machine with the new parameters, as a result, is possible to determine optimal values of kΔU.

As a result of the installation of optimal adjustment parameters of AVR, damping time was 10 seconds. In Figure 6 shows the results of researches demonstrating the effectiveness of methods.

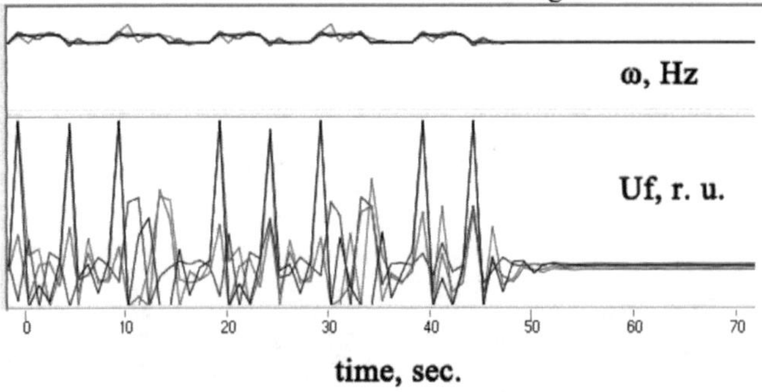

Figure 6. Results of research

The optimal parameters of AVR are determined for a particular scheme-modal state modeled power system as a result of automated scripts. Accordingly, when it changes the process of finding the optimal values is repeated which gives statistical material to determine the optimum parameters of AVR across the spectrum of modes of power system and development of appropriate recommendations.

4. Conclusions

Results of the practical experiments show the linkages adjustment parameters AVR of different synchronous machines included in the simulated power system. Consequently, when setting AVR for a particular generator or stations necessary to consider the current parameters of AVRs other synchronous machines, especially those with strong electrical interconnections.

To determine the optimal parameters AVRs necessary to consider all the features of the power system (or the local power district) as detailed as possible, otherwise the numerical values of the parameters will be defined incorrectly. The extent and nature error may lead to a radical difference between the processes at work of AVR in the real grid.

Only one criterion is used in the work as the optimal parameter settings to minimize the time of AVR. This is minimization of time of electromechanical oscillations. However, such criteria may be set (range of voltage and reactive power generation and other user-defined requirements).

References

[1] Merkuriev G 2008 *Stability of power systems* (St. Petersburg: Training Center of Energy)
[2] Portnoy M *Methodical instructions on determination of stability of power systems* Part 2 (Moscow: Soyuztehenergo)
[3] Lirin V 1968 *Setting and experimental works ORGRES* **35** 255-262
[4] Yurganov A and Kozhevnikov V 1996 *Excitation control of synchronous generators* (St. Petersburg: Science)
[5] Zhdanov P 1979 *Questions of the stability electrical* (Moscow: Energy)
[6] Venikov V 1985 *Transitional electromechanical processes in electrical systems* (Moscow: Higher school)
[7] Chernikh V 2008 *Simulation of electrical devices on Matlab, SimPowerSystem and Simulink* (Moscow: DMK Press)
[8] Gusev A, Khrushchev A, Gurin S, Svechkarev S and Plodisty I 2009 *Electricity* **12** 7-9
[9] Gusev A, Svechkarev S and Plodisty I 2004 *Electricity and the Future of Civilization : Materials of the International Scientific and Technical Conference* pp 329-330
[10] Ilina A and Svechkarev S 2014 *Electrical power industry by eyes of youth: collection of reports of V International youth Scientific and Technical Conference* pp 503-507
[11] Kosuke N, Kiyosi T and Masuo G 1997 *Elec. Rev* 82 **5** 67-71
[12] Akihiro A 1998 *Trans. Ing. Elec. Eng* 118 **4** 360-363

IOP Conf. Series: Materials Science and Engineering **93** (2015) 012014 doi:10.1088/1757-899X/93/1/012014

Correction of gasoline blending recipes with the use of computer modelling system

M Kirgina[1], E Sviridova[1], B Sakhnevich[1], N Chekantsev[1] and E Ivanchina
[1] National Research Tomsk Polytechnic University
30 Lenin av., Tomsk, 634050, Russia

E-mail: lizasvir@mail.ru

Abstract. The process of gasoline blending is a difficult multistage industrial technology. In this paper gasoline blending recipes for one of the largest refineries in Russian Federation were corrected by using of developed computer modelling system «Compounding». As the result of correction, production volume of high-octane and high-quality gasoline was increased on 47 wt. %.

1. Introduction

Since the January, 1-st, 2015, all refineries in Russian Federation have to switch their technologies to produce automobile gasoline of ecological class no below Euro-4 quality standard. One year later, only Euro-5 gasolines can be produced and presented in the domestic market. In order to solve this problem, trade gasoline producers perform upgrades of existent refineries by inputs of the new secondary refinery units and reconstruction of the old ones. As a result, the total production volume of oil-products increases, along with the increase of oil conversion ratio; in turn, it leads to upgrades in ecological and operational qualities of automobile gasolines.

From year to year, the total production volume of low-octane gasoline brands such as Normal-80 (RON 80) reduces, at the same time, high-octane gasoline (for example, brands Premium-95 and Super-98 have RONs 95 and 98, respectively) production increases permanently. In order to decrease the percent of low-octane gasoline brands production, it is necessary to revise existent gasoline blending recipes, redistribute the raw between the secondary refinery units, and their products – between different gasoline brands.

Solution of this problem is extremely difficult due to list of technological features, which take place at industrial gasoline production, as this process reflects the idea of blending of different hydrocarbon streams, such as: products of catalytic reforming, isomerization, catalytic cracking, alkylation with octane boosters and special anti-knock additives.

This multi-staged process is one of the most sophisticated technologies from the economical optimization point of view. The key point lies in complexity of mixtures, consisting of large quantities of individual components, in conditions of constant changes in feedstock composition. In addition, detonation resistance does not subject to the law of additivity [1-4], so it makes it more difficult to optimize the process. All above mentioned factors counteract developing of a universal blending recipe; existent recipes require for revisions and constant corrections due to changes in external factors.

IOP Conf. Series: Materials Science and Engineering **93** (2015) 012014 doi:10.1088/1757-899X/93/1/012014

Solution of such a multi-factorial and multi-criteria optimization task could be carried out in the most effective way by means of a mathematical modelling approach, and applying of computer modelling systems using a physic-chemical basis of the blending process [5, 6].

2. Computer modeling system «Compounding»

Previously, at the Chemical Engineering of Fuels and Chemical Cybernetics Department of Tomsk Polytechnic University, Russia, the mathematical model for gasoline blending process was developed. It described the nature of blending taking into the account deviations from the law of additivity of octane numbers caused by intermolecular interactions in a mixture. It provides calculations for main detonation characteristics of gasoline streams and anti-knock additives influence on octane number values. On a basis of this model, the computer modeling system «Compounding» was developed [7-9]. It provides calculations for:

1) Octane numbers of hydrocarbon stream involved into the blending process taking into account their non-additivity by the research and motor method (RON, MON);

2) Mixture density, by the Mendeleev formula:

$$\rho_4^t = \rho_4^{20} - \Delta t \cdot (t - 20) \qquad (1)$$

where ρ_4^t – density for temperature t, kg/m^3; ρ_4^{20} – density for temperature of 20 °C, kg/m^3; Δt – temperature correlation for density by 1 °C.

3) Mixture viscosity, by the Orrick and Erbar formula:

$$\ln(\eta/(\rho \cdot M)) = \alpha + \beta/T \qquad (2)$$

where η – viscosity, sP; T – temperature, K; ρ – density for T = 20 °C; M – molar mass; α, β – constants depending on the substance nature.

4) Saturated vapor pressure (SVP) by the Antuan equation:

$$\ln P_T = A - B/(T + C) \qquad (3)$$

where T – temperature, K; A, B, C – physic-chemical constants.

5) Aromatics, olefins hydrocarbons and benzene percentage;

The main module is developed in Borland «Delphi 7» workspace combining a user-friendly interface, coordination, integrity of sub-components, and stable functioning of system in general. The module has open database for detonation and physic-chemical characteristics of individual components.

In this modeling system, the input data for calculations are the actual industrial data and chromatography data of streams involved in the blending process. For unification of an experimental chromatography data, the module of chromatographic data systematization has been introduced into the system. During the chromatographic data systematization, all chromatograms are basically considered sets of 110 key components, and this set is a baseline for the high accuracy calculations of the blending gasoline octane numbers.

There is a possibility to calculate detonation characteristics of hydrocarbon mixtures by inputting the volumetric ratios of components, and calculates gasoline blending recipes with required brand and quality.

3. Results and discussion

For this study, by using of computer modelling system «Compounding», gasoline blending recipes for one of the largest refineries in Russian Federation were investigated. We explored 19 real hydrocarbon streams involved into the blending process for different brands of gasoline. During the research, blending recipes were reproduced, and all detonation characteristics including RON and MON and composition of streams, were calculated and analyzed.

Oil refinery produces such gasoline brands as Normal-80, Regular-92, Premium-95 and Super-98 alongside with gasoline for industrial purposes (GIP), which is semi-finished product.

Table 1.Volume of gasoline production.

Brand	Production volume, tons
Normal-80	12742
Regular-92	145592
Premium-95	70656
Super-98	7006
GIP	35814

As it can be seen from Table 1, the main part of produced gasoline is Regular-92 (53 wt. %), Premium-95 brand amounts to the quarter part, and remaining volume relates to Super-98, Normal-80 gasolines and GIP. In addition, the results shown in Table 1 demonstrate large amounts of produced gasoline falls up to GIP, which is not a trade product, wherein amount of high-octane Super-98 gasoline is small. This ratio doesn't represent a resource saving strategy, thus upgrading the quality of trade oil products is a top-priority aim for all producers. The main properties of gasoline produced at oil refinery by existing recipes are shown in the Table 2.

Table 2. The main properties of gasoline produced at oil refinery by existing recipes.

Characteristic	**GIP**	**Normal-80**	**Regular-92**	**Premium-95**	**Super-98**
RON	56.3	80.8	93.5	96.4	99.0
SVP, kPa	24.1	40.7	46.7	65.3	58.8
Density, kg/m^3	696.1	732.0	728.0	722.3	721.8
Viscosity, sP	35.9	44.7	41.3	39.6	38.6
Olefins, wt. %	0.8	11.3	13.6	7.4	3.0
Benzene, wt. %	0.7	0.9	1.0	0.9	1.0
Aromatic hydrocarbons, wt. %	1.1	27.4	35.3	38.0	34.7

As it can be seen from Table 2, several properties of producing gasoline do not comply with the demands of the Russian Technical Regulations and State Standard R 51866-2002. It can be observed that gasoline octane number is always higher than it is required. It is not economically profitably, that is why in this case there is a high-octane feedstock over expenditure. That feedstock can be used for production of other brands of gasoline. At the same time, in Regular-92 and Premium-95 gasolines the aromatic hydrocarbons content exceeds the maximum allowable standards (according to the Russian Technical Regulations, the maximum of aromatic content is 35 wt. %), thus they do not correspond the established demand. Thus, according to the data from Table 1 and Table 2, the existing recipes used at oil refinery do not allow obtaining trade product, corresponding with the Russian Technical Regulations demands and State Standard R 51866-2002. Moreover, the ratio of volumes in gasoline production is not reasonable, thus it is necessary to correct existing recipes for increasing the quality of trade product and the part of high-octane brand.

The aim of correction was to decrease the volumes of is semi-finished product (GIP) and to increase the volumes of high-octane and high-quality Super-98 brands production. Table 3, 4 show corrected recipes for gasoline blending. GIP is semi-finished product and it is used for further usage on the filling station, thus the task of correction recipes is to decrease its production volumes in favor of increasing production of Normal-80 gasoline, which has higher operational and ecological properties. As it can be seen from Table 3, the corrected recipe for GIP blending is practically the same as the existing one. The main part of feedstock consists of straight run gasoline No.2 streams and raffinate of aromatics production. It can be noticed that in existing recipes there are streams of straight run gasoline No.1, and aromatics production feedstock streams, but in corrected recipes they were excluded.

Table 3. Corrected blending recipes of Normal-80 and GIP gasoline brands, wt. %.

Feedstock streams	Normal-80 (cor.)	Normal-80 (ex.)	GIP (cor.)	GIP (ex.)
Catalytic cracking gasoline No.1	17.17	55.72	–	–
Catalytic cracking gasoline No.2	3.90	1.09	–	–
Reformate (fixed-bed)	8.05	6.25	–	–
Reformate (moving-bed)	5.09	–	–	–
Izomerizate	–	0.97	–	–
Isopentane No.1	6.36	–	–	–
n-butane	0.53	–	–	–
Straight run gasoline No.1	0.31	0.09	–	0.29
Straight run gasoline No.2	23.84	15.89	49.39	51.72
Aromatics C_8	3.97	–	–	–
Aromatics C_9	1.19	–	–	–
Feedstock for aromatics production	0.08	0.03	–	0.08
Raffinate of aromatics production	19.85	14.72	50.61	47.91
Condensate of aromatics production	1.01	3.01	–	–
Toluene	8.66	2.24	–	–

Table 4. Corrected blending recipes of Regular-92, Premium-95 and Super-98 gasoline brands, wt. %.

Feedstock streams	Regular-92 (cor.)	Regular-92 (ex.)	Premium-95 (cor.)	Premium-95 (ex.)	Super-98 (cor.)	Super-98 (ex.)
Gasoline of catalytic cracking No.1	4.54	3.93	7.41	7.21	4.08	4.01
Gasoline of catalytic cracking No.2	48.77	45.46	18.72	21.05	7.78	7.65
Reformate (fixed-bed)	15.09	16.73	8.49	6.85	15.05	14.80
Reformate (moving-bed)	3.81	4.66	14.73	15.47	13.36	13.13
Izomerizate	11.76	13.59	6.80	1.24	1.57	1.54
Isopentane No.1	4.00	3.82	20.10	23.82	20.33	20.00
Isopentane No.2	0.10	0.24	1.44	1.53	3.85	1.28
n-butane	0.34	0.49	1.67	1.63	1.39	1.38
Alkylate	0.82	1.30	5.02	4.60	16.54	16.26
Aromatics C_8	3.27	2.70	5.87	8.97	7.75	7.62
Aromatics C_9	0.88	1.22	4.15	4.04	3.48	3.43
Straight run gasoline No.2	0.77	–	2.04	–	–	–
Platformate of aromatics production	–	0.40	2.31	1.32	–	1.11
Raffinate of aromatics production	1.63	–	–	–	–	–
Toluene	4.12	5.47	–	0.96	0.80	0.79
MTBE	0.10	–	1.26	1.31	4.00	6.99

According to the data from Table 3, in correct recipes amounts of raffinate of aromatics production and straight run gasoline No.2 were significantly increased due to increase in production volumes of this gasoline brand. That streams were saved at the expense of reducing volume of GIP production. In this regard for enhancing the main property (RON) of gasoline, the following streams were involved: reformate (moving-bed), aromatics C_8, aromatics C_9; contents of the following streams were increasing: reformate (fixed-bed), catalytic cracking gasoline No.2, toluene. The isopentane stream No.1 and n-butane were involved in the blending scheme in order to increase SVP.

The aim of the recipes correction for Regular-92 gasoline blending was to reduce RON to the desired (from 93.5 to 92 points) and, in the same time, to decrease amount of aromatic hydrocarbons. Corrected volume of Regular-92 gasoline production affects the corrected recipes. As the part gasoline, that was previously used to produce Regular-92 gasoline, was switched to produce Normal-80 gasoline, the production volume of this brand was reduced to 10 000 tons (6.23 wt. %).

The amount of the next streams, such as toluene, reformate (moving-bed), reformate (fixed-bed) was decreased and streams of raffinate of aromatics production and straight run gasoline No.2 was involved to reduce contents of aromatics hydrocarbons and ON.

The task of blending recipes correction was to decrease the ON to the desired values and to decrease high contents of aromatics (from 38 wt. % to 35 wt. %). For reducing of aromatics hydrocarbons amount mainly the contents of aromatics C_8 streams were decreased (from 8.97 wt. % to 5.87 wt. %), that surplus was used to the Regular-92 brand production. Also, the toluene stream was excluded, but the content of not-aromatics isomerizate stream was increased. For increasing SVP of Normal-80 gasoline the isopentane No.1 stream taken from recipes Premium-95 gasoline was used, thus the content of isopentane No.1 stream was decreased. At the expense of increasing volume production Super-98 gasoline volume production Premium-95 gasoline was reduced to the 3 wt. %.

As it can be seen from Table 4 change of Super-98 gasoline recipes is insignificant in spite of increasing volume production on 47 wt. %. ON in existing recipes was enhanced, thus amount of supplements-oxygen was decreased in corrected recipes. That surplus was used to increase ON of other brand gasoline.

Table 5 shows the main properties of gasoline producing by correct recipes.

Table 5. The main properties of gasoline produced at oil refinery by corrected recipes.

Characteristic	Normal-80	Regular-92	Premium-95	Super-98	GIP
RON	80.0	92.0	95.1	98.0	56.1
SVP, kPa	37.1	46.6	63.5	62.1	24.1
Density, kg/m^3	733.17	725.49	718	717.85	695.03
Viscosity, sP	41.38	41.11	39.47	38.38	35.55
Olefins, wt. %	4.74	14.66	6.77	3.05	0.84
Benzene, wt. %	0.9	1.0	1.0	0.9	0.7
Aromatic hydrocarbons, wt. %	30.4	33.2	35.0	34.5	1.1

As it can be seen from Table 5 all properties of gasoline comply the Russian Technical Peculations demands and State Standard R 51866-2002. It means that recipes can be used for gasoline preparation. The amount of aromatic hydrocarbon was decreased in Regular-92 and Premium-95 gasoline as a result of correction. Also it can be noticed that there is not a surplus of primary streams, because ONs consistent with the stated. It allows save high-quality streams to using them for high-octane gasoline production.

Table 6 shows the data about gasoline volume changes, which is produced by corrected recipes.

As it can be seen volume of production GIP was decreased on 48 wt. % while the volume of production of Super-98 gasoline was increased on the 47 wt. %.There are a volume changes of production of GIP due to the involving in the production part of Normal-80 gasoline, because of the volume of production Normal-80 gasoline increases on the 196 wt. %. As in the production of Normal-80 gasoline low-octane streams, such as straight run gasoline No.1 and raffinate of aromatics production were involved, it was necessary to use high-octane streams from Regular-92 gasoline production. In this regard there is a reduction volume of production of Regular-92 gasoline on the 6 wt. %. Volume changing of production Super-98 gasoline was due to decreasing high-octane components in existing recipes and involving the part of Premium-95 gasoline. Because of this there is a reduction volume of production Premium-95 gasoline on the 2 wt. %.

Table 6. Changing volumes of gasoline production.

Brand	Existing recipes, tons	Correct recipes, tons	Volume changing, wt. %
Normal-80	12 742	37 750	+196.26
Regular-92	145 592	136 526	-6.23
Premium-95	70 656	68 791	-2.64
Super-98	7 006	10 332	+47.47
GIP	35 814	18 411	-48.59

4. Conclusion

In this paper, a novel computer modelling system «Compounding» was developed. Practical importance of this system consists in increasing of the resource efficiency of gasoline blending process, by maximizing the yield and quality of high-octane trade gasoline. Such an effect is achieved by using only internal oil refinery resources, and this is beneficial in terms of economy and gasoline production technology.

Acknowledgements

The current study was performed in the frame of the project part of the State Task in the field of scientific work 1.1404.2014, The Ministry of Education and Science of The Russian Federation.

References

[1] Albahri T A, Riazi M R and Alqattan A A 2002 *Fuel Chem. Div. Preprints.* **47 (2)** 710–11
[2] Albahri T A 2003 *Ind. and Eng. Chem. Res.* **42 (3)** 657–62
[3] Twu C H and Coon J E 1996 *Hydrocarbon Processing.* 75 (2) 51–56
[4] Ghosh P, Hickey K J and Jaffe S B 2006 *Ind. and Eng. Chem. Res.* **45 (1)** 337–45
[5] Gyngazova M S, Kravtsov A V, Ivanchina E D, Korolenko M V and Chekantsev N V 2011 *Chem. Eng. J.* **176** 134–43
[6] Chekantsev N V, Gyngazova M S and Ivanchina E D 2014 *Chem. Eng. J.***238** 120–28
[7] Kirgina M V, Ivanchina E D, Dolganov I M, Chekantsev N V, Kravtsov A V and Fu F 2014 *Chem. and Tech. of Fuels and Oils.* **1** 12–18
[8] Kirgina M V, Maylin M V, Ivanchina E D and Sviridova E V 2014 *Adv. Mat. Res.* **880** 121–27
[9] Kirgina M, Gyngazova M and Ivanchina E 2012 *Proc. 7th International Forum on Strategic Technology IFOST-2012.* **1** 30–33.

Energy efficiency by use of automated energy-saving windows with heat-reflective screens and solar battery for power supply systems of European and Russian buildings

V M Zakharov[1], N N Smirnov[1], V V Tyutikov[1] and B Flament[2]
[1] Ivanovo State Power Engineering University, Ivanovo, Russia
[2] National Institute of Applied Sciences in Strasbourg, Strasbourg, France

E-mail: nsmirnov@bk.ru

Abstract. The new energy saving windows with heat-reflecting shields have been developed, and for their practical use they need to be integrated into the automated system for controlling heat supply in buildings and the efficiency of their use together with the existing energy-saving measures must be determined. The study was based on the results of field tests of windows with heat-reflective shields in a certified climate chamber. The method to determine the minimum indoor air temperature under standby heating using heat-reflective shields in the windows and multifunctional energy-efficient shutter with solar battery have been developed. Annual energy saving for the conditions of different regions of Russia and France was determined. Using windows with heat-reflecting screens and a solar battery results in a triple power effect: reduced heat losses during the heating season due to increased window resistance; lower cost of heating buildings due to lowering of indoor ambient temperature; also electric power generation.

1. Introduction

Pursuant to Russian legislation, annual specific consumption of energy in buildings as of 1 January 2020 shall be reduced by 40% of the basic level. In France, the thermal regulations RT, which set reduction of energy consumption in buildings, were adopted. Annual energy consumption per 1 square meter of building Q, kWh/(m^2·year) is the regulated value. The requirements of the Act Grenelle (Loi Grenelle), dated August 03, 2009 and the thermal regulation RT 2012 allowed to build buildings with low energy consumption (BBC, Q <50 kWh / (m^2·year) depending on region of France) with January 1, 2013 and with January 1, 2020 - only buildings with "positive energy» (Bâtiment à énergie positive, BEPos), i.e. with positive balance, such as the production of electricity (often due to photovoltaic method). Energy efficiency class "A" is assigned to such buildings.

Existing European and Russian energy efficiency regulations stipulate strict requirements regarding annual energy consumption, and particularly, the heat transfer resistance coefficient of translucent structures. Thus, European Union legislatures stipulate a coefficient of heat transfer resistance for windows by 2020 of 1,67÷2,0 m^2·°C/W, while Russian official bodies stipulate 1,0 ÷ 1,05 m^2·°C/W by 2016.

The scientists from Ivanovo State Power Engineering University (ISPU), Scientific and Technical Committee for Industrial Climatology (COSTIC) and National Institute of Applied Sciences in Strasbourg (INSA de Strasbourg) [1] developed and patented window designs with panel, roll and louvre type heat-reflecting screens, which are made of metal and significantly reduce heat losses. The

IOP Conf. Series: Materials Science and Engineering **93** (2015) 012015 doi:10.1088/1757-899X/93/1/012015

use of screens is desirable during nighttime or in the absence of people. Screens may be placed inside or outside buildings, or between window panes. The use of screens not only reduces losses related to heat transfer but also permits ambient temperature reduction in setback heating mode. There is a need for creating a system of automated window screen control, and also for methods of determining ambient temperature in setback heating mode.

2. Determining Minimal Ambient Temperature in Setback Heating Mode

A certified environmental chamber belonging to "Ivanovostroiispytaniya" (an independent non-profit making organization engaged in construction engineering) was used [1] to study the impact of using heat reflecting window screens on raising heat transfer resistance in windows and reducing heat losses. For control purposes, two double-glazed windows, made according to formula 4M1x10x4M1x10x4M1 and 4M1x10x4M1x10x4i, respectively, with low-emission coating, were used.

According to the given data (refer with Table 1), the use of low-emission I-glass increased heat transfer resistance from 0.47 $m^{2.0}C/W$ to 0.61 $m^{2.0}C/W$ (by 29 %), while adding a panel consisting of two metal screens positioned at a distance of 10 mm from each other on the cold side of the environmental chamber increased resistance to 1.76 $m^{2.0}C/W$ (by 274% of control value 1, or 189 % of control value 2).

Table 1. Results of tests on heat-reflecting screens in windows with low-emission coating.

Item	Heat flow (W/m^2)	Inner glass temperature t_w (^0C)	Heat transfer resistance R ($m^2 \cdot {}^0C/W$)	$R/R_{control\ 1}$	$R/R_{control\ 2}$
Control 1 4M1x10x4M1x10x4M1	76,1	12,5	0,47	-	-
Control 2 4M1x10x4M1x10x4i (coated)	58,9	14,15	0,61	1,29	-
2 cold side screens	20,77	18,65	1,76	3,74	2,89

The current regulations (SP 60.13330.2012. "Heating, ventilation and air conditioning") with reference to buildings during the cold season when not in use or during non-working hours permit reducing indoor ambient temperature below the regulation value, but not below 15 ^0C in residential buildings, 12 ^0C in public buildings and 5 ^0C in the case of industrial premises. In France, according to the regulations it is not allowed to decrease indoor ambient temperature below 16 ^0C for residential, public and administrative and service buildings.

Indoor temperature reduction in the setback heating mode creates a high energy saving potential. Minimal ambient temperature in setback heating mode is to a large extent determined by the conditions necessary to prevent condensation on enclosing structure surfaces. We note the high probability of condensation on windows, because translucent structures are the weak spot in heat insulation of buildings.

It should be noted that the appearance of moisture on glass surfaces is not only an esthetic defect, as continuing condensation may eventually result in moistening of the structures [2], with possible formation of fungi and mildew on window sills. Particular attention should be paid to horizontally positioned and slanting windows, also skylights, as pursuant to Current regulations (SP 50.13330.2012 "Buildings Heat Insulation"), the inner surface temperature should not be lower than indoor ambient dew-point temperature at the assumed outdoor ambient temperature during the cold season.

Relative humidity inside buildings *RH* is a regulated value (30 % to 65 % for public and residential buildings), while air humidity, window heat transfer resistance and outdoor ambient temperature are the factors determining dew-point temperature along internal glazing, and, therefore, the minimal ambient temperature in setback heating mode (at a given heat transfer coefficient for inner window surfaces). As shown by calculations and experimental data (refer to Table 1), the use of external heat-reflecting screens increases heat transfer resistance up to 1.76 $m^{2.0}C/W$ with a significant increase in internal glass temperature, thereby enabling us to further lower indoor ambient temperature (depending on indoor humidity) in the absence of people.

We were interested to know how to determine the value (refer with Fig. 1) to which an automated window control system may lower the ambient temperature in the setback heating mode while excluding condensation on internal window glass surfaces with enhanced heat insulation properties (using heat-reflecting screens).

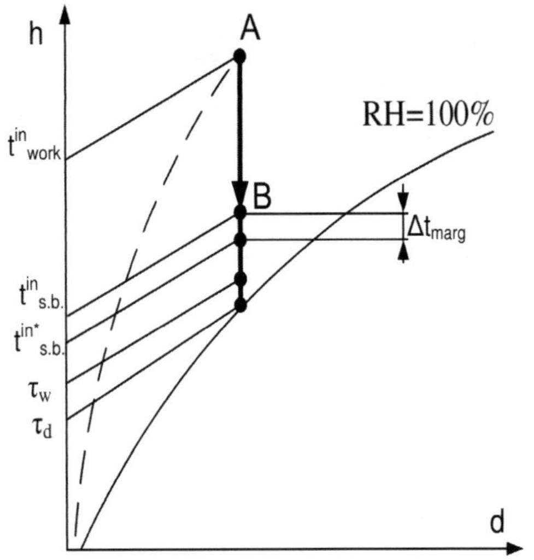

Figure 1. Determining ambient temperature in setback heating mode.

In the course of mathematical transformations, it was deduced that minimal indoor ambient temperature in setback heating mode may be determined by:

$$t_{s.b}^{in} = \frac{\tau_d \cdot \alpha_{\text{in}} \cdot R_{\text{window}} - t_{out}}{\alpha_{\text{in}} \cdot R_{\text{window}} - 1} + \Delta t_{\text{marg}}, \tag{1}$$

where τ_d - dew-point temperature at working hour ambient conditions, may be determined by the Magnus-Tetens formula [3], ^0C;

Δt_{marg} - temperature margin for preventing condensation (assumed to be 1 ^0C);

t_{out} - outdoor ambient temperature, ^0C;

R_{window} - heat transfer resistance of window translucent area, $m^{2.0}C$ /W;

α_{in} - coefficient of heat transfer from indoor air to glass, assumed to be 8.7 W /$m^{2.0}C$.

Thus, knowing the window resistance value R_{window} during non-working hours, outdoor ambient temperature, ambient temperature and humidity during working hours we can determine minimal ambient temperature in setback heating mode.

In accordance with formulae obtained in Mathcad and Excel computing environments, we prepared a programme for calculating the minimal ambient temperature in setback heating mode and using windows with the heat-reflecting screens.

We simulated changes in the resistance of the translucent part of Window R from 0.5 to the resulting experimental value of 1,757 м$^{2.0}$C/W, while setting outdoor ambient temperature t_{out} from – 10 to - 30 ^{0}C, and ambient temperature during working hours at 20 ^{0}C. The calculation was made for relative humidity RH equal to 35 and 50%. The results of these calculations may be seen in Fig. 2.

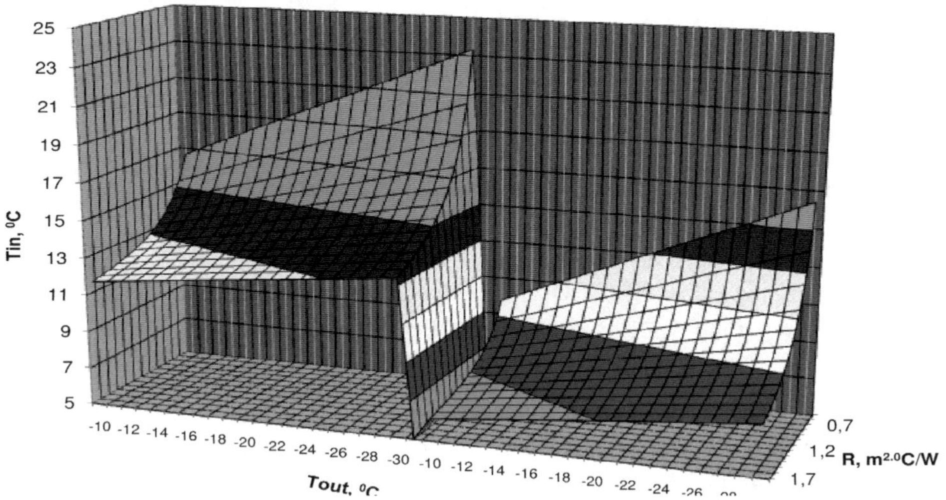

Figure 2. Dependence of changes in minimal permissible ambient temperature tin in setback heating mode on R and t_{out}., given RH=50 % in the left half of the graph and RH=35 % in the right half.

As it is expected, the $t^{in}_{s.b}$ will be significantly higher with higher indoor humidity. A significant difference also results from the use of screens. Thus, according to calculations, with relative humidity of 35%, outdoor ambient temperature of - 30^{0}C and the use of I-glass windows with two heat-reflecting screens, minimum permissible ambient temperature in setback heating mode falls from 16.4 ^{0}C to 7.8 ^{0}C, that is, by more than 8 ^{0}C, while with relative humidity of 50% the difference in temperature is more than 10 ^{0}C.

The authors also developed and patented a panel design featuring heat-reflecting screens using a solar battery on the external side, consisting, for example, of thin-film photocells. The use of a solar battery permits generation of electric power which not only ensures the independent operation of windows with heat-reflecting screens but also partially or completely serves the consumer's electrical load, and in the event of large-scale generation may be fed into the exterior electrical network. Moreover, the conversion efficiency of solar batteries reaches 20%.

Thus, using windows with heat-reflecting screens and a solar battery results in a triple power effect: reduced heat losses during the heating season due to increased window resistance; lower cost of heating buildings due to lowering of indoor ambient temperature; also electric power generation.

3. Determination of Thermal Energy Saving

After processing data received from the Russian and French Meteorological Services, the authors determined the duration of daylight hours for each month in various cities of the Russian Federation. As it is well known, it is desirable to use screens at night and in the absence of people. Hours of daylight per day (24 hours) during a calendar year in the city of Norilsk are presented in diagram form in Fig. 3. Efficiency is determined without the inertial component of walls, windows, etc, as well as power supply systems.

Staff working hours are assumed to be from 8:00 up to 17:00. Significantly, in Norilsk, December marks the onset of polar night, when the sun does not rise above the horizon, hence screens are used around the clock.

IOP Conf. Series: Materials Science and Engineering **93** (2015) 012015 doi:10.1088/1757-899X/93/1/012015

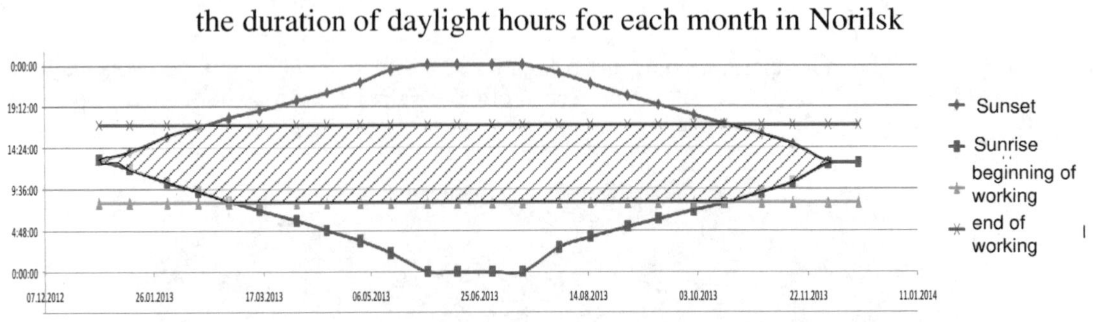

Figure 3. Duration of heat reflecting screen use on windows (unshaded area) during a 24-hour day in Norilsk conditions.

We also analyzed the use of screens in Russian and French cities representing different climatic zones.Determining thermal energy saving also required additional city data, including geographical latitude S, number of heating period days $n_{\text{heat.per}}$, hours during which screens were used and not used during the heating period (n_1 and n_2, respectively), design ambient temperature tdes and mean ambient temperature during the heating period tmean heat. Data for the cities under consideration are given in Table 2. Despite the fact that France is located south of 52 degrees north latitude (south of the Russian Belgorod) and the climate is much warmer, heating period longer than many cities of the Central Federal District in Russia.

Table 2. City climatic data.

City	S (Degrees N)	$n_{\text{heat.per}}$ (days)	$t_{\text{mean heat}}$ (^{0}C)	t_{des} (^{0}C)	n_1 (h)	n_2 (h)
Norilsk	69	365	-11,5	-40	1786	6974
Moscow	55	205	-2,2	-25	1162	3758
Strasbourg	48	232*	5,8	-15	1410	4158
Lyon	45	232*	7,4	-8	1431	4137
Marseille	43	232*	11,1	-5	1431	4137

* - In France, the heating season for all cities is approved from October 01 to May 20

Calculations were made for a window comprising a multiple heat reflecting I-glass unit (4M1x10x4M1x10x4I), using a panel screen consisting of two metal sheets separated by a low heat conductivity frame (Table 1). Design indoor temperature during working hours was assumed to be 20^{0}C, with relative humidity of 40%. For non-working hours (nighttime) in setback heating mode two levels of ambient temperature reduction were assumed – to standardized 12 ^{0}C (16 ^{0}C – for France), and to the minimal permissible ambient temperature satisfying the requirement of preventing condensation on translucent structures. Heat losses for the heating period were determined per 1 m^2 of window surface. Calculation data are given in Tables 3 and 4. Thus, the maximum thermal energy saving was achieved in all cities using panel heat reflecting screens and automation systems intended to maintain setback heating mode during non-working hours at the minimal permissible ambient temperature, and in the absence of condensation. Maximum saving (0.235 Gcal) was actually achieved for conditions in the city of Norilsk.

For the southern cities of Russia and most of French cities in the case of using these proposed energy-saving measures annual consumption of thermal energy for compensating transmission losses will be minimal (from 0.11 to 0.28 Gcal / m^2).

Table 3. Heat losses during the heating period per square metre of translucent structures (Gcal).

City	Check value, Q_k	Using setback heating mode ($t_{setback}$=12^{0}C and 16^{0}C), $Q_{setback\ (12/16)}$	Using a screen, $Q_{scr.}$	Using a screen in setback heating mode ($t_{(setback)}$=12^{0}C and 16^{0}C), $Q_{setback\ 12/16+scr}$	Using a screen in setback heating mode at lowered temperature, $Q_{setback\ (min)+scr}$
Norilsk	0,217	0,162	0,124	0,098	0,084
Moscow	0,154	0,112	0,096	0,075	0,062
Strasbourg	0,111	0,088	0,057	0,049	0,028
Lyon	0,099	0,076	0,052	0,044	0,026
Marseille	0,07	0,046	0,036	0,028	0,018

Table 4. Heat saving during the heating period per square metre of translucent structure (Gcal/%).

City	Check value, Q_k	Using setback heating mode ($t_{setback}$=12^{0}C and 16^{0}C), $Q_{setback\ (12/16)}$	Using a screen, $Q_{scr.}$	Using a screen in setback heating mode ($t_{(setback)}$=12^{0}C and 16^{0}C), $Q_{setback\ 12/16+scr}$	Using a screen in setback heating mode at lowered temperature, $Q_{setback\ (min)+scr}$
Norilsk	-	0,055 / 25,4	0,093 / 43	0,119 / 55	0,133 / 62
Moscow	-	0,042 / 27	0,058 / 38	0,079 / 51,4	0,092 / 60
Strasbourg	-	0,023 / 21	0,054 / 49	0,062 / 56	0,083 / 75
Lyon	-	0,023 / 23	0,047 / 47	0,055 / 56	0,073 / 74
Marseille	-	0,024 / 34	0,034 / 49	0,042 / 60	0,052 / 74

4. Conclusion

The application for energy-efficient purposes is not only of heat-reflecting screens with solar cell panels, automated heating systems using setback heating mode, heat recovery in exhaust air with the aid of regenerative heat exchangers [4], but also of new designs and the results of research by COSTIC and INSA staff in the fields of "hybrid" ventilation, improved building air tightness and controlled fresh air supply depending on carbon dioxide levels in buildings has created a big energy-efficiency potential for systems supplying energy to both residential and industrial buildings. Using the recommended window designs will help to achieve the energy consumption levels stipulated for buildings by Russian and European regulations [1].

References

[1] Beaumont C, Holtzweiler E, Zakharov V M, Smirnov N N, Yablokov A A and Lapateev D A 2013 *Vestnik IGEU* **5** 13-24
[2] Merrill J and TenWolde A 1989 *ASHRAE Transactions* **95** 405-414
[3] Barenburg A 1974 *Psychrometry and Psychrometric Charts* (Cape Town, SA: Cape and Transvaal Printers Ltd.)
[4] Missenard M, Blandin C and Feldmann C 2010 *Indoor microclimatology* (Paris: COSTIC)

IOP Conf. Series: Materials Science and Engineering **93** (2015) 012016 doi:10.1088/1757-899X/93/1/012016

Using Controlled Shunt Reactors for Voltage Stabilization on the Example of Real Electric Power System

A S Gusev[1], A A Suvorov[1] and A O Sulaymanov[1]
[1]National Research Tomsk Polytechnic University, Lenin Avenue 30, Tomsk, 634050, Russia

E-mail: lejkasvr@yandex.ru

Abstract. The article is devoted to actual task of real-time simulation controlled shunt reactors to use in the appropriate electric power system. Such development allows fully and reliably reproducing the processes running in controlled shunt reactors and electric power systems as whole. As an example of such task solution the working results of controlled shunt reactors simulation and its application for voltage stabilization are presented.

1. Introduction

One of the main operation conditions of electric power systems (EPS) is a voltage level. The values of this are connected to reactive power (RP) flow. The main voltage stabilization and generation RP devices are located on power stations, so located in electrical network devices is not provide necessary voltage level and RP regulation. For example, in Russian networks, the most widespread devices for reactive power compensation (DRPC) are providing only single-stage voltage regulation shunt reactors. An effective way of solving this problem is using Flexible Alternative Current Transmission Systems (FACTS) technologies, such as: controlled shunt reactors (CSR) [1], static reactive power compensator (SPC), static synchronous compensator (STATCOM) and others. Nowadays, the CSR is the most widespread FACTS devices, representing a device with continuously variable inductive reactance [1-3].

Design, research and exploitation of EPS, including CSR, are based on the analysis and using information about a full spectrum of various processes in CSR and EPS as a whole. The known specificity of electrical power systems is excluded the possibility of obtaining this information by full-scale test, so the main way of getting information is simulation, mostly mathematical because of the difficulty of the real EPS and impossibility of full physical simulation [4, 5]. As a result, for reliable and efficient operation of CSR as EPS part is necessary to create a mathematical model CSR, and software and hardware are adapted for using in the appropriate EPS.

2. Creating a CSR mathematical model

The describing the running processes in the CSR equations are taken into account the interaction of each phase coil with its own main magnetic flux and stray flux, so created mathematical model have to include a system of three phase three-circuit equations:

1) Equations magnetically coupled by magnetic flow phase each winding:

$$w_{PWi}\frac{d\Phi_{0i}}{dt}+L_{PWi}\frac{di_{PWi}}{dt}+r_{PWi}i_{PWi}=u_{PWi} \quad (1)$$

$$w_{CWi}\frac{d\Phi_{0i}}{dt}+L_{CWi}\frac{di_{CWi}}{dt}+r_{CWi}i_{CWi}=u_{CWi} \quad (2)$$

$$w_{CONWi}\frac{d\Phi_{CONWi}}{dt}+L_{CONWi}\frac{di_{CONWi}}{dt}+r_{CONWi}i_{CONWi}=u_{CONWi} \quad (3)$$

where: j – power, control and compensation winding; w_{ji} – number of turns; i – A, B, C; Φ_0 – main magnetic flux i; Φ_{VO} – control winding magnetic flux i; L_{ji} – winding-leakage inductance i; i_{ji} – current in winding; r_{ji} – active resistance; u_{ji} – line voltage.

2) Equations magneto-motive forces for each phase:

$$w_{PWi}\cdot i_{PWi}+w_{CWi}\cdot i_{CWi}+w_{CONWi}\cdot i_{CONWi}=F_{MMFi} \quad (4)$$

where F_{MMFi} – magneto-motive force for phase i, MMF is determined by taking into account the possible saturation of steel:

$$F_{MMFi}=K_{\mu}\Phi_{0i}^{p}. \quad (5)$$

The resulting system of equations allows synthesizing CSR mathematical model reproducing running processes in the windings, with the non-linearity of magnetic circuit, and hardware and software of models implementation. It allows solving the nonlinear system of differential equations without any decomposition and limitation on their duration in real time with guaranteed accuracy [6, 7]. Structural diagram of the specialized reactor processor (SRP) implements CSR mathematical model, see Figure 1.

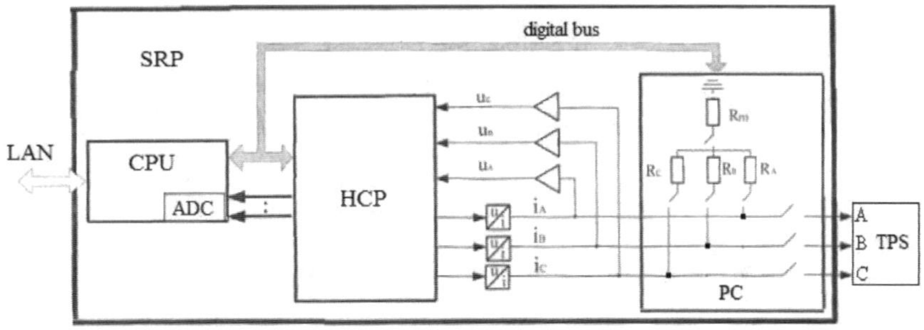

Figure 1. The structural diagram of the hybrid software and hardware simulation of CSR.

where: SRP - specialized reactor processors, LAN – local area network, TPS – three-phase switch, CPU - central processing unit, ADC –analog-to-digital converter; HCP – hybrid coprocessor; PC - switching processor; u/i - voltage-current converter.

This SRP realized in the Hybrid Real Time Power System Simulator (HRTSim), which is developed in Institute of Power Engineering of Tomsk Polytechnic University [8]. It represents multiprocessor software and hardware system that provide the real time continuous implicit integration of differential equations system describing processes in three phase power system. HRTSim allows avoiding decomposition of power system processes and simplification of power equipment models independently on the size of modeled EPS. Besides, the capability to provide any longitudinal and transversal commutations this system allows simulating any abnormal regime of power system [9, 10].

3. Research the CSR effectiveness in real EPS

A current pilot project on creation of system with FACTS devices for 'Elgaugol' energy cluster which is a part East Power System is being implemented in Russia. A scheme of simulated East Power System including Elgaugol energy cluster and adjacent network is given in Figure 2 [11].

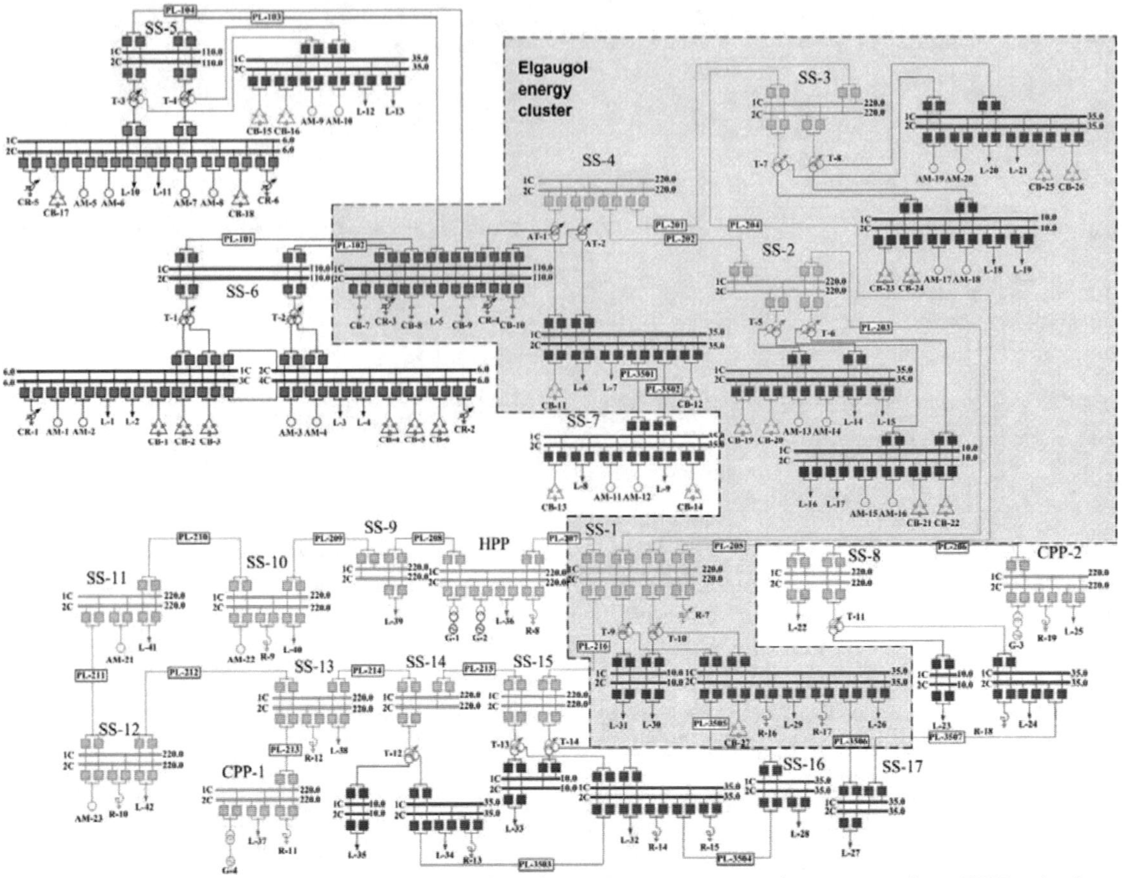

Figure 2. Simulated power system: SS – substation, CPP – co-generation power plant, HPP – hydro power plant, PL – power line, CB – capacitors bank, CR – controlled shunt reactor, R – switched reactor, L – load, AM – asynchronous motor, G – generator, (A)T – (auto) transformer.

The major components of energy cluster power equipment controlled shunt reactor (CSR): 100 Mvar reactor is expected to be installed at Prizeyskaya substation that powers energy cluster and two 25 Mvar reactors are planned to be installed at Elgaugol substation. Automatic control system (ACS) of CSR contains hysteresis regulator which switches-on/off one or several banks of capacitors (BC) in case CSR experiences maximum/minimum reactive power load level and this level is being hold for certain period of time. The purpose of researching is check CSR effectiveness for voltage and RP regulation, it allows: to support acceptable voltage level in EPS; to reduce electricity losses during transportation and distribution; to provide positive effect on operational conditions of energy cluster and EPS as whole.

The purpose of the first experiment was to investigate the CSR application for stabilization voltage level in load node when operating electrical equipment for real the daily schedule of active and reactive loads mining enterprise in three-shift work in winter. The example of real time simulation results of self-starting process of energy cluster consumer's asynchronous motors and CSR actions in the case of voltage dip occurring when total loads are increasing according to the daily schedule is performed in Figure 3 and Figure 4.

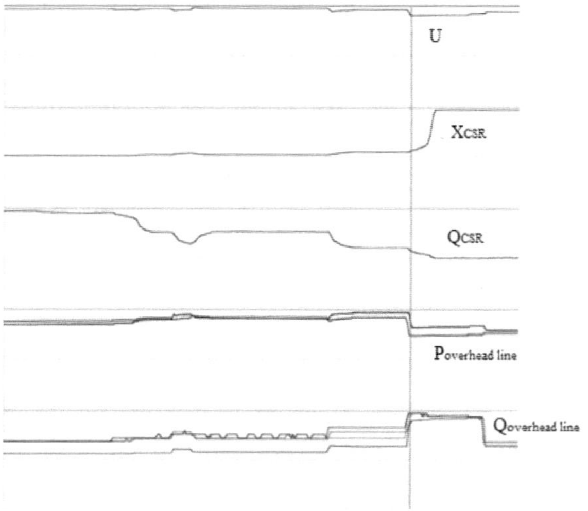

Figure 3. Real time data oscillogram of changing loads, CSR and banks of capacitors switching-on.

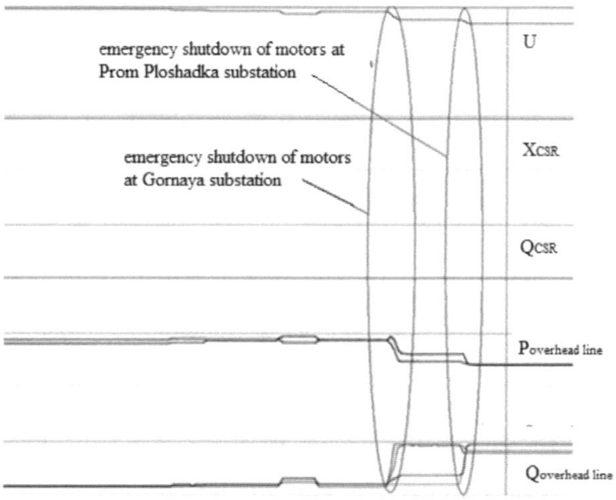

Figure 4. Real time data oscillogram of changing loads, CSR and banks of capacitors switching-off.

ACS CSR runs with increasing load and unloads the reactor power, because of this bus bar voltage level does not leave the permissible 5 percent level from the set point, and a power failure consumer does not occur, see Figure 3. On the next diagram Figure 4, a significant voltage dip occurred during increasing the load, it led to stop the located on the SS Prom Ploshadka and SS Gornaya motors are powered by bur buses 110 kV Elgaugol with installed DRPC. As a result, active power connected SS Elgaugal and motor consumers overhead lines sharply decreased and reactive power increased. Obtained diagrams shows that continuity of power supply main customers Elgaugol energy cluster depends on the functioning CSR.

The following are the results of experiments demonstrating the CSR influence on the operation stability consumer when one of the energy cluster supplying power overhead line is suffered a short circuit is performed in Figure 5, 6. At some moment of time a short-sscircuit is simulated on the supplying power overhead line, it is followed to activate of relay protection (RP) and the successful automatic reclosing. Thus, the load stability is preserved, see Figure 5. Figure 6 shows the results of a similar experiment when CSR and banks of capacitors are disabled. Obviously, motors of Elgaugol energy cluster have stopped after a short circuit in the EPS.

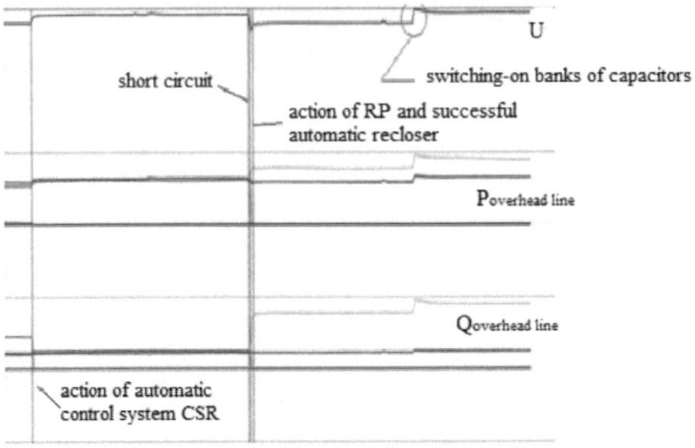

Figure 5. Real time data oscillogram of changing loads on time feeder short-circuit, CSR and banks of capacitors switching-on.

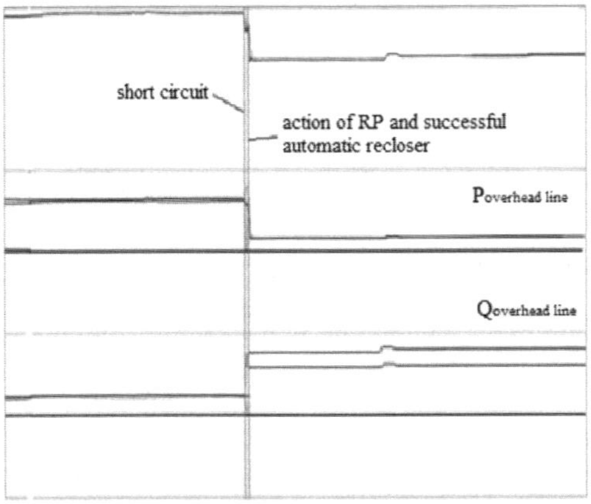

Figure 6. Real time data oscillogram of changing loads on time feeder short-circuit, CSR and banks of capacitors switching-off.

Losses of active and reactive power overhead lines and transformers of Elgaugol energy cluster are taken readings for researching the power losses, see Figure 7. Presented results show that the active and reactive losses are higher when CSR and banks of capacitors are disabled than with their using.

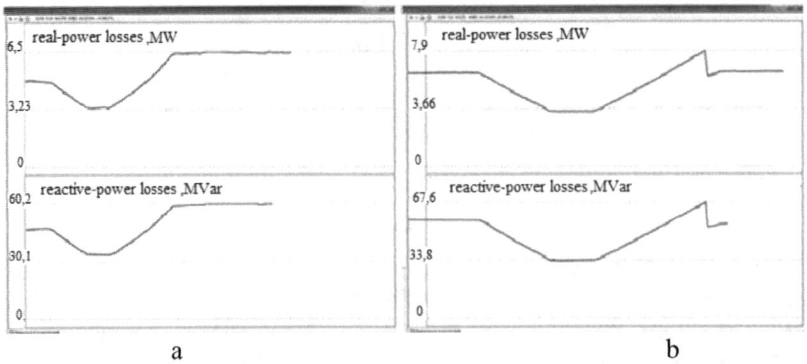

Figure 7. Real time data oscillogram of total power losses in Elgaugol a - CSR and banks of capacitors switching-on, b - CSR and banks of capacitors switching-off.

4. Conclusion

Currently, the voltage stabilization and RP generation in EPS are carried out by means of shunt reactors, capacitor banks, transformer and autotransformer ratio are can only be provided step regulation and such regulation is proved to be insufficient to maintain the voltage level within permissible limits. In this regard, approach is to use DRPC becomes justified and useful. It allows carrying out the automatic continuous voltage stabilization and RP generation. The finished research shows that the combined using of CSR and banks of capacitors allow solving the tasks formulated by above. However, CSR is not always able to provide the stability of consumers in emergency conditions because of its response time, it was established. It depends on many factors, such as the fault location, the running time of relay protection, the duration automatic recloser et al, so in some cases, to use a high-speed device is more profitable, such as for example a static synchronous compensator (STATCOM).

Acknowledgment

This work was supported by mega-grant "Hybrid simulation and control of smart grids".

5. References

[1] Canizares C A 1999 Modeling and Implementation of TCR and VSI Based FACTS Controllers *ENEL Int. report* (Milan: Polytechnics di Milano) p 43

[2] Borovikov Yu S, Gusev A S, Sulaymanov A O and Ufa R A 2014 Hybrid real-time simulator of power system for advanced simulation of the FACTS and HVDC system based on Voltage Source Converter *Proc. Int. Conf. Systems and Informatics (ICSAI 2014) (Shanghai)* (Beijing: IEEE) pp 148–52

[3] Borovikov Yu S, Gusev A S, Sulaymanov A O and Ufa R A 2014 Synthesis of a hybrid model of the VSC FACTS devices and HVDC technologies *Proc. Int. Conf. Modern Techniques and Technologies (MTT 2014) (Tomsk)* (London: Institute of Physics Publishing) pp 1–6

[4] Mahseredjian J, Dinavahi V and Martinez J 2009 Simulation tools for electromagnetic transients in power systems: overview and challenges *IEEE Transactions on Power Delivery* **24(3)** 1657-68

[5] Woodruff S Complexity in power systems and consequences for real-time computing 2004 *IEEE PES Proc. of the Power Systems Conf. and Exposition* (Atlanta: IEEE) pp 1770-75

[6] Gusev A S, Gurin S V, Svetchkaryov S V and Plodisty I L 2004 Hybrid methods as a way out of electric power system modeling problem *Proc. Int. Symp. KORUS 2004 (Tomsk)* vol 1 (Tomsk: Tomsk Polytechnic University) pp 215–19

[7] Gusev A S, Svechkarev S V and Plodisty I L 2005 Basic aspects of modeling problem for electrical power systems, perspectives and methods of their solution *Proc. Int. Symp. KORUS 2005 (Novosibirsk)* (Novosibirsk: Novosibirsk State Technical University) pp 322–26

[8] Borovikov Yu and Sulaymanov A 2013 Information and control system for use in the real time multiprocessor simulation of power equipment *Russian Electrical Engineering* **84** 290–95

[9] Borovikov Yu S, Sulaymanov A O, Gusev A S and Andreev M V 2014 Simulation of automatic exciting regulators of synchronous generators in hybrid real-time power system simulator *Proc. Int. Conf. Systems and Informatics (ICSAI 2014) (Shanghai)* (Beijing: IEEE) pp 153–58

[10] Ruban N Yu, Borovikov Yu S and Sulaymanov A O 2014 Simulation of transmission lines phase-comparison protection for optimization its settings *Proc. Int. Conf. IFOST 2014 (Cox's Bazar)* (Bangalore: IEEE) pp 264–67

[11] Borovikov Yu S, Prokhorov A V and Andreev M V 2012 Application of hybrid real time simulator for solution of smart grid tasks on the example of Elgaugol energy cluster pilot project *Proc. Int. Conf. IFOST 2012 (Tomsk)* vol 2 (Tomsk: Tomsk Polytechnic University) pp 604–08

The effectiveness of the small-tonnage solid composite fuel production from biomass

R B Tabakaev[1], A V Astafev[1], A V Kazakov[1], A S Zavorin[1] and M Polsongkram[2]
[1] National Research Tomsk Polytechnic University, Tomsk, Russia
[2] King Mongkuts University of Technology, Nakhon Ratchasima, Thailand

E-mail: TabakaevRB@tpu.ru

Abstract. The relevance of the work is caused by necessity of the involving of local low-grade raw materials in the fuel energy balance. The purpose of the work is technical and economical evaluation to implementation possibility of the solid composite fuel production from peat as an example of the Tomsk region. The results of a processing of the low-grade raw materials at certain types from Tomsk region into the solid composite fuel are shown, their competitiveness is evaluated, the process line to production of this fuel is suggested and the economical calculation of the production organization by its basis is made. As a result, the prime cost of solid composite fuel and technical and economical parameters of investments efficiency are determined.

1. Introduction

Questions of the biomass implication (peat, wood, agricultural wastes, etc.) in the fuel energy balance are actual for several decades. At first, biomass is a renewable resource that allows is got energy fuel based on this for the long term. Secondly, the low sulfur content in biomass does not result to the formation of oxides SO_x during combustion. In addition, CO_2 emissions are not counted during combustion of biomass in accordance with the Kyoto Protocol, as amount of CO_2 which absorbed by growth of biomass under the influence of photosynthesis from the environment is equal to the amount of CO_2 which generated during its combustion [1, 2]. Thirdly, biomass is perspective raw from economic point because its reserves locate on the earth surface and do not require special conditions and huge costs for extraction.

However, low thermotechnical characteristics and local of application hinder its extensive energy use. Biomass during transportation clod, stick together and frozen in winter, because it has high moisture, thus unloading process is complicated, and biomass is made nontransportable raw.

Question of biomass implication is especially acute in remote and inaccessible areas, heat supply of which is currently provided by imported coal mainly. Long multielement distances of transportation and seasonality of delivery lead to a substantial increase in the cost of imported fuel that consequently affects to the cost of heat supply.

In view of the above, search of ways to effectively biomass implication in the fuel energy balance is an actual problem of scientific research.

2. Main provisions of the heat-technology

Replacement or upgrading of the boiler equipment park, which used in conditions of the exploitation of low power heating systems, will lead to a prolonged payback periods. According to this implication of biomass should be carried out by its processing into a solid composite fuel (briquettes, pellets, granules), which suitable for the energy use in an extensive range of the fuel combustion systems of layer type. The main part of the Russian composite fuel market is process lines of thermal biomass pressing Ruf, Pini&Kay, Nestro and etc. [3–5]. However, pressing equipment is quite expensive, and the pressing process requires high energy costs for its implementation. As a result, produced fuel has a high prime cost and, usually, can't compete with imported coal.

The authors have developed heat-technology of the small-tonnage solid composite fuel production [6–9], forming of which is based on products of low-temperature initial biomass pyrolysis. The heat technology carried out in three stages: the thermal processing of biomass, forming and drying the composite fuel. At the first stage, the biomass is exposed to low-temperature pyrolysis at a temperature 400-450 °C, resulting in charcoal (semi-coke), pyrolysis condensate (mixture of pyrolysis resin and pyrogenetic water) and gas. At stage of forming, used for this pyrolysis condensate is heated to a temperature 50-70 °C, after that in it dissolves dextrin. According to researches [6], the ratio of the pyrolysis condensate and dextrin is recommended 10:1. The resulting forming solution is mixed with a charcoal (semi-coke). The solid composite fuel is formed from mixture and draw at a temperature 20-80 °C for pellets and no more 40 °C for briquettes.

Binder based on pyrolysis condensate is forced out from stamping form when exposure to pressure: dry semi-coke without binder is remained in the form, a stable solid form and strength of which is impossible given by force of pressing up to 30 MPa. Accordingly, the production of the solid composite fuel should be ensured by forming unit of stamping type, energy consumption of which is much lower than pressing equipment.

3. Results of experimental research

The results of the processing of some type's biomass in Tomsk region to solid composite fuel as described in heat technology are shown in Table 1.

Table 1. Burning characteristics of fuels under study.

Thermotechnical characteristics	Kandinsky peat	Sukhovsky peat	Arkadjevsky peat	Woodchips
Feedstock / solid composite fuel				
Humidity W_t^r, %	72.8/0	59.6/0	38.2/0	45/0
Ash on dry basis A^d, %	9.1/19.1	39.5/40.7	31.5/51.6	0.6/2.9
Volatile yield V^{daf}, %	71.6/19.4	69.3/20.9	71.0/15.5	91.8/13.1
Lower calorific value Q_i^r, MJ/kg	3.1/21.1	4.2/14.7	7.3/13.1	9.3/29.4

The lower calorific value of fuel increases despite the increase of composite fuel ash content in comparison with the primary biomass.

Processing of biomass can be carried out without the addition of dextrin at high resin content in the pyrolysis condensate (comparable with yield of semi-coke) according to the method [10], wherein the pyrolysis resin serves as the binder. The high calorific value of the resin (for kandinsky peat – 29.4 MJ/kg, sukhovsky peat – 34.2 MJ/kg, talovsky coal – 33.1 MJ/kg) and the mineral component absence allows significantly improve the calorific value of the produced fuel. For example, the composite fuel from sukhovsky peat, which is obtained by this technology [10], has a calorific value of 17.7 MJ/kg and ash content of dry mass – 32.2 %. Woodchips can't be processed according to this method, because the pyrolysis condensate does not contain the resin.

Most of the solid fuel loss is observed during loading and unloading under transportation, as well as under fuel supply to the furnace of boiler, when it subjects to multiple falls and the subsequent destruction of entirety. As such, proving to dump according to GOST 21289-75 «Coal briquettes. Methods of the mechanical strength evaluation» is carried out as the composite solid fuel testing for mechanical strength. As a result, all test samples of the composite fuel showed 100 % of resistance to fracture, that a fully meets the requirements of GOST 54248-2010 «Peat briquettes arid pellets for heating purposes. Specifications».

The mechanical compressive strength parameters of solid composite fuel (more than 0.3 MPa) show the possibility of stack storing and warehousing with small losses.

4. Technical and economical evaluation of the implementation (as an example of the Tomsk region, Russia)

The translation of remote areas boiler-houses to local peat fuel registers in departmental plans, its will reduce the cost of energy generation and ensure energy security of settlements in the winter and the off-season, when transport accessibility misses. The total area of the Russian Federation peat fields is 80 million hectares, when concentrated is more than 186 billion tons of peat. Question by use of peat fields for Tomsk region is particularly relevant. The Tomsk region ranks second place in Russia by peat reserves. 1300 fields, constituted up 30 % of Russia peat reserves, account for region share (29 billion tons when rate of annual growth is about 1.2 million tons) [11], some of which is early drained and suited for industrial use.

The amount of the Tomsk fuel and energy market for the solid fuel boiler-houses is more than 162 thousand tons of referent fuel every year [12], but imported coal occupies 80 % of the market, and the rest share is firewood. Potentially local composite fuels, which surpass to the firewood and comparable to the imported coal by their characteristics, can cover most of the market at a comparable cost.

The Kandinsky field was chosen for cost-efficiency evaluation of thermal technological by processing peat into a solid composite fuel, in this field is currently produced mining of non-metallic materials [13–16]. The peat is stored in the field at mining without further use. If peat moisture and special aspects of thermal technology will be considered [6] then 40 thousand tons of composite fuel may be received from this field every year. However, the practice shows that the enterprise can be considered successful if its production occupies 5-10 % of the market. As such, the amount of made production was taken 45 tons of composite fuel every day at the cost-efficiency calculation. Two scripts of calculation were reviewed: optimistic and pessimistic. The way of optimistic script assumes that the initial investments are equity capital and the peat is a by-product at the mining of non-metallic materials with its cost equalled 0; at the pessimistic script – the initial investments are deposits by customers (credit for three years at 20 % per annum), the peat cost amounts 100 rub/m^3 (break-even value of peat mining for industrial use).

One example of the technic realization of heat technology is process line, layout diagram of which shows in Figure 1.

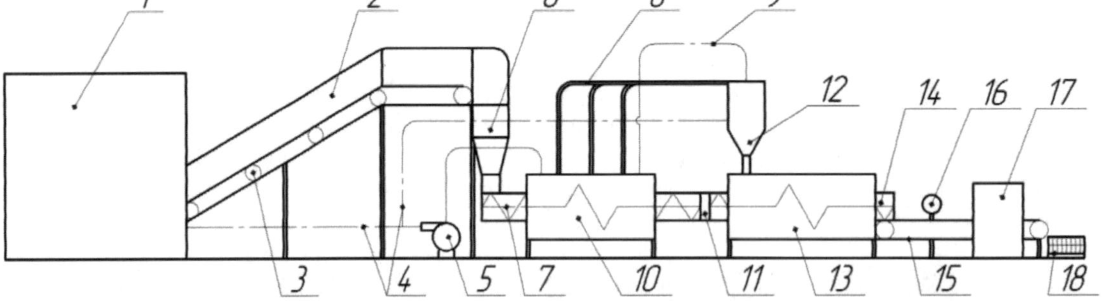

Figure 1. Layout diagram of the process line at small-tonnage solid composite fuel production: 1 – storeroom, 2 – dry of raw, 3 – conveyer, 4 – conduit of flue gases, 5 – exhauster, 6 – grinder, 7 – conveyer screw, 8 – conduit of volatile products, 9 – conduit of gas, 10 – combustion chamber, 11 – box-coupling, 12 – unit of binder preparation, 13 – heat-exchange unit, 14 – conveyer screw, 15 – conveyer, 16 – press roll, 17 – molder of stamping, 18 – dilly.

If we considered the power inputs of this process line with inputs of the composite fuel production lines from other manufacturers, which use press equipment or forming extruder, then should be noted that the specific power consumption of the equipment of this line is lower with comparing to other manufacturers.

The calculation of the production cost-efficiency for solid composite fuel (45 tons every day) was performed by the software "Project Expert 7" (Table 2). Its cost was chosen from condition that modified internal revenue rate equal 30 %. Calculation is made for 5 years.

Table 2. Parameters of investments efficiency.

Parameter	Script of calculation	
	Optimistic	**Pessimistic**
Cost of solid composite fuel ton, rubles	3 400	4 180
Pay back (PB), months	7	6
Average rate of return (ARR), %	72.30	73.36
Net present value (NPV), rubles	11 783 301	14 180 230
Productivity index (PI)	3.61	3.67
Modified internal revenue rate (MIRR), %	30	30

The calculations showed that composite fuel cost should be 3 400 rubles at optimistic script and 4 180 rubles at pessimistic script for the achieve of modified internal revenue rate (MIRR) value, which equal 30 %. If taking into account that cost of coal in Tomsk is 2 100-2 800 rubles/ton, then produced composite fuel at pessimistic script will be difficult to compete with it. In this case the cost of composite fuel is lowed by reducing the modified internal revenue rate (MIRR) value.

It is fair to assume that demand for composite fuel, which produced to the optimistic script and had additional advantages (ease of unloading, deficiency of dust, low moisture), will rise. In consequence of which it is possible of the production expanding, result of which is the increase of net present value (NPV).

The cost of solid composite fuel from other manufacturers, which are representing on the market now (Pini&Kay, Nestro, Ruf, pellets and other), found in range from 5 500 to 12 000 rubles per ton. The cost-efficiency of the heat-technological processing of law-grade raw materials into composite fuel according to the developed technology is obvious, if foregoing manufacturers will be regarded as the main competitors.

5. Summary

1. The new heat-technology by processing of the biomass into high calorific solid composite fuel was considered.

2. The results of the biomass processing in Tomsk region into solid composite fuel with according to this heat-technology are showed. Solid composite fuels, lower calorific value of which equal 31.9 MJ/kg, are obtained when processing of woodchips, and solid composite fuels with lower calorific value, which found in range 10.1-21.5 MJ/kg, are obtained from the least remote fields of peat.

3. Technical and economical evaluation of the heat-technological processing of peat from one Tomsk region field is conducted. The economic expediency of processing occurs when by-products or wastes of production (peat at mining of non-metallic materials, wood wastes and other) use as a feedstock.

Acknowledgments
The reported study was partially supported by the Ministry of education and science of The Russian Federation (Government Order No. 2069 (2.1322.2014)).

IOP Conf. Series: Materials Science and Engineering **93** (2015) 012017 doi:10.1088/1757-899X/93/1/012017

Reference

[1] Sonar D, Soni S L and Sharma D 2014 *Applied Thermal Engineering* **71** 790
[2] Ryabov G A, Litun D S, Dik E P and Zemskov K A 2006 *Thermal engineering* **53** 559
[3] Perederiy S 2010 *LesPromInform* **72** 162 [in Russia]
[4] De Oliveira Maia B G, Souza O, Marangoni C, Hotza D, De Oliveira A P N and Sellin N 2014 *Chemical Engineering Transactions* **37** 439
[5] Srivastava N S L, Narnaware S L, Makwana J P, Singh S N and Vahora S 2014 *Renewable Energy* **68** 270
[6] Tabakaev R B, Gergelizhiu P S, Kazakov A V and Zavorin A S 2014 *IOP Conference Series: Materials Science and Engineering* **66** 012052
[7] Kazakov A V, Tabakaev R B, Novoseltsev P Y and Astafev A V 2014 *MATEC Web of Conferences* **19** 01014
[8] Zavorin A S, Tabakaev R B, Novoseltsev P Y and Astafev A V 2014 *MATEC Web of Conferences* **19** 01015
[9] Gergelizhiu P S, Khaustov S A, Tabakaev R B, Novoseltsev P Y, Kazakov A V and Zavorin A S 2014 *Proc. of 2014 Int. Conf. on Mechanical Engineering, Automation and Control Systems* (Russia: Tomsk/Institute of Electrical and Electronics Engineers Inc.) 109804
[10] Zavorin A S, Kazakov A V and Tabakaev R B 2012 *Bulletin of the Tomsk Polytechnic University* **320** 18 [in Russia]
[11] Kasitskaya L V, Sarkisov Yu S, Gorlenko N P and etc. 2007 *Peat resources of Tomsk area and the ways of their use in civil engineering* (Tomsk: Scientific & Technical Translations) p 292 [in Russia]
[12] Renewable energy in the Tomsk region, http://green.tsu.ru/tomres/?cat=11
[13] Suglobova N V, Plakhova T M, Kazakov A V and Zavorin A S 2014 *9th Int. Forum on Strategic Technology "IFOST 2014"* (Bangladesh: Cox's Bazar/ Institute of Electrical and Electronics Engineers Inc.) p. 386
[14] Kazakov A V, Plakhova T M and Popov R I 2014 *MATEC Web of Conferences* **19** 01019

IOP Conf. Series: Materials Science and Engineering **93** (2015) 012018 doi:10.1088/1757-899X/93/1/012018

Simulation of automatic frequency and power regulators

Y S Borovikov, A Y Pischulin and R A Ufa
National Research Tomsk Polytechnic University, Lenin Avenue, 30, Tomsk, 634050, Russia

E-mail: hecn@tpu.ru

Abstract. The motivation of the presented research is based on the need for development of new methods and tools for adequate real time simulation of automation control frequency and power regulators of generator played an important role in the planning, design and operation of electric power system. This paper proposes a Hybrid real time simulator of electric power system for simulation of automation control frequency and power regulators of generator. The obtained results of experimental researches of turbine emergency control of generator demonstrate high accuracy of the simulator and possibility of real-time simulation of all the processes in the electric power system without any decomposition and limitation on their duration, and the effectiveness of the proposed simulator in solving of the design, operational and research tasks of electric power system.

1. Introduction

According to the well-known reasons, the simulation of electric power system (EPS) is the only effective tool for research of processes in EPS as a whole and solving of the design, operational and research tasks, including adequate settings of relay protection and automation system (RPA) of EPS [1, 2]. Therefore, accurate modeling of RPA is one of the conditions for achieving a high adequacy of simulation EPS. At the same time simulation of RPA is quite complex problem that has significant practical value and can to solve a number of tasks. There are most urgent tasks [2, 3]:

- The analysis and optimization of calculated settings of RPA;
- The formation of methods and means of adequate settings of RPA (including automatic voltage regulator (ARV), Automatic frequency and power control (known as automatic generation control (AGC)), and others;
- The development of new emergency control systems of EPS, including multi-agent systems;

In particular, the simulation of AGC of generator is one of the most urgent tasks [4, 5]. Especially important from a researcher's viewpoint is the depth of the simulation, as well as impact of the simulation depth to adequate modeling of the processes in EPS. The main criterion for the solution of this task is ability of realization of the AGC model in the particular simulated EPS with the maximum possibility of simulation depth of AGC model.

Several approaches to solving the task of adequate simulation of AGC model have been implemented during the development of Hybrid real time simulator (HRTSim) of EPS, based on the concept of hybrid simulation and developed in Tomsk Polytechnic University [6, 7]. The first version of AGC model created in 1988 was developed at analog simulation level via operational amplifiers and analog multipliers. In the current version, the AGC model has been realized at digital simulation level via microcontroller, which is a coprocessor of specialized hybrid processor (SHP) of electric

IOP Conf. Series: Materials Science and Engineering **93** (2015) 012018 doi:10.1088/1757-899X/93/1/012018

machine. As a result the depth of mathematical models and flexibility of parameter controlling have been greatly increased.

According the concept of hybrid simulation [6, 7], the features and properties of the HRTSim can be used for effectively solving a lot of problems, including the real-time simulation of all the processes in the electric power system without any decomposition and limitation on their duration, and testing of AGC under all possible normal, emergency and post-emergency modes of EPS.

The issue of adequate modeling of AGC and turbine emergency control of generating unit are presented are presented in the article.

2. A mathematical model of AGC implemented in HRTSim

The task of synthesis implemented in the microprocessor unit of SHP of HRTSim mathematical model of AGC was the creation a library of universal mathematical model of AGC, which can be adapted by setting the parameters of the transfer functions (coefficients of differential equations) for adequate modeling of currently used in Russian EPS of various hydraulic-mechanical and electro-hydraulic model of AGC.

The mathematical model of AGC was created by analyzing of well-known mathematical units and structures of existing AGC [8, 9]. The developed model of AGC allow us to provide fully and accuracy modeling of AGC operation at primary and secondary regulation, as well as emergency control of generating unit by centralized emergency control system (CECS) of EPS, including automatic pulse unloading of turbine (APU) and post-emergency unloading of turbine (PEU).

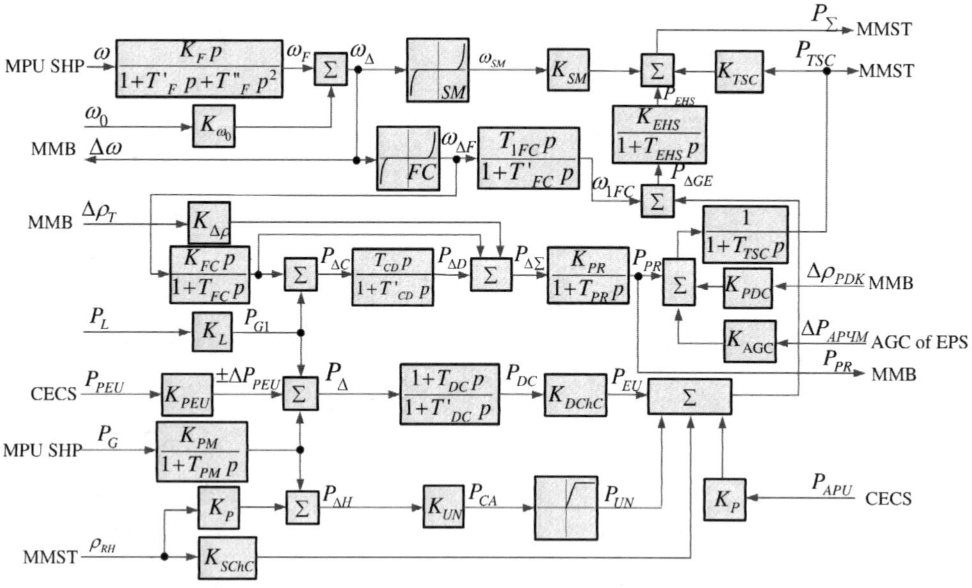

Figure 1. Structural diagram a mathematical model of AGC.

According to Figure 1:

- K_L – programing factor of generator unit load (P_L), determining in AGC active power of generator unit (P_{G1}), which in functional changes of the K_L can display a required load curve;
- K_{PEU} – programing factor of post-emergency unloading (additional loading) of generator unit $\pm\Delta P_{PEU}$;
- Differential equation of generator power meter

$$\frac{dP_{GE}}{dt} = \frac{1}{T_{PM}}(K_{PM}P_G - P_{GE});$$

- Differential equation of frequency corrector of slow-speed circuit of AGC

$$\frac{d\omega_{FC}}{dt} = \frac{1}{T_{FC}}(K_{FC}\omega_{\Delta F} - \omega_{FC});$$

- Equations system of the dynamic correction of the slow-speed circuit of AGC

$$P_{\Delta D} = \frac{1}{T'_{DC}}(T_{DC}P_{\Delta C} - Z_{\Delta C}), \frac{dZ_{\Delta C}}{dt} = P_{\Delta C};$$

- $K_{\Delta\rho}$ – programing factor of the frequency corrector of the slow-speed circuit of AGC at change of pressure of live steam $\Delta\rho_T$ (from MMB – mathematical model of boiler);
- Differential equations of general slow-speed circuit of AGC

$$\frac{dP_{PR}}{dt} = \frac{1}{T_{PR}}\left(K_{PR}P_{\Delta\Sigma} - P_{PR}\right);$$

- K_{PDC} – programing factor of common channel of power control for condensing turbine in sliding movement modes and regulating the steam pressure regulation, including stand-by mode, as well as power control with backpressure, turbines with industrial and cogeneration steam extraction;
- K_{AGC} – multifunctional programing factor, which can be used for modeling of external control actions of AGC from EPS;
- K_{TSC} – transfer ratio of turbine speed changer (TSC);
- Differential equation of TSC

$$\frac{dP_{TSC}}{dt} = \frac{1}{T_{TSC}}(P_{\Delta M} - P_{TSC});$$

- K_P и K_{UN} – programing factor of the corrector of the high-speed circuit of AGC, which, together with the current value of power of generator unit (P_G) from microprocessor unit of specialized hybrid processor (MPU SHP) of electric machine and the steam pressure of reheater (ρ_{RH}) from the universal mathematical model of steam turbines (MMST), forms a corrective action (P_{CA}) according to the equation

$$P_{CA} = (P_{GE} - K_P\rho_{RH})K_{UN}$$

- K_{SChC} – programing factor of the static correction channel of unevenness of the high-speed circuit of AGC;
- Differential equation of the dynamic correction channel of unevenness of the high-speed circuit of AGC

$$P_{DC} = \frac{1}{T'_{DC}}(T_{DC}P_{\Delta} - Z_{D\Delta}), \frac{dZ_{D\Delta}}{dt} = P_{DC} - P_{\Delta},$$

- K_{DChC} – transfer ratio of the dynamic correction channel of the high-speed circuit of AGC;
- K_{EU} – programing factor of the unloading channel of the high-speed circuit of AGC, including channel of the APU, and forms the ΔP_{APU} by shown in figure 2 graphics:

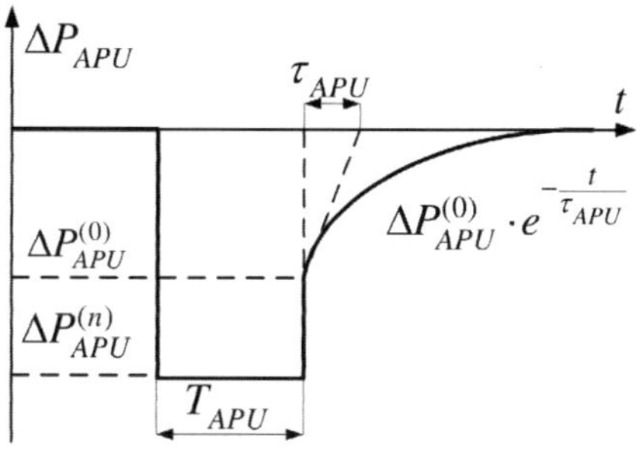

Figure 2. Control actions of the turbine emergency control of generating unit, where $\Delta P_{APU}^{(n)} =$ var – depth of the APU, n = var – index of the depth; T_{APU} = var – duration of the APU; $\Delta P_{APU}^{(0)}$ = var – part of ΔP_{APU}, exponential removing $\Delta P_{APU}^{(0)} \cdot e^{-\frac{t}{\tau_{APU}}}$, τ_{APU} = var.

IOP Conf. Series: Materials Science and Engineering **93** (2015) 012018 doi:10.1088/1757-899X/93/1/012018

- Differential equation of speed turbine meter
$$\frac{d^2\omega_F}{dt^2} = \frac{1}{T''_F}\left(-T'_F\frac{d\omega_F}{dt} - \omega_F + K_F\omega\right);$$
- Differential equation of the dynamic frequency correction of the high-speed circuit of AGC
$$\omega_{1FC} = \frac{1}{T'_{1FC}}(T_{1FC}\omega_{\Delta F} - Z_{1\Delta F}), \frac{dZ_{1\Delta F}}{dt} = \omega_{1FC} - \omega_{\Delta F};$$
- Differential equation of the common channel of the high-speed circuit of AGC
$$\frac{dP_{EHS}}{dt} = \frac{1}{T_{EHS}}(K_{EHS}P_{\Delta GE} - P_{EHS});$$
- $K_{SM} = \frac{1}{\sigma}$ – transfer ratio of the turbine governor, where $\sigma = var$ – drop of the turbine governor;
- K_{ω_0} – coefficient of setting of the turbine rotor speed (ω_0).

The developed mathematical model of AGC and mathematical model of the generator unit are implemented via SHP, which is the basic element of HRTSim [7].

3. Experimental research of the AGC model

Experimental researches of the AGC model have based on the simulation of the APU and PEU of the generator unit.

The part of the created intelligent scheme of power cluster "Elgaugol" implemented in the HRTSim has been used for modeling and analysis of AGC model based on the generation unit G-1. (figure 3).

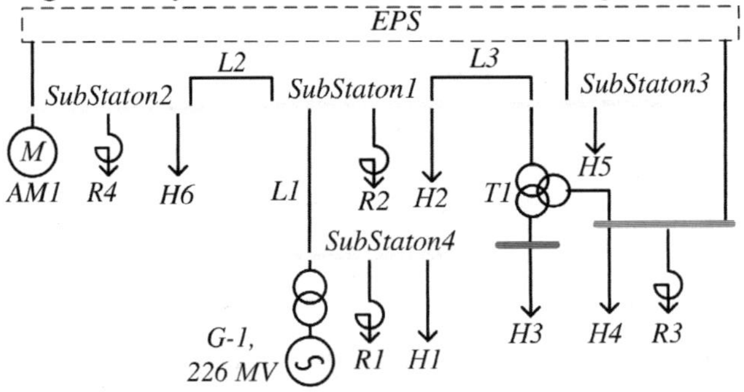

Figure 3. Part of analysis scheme of power cluster "Elgaugol".

The control of APU and PEU is carried out by its preset, which includes:

- Determination and parameter setting of the APU;
- Formation of signal in central system emergency control (CSEC) for PEU as a values of surplus power (P_{SP}), used for deterring a value of generator unit power after PEU (P_{EU}) according to the following equation:
$$(P_{EU} = P_T - P_{SP}),$$
where P_T – current value of load of the steam turbine.

There are parameters of the APU and PEU model, whose value can be changed both stationary and interactively:

- A_P – the value unloading pulse of turbine;
- P_{EU} the value of the unloading pulse of turbine set between A_P to 0, in according to an algorithm of emergency control of generating unit;
- A_0 – the value of the residual control action of the unloading pulse of turbine значение;
- t_P – time duration of the applied pulse;
- T_{EU} – time constant of exponential removal of the unloading pulse A_0 ;
- P_{EU} – the value of generator unit power after PEU;
- K_{EU} –gain factor of the channel of the PEU.

Interactive changing of parameters is carried out using the corresponding table "Prime mover" in the dynamic monitoring and control panel (DMCP) of generator unit. In automatic mode, it is possible to carry out with the specially prepared scenarios that run from the Client software "Script of APU and PEU" from DMCP.

Analysis of APU and PEU model operation in the HRTSim includes the following steps:

1. The performing the dynamics scenario simulated the occurrence of emergency mode and the corresponding operation of the RPA.
2. The modeling the CSEC and exporting the unloading pulse of turbine to the model.
3. The operation of APU and PEU model, taking into account data from CSEC:
   a. Assigning P_{EU} to A_P during the time t_P ;
   b. Changing P_{EU} to A_0 after t_P and removing of the residual control action $P_{EU} = A_0$ with the time constant $T_{EИ}$;
   c. Assigning the PEU settings: P_{EU}, according to the equation (1), and $K_{EU} = (1 - 10)$, (particular value is determined as a result of research);
   d. Zeroing K_{EU} after control time.

The following research results for determining the effective settings of automatic emergency power control of generator unit by means of a thermal power station of APU and PEU are presented below and modeled at the original operating mode of the simulated EPS:

- Generation unit G-1 produce around 226 MV;
- Overhead line (L1) performs communication G-1 with EPS model and transmit the 136 MV of active power;
- This transmitted active power distribute through two overhead lines (L2 transmit the 105 MV and L3 transmit the 31 MV).

According mentioned scenario:

1. The single-phase short circuit is simulate on a L2;
2. The action of RPA, non-successful reclosing and repeated action of RPA with acceleration simulated.

The consequence of these actions is the weakening of the link G-1 and EPS (L2 tripping and increasing power flow via L3). As a result, CSEC sends action and data to perform turbine emergency control of the G-1. In figure 4a the waveform illustrating the significant power oscillations that lead to the destruction of the stability of EPS, in the operation of relay protection without the use of turbine emergency control is shown. In figure 4b the waveform corresponding to operation of APU without smooth removing of the residual control action is shown. There are some overshoot and long power oscillations in a simulated EPS.

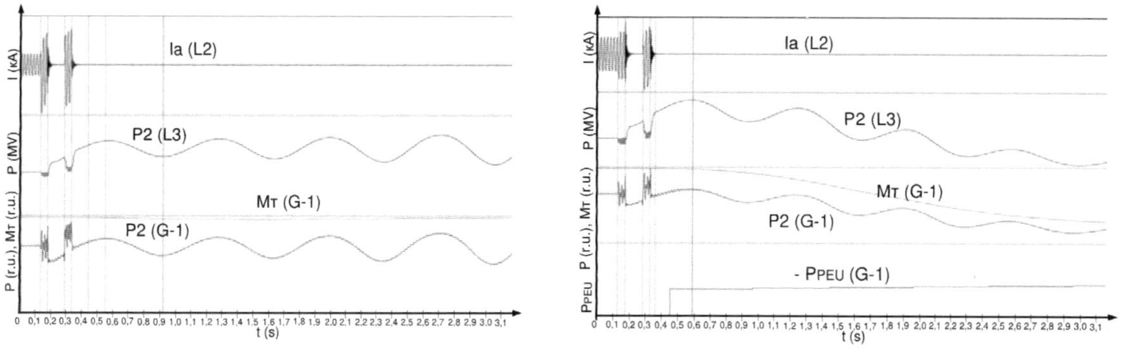

(a) (b)

Figure 4. Waveform in the operation of relay protection without the use of turbine emergency control (a) and with the use of APU and PEU of G-1 ($A_P = 4$ u. un., $A_0 = 1$ u. un., $t_P = 0.1$ s, $T_{EU} = 3{,}5$ s) (b).

The waveform of processes, when APU and PEU are configured, is shown in figure 5 and 6. As can be seen, the power oscillations rapidly damped and EPS stability is saved for the given setting of APU and PEU. It should be noted that the when pulse is equal of 2,5 u. un. (unit of unevenness) the dynamic impact on the turbine valves system is less than pulse is equal of 4 u. un.

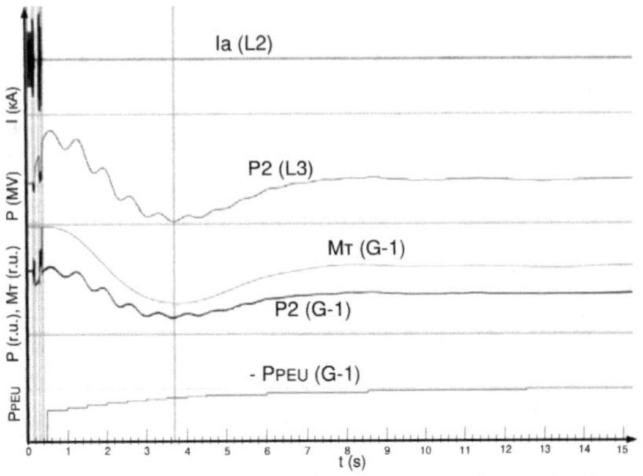

Figure 5. Waveform in the operation of relay protection with the use of APU and PEU of G-1 (A_P = 2,5 u. un., A_0 = 1 u. un., t_P = 0.1 s, T_{EU} = 3,5 s).

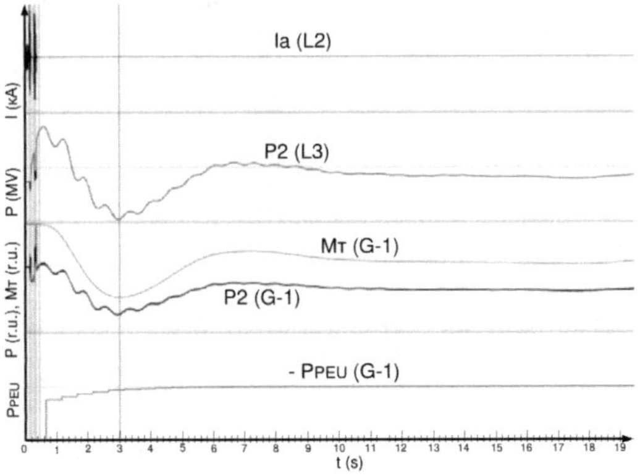

Figure 6. Waveform in the operation of relay protection with the use of APU and PEU of G-1 (A_P = 4 u. un., A_0 = 1 u. un., t_P = 0.1 s, T_{EU} = 1,5 s).

As a result of the analysis the following results have been obtained:

- The adequate simulation of developed APU and PEU model in real time is confirmed;
- The optimal settings of developed APU and PEU model for simulated EPS are determined.

4. Conclusion

- The simulation of RPA is an essential part in the simulation of EPS, to achieve the adequate modeling a processes in equipment and EPS as a whole;
- The developed model of AGC implemented in the HRTSim, taking into account the depth of mathematical models sufficient for adequate modeling a processes in EPS;
- The experimental results presented in this paper confirm the adequacy of obtained information, as well as the possibility of using the developed model for analysis and effective settings of AGS to provide a dynamic and static stability of EPS.

Acknowledgements
This work was supported by mega-grant "Hybrid simulation and control of smart grids".

References

[1] Chen Y and Dinavahi V 2012 *IEEE Transactions on industrial electronics* **59 (2)** 1300–09
[2] Nayak O, Santoso S and Buchanan P 2002 *IEEE Computer Applications in Power* **15** 37-44
[3] Chusovitin P and Pazderin A 2012 *IEEE PES Innovative Smart Grid Technologies European* (Germany: Berlin) pp. 1-5
[4] Usman A and Divakar B 2012 *IEEE Global Humanitarian Technology Conf.* (USA: Seattle) pp. 1-6
[5] Prytkov N, Fedorov V and Mosquitoes K 2007 *Thermal Engineering* **10** 2-7
[6] Ruban N, Borovikov Y and Sulaymanov A 2014 *International Forum on Strategic Technology* (Bangladesh: Cox's Bazar) pp. 264-267.
[7] Borovikov Y, Sulaymanov A, Gusev A and Andreev M 2014 *International Conf. on Systems and Informatics* (China: Shanghai) pp. 153- 158.
[8] Ivanov V 1982 *Regulation of generator unit* (Moscow: Mechanical Engineering)
[9] Rabinovich R and Polonsky M 1983 *Elektrichestvo* **3** 11-19

IOP Conf. Series: Materials Science and Engineering **93** (2015) 012019 doi:10.1088/1757-899X/93/1/012019

Plasma dynamic synthesis of ultradispersed copper oxides

Yu Shanenkova[1], A Sivkov[1], A Saygash[1] and I Shanenkov[1]
[1] National Research Tomsk Polytechnic University
30 Lenin av., Tomsk, 634050, Russia

E-mail: julia_kolganova@mail.ru

Abstract. Copper oxide is necessary material for production of superconductors. The issue of obtaining high purity and nanosides CuO is actual. This article shows the results on the obtaining of nanodispersed copper oxide by plasma dynamic method in system based on coaxial magneto plasma accelerator with copper electrodes. Such analyses of ultradispersed synthesized products as X-Ray diffractometry, IR-spectroscopy and thermal analysis were carried out. According to XRD such phases as copper Cu, copper oxide (I) Cu_2O, copper oxide (II) CuO, and copper hydroxide hydrate $Cu(OH)_2•H_2O$ were identified in the product. It was found that with the gradual heating of the initial product up to 800 °C the phase content changed dramatically in terms of enhancing copper oxide phase (up to 97%).

1. Introduction

Nanodispersed copper oxide (CuO) is widely used in catalysis [1], high critical temperature superconducting [2] applications due to its excellent physical and chemistry properties and cost effectiveness [3-6]. Copper oxide is one of the main components of yttrium-barium cuprates composition. This material has unique property to reach zero electrical resistance at certain temperature and convert to superconductive state. Using this property it is possible to create effective systems of production, accumulation and transfer of electrical energy over long distances, super-power generators, high-speed magnetic levitation transport, and powerful magnet systems for thermonuclear reactors, particle accelerators, super-fast computers and ultrasensitive diagnostic devices. The main advantage of CuO in comparison with Cu_2O is the stability. The CuO is stable at room temperature (Cu ion is in the +2 valance state) whereas Cu_2O (+1 valance state) is only stable above 1020 °C.

In general, all methods of obtaining nanodispersed copper oxide (II) are chemical-based techniques. In this paper, the possibility of obtaining nanodispersed copper oxide by plasma dynamic method is shown.

2. Materials and Methods

The basis of plasma dynamic synthesis is the use of a high-current pulsed coaxial magneto-plasma accelerator (CMPA) of erosion type with copper central electrode and barrel-electrode [7-8]. Pulse power supply is provided to CMPA from the sectioned capacitive energy storage with the total capacity of 57.6 mF and the charging voltage $U \leq 5.0$ kV. The CMPA consists of two main components: the removable one is a classic coaxial Z-pinch accelerator and the stationary one is an external induction system. The use of copper electrode and the copper accelerating channel (barrel) allow to realize the synthesis of nanodispersed copper oxides.

[1] To whom any correspondence should be addressed.

The principle of the accelerator operation is as follows. Initially, the CMPA is included in the discharge circuit of capacitive energy storage. With the closure of the power keys the discharge current begins to flow from the capacitor bank to the inductive load. At achievement of a certain level by the discharge current, there is an arc discharge with plasma structure of Z-pinch type with circular plasma bridge. The plasma structure stable state is provided by the magnetic field pressure of its own current and the magnetic pressure from axial component of an external magnetic field from the solenoid induction system.

Electro erosive recovery of material from the copper surface of the acceleration channel occurs when the plasma flows through this channel. Copper is involved into the discharge structure, converted into the plasma state and removed from the accelerating channel at hypersonic speeds in a closed space of the chamber. The chamber is filled with a mixture of gaseous oxygen. Thus, the main precursors of oxidative processes interact in the chamber.

The registration of energy parameters u(t) and i(t), was carried out using active voltage divider and Rogowski transformer. Their signals were recorded using a digital oscilloscope Tektronix. Typical oscillograms of discharge current i(t) and voltage u(t) are shown in figure 1. Using these oscillograms the discharge power curve was built. By its integrating the value of input to the CMPA energy (figure 1) was found. It should be noted the whole process is no much longer than 500 μs.

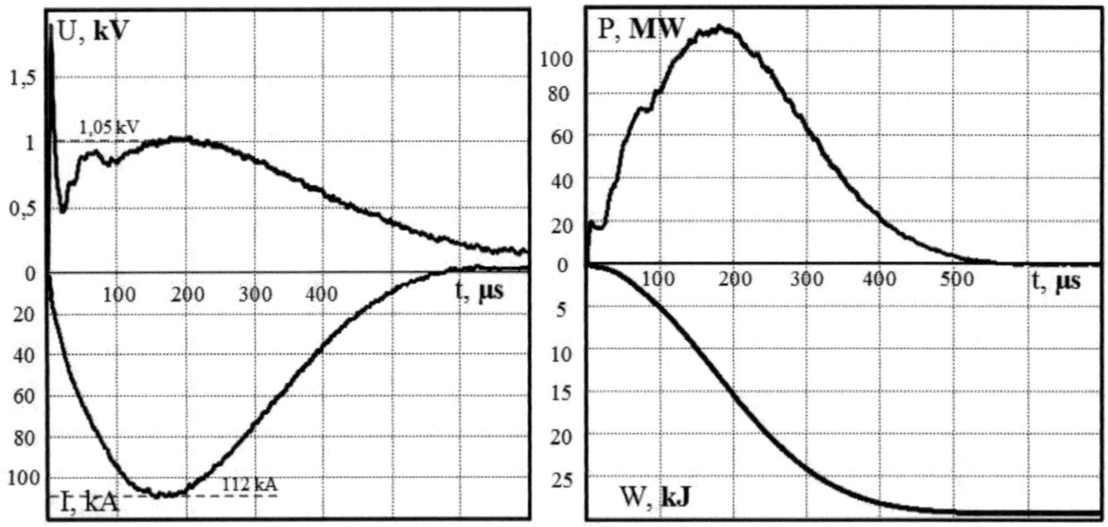

Figure 1. Typical oscillograms current i(t), voltage u(t) and curve power p(t) and energy w(t).

The collection of the resulting product was made after waiting few hours. The phase composition of the resulting plasma dynamic synthesis product was studied by X-ray diffractometry (XRD) with using diffractometer Shimadzu XRD7000 having copper cathodes (CuKα-radiation). The initial product was annealed using high-temperature oven Nabertherm 1800/3 at temperature of 420 °C, 600 °C and 800 °C. Also the synthesized product was analyzed by IR-spectroscopy (Nicolet 5700), mass spectrometry (TRACE DSQ (GC / MS)) and thermal analysis (combined TGA / DSC / DTA analyzer SDT Q600).

3. Results and discussion

Figure 2 shows XRD patterns of initial and annealed products. The software PowderCell 2.4 with the structural database PDF2+ was used to analyze XRD-patterns. According to the diffraction pattern (Figure 2a)) of initial powder it is the heterophase product. The dominant phase is copper oxide CuO with an average size of coherent scattering regions (CSR) of about 14 nm. It should be noted the final product also has several nanocrystalline phases (copper oxide (I) Cu_2O, copper Cu, copper hydroxide hydrate $Cu(OH)_2{\bullet}H_2O$). Table 1 shows the percentage of crystalline phases in the product. XRD-pattern of the powder, heated in air up to 420 °C at a rate of 80 °C/min with the exposure of 30 min and followed natural cooling down to the room temperature is presented in figure 2b. It unambiguously indicates the fully disappearance of strong small angle reflection and other reflections corresponded to the phase of copper hydroxide hydrate $Cu(OH)_2{\bullet}H_2O$. The formation of this phase can be explained by the presence of air humidity in the chamber. After being heated, it is decomposed to copper oxide CuO and the water that evaporates, ensuring irreversibility of process. Due to this, the percentage of CuO is increased up to 86.0%, according to the change in the ratio of the coherent reflection intensities of the remaining phases. However, X-ray amorphous component is still present in the product.

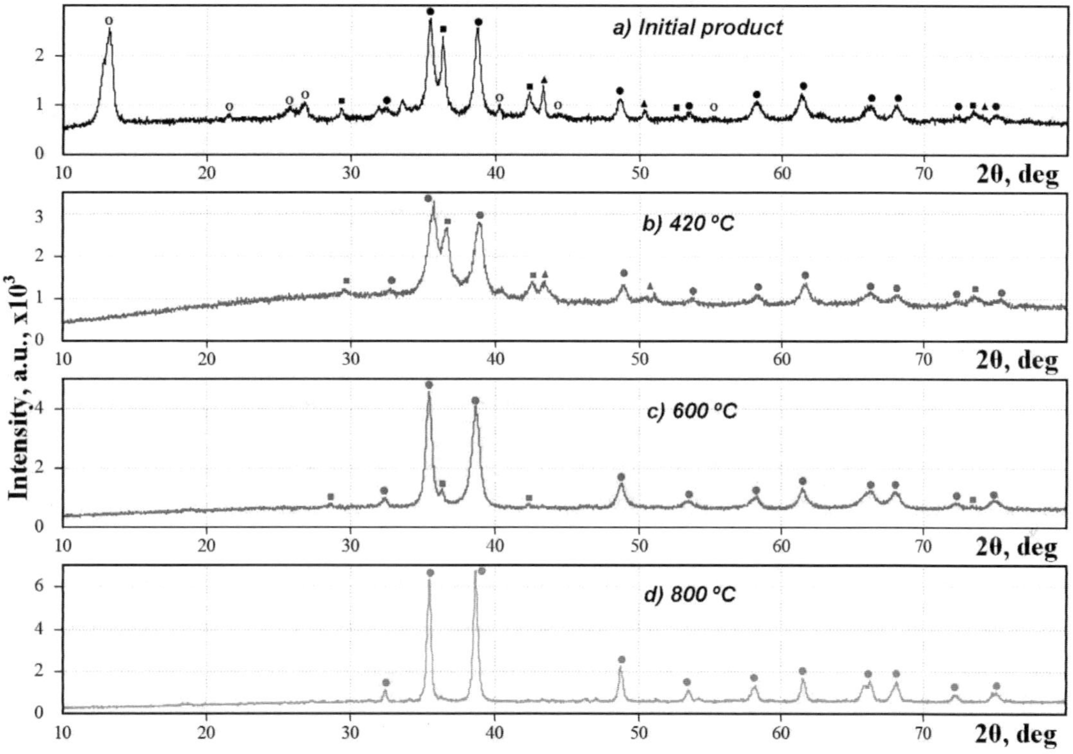

Figure 2. XRD patterns of the plasma dynamic synthesis product and its evolution under the temperature influence.
● CuO ; ■Cu_2O; ▲ Cu; **O** $Cu(OH)_2{\bullet}H_2O$

Table 1 - Ratio in weight % of crystalline product components.

Product	CuO (C2/c (15), monoclinic)	Cu_2O (Pn-3m (224), cubic)	Cu (Fm-3m (225), cubic)	$Cu(OH)_2{\cdot}H_2O$ (P^*, -1, anorthic)
Initial	48%	25%	7%	20%
420 °C	86%	9%	5%	-
600 °C	91%	7%	2%	-
800 °C	97%	2%	~1%	-

The figures 2c and 2d show XRD-pattern of product annealed with the same parameters (rate and exposure) to the temperature of 600 °C and 800 °C, respectively. The XRD-pattern of the resulting material is characterized by the almost complete disappearance of the reflections of the crystal copper and consistent reducing the intensity of reflections of Cu_2O to the level of "traces". In addition, there is a significant reduction of x-ray amorphous component after the high-temperature heat treatment. A significant decrease in the broadening of the strongest reflections CuO indicates the recrystallization process in the temperature range of about 400-800 °C, the growth of CSR and hence the crystallite sizes of this phase. Calculations show an increase of the CuO mass content up to 97.0% (table 1), increase of this phase CSR up to ~ 100 nm.

Using IR-spectroscopy the initial copper-containing plasma dynamic synthesis product and the sample after heat treatment in air at 420 °C were studied. Initial sample IR-spectrum is shown in the figure 3. This spectrum is characterized by an intense absorption band with a maximum at 3471 cm^{-1} and a shoulder at 3253 and 3186 cm^{-1} due to stretching vibrations of water molecules and the structural hydroxyl groups [9, 10]. The absorption bands with maxima at 2958, 2925 and 2855 cm^{-1} can be attributed to the stretching vibrations of OH-groups. The absorption band of low intensity at 1660 cm^{-1} is explained by the deformation vibrations of the OH group. According to the literature [11], even momentary contact of freshly prepared nanoparticles of copper-containing powder with air environment with relative humidity of 58-60% is accompanied by the formation of hydroxyl hydrate cover. The sorption of water molecules on the copper containing particles is accompanied with both the dissociation with the formation of surface hydroxyl groups and the localization of the water molecules on the surface of the nanoparticles in the coordinated state. In the range of 2516-2196 cm^{-1} of the initial sample IR-spectrum there are absorption bands of adsorbed molecules of CO_2. As a result of the carbon dioxide sorption on the surface of the nanoparticles hydroxycarbonate cover $Cu_2CO_3(OH)_2$ can be formed [11]. Its presence is manifested in the range of 1582-1247 cm^{-1} with the intensity absorption bands that are typical for carbonate structures of monodentate type. The absorption band at 738 cm^{-1} is also can be explained by vibrations of the carbonate ion. Absorption bands in the 592-430 cm^{-1} indicate the presence of copper oxide CuO, absorption bands at 874 and 678 cm^{-1} are due to deformation vibrations of Cu-O-H [12].

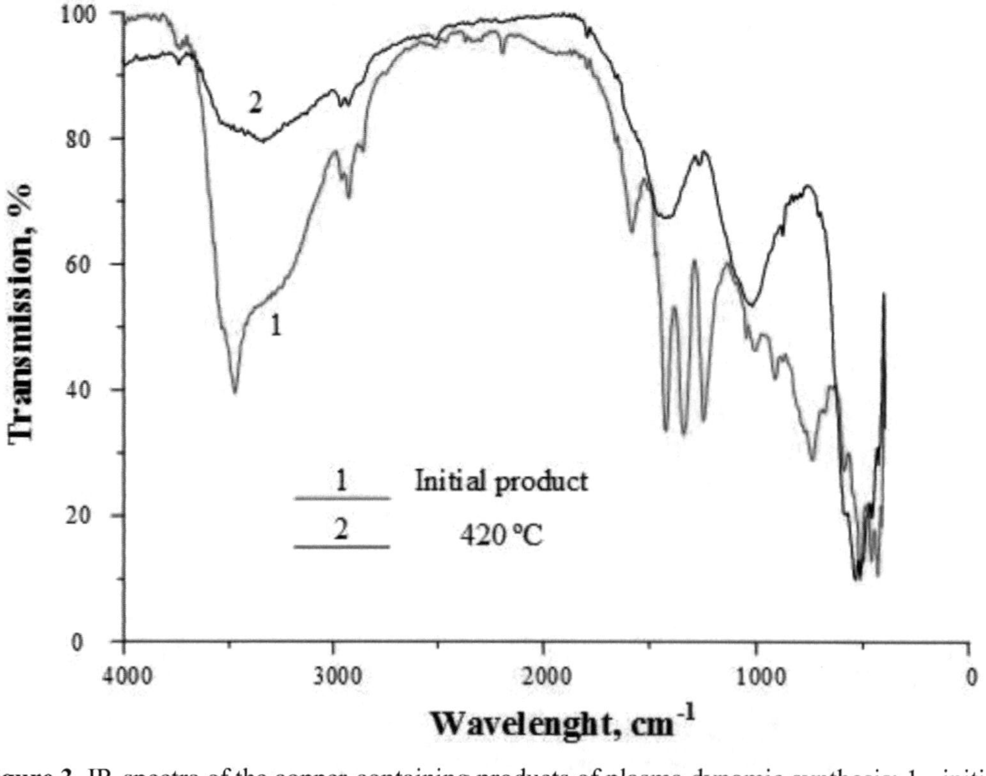

Figure 3. IR-spectra of the copper-containing products of plasma dynamic synthesis: 1 - initial powder, 2 - powder after heating treatment in air at 420 °C.

The results of thermographic investigations of the initial dispersed synthesis product obtained by linear heating of sample with mass of 5.478 g in air at 10 °C/min are shown in Figure 4. The character of TG-curve shows that at heating up to 1000 °C there is a decrease in sample mass by 19.5%. At a temperature range up to 650 °C several intervals of consistent reduction of the derivative changes in the mass of the sample are clearly distinguished. Such uneven weight reduction of the sample with increasing of temperature occurs simultaneously with the appearance on the DSC-curve of exothermic peaks at 242 °C extremes, 321 °C, 431 °C and 500 °C, coincides with the boundary of said intervals. Results of mass-spectrometric synchronous measurements show that after heating of the sample up to 800 °C there is a removal of the adsorbed on the particle surface gases of nitrogen and oxygen: in mass spectrum there are peaks with mass numbers of m/z equal to 28 and 32. Upon further heating the crystallization water is removed.

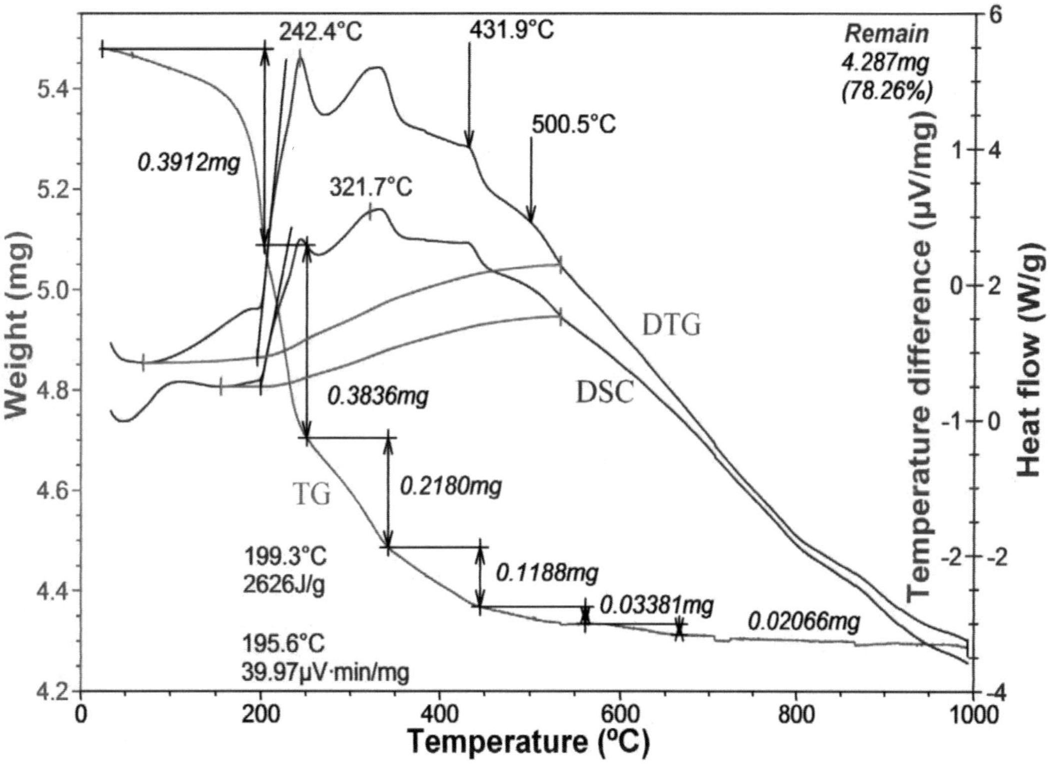

Figure 4. The data of thermographic studies

The main weight reduction upon heating the sample to about 450 °C is connected with the decomposition of hydrated copper hydroxide, as shown by XRD. In the temperature range of 200-280 °C in parallel with the endothermic decomposition reaction of $Cu(OH)_2$•H_2O flows the similar process of decomposition of copper hydroxycarbonate $Cu_2CO_3(OH)_2$. Its presence in the plasma dynamic synthesis product is determined according to IR-spectroscopy (Figure 3) and mass spectrometric measurements in the temperature range of 200-250 °C with the presence of a peak with a mass number m/z=44, corresponded to the carbon dioxide emissions. Small endoeffects caused by the decomposition reaction of the copper compounds, completely overlap by exothermal effects due to the oxidation of copper metal to oxide Cu_2O and further oxidation of the formed and present in the initial product Cu_2O to CuO. The growth of mass of the sample due to the oxidation process is very small in comparison with the loss by the above mechanism.

IOP Conf. Series: Materials Science and Engineering **93** (2015) 012019 doi:10.1088/1757-899X/93/1/012019

4. Conclusions

The paper shows the possibility of direct plasma dynamic synthesis of nanodispersed copper-containing powders using a high-current coaxial magneto plasma accelerator of erosion type. Components of the powder obtained by plasma dynamic method are metallic copper, copper oxides Cu_2O and CuO, as well as hydrated copper hydroxide $Cu(OH)_2 \cdot H_2O$. Using XRD analysis, IR spectroscopy and thermal analysis the changes in the composition of the synthesis product are investigated by heating it in air. The heat treatment of the synthesized powder at 800 ° C in air allow to yield the nanodispersed product with the main crystal phase of copper oxide CuO (97 %).

Acknowledgements

The work was supported with the Russian Science Foundation (project # 15-19-00049).

References

[1] Umadevi M and Christy A 2013 *Spectrochim. Acta* A **109** 133–137
[2] Oku T, Motoyoshi R, Fujimoto K, Akiyama T, Jeyadevan B and Cuya J 2011 *J. Phys. Chem. Solids* **72** 1206–1211
[3] Hassan M S, Amna T, Yang O B, El-Newehy M H, Al-Deyab S S and Khil M-S 2012 *Colloids Surf.* B **97** 201–206
[4] Pandiyarajan T, Udayabhaskar R, Vignesh S, James R A and Karthikeyan B 2013 *Mater. Sci. Eng.* C **33** 2020-2024
[5] El-Nahhal I M, Zourab S M, Kodeh F S, Selmane M, Genois I and Babonneau F 2012 *Int. Nano Lett.* **2** 14–18
[6] Ananth A, Dharaneedharan S, Heo M and Sun Mok Y 2015 *Chemical Engineering Journal* **262** 179–188
[7] Sivkov A A, Saigash A S and Kolganova Yu L 2012 *Russ. Electr. Eng.* **84 (8)** pp 418-421
[8] Sivkov A, Pak A, Shanenkov I, Kolganova J and Shatrova K 2014 *IOP Conference Series: Materials Science and Engineering* **66 (1)** 012001
[9] Dedov N V and Ivanov Yu F 1991 *Glass and ceramics* **10** p 17
[10] Nakamoto K 1991 IR and Raman Spectra of Inorganic and Coordination Compounds *Moscow. Mir*
[11] Larkin P 2011 Infrared and Raman Spectroscopy. Principles and spectral Interpretation *Elsevier*
[12] Harchuk V G, Korykova O V and Hmara E F 2005 *Analytics and Control* **9 (4)** 423–427

IOP Conf. Series: Materials Science and Engineering **93** (2015) 012020 doi:10.1088/1757-899X/93/1/012020

Analysing the thermal state of voltage transformer based on resistive voltage divider

V Lebedev[1,a], **V Zhukov**[1,b] **and A Yablokov**[1,c]
[1] Ivanovo State Power Engineering University, Ivanovo, Russia

E-Mail: [a]vd_lebedev@mail.ru, [b]zhukov-home@yandex.ru, [c]AndrewYablokov@yandex.ru

Abstract. We performed a simulation of the thermal state of a resistive voltage divider based on an equation of heat conductivity with internal sources of heat, solving this equation by using two numerical procedures. We also conducted experimental research regarding transformer thermal state on a laboratory stand. We obtained numerical results of the above heat conductivity equation, taking into account the supply of heat energy from internal sources, and compared the results of our calculations with our experimental data. Transformer thermal state simulation and numerical solution procedures enable us to formulate and resolve the problems of choosing optimal transformer design and operating modes, ensuring maximum measuring accuracy when limiting the thermal state of resistive elements.

1. Introduction

As part of the creation of electronic measuring instruments for Smart Grids it is vital to develop a high-voltage measuring transformer based on a resistive voltage divider. The use of solid state insulation material creates additional heat resistance, hindering heat removal from resistive elements. Warming of resistive elements may lead to changes in their active resistance and therefore reduced accuracy in voltage measurement using a resistive divider. Increased resistance leads to reduced self-heating of resistive elements and thus smaller errors. However, increasing active resistive element resistance reduces total current passing through the voltage divider, so that with significant resistive element resistance measuring accuracy becomes subject to displacement volume current in insulator and environment dielectrics, and also leakage current along the insulator's surface. Most significantly, the impact of volume (displacement) current increases the phase angle.

Thus, to achieve a compromise solution it will be necessary to increase current passing through the resistive divider, resulting in lower active resistive element resistance, which reduces the current's capacitive component. To reduce the impact of resistive element self-heating on accuracy, resistive elements with a minimal temperature resistance coefficient (TRC) should be selected.

It should be noted that this approach leads to greater heat release in the resistive elements, with possible overheating and damage. To limit maximum temperatures as required by guidelines for heat testing [1] and use of resistive elements, we selected a maximum positive temperature value for resistive element surfaces of 100° C.

The purpose of this study is to analyse the thermal state of resistive voltage divider transformers, to choose the optimal position of resistive elements and their characteristics, the dielectric material for the support insulator and filler having the appropriate thermo-physical and dielectric properties.

To achieve this goal the following tasks must be fulfilled in the given order:

- conduct experimental studies of non-steady temperature fields inside a transformer;
- develop mathematical models of non-steady heat conductivity in a cylinder with internal sources of heat energy due to heat release from resistive elements;
- identify mathematical models with the help of obtained experimental data;
- estimate transformer thermal state within the framework of the developed model, and compare the obtained data with the results of experimental studies;
- find the optimal position of resistive elements in the developed model.

2. Experimental examining the thermal field and mathematical simulation of resistive voltage divider transformers

For experimental research purposes an experimental prototype resistive divider transformer (see Fig.1, a) was designed and made. The experimental studies were carried out with real time temperature measurements, using a measuring system comprising a National Instruments Compact RIO 9074 controller with NI 9223 enter-release modules, thermo-resistive temperature sensors and a source of direct current for those sensors. The temperature sensors for temperature field measurement were placed in the insulator body. The experimental model was heated by applying voltage to the resistive divider until a steady-state mode was achieved.

The results of the experimental studies are shown in figure 2 as dependence of maximum resistor surface temperature on time, and in figure 3 as a steady-state temperature distribution along the transformer's radius.

For mathematical simulation of transformer thermal state we used the thermal conductivity equation recorded in the cylindrical coordinates system. The directions of its axes are shown in figure 1, c.

$$\frac{\partial T}{\partial t} = a\left[\frac{1}{r}\frac{\partial}{\partial r}\left(r\frac{\partial T}{\partial r}\right) + \frac{1}{r^2}\frac{\partial^2 T}{\partial \varphi^2} + \frac{\partial^2 T}{\partial z^2}\right] + \frac{q_v}{c \cdot \rho}, \tag{1}$$

where T – temperature, r, φ, z - coordinates (axis z coincides with cylinder axis), t – process time, a – temperature conductivity coefficient, q_v – heat source related to heat release in resistive elements, c – heat capacity, ρ – density. The parabolic equation (1) is resolved by specifying boundary single-valuedness conditions. Transformer operation is planned on an open distribution device, therefore third genus boundary conditions are accepted on the outer surface, as

$$q_s = \alpha\left(T_w - T_a\right), \tag{2}$$

where α –coefficient of heat release from wall to air, indexes "w" and "a" refer to wall and air temperatures, respectively, q_s – heat flow via cylinder outer surface.

Equation (1) is solved using two methods: modified finite volume method and finite element method.

For the finite volume method [2, 3] of solving equation (1) we developed an algorithm and software package using Markov's chain theory mathematical apparatus. The method of calculating the process of heat conductivity with a three-dimensional approach to the task is based on cell methodology. Heat distribution among cells in a three-dimensional space was chosen as the required function. The obtained multidimensional cell array is compressed into a single-dimensional matrix (vector-line) from the same cells. Heat distribution among the cells is represented by vector S ={S_i}, where index "i" corresponds to the cell number. The algorithm for calculating the required distribution value S includes the following stages. First, for each space cell we determine the numbers of the cells with which its may interact, then prepare balance equations to determine energy flows between the cells. Certain energy flows allow us to determine the shares of heat transferred to neighboring cells for the time step. To describe the evolution of the state of the system we summarize heat energy transferred from cell j to cell i:

$$S_i^{k+1} = \sum_{j=1...n} S_i^k p_{ij} , \tag{3}$$

where p_{ij} - share of heat transferred from cell j to cell i; index k corresponds to the time step number.

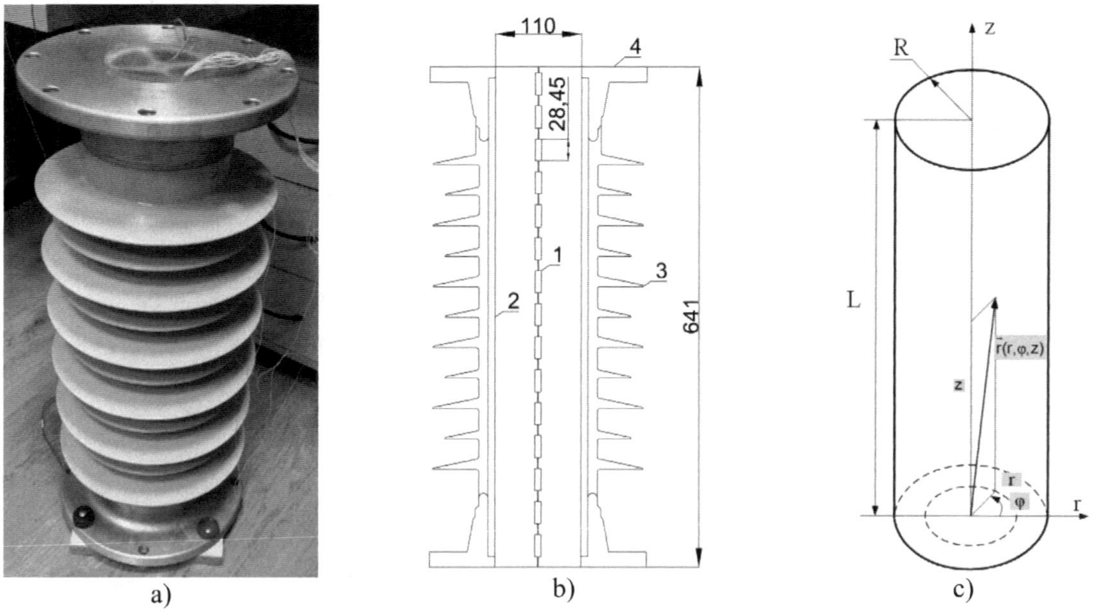

Figure 1. Resistive voltage divider transformer: a – experimental model photograph; b – drawing; c – cylindrical coordinates system model; 1 – resistive elements; 2 – glass textolite tube; 3 – silicone cover; 4 – flange.

The second numerical method of solving heat conductivity equation is the finite elements method, using COMSOL Multiphysics software package [4] in axially symmetric two dimensional and three dimensional versions. The results of our calculations of transformer thermal state in non-steady and steady-state modes are given in figures 2, 3.

Figure 2 shows the results of numerically solving equation (1) using two methods involving time changes in the maximum resistive element surface temperature. The same chart shows the results of the experimental studies described above. Temperature changes along the transformer radius in steady-state mode are given in figure 3.

The adequacy of describing experimental data by the results of simulation is assessed according to Fisher's criterion F [5]. The value of criterion F is calculated by the ratio of mean and residual dispersion. The calculated value of criterion F for the finite elements method with COMSOL Multiphysics is 0.7; for the modified finite volumes method it is 1.9. Comparison of obtained values with table criterion value ($F_{кр}$=3.04) for a significance level of q=0.05 has shown that both obtained values are less than the table figure, which confirms the adequacy of the proposed ways of analysing thermal states in the given time domain.

Comparison of time spent on dynamic daily calculations has shown the advantage of the modified matrix method [2] over the finite elements method using COMSOL Multiphysics: required time for calculating one model by the above methods was 100 s and 1000 s, respectively.

For design calculations and analysis of equipment operating modes in the electric power industry, also for the purpose of optimization calculations it is convenient to use the modified matrix method [2], with subsequent itemization of results using the more accurate and resource-intensive (data hungry) COMSOL Multiphysics package method.

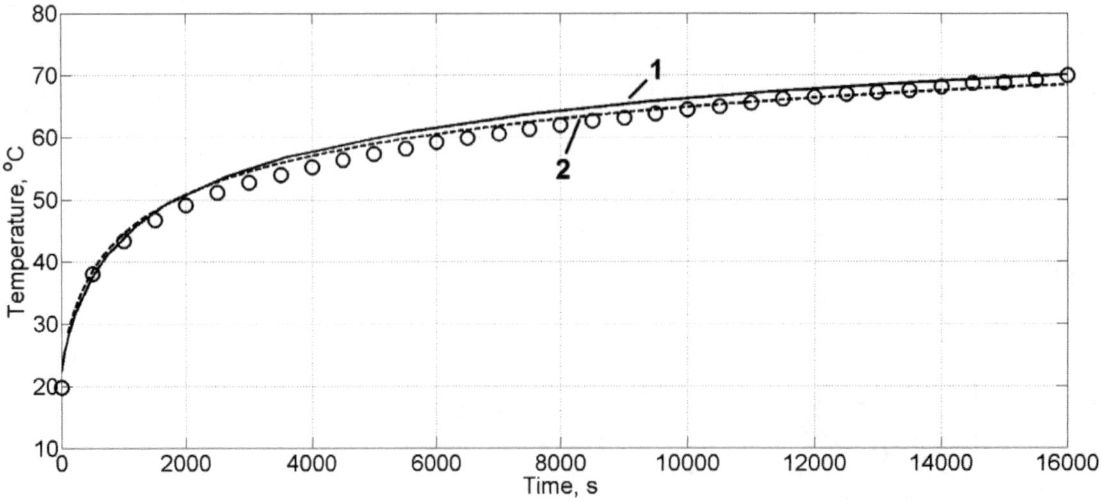

Figure 2. Calculated (1 – COMSOL Multiphysics, 2 – recommended matrix method [2]) and experimental (points) of dependence of maximum resistive element surface temperature on time.

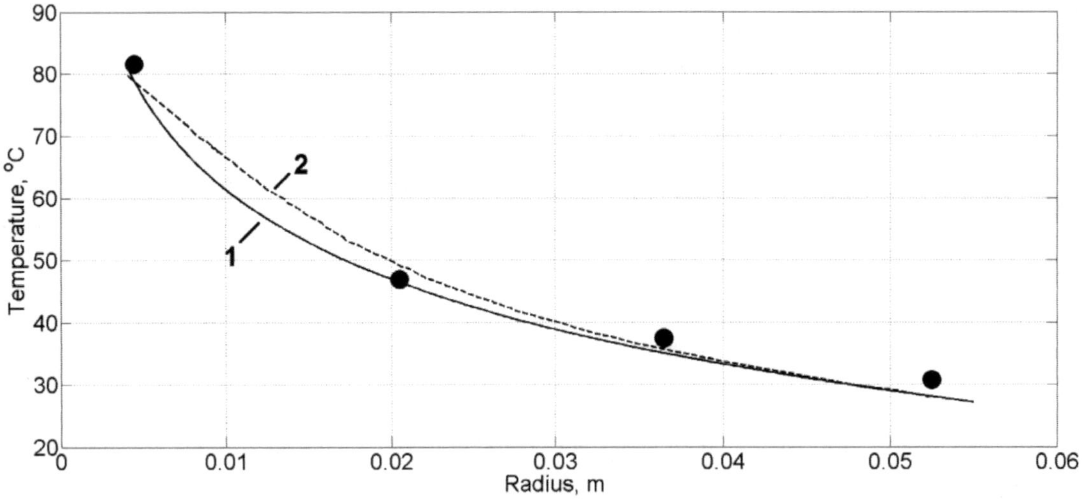

Figure 3. Dependence of steady state radial temperature distribution in transformer's central part; 1 – results of COMSOL Multiphysics calculations; 2 – results of modified method calculation [2], points representing results of experimental measurements.

3. Search of the most optimal position of resistive elements

Experimental research regarding the thermal field of resistive voltage divider transformer was conducted on a verified mathematical model (figure 4) in order to determine the position of resistive elements where their self-heating is minimal.

The research has shown that resistive divider moving to inner wall of the insulator allows decreasing resistive elements temperature from 81 °C to 54 °C, and spiral position of resistive elements reduces the temperature by 3.4 °C.

According to Russian State Standard GOST 1983-2001 voltage transformer should withstand voltage 1.2 U_n (of design) and 1.9 U_n for 8 hours. The research of self-heating rate of resistive elements was conducted on the developed mathematical model of transformer with spiral position of resistive elements when applied voltage to the resistive divider was equal to 1.9 $U_{n.}$ The research has shown resistive elements of spiral divider warm up to 170 °C at voltage 1.9 U_n and an ambient temperature of 70 °C (the most severe conditions according to Russian State Standard GOST 15150-69).

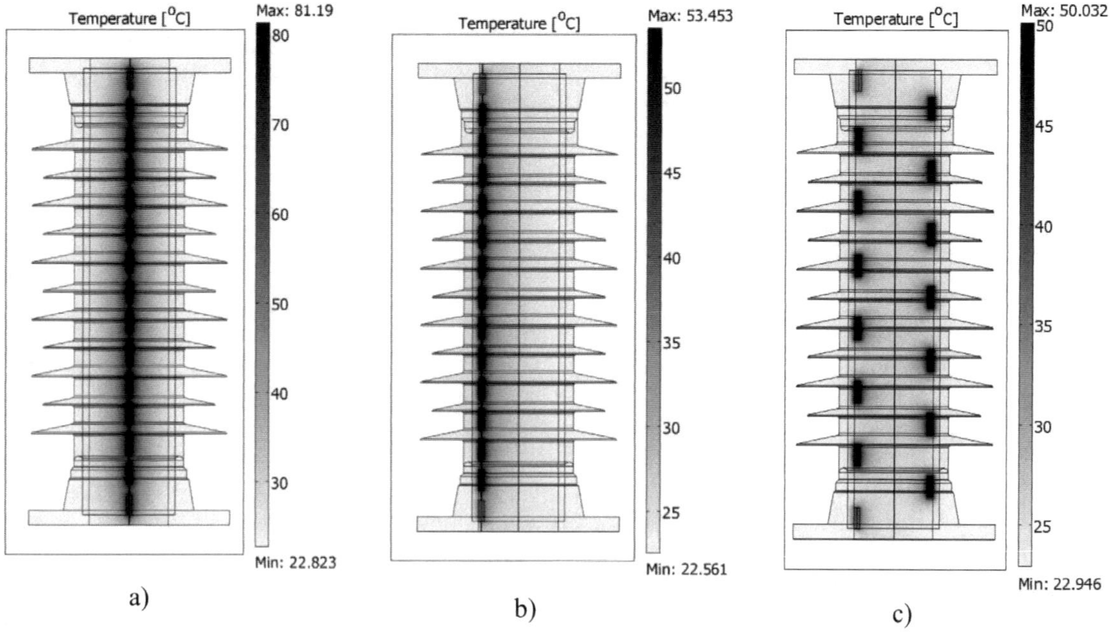

Figure 4. The comparison of thermal fields of resistive voltage divider transformer with axial (a), boundary (b) and spiral (c) position of resistive elements.

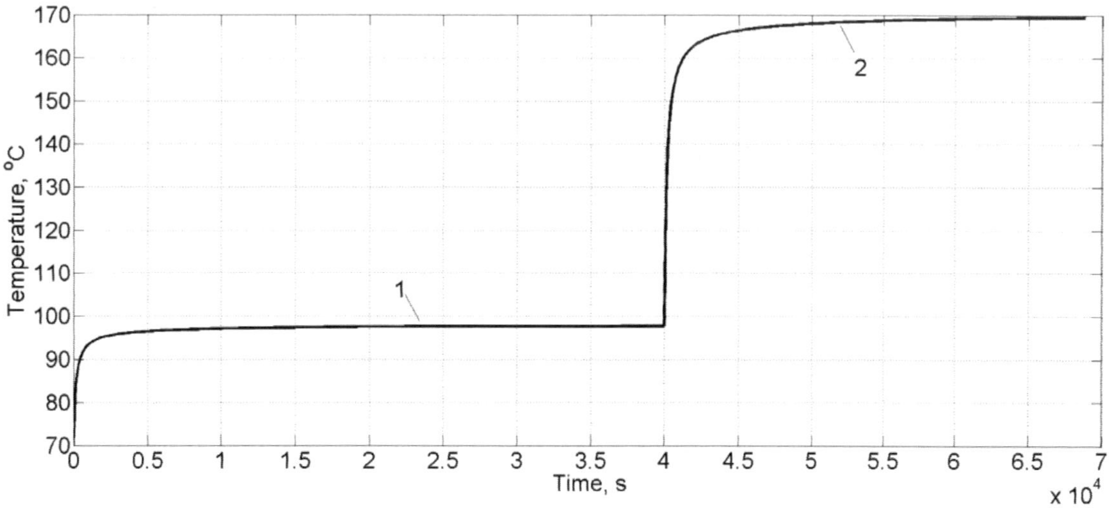

Figure 5. Dependence of maximum resistor surface temperature on time: 1 – $U=U_n$; 2 – $U=1.9\ U_n$.

4. Conclusions

The mathematical models of resistive voltage divider transformer developed and verified according to experimental data allow determining the most optimal position of resistive elements and their characteristics in compliance with temperature criterion. According to the temperature criterion resistor surface temperature should not exceed 100 °C (at an ambient temperature of 70 °C in conformity with Russian State Standard GOST 15150-69) at voltage 1.9 $U_н$. The given methods of simulation and calculation may be used in the design of measuring resistive voltage divider transformers.

Acknowledgements
This research was conducted at Ivanovo State Power University within the framework of a federal target programme, "Research and Development in Priority Areas of Developing the Russian Scientific and Technological Complex in 2014-2020" concerning "Development and Research Relating to Digital 110 kV Transformers Based on Basic Physical Laws with Optoelectronic Interface for Metering Electric Power in an Intelligent Electric Power System with a Smart Grid" (Agreement No. 14.574.21.0072 on Granting Subsidies, dated 27 June 2014, Unique Identifier for applied scientific research (projects) RFMEFI57414X0072).

References
[1] *Russian State Standard GOST 15150-69. Machines, Instruments and Other Technical Products. Versions for Different Climatic Regions. Categories, Operating Conditions, Storage and Transportation, as Influenced by Environmental Climatic Factors (with Emendations Nos. 1,2,3,4,5)* (Moscow: Standards Publishing And Printing Centre)
[2] Zhukov V P, Belyakov A N and Zhukov P V 2012 *Certificate of State Registration of Computer Software No. 2012611546*
[3] Royak M E, Soloveichik Yu G and Shurina E P 1998 *Grid-Based Approaches to Solving Boundary Problems in Mathematical Physics* (Nizhny Novgorod: Nizhny Novgorod State Technical University Press)
[4] Lebedev V D and Yablokov A A 2013 *Simulation of Physical Processes in Technical Devices Using COMSOL Multiphysics* (Ivanovo: Ivanovo State Power Engineering University Press)
[5] Gaidyshev I 2001 *Data Analysis and Processing: Special Reference Book* (Saint Petersburg: Piter)

Increasing accuracy of high temperature and speed processes micropyrometry

M P Boronenko, P Yu Gulyaev, A E Seregin and A G Bebiya
Ugra State University, Chehov st. 16, Khanty-Mansiysk, 628012 Russian Federation

E-mail: m_boronenko@ugrasu.ru

Abstract. The correction factor introduction in the method of measuring the brightness temperature of individual hot particles moving at speeds of 100-1140m/s with diameter above the diffraction limit of the OES, can solve the problem of the moving objects brightness pyrometry, increasing accuracy of at least 2.5 %.

1. Introduction

The intensive development of the optical-electronic systems (OES) element base allow effectively use them for non-destructive control of high-speed and high-temperature processes [1-4]. The review [5] of modern noncontact temperature control methods has revealed brightness pyrometry moving objects problem due to introduction of more methodological errors in the measurement [1, 6-7]. High speed camera been calibrated at stationary blackbody this introducing error to the moving particles temperature measurement. In this approach temperature is calculating as the average value over the track length, ignoring their size and spatial movement. Therefore radiation superposition of registering particles during their spatial moving doesn't taking into account. The purpose of research is to develop the methodology revised for brightness micropyrometery.

2. Experimental part

We measured the brightness temperature of the condensed phase sprayed particles. The brightness of the particles registering by the OES, based on multielement matrix photodetector [5-6]. We know that all condensed phase heated particles have a continuous spectrum of thermal radiation. The temperature of the object is determined in the absence of heating current I the filament of standard lamp on the stored value of the current intensity, corresponding to the EOS current output gradation. Thus, brightness of measured object and brightness of standard lamp digital equivalent are compared. The digital equivalent is representing by conformity spreadsheet between filament viewing area pixel brightness of standard lamp TRU 1100-2350 and current passing thought the filament. Digital equivalent is realized on the basis of macro program ImageJ. For determining the particles temperature it is necessary to take into account their linear dimensions. For this purpose the following assumptions are made: all particles are spherical with radius r, l is the particle track length. The particle radiation energy during one EOC shutter time, ie during the exposition time [8]:

$$E = \int_0^{t_{exp}} \phi(t)dt. \qquad (1)$$

During this time, the absorbed light radiation creating the particles image on the photomatrix surface. Replacing the analog signal *x (t)* by the sequence, represented by the signal expansion coefficients on an orthogonal basis, is the most common discretization method. Instead of considering the functional dependence of *x(t)* in uncountable set of points, we can characterize the signal counting system coefficients xn. Basis is chosen from the convenience of the physical realization, simplicity of coefficient calculation and accuracy of approximation. Input optical image signal *E(x, y)* represents a two-dimensional generally continuous function of continuous spatial arguments (coordinates *x, y*) is converted into an electrical signal *I(x, y)* described the two-dimensional discrete continuous function of spatial arguments *(x, y)* and weight function *g(x, y)*. Converting the input signal by optical system and the image formation on the sensitive area of the photodetector device can be represented as follows [9]:

$$I = \int_0^l g(x, y) l(x, y) dx dy. \tag{2}$$

If the particle was unmovable, then charge accumulation was continuous on the same photosensitive pixels area during the exposition time. Track of the moving particle has less brightness.

Suppose that on the image particle has size n × m pixels and also suppose that radiation of the particle is constant by their surface. In determining the temperature by thermal radiation, the particle brightness is associated with an average brightness of the particle image:

$$\bar{I} = \frac{\sum_{i,j}^{n,m} I_{i,j}}{n \times m} \tag{3}$$

During the time of particle moving the radiation illuminate area with n x m size. On the track length the l/m of such particles is fit into (Figure 1) [10]:

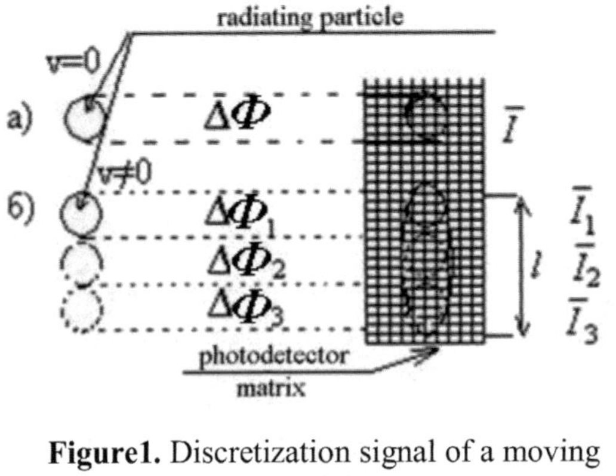

Figure1. Discretization signal of a moving particle.

Therefore, taking into account the superposition principle, integrated brightness at which we can calculate moving particle temperature in a discrete representation:

If the total brightness of the track pixels at a given exposition time does not exceed 255, then with using the spreadsheet conforming the brightness temperature of the standard lamp and brightness gradations of pixels is possible to determine the temperature of the moving particles. The brightness of a luminous object on the short exposition time is less than brightness of same objects at high exposition time.

During different exposition times particles flies different distances, and consequently, different number of pixels will forming the track. As seen from the graphs in Figure 1, for the same temperature, and hence the value of the brightness gradations, will correspond different exposition times.

Consequently, it is possible to find the coefficient for translation of image brightness from one exposition to another, and measure the temperature corresponding to the pixel luminescence gradation of the image with different exposition time.

For example, during registering the radiation flux at 150 microseconds we obtain an image brightness of 200 gradations. If the exposition time of the same signal was 26 microseconds, the image had a brightness of 50 gradations. The conversion coefficient is equal to:

$$k = \frac{I_{26}}{I_{150}} = \frac{50}{200} = 0{,}4 \tag{4}$$

For moving particles temperature measurement is necessary to reduce exposition time, calculated with assumption of particle speed constancy during the track registration time.

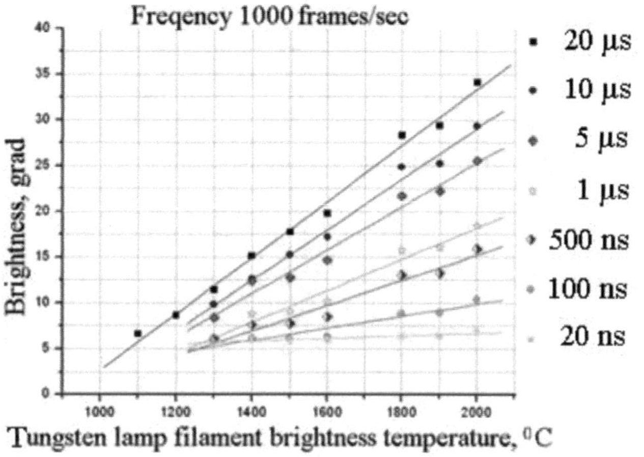

Figure 2. Calibration graphs for the OES based on camera with EOC.

Method of converting the image brightness from one exposition time to another with moving particle observation area of one photosensitive cells to the other observation area, consists in follows:

1. Suppose that the emitter is unmovable.

The charge of each detector (pixel) accumulated during the exposition time. It is convenient what most probable value of signal is equal to the average brightness registered on the light absorption zone. Most probable value of the signal is equal to the average brightness of the illuminated area. The most probable brightness value of illuminated area is associated with the temperature of radiated object. During all exposition time the same detectors is absorbing the light.

2. The object speed is not equal to zero.

If object is moving, than at different times, different detectors are illuminating by different parts of the object. Thus, same detectors that were used in the case of a stationary emitter, will be irradiated only during the time where particle will take place in the projection sector of the pixel. Therefore, the scheme of moving particle Is equivalent to stationary emitter but exposition time is less and equal to the particle transit time along the same detector witch used for the case 1The particle velocity at all time of track registration is considered constant. Then:

$$\frac{L-D}{t} = \frac{D-d}{t_1}, \tag{5}$$

$$t_1 = t \cdot \frac{D-d}{L-D}. \tag{6}$$

where: L-track length, D-particle diameter, d-pixel size, t1-passage time of pixel projection zone, t- exposition time.

Thus, it is possible to measure the temperature of a moving emitter by comparing the brightness of pixels corresponding to the image gradation in another exposition time:

1. Measure the average brightness (grayscale) in the selected track, resulting In the frame sequence with exposition time t;
2. Measure the transverse and longitudinal dimensions of the track (in pixels);
3. Accept the diameter of the moving emitter as transverse dimension of its clear picture. We accept that track length contains a minimum L = 2D pixels;
4. According to the formula 6 providing exposition time calculation, corresponded to the passage time of pixel projection zone;
5. According to calibration curve finding the brightness temperature corresponding to the measured average brightness of the image on the exposition time t1.

For example, when the particle diameter D = 2d, and the track length L = 4d exposition time is reduced by 2 times. So, if the average brightness (Fig. 2) in the track on the exposition time 20 microseconds was 29 shades of gray, this corresponds to a brightness temperature at 2000 C 10 microseconds exposition. If a correction coefficient has not been entered, the additional error amounted 100C, which is 5% of the measured temperature. At lower temperatures, the introduced error about 50 C.

3. Results and conclusion

The introduction of a correction coefficient in the brightness micropyrometery method allows increasing the result accuracy not less than 2.5%. Future developing will allow to create a method the OES using for two-phase flows diagnostics, estimates the intensity of the flows independently from temperature and dispersion composition [11].

References

[1] Novitsky P V 1968 *Fundamentals of measurement devices information theory* (Leningrad: Energy) p 248
[2] Korotaev V V and Krasnyaschih A V 2008 *Television measuring systems* (SPb: Gos. Univ ITMO) p 108
[3] Berkin A B and Vasiliev A B 2007 *A new approach to modeling the current amplification channel MCP* (Technical Physics Letters vol 33№ 15) pp 664-666
[4] Jespersen P 1979 *Semiconductor image sensor* (Moscow: WORLD) p 573
[5] Boronenko M P 2014 *Bulletin of Ugra State University* **2(33)** 43-55
[6] Tewes J, Hertter M, Voggenreiter D H, Stuttgart D, Biermordt T, Zimmermann S, Landesand K and Schein J 2013 *Electronic materials Journal of Thermal Spray Technology* **22** 892-900
[7] Vattulainen J 2001 *Journal of Thermal Spray Technology* **10** 94–104
[8] Epstein M 1990 *Measurements of optical radiation in electronics. Production Edition* (Moskow: Energoatomizdat) p 253
[9] Andreev A L 2012 *Modelling and calculation of automated video information systems for objects monitoring. Methodical instructions to laboratory works* (St. Petersburg: ITMO) p 82
[10] Boronenko M P, Gulyaev I P, Gulyaev P Yu and Seregin A E 2014 *Proceedings of the universities. Physics* **57** 73-77
[11] Boronenko M P, Gulyaev I P, Gulyaev P Yu, Demyanov A I, Dolmatov A V, Jordan V I, Korzhik V N, Krivtsun I V and Kharlamov M Yu 2014 *Fundamental Research* **10** 2135-2140

Experimental facility control system for optical studies in the frame of problem solving of brain hematoma diagnostics

A Novoseltseva[1], A Aristov[1] and K Timchenko[1]
Tomsk Polytechnic University, Tomsk, Russia

E-mail: ania.minis@yandex.ru

Abstract. The article describes a system to control a unit moving to find the dependence of the photodetectors signals from the distance to the hematoma. It considers existing devices that operate in accordance with the principle of near infrared spectroscopy and reveals their limitations the illuminations of which are our key objective. The limitations of the designed unit are described and possible ways of upgrading the system are outlined.

1. Introduction

Identification of intracranial hematoma is extremely important in traumatic brain injury. For now the most accurate methods for detecting such injuries are computed tomography and magnetic resonance imaging, however it is almost impossible to perform the tomographical scanning immediately after the injury. Therefore, the problem of creating a device that could quickly identify the intracranial hematomas is important. Possible solution of this problem is to use a diagnostic spectroscopy in the optical wavelength range. This method is based on the fact that the substance has characteristic absorption spectra of molecules, and the intensity of the absorption is specified by the content of the absorbing component within the irradiated object. The spectral region used in this work is not harmful both for an operator and a patient. With proper technical support is possible to investigate the organs and tissues of animals and humans, without violating their integrity and without causing any damage to the body [1].

2. Problem definition

InfraScan company has developed a device Infrascanner [2, 3]. It is a portable detector of intracranial hematomas operating in its near infrared range (808 nm). According to the published experimental studies performed on models of intracranial hemorrhage, clinical trials have shown a high diagnostic sensitivity of the device. This scanner allows you to detect the presence of hematomas with 91 % probability (in the case when hematoma is not deeper than 2.5 cm from the brain surface and is greater than 3.5 cm) [4, 5].

However, this device does not allow to receive data on hematoma such as its size, density, exact location and etc. that are of great diagnostic value for further resuscitation and rehabilitation medical treatment. Since the diagnostics is performed by comparing the optical properties of symmetric sections on the head surface in exact locations, there is a high probability of missing small or deeplying hematomas outside of these zones. Therefore, it is necessary to improve the method and upgrade the device for getting better medical and technical characteristics.

We design a model of the device that allows to carry out a laboratory research which will help to improve optical detection diagnosis of hematomas. For this purpose we are facing many challenges, such as, design and engineering implementation of the optical part of the device, processing of the analog signal received from the photodetectors, digital processing and visualization of information read from the device.

3. Experimental unit

Construction of the laboratory model of the device is shown in Fig.1 [6]. Measuring optical part of the device consists of a carcass and movable fastenings (4). The optical block (2) is fixed on a carcass using a movable fastening. This optical unit is placed above the phantom of human head (3) and consists of a radiation source and receivers. A movable fastening (4) is used to fix a model object (5) that imitates the hematoma. The optical unit and a fastening for “hematoma” are fixed to the carcass using flexible and sliding joints, which allows us to move them both vertically and horizontally with respect to the phantom of the head.

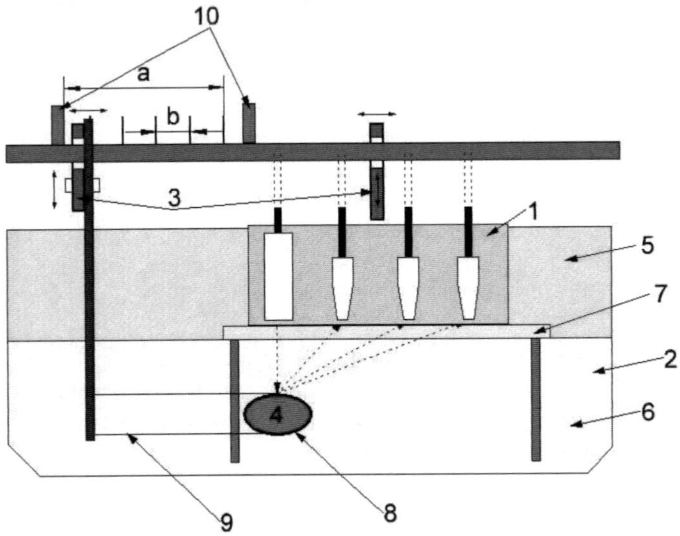

Figure 1. Construction of the laboratory model of the device. 1 – optical block; 2 – phantom of head; 3 – movable fastening; 4 – hematoma; 5 – volume; 6 – liquid imitating brain tissue; 7 – fluoroplastic plate; 8 – silicone transparent bag; 9 – wire grid; 10 – switches.

The phantom of the head consists of a container (6) filled with a dispersed viscous fluid (7) simulating the brain tissue. Above the liquid surface a fluoroplastic plate (8), which simulates the cranial bone, is placed. The hematoma is modulated by the human blood placed in a transparent small bag (9) fixed on a wire grid (10) of the movable fastening (4) and is set into the liquid under the fluoroplastic plate.

Varying the position of ‘hematoma’ relative to the optical measuring unit we can determine how its movement change the data received from photodetectors and whether on the basis of these data to determine the size of the hematoma and its exact location. For this purpose it is necessary to develop a device that would move the hematoma by a predetermined distance a with a given step b. In those moments when the unit reaches a given step b the values must be read from the photodetectors. Thus we get the dependence of the photodetectors signal from the distance to the hematoma.

The stepper motor is used to move the fastening with hematoma. Switches (11) limit the distance that the stepper motor can move the optical part of the device. The switch located to the left corresponds to the starting point that begins data reading. The switch to the right corresponds to the maximum distance at which hematoma can move.

Block diagram of the device for controlling the stepper motor is shown in Fig. 2. A program that controls the data acquisition device NI USB-6009 runs on a personal computer. This device has 12 channels of digital input / output, these pins are used for transmitting control signals to the microcontroller, and for transmitting data - values of "distance a» and «step b». The microcontroller STM8S003K3 sends control signals to the stepper motor (SM) TM100 and receives signals from the switches. Fig. 3. shows the algorithm of the microcontroller program. Chip ULN2003 is used as a SM-driver for coordination of voltage levels to control the stepper motor.

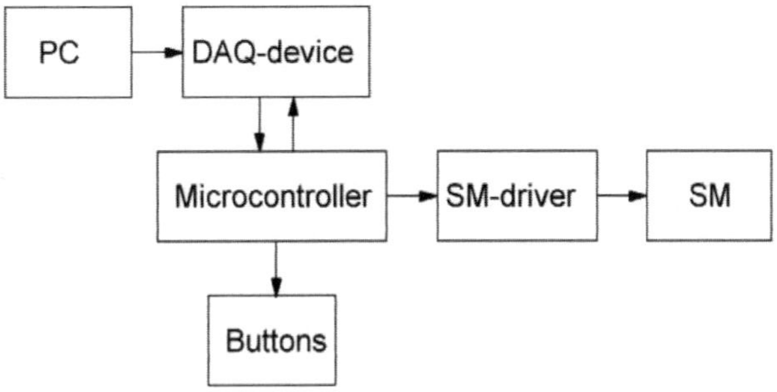

Figure 2. Block diagram of the device. PC – personal computer, DAQ-device - data acquisition device, Microcontroller- microcontroller STM8S003K3, SM – stepper motor, SM-driver – driver for stepper motor, Buttons – switches are limit the distance that the stepper motor can move.

Block diagram of the device for controlling the stepper motor is shown in Fig. 2. A program that controls the data acquisition device NI USB-6009 runs on a personal computer. This device has 12 channels of digital input / output, these pins are used for transmitting control signals to the microcontroller, and for transmitting data - values of "distance a» and «step b». The microcontroller STM8S003K3 sends control signals to the stepper motor (SM) TM100 and receives signals from the switches. Fig. 3. shows the algorithm of the microcontroller program. Chip ULN2003 is used as a SM-driver for coordination of voltage levels to control the stepper motor.

The program on a PC for measuring the position of the hematoma depending on the value from photodetectors is implemented in software development environment LabView. The program works as follows. First, the user enters the depth of the hematoma, then enters the distance – a that the hematoma must pass, and enters the value of step – b. After clicking the "Run" button the measurement starts. First values from the photodetectors are read. Then a control signal is sent to the microcontroller, and the unit begins to move at the given step. At this time, a microcontroller puts a prohibition to read data from the photodetectors. Once the hematoma has moved a predetermined step, the microcontroller sends the enable signal to read data from the optical part of the device. Now the data of the distance b is read. Then the device goes another step, data of a distance $2 \cdot b$ is read. And this process occurs until the unit passes the distance a specified by the user. Thus, after completion of reading data, we get the dependence of the photodetectors output signals from the distance to the hematoma.

4. Conclusion

As a result of this work a stepper motor has been implemented to automate data reading from the optical part of the device. The programs have been debugged. Further the experiments to determine hematoma location will be carried out using the device.

Further it is planned to develop our own software for the visualization of information read from the device. To get more high-speed data retrieval and to reduce the mass-dimensional parameters of the device it is planned to substitute the data acquisition device NI USB-6009 for analog-to-digital converter chip.

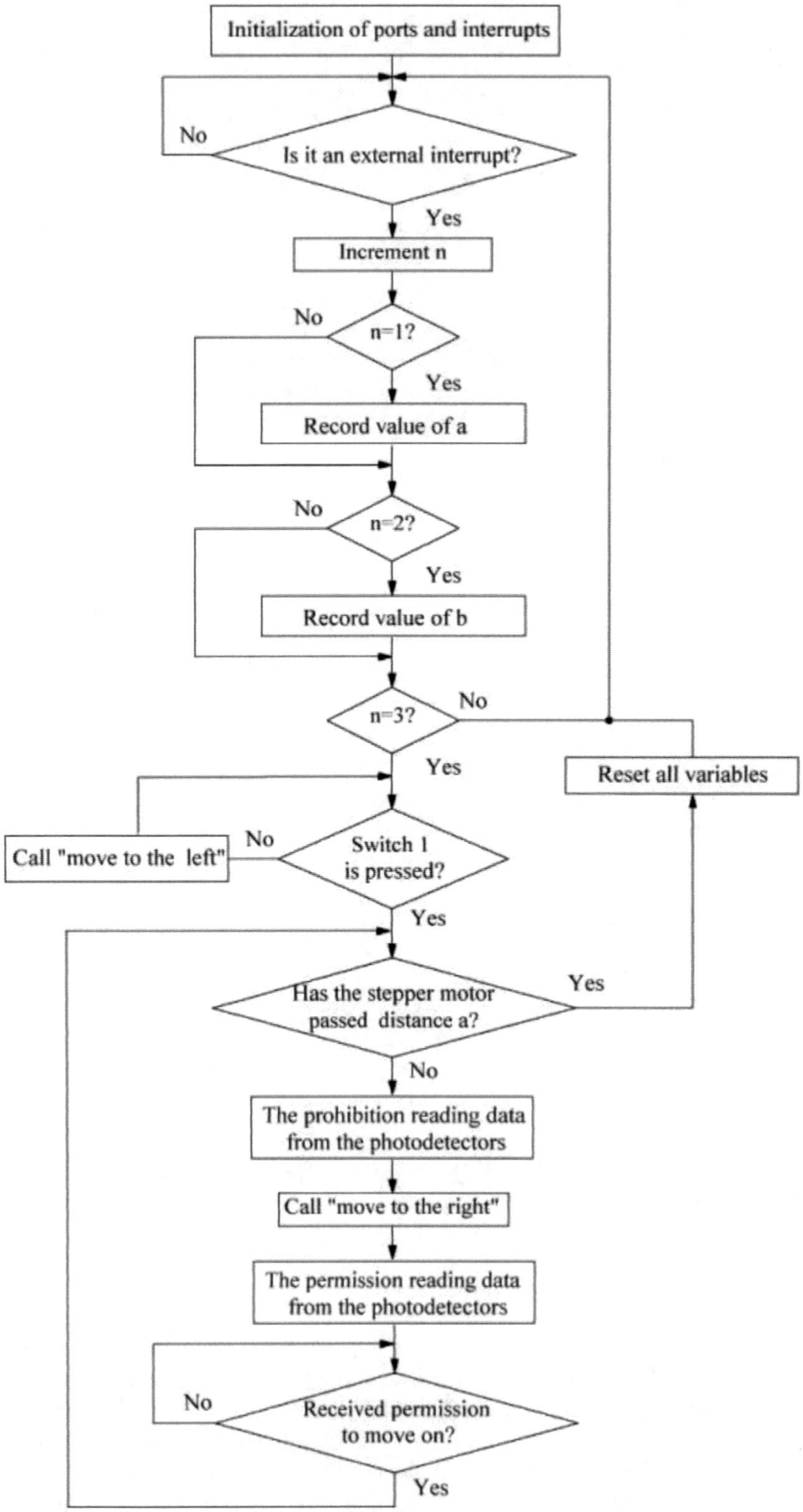

Figure 3. Flow chart of program algorithm embedded into the microcontroller.

IOP Conf. Series: Materials Science and Engineering **93** (2015) 012022 doi:10.1088/1757-899X/93/1/012022

References

[1] Zimnyakov D A and Tuchin V V 2002 *Quantum Electronics* **32(10)** 849-867

[2] [Electronic resource]/The *Infrascanner Model 2000* - access mode:. http://www.infrascanner.com/solutions/the-infrascanner-model-2000/, (date of treatment 22/03/2014.).

[3] [Electronic resource]/The *Infrascanner Model 1000* - access mode: http://www.infrascanner.com/solutions/the-infrascanner-model-1000/,(date of treatment 22/03/2014.).

[4] Bressan S, Daverio M, Martinolli F, Dona D, Mario F, Steiner I P and Da Dalt L 2013 *Childs Nerv. Syst.* **30** 477-484.

[5] Robertson C S , Zager E L , Narayan R K , Handly N , Sharma A , Hanley D F, Garza H, Maloney-Wilensky E, Plaum J M, Koenig C H, Johnson A and Morgan T 2010 *Journal of Neurotrauma* **27** 1597-1604.

[6] Timchenko K A , Aristov A A , Musorov I S and Evtushenko T G 2014 *Proceedings of the 2014 15th International Conference of Young Specialists on Micro/Nanotechnologies and Electron Devices (EDM)* 319-322

IOP Conf. Series: Materials Science and Engineering **93** (2015) 012023 doi:10.1088/1757-899X/93/1/012023

Simulating magnetic field of a ferromagnetic pipe underwater in COMSOL Multiphysics

D Soltanova, P Baranov, V Baranova and A Chudinova
Tomsk Polytechnic University, Tomsk, Russia

E-mail: bpf@tpu.ru

Abstract. Nowadays ecological situation in seas and oceans requires permanent supervision and control. Carrying out building activity such as building hydraulic structures, oil- and gas-pipes in areas of past warfare is the reason for the active usage of geophysical methods to search method of the objects underwater. The paper examines the classification of magnetic search methods and theoretical base statements of electromagnetics. The work represents the investigation of an object influence on geomagnetic field in problem-solving environment "COMSOL Multiphysics". The article also contains the results of simulating for variations of different object parameters. This paper is connected with the magnetometric

1. Introduction

Nowadays ecological situation in seas and oceans requires permanent supervision and control. Carrying out building activity such as building hydraulic structures, oil- and gas-pipes in areas of past warfare is the reason for the active usage of geophysical methods to supply safety and environmental security. So, there is the topical problem of developing precise methods and means for search, supervision and investigation of underwater pipelines, cables and other objects [1-6].

As a rule, these objects are made of ferromagnetic materials and as they are in the magnetic field of the Earth they influence it, thus, creating magnetic anomalies. Underwater search means need this information.

There are two types of search magnetic methods: active and passive. Active methods consist of the inductive method that performs the search of metal objects and the radio-locating method that performs the search of both metal and nonmetal objects. Passive method is the magnetometric method that allows to estimate the geomagnetic field distortions induced by ferromagnetic objects and magnetic field sources.

2. Magnitometric method

If current passes not through an object the object location can be determined via magnetic flux density measurement. Magnetic flux density depends on the object size and the distance on it. It is necessary to know the character of magnetic flux density variations to create precise search means.

Modelling such conditions is available with the help of program environment "COMSOL Multiphysics" (COMSOL). Calculations of magnetic fields in COMSOL are based on Maxwell equations [7]:

$$\oint \bar{E}dl = -\int \frac{\partial \bar{B}}{\partial t} dS,$$
$$\oint \bar{D}dS = \int \rho dV,$$
$$\oint \bar{H}dl = \int \left(j + \frac{\partial \bar{D}}{\partial t} \right) dS, \quad (1)$$
$$\oint \bar{B}dS = 0,$$

where $\bar{H}$ – the magnetic field vector, $\bar{B}$ – the magnetic flux density vector, $\bar{D}$ – the bias current vector, ρ – the volume density of exterior charges, j – the current density, S – the surface square, l – the distance.

Except Maxwell equations the system of equations for electromagnetic fields includes the material equation. The material equation in its simple form has the following view (2):

$$B = \mu_0 \mu_r H , \quad (2)$$

where B – the material magnetic flux density, T; μ_0 – the vacuum magnetic permeability H/m, μ_r – the medium (material) relative magnetic permeability.

The object magnetic field can be defined by the following more complete relation (3) [8]:

$$B_x = \frac{B_i V}{4\pi x^3} , \quad (3)$$

where B_x – the object magnetic flux density, T; B_i - the material internal flux density, T; x – the distance to the object, m.

The material internal flux density B_i is determined by the formula (4) [3]:

$$B_i = (\mu_r - 1) B_E , \quad (4)$$

where B_E – the value of the ambient flux density equal to 50 mkT.

3. Modelling

It was a ferromagnetic pipe in water simulated in COMSOL to investigate the theoretical dependence (3) and the object magnetic field dependence on its geometric parameters and the relative magnetic permeability of pipe material. The model consists of the ferromagnetic hollow cylinder (pipe) and the sphere that imitates water space. Modelling process includes the following steps: Geometry building, material assignment, the application of Physics, partitioning into the finite elements, calculations and visualization of the results.

3.1. Geometry building

Geometry building was implemented with the function "Work Plane". Two cylinders of different diameters were created to simulate the pipe. The pipe thickness was assumed equal to 1 % of the external diameter and the length was chosen equal to 10 m. The first meaning of the external pipe diameter D_1 was defined as 1 m. The internal pipe diameter D_2 was assumed equal to 0.98 m, then. Three-dimensional imagine of the observed pipe geometry is shown in figure 1.

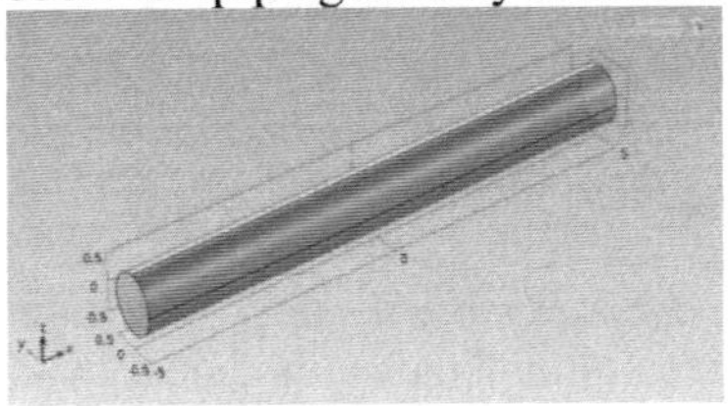

Figure 1. Pipe Geometry.

To calculate the magnetic field induced by the pipe it was necessary to place it into the sphere that simulates water space. The sphere was created around the pipe and had the diameter equal to 10 m. The model of the pipe in sphere water space is presented in figure 2.

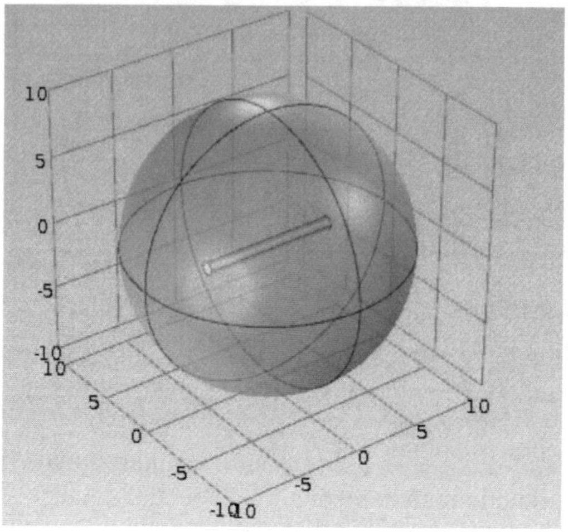

Figure 2. Model Geometry.

3.2. Material assignment

Material assignment in COMSOL is carried out via "Material Browser". It is convenient for quick system simulating with the real characteristics. The pipe material was defined as "Iron" and the sphere material – as "Water". User can define the relative magnetic permeability for some materials in the category "Material Contents". The initial value of magnetic permeability for iron was defined as $\mu_r = 1000$. Relative magnetic permeability for water is equal to 1.

3.3. The Application of Physics

The function "Magnetic Flux Density" in the category "Magnetic Fields, No Currents" allows assigning the value of the background magnetic field and the one of the magnetic field of the object. Background magnetic field corresponds to the geometric field of the Earth and is equal to 50 mkT along the axis z and zero values along the axes x and y. Magnetic field of the pipe is determined through the relative magnetic permeability μ taken from the information about material.

3.4. Partitioning into the finite elements

The model partitioning into the finite elements was carried out after Geometry building, material assignment and the application of Physics. Two types of mesh with different element size were used for partitioning. "Coarser" mesh was chosen with the maximum element size 3.8 m and the minimum element size 0.8 m for the sphere. "Normal" mesh was chosen with the maximum element size 2 m and the minimum element size 0.36 m for the pipe.

3.5. Visualization of the modelling results

Visualization of simulating results after calculations is implemented with the help of the category "Results". Firstly, it was necessary to create the *zx*-plane for mapping the distribution of magnetic flux density near the pipe. Secondly, we chose the expression for the magnetic flux density absolute magnitude: "mfnc.normB". Figure 3 demonstrates the distribution of magnetic flux density along *zx*-plane.

The graph of magnetic flux density dependence on the distance to the pipe (along the axis z) was also built in (figure 4). The expression for magnetic flux density variation in per cent in COMSOL has the following view: "((mfnc.Bz+50e-6)/50e-6)*100".

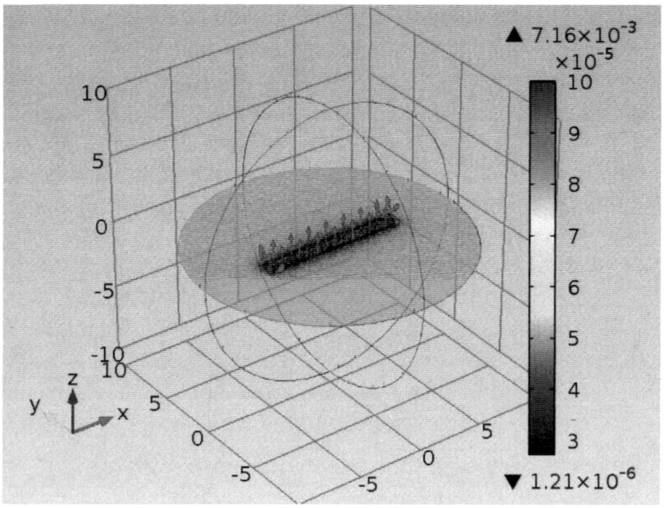

Figure 3. Distribution of magnetic flux density.

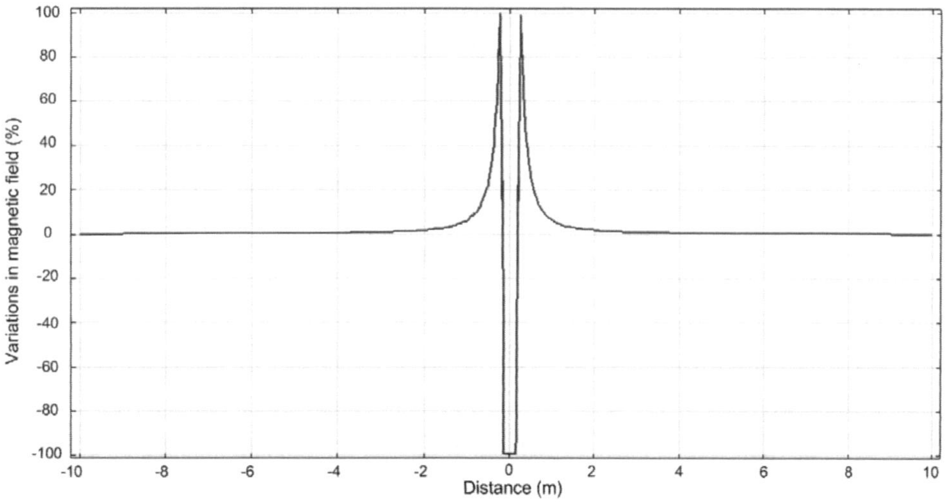

Figure 4. Magnetic flux variation.

The graph in figure 4 demonstrates that the magnetic field near the pipe doubles. While the distance to the pipe increases the magnetic field variation decreases. Thus, the magnetic field variation is equal to zero with the distance of 4 m.

4. Model investigation

Magnetic flux variation was also calculated in the program "Mathcad" with the theoretic formula (3). Magnetic field variation in per cent γ was evaluated by the formula (5), then:

$$\gamma = \frac{\left((B_x + B_E) - B_E\right)}{B_E} \cdot 100\,\% \tag{5}$$

The calculating results are represented graphically in figure 5.

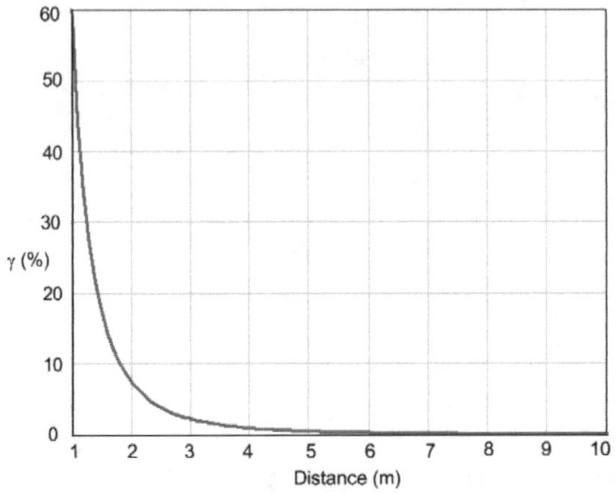

Figure 5. Magnetic flux density variation in "Mathcad".

It is evident that the theoretical law (figure 5) and the practical law (figure 4) are similar. The table 1 compares the meanings of the laws according to different z-coordinates.

Table 1. The theoretical and practical values of magnetic flux density variation.

z-coordinate (m)	The theoretical value of magnetic flux density variation (%)	The modelling value of magnetic flux density variation (%)
1.5	18.5	12
3.0	2.3	2.8
5.0	0.5	0.7

The results obtained by the formula (5) and by the simulating are nearly identical. Thus, it can be said that the computer model of the pipe under water is adequate and can be used to evaluate magnetic field variations.

It is interesting to establish the magnetic field variation dependence on pipe geometric parameters (diameter, length) and the relative magnetic permeability of pipe material.

The pipe length took the values from 1 to 10 m. There were observed the graphs of magnetic flux density variations for each case. All the graphs have similar view as shown in figure 4 and contain not completely different meanings of variation.

The next step was the variation of pipe diameter. Diameter variation range was from 0.5 m to 1.5 m. The graphs observed for the diameters and equal to 0.5 m and to 1.5 m are presented in figure 6 and figure 7.

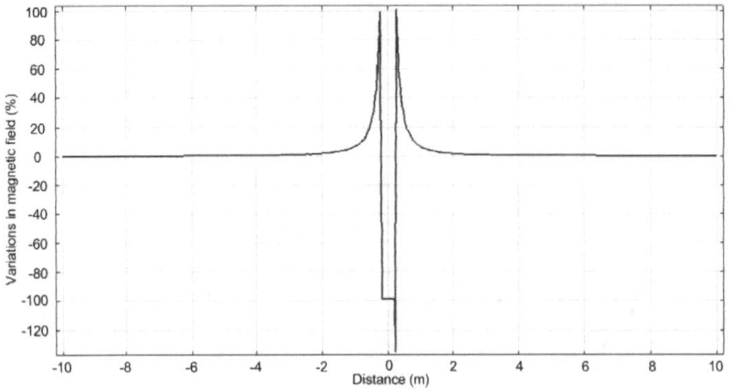

Figure 6. Magnetic flux density for pipe diameter 0.5 m

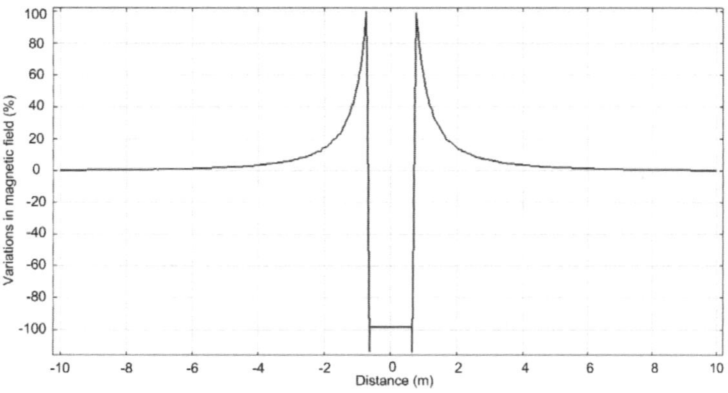

Figure 7. Magnetic flux density for pipe diameter 1.5 m

The comparison of the graphs makes it possible to conclude the following: while the diameter increases the magnetic flux density variation increases, too. For example, the magnetic flux density variation with the distance 1.5 m (x-coordinate is equal to 2 m) for diameter equal to 0.5 m is 1 % or ≈500 nT. Magnetic flux density variation equal to 1 % for diameter 1.5 m corresponds to the distance 5.25 m (x-coordinate is equal to 6 m). Consequently, the distance of magnetic field variation with 1 % increases by a factor of three as the diameter increases in three times. So, direct dependence between diameter and distance is observed.

5. Conclusion

Simulating results make it possible to conclude the following: Earth geomagnetic field variations increase with the object diameter increase because of the influence of the object and they do not depend on the length or relative magnetic permeability of the material for the object. Magnetic field variation near the object directly depends on its diameter.

Acknowledgments

The reported study was partially supported by RFBR, research project No. 15-08-01007.

References

[1] Hirota M, Nanara K 1997 *IEEE Transaction on Applied Superconductivity* **7** pp 2327-2333.
[2] Scheinker A, Lerner B, et al 2007 *Meas.Sci.Technol.* **18** pp 64-65.
[3] Scheinker A, Salomonski N, et al 2008 *Meas.Sci.Technol.* **045205** pp 1-5
[4] Huang Yu, Li Chao, Sun Feng 2010 Research on correction of tri-axial magnetometer and application in underwater magnetic field localization *Information and Automation: Proc. IEEE International Conference on* pp1919-1923.
[5] Huang Yu, Yan-ling Hao 2010 Method of separating dipole magnetic anomaly from geomagnetic field and application in underwater vehicle localization *Information and Automation: Proc. IEEE International Conference on* pp.1357-1362.
[6] Vaizer L, Lathrop J and J Bono J 2004 Location of Magnetic Dipole Targets *Proc. OCEANS'04* **2** pp 9-12.
[7] Stratton J 1941 *Electromagnetic Theory.*
[8] Sea-SPY Overhauser magnetometer 2007 *Technical appliction guide* p 13.

IOP Conf. Series: Materials Science and Engineering **93** (2015) 012024 doi:10.1088/1757-899X/93/1/012024

The Centerless Measurement of Roundness with Optimal Adjustment

O A Yalovoy[1], O V Zakharov[1] and A V Kochetkov[2]
[1] Yuri Gagarin State Technical University, Saratov, Russia,
[2] Perm National Research Polytechnic University, Perm, Russia

E-mail: zov17@mail.ru

Abstract. A new metering and data-processing method for the centerless roundness measuring, using corrective adjustment, is scientifically developed. In this paper, a mathematical model based on numerical (iterative) method and harmonic analysis, which additionally considers the reposition of the prism planes contact points with the component part by rotation, is proposed. In accordance with practical measurements and modeling, a possibility of consistent error minimization and improving accuracy of the centerless roundness measuring is found.

1. Introduction

The form testing of the machine or mechanism components surface is an important task today which has always attracted the attention of the specialists. Rotary bodies form accuracy in the cross-section is normalized by a special index "roundness" which indicates to what extent actual profile of the product differs from the reference circle. Standard ISO 12181-1:2011 recommends Minimum Zone Tolerance Circle as a basic variant of the measurement reference, but the usage of the identified derived elements in the form of superimposed external and inner circles (Maximum and Minimum Inscribed Circles) or central Circle Least Squares Estimate (LSE) is also considered acceptable. The choice of the specified reference circles depends on the functional purpose of the machine or mechanism components.

Different accuracy, efficiency and control condition requirements caused the development of a wide range of measurement methods and instruments each of which cannot be universal [1]. Methods of radius, coordinate and differential roundness measuring are well-known. Each of these methods has its own peculiarities and involves the use of different measuring instruments and results processing algorithms. The following is a short review of the methods from the point of view of their application to the specific control conditions.

The radius method means the usage of the precise rotation of roundness measurement instrument spindle with the small linear displacement sensor which contacts with the measurable component. Currently, high accuracy of roundness measuring is promoted by the roundness measurement instruments produced by Taylor Hobson (UK) under a trademark «Talyrond». If there is a misfit of spindle and workpiece axes of rotation then systematic inaccuracy appears and it is corrected with pre-centering and statistical analysis of the results based on harmonic analysis [2, 3]. The characteristic of the precise roundness measuring instruments is special placement and highly qualified staff necessity; there are also size and weight restrictions for the measured parts; and the high cost of the equipment of this type.

The coordinate measuring method is performed by Coordinate Measuring Machines (CMM) using special mathematical apparatus for result processing. Typically, the geometric image of the ideal components is created, then this image is corresponds to the real coordinates and the firm deviations are calculated. The important feature of the coordinate measuring method involves the irregularity of measured points position at the roundness chart, as it was mentioned in the research paper [4]. Using contact measuring method, accuracy and efficiency of CMM is considerably inferior to roundness measurement instrument with precise rotation.

When using the differential method, not absolute values of the investigated variable are defined but the difference between its consecutive values pulled apart with the definite interval. Moreover, analytic dependence between measured and desired quantities is unknown. Under the factory conditions for roundness measuring the devices with two-point contact (micrometers and gages) and three-point contact (different combinations of the prisms and small linear displacements sensors) are mainly used. It is the latter type of measuring that the centerless method under review is relevant to.

Instruments for the centerless measuring are generally like different combinations of prisms and small linear displacements sensors. The instruments with prism angle of 60, 90 or 120 degrees or the measurement direction of which is the same as prism bisectrix of angle are most common. The prisms with regulated expansion angle and sensor change position are also known. Such devices have a simple construction, high reliability and small size.

The main point of the centerless measuring method is that the product rotates and locates on the prism planes actually with the measured surface, and the small linear displacements detector fixes the roundness. The main disadvantage of this scheme is the systematic measurement inaccuracy which is up to 100 percent according to the different estimates [5]. Despite of the obvious advantages of this device – the simplicity and high efficiency – it is this reason that limits its operation.

The certain attempts to minimize the systematic measurement inaccuracy in terms of device construction solutions and the development of special measurement result processing procedures only partially solved this problem [6]. The main difficulty involves development of a mathematical model for measuring with high degree of adequacy. Therefore, a new centerless measuring model using numerical techniques and harmonic analysis is proposed in this paper. Due to this model the modeling for the actually measured details is done and the opportunity for systematic measuring inaccuracy minimizing is proved.

2. Argumentation and development of the mathematical model of the roundness centerless measuring.

The main feature of the centerless measuring is that the device sensor reading data and actual circular deviation value are non-linearly related, that is, there is a systematic inaccuracy. This inaccuracy is due to the measuring scheme itself and therefore it cannot be entirely precluded by the constructive solution or mathematical data processing. The principle of the measuring process is that the item locates on the prism planes actually with the measured surface. Consequently, the error of locating leads to repositioning of detail profile center and thus, changes the distance to the measuring sensor. The device detects these changes as well as surface form deviation. A specific device with the fixed prism angle value and sensor position has a different measurement error for different harmonic components in the profile. Since the product profile is described by superposition of harmonicas with different amplitudes and initial phases, systematic measuring inaccuracy minimizing is a complex mathematical problem.

The known mathematical model of centerless measuring presented in the paper [7] means invariability of the detail contact points with prism planes in the course of control. These dependences are true only for individual harmonicas since they do not consider their initial phases and that prevents from superposition principle proper using. The results obtained from this model are special and that makes their application challenging.

A new roundness centerless measuring model which is more adequate to a real process and focused on numerical methods and harmonic analysis is developed. Mathematical description of a measuring process is considered in three stages: detection of the profile center circle center after locating (for each detail current position); definition of the radii of the profile points measured by the sensor; resulting roundness estimation by the measured points.

In the paper [8] it is shown that the detail crossing profile is most accurately described by a trigonometric polynomial. Values of amplitudes and harmonics initial phases are obtained with harmonic analysis of real parts profiles. Thus, it is possible to model both a specific part profiles and profiles of specific kit of parts. In polar coordinates the profile is defined in the following way:

$$r = R + \sum_{n=2}^{p} a_n \cos(n\varphi - \varphi_n),$$

where, R is central circle radius; a_n, φ_n is the n-harmonic's amplitude and initial phase; p is a maximum number of relevant harmonics; φ is a vectorial angle.

At the first stage locating inaccuracy is defined, which makes for departure of actually achieved part location from nominally cylindrical part of radius R without form deviation. In such a case, in a part cross-section the required center position (point 0) is uniquely determined by the radius R and the prism angle α (figure 1).

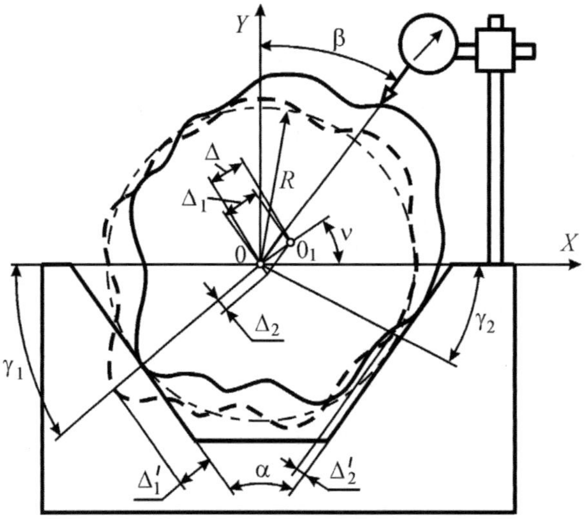

Figure 1. Measurement calculating model.

At the second stage we define radii r_2 of the measured profile points after locating. The initial values are coordinates of the profile central circle centre (Δ,ν), obtained at the first stage, and the radius r_i of the part profile points.

The third stage concerns with the roundness deviation determination – it is the maximal distance from profile points to the reference circle. The easiest is to find the roundness deviation for the Least Squares Circle (LSC) [8]. If the center of the LSC of the out-of-roundness coincides with the origin of the coordinate system, the roundness deviation is the difference between the maximum and minimum radii. Otherwise, it is required to determine additionally the coordinates of the LSC center and then the roundness deviation. To find the roundness deviations of the Minimum Zone Tolerance Circle or Maximum Inscribed Circle and Minimum Circumscribed Circle, it is possible to apply the techniques stated in works [9, 10].

3. The measuring and simulating of the centerless roundness measuring.

The proposed strategy for the centerless roundness measuring of the product with corrective adjustment assumes repeated measurement of one product. The first measurement allows revealing harmonic structure of the profile to a first approximation. These data are used for simulation by the developed model, which result becomes a new device setting. After re-measuring to the accepted setting the received values of the harmonic structure and roundness are comparing to the original ones. If the roundness does not decrease, it is considered that setting parameters are optimal. Otherwise, the process of simulating and corrective setting repeats. Generally, in practice, the process of the corrective setting is limited by two or three measurements.

On the basis of mathematical simulation of the centerless measuring is developed the algorithm and the program in C ++. The algorithm includes two cycles: to set a profile by the angle φ with the variable j distributing from 1 to m (m – number of points on the profile); product rotation by measurement by the angle with the variable j distributing from 0 to 2π. During the first cycle the cross-section profile of the product is formed as a set of radius-vectors r_i based on the original data (n, a_n, φ_n) concerning the roundness harmonic structure. The second cycle calculates basing error and measures radii r_2 of the profile. At that the number of initially set points of a profile and number of the measured points may not coincide. According to the simulating results it is possible to find in an interactive mode the optimum device settings on α and β angles. The harmonic analysis, centering of the out-of-roundness and calculation of the roundness deviation by various techniques are considered to be known and the scheme shows them conditionally.

Made numerical experiments demonstrate that by using of the single prism with a constant angle and the angular position of the sensor it is not possible to provide both an acceptable measurement error for all harmonics and a constant measurement error for the first 20 harmonics. This result corresponds with similar data presented in the study [6]. The systematic error is about 5% for the 9th harmonic and up to 105% for the 8th harmonic. The received results can be used by the device setting only by the single dominant harmonic in the profile but this is rare state in the contemporary manufacturing technology. The superposition principle by the centerless measurement simulating is not observed due to system nonlinearity. Therefore the recommendations for the certain harmonics do not allow reducing the measurement error in the real production.

The measurement and simulating for 50 external racers with a diameter of 30 mm were made with the tolerance roundness of 0.002 mm with the purpose to research the systematic error and its reduction methods. Sample measurements which set the valid profile, were carried out on a micrometric out-of-round gage Talyrond 30. Prism angle varied of α = 60, 90, 120° and the sensor angle varied of β = 0, 15, 30, 45, 60, 75°. Figure 2 presents an average ratio error and figure 3 reflects results for a standard measurement error deviation.

The graphs show that the prism angle and the sensor angle affect the measurement error. For individual product and device settings ratio error reached 120%. For party of product the worst ratio error is 62 % and corresponds to the parameters α = 120°, β = 75°, the best ratio error is 17.5% for the parameters α = 90°, β = 15°. The standard deviation for these setting variants was 26% and 14.5% correspondingly. Thus, the optimal setting for the product kit can simultaneously reduce the average systematic error and its range. The minimum value of the standard deviation does not always correspond to the minimum average error. However, this is important only by the optimal device setting for the specific product kit.

The above presented measurements were performed for precise ball bearing product of the small roundness compared to its diameter. Additional measurements for the less precise product with a higher roundness deviation established that for them the systematic measurement error increases slightly. However, the general pattern of the optimal setting for measurement process stays correct.

Not only device setting for product kit was carried out for the data, but the setting for every product separately was made accordingly to the reasonable strategy for corrective setting. Optimal setting provided an average error about 4.11% and the standard deviation about 4.12% for 50 components. The choice of the prism angle was also limited only by three values of α= 60, 90, 120°. Obviously, the finer device setting according to the prism angle and the sensor angle will allow reducing the systematic measurement error more effectively. The complexity by such measurements increases, but the simulation by the modeling program develops much faster than the device measurement itself.

Therefore, centerless measurement in its efficiency will hold on to the Rotational datum method and the Coordinate Measuring Method.

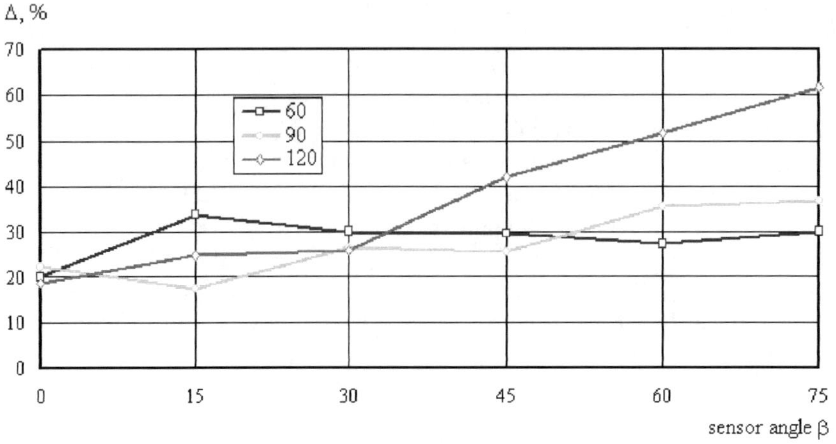

Figure 2. Average ratio error.

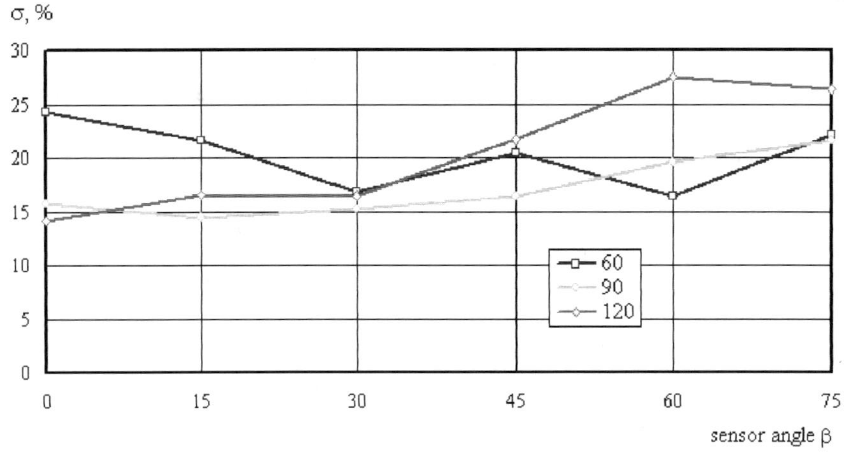

Figure 3. Ratio error, standard deviation.

4. Abstract

The analysis showed that the measured roundness value in most cases exceeds the actual value. When monitoring the probability of making α- error increases. At the same time, if the measured roundness value is placed in the tolerance, we can guarantee a minimum probability of β-error. This is particularly true when monitoring mass-produced product, e.g., in car and ball bearing industry.

Our research suggests that the centerless roundness measurement of the prism with small linear motion sensor has a systematic error. It depends on the harmonic content of the product form deviations. The superposition principle for the harmonics is not observed due to system nonlinearity.

Ratio error in the experiments reached the maximum 120%. On the other side the device settings such as the prism angle and the sensor position also contribute to the systematic error. The optimal device setting helps to reduce the measurement error which will not exceed the amount of 5%. It is reasonable to complete settings on the basis of simulating by the proposed algorithm in several stages. Optimal setting parameters can be determined for a separate product and for the product kit.

5. Conclusion

The new method of the centerless roundness measuring with corrective adjustment using a new mathematical model can effectively reduce systematic error. In this case, centerless measurement is a good alternative to the precise roundness measuring indicators or to the Coordinate Measuring Machines due to its accessibility to the users. Being inferior to a certain degree to these methods in precision, centerless measuring has high efficiency, low cost and does not require highly qualified staff. It is recommended for the control of the mass-produced product with the high and medium accuracy, both in labs and at factories.

Acknowledgments

The reported study was partially supported by RFBR, research project No. 15-08-08411\15 a.

References

[1] Avdulov A N 1974 Monitoring and evaluation of roundness of machine parts Moscow
[2] Palei M A and Chudov V A 1972 *Measurement Techniques* **15** 532-534
[3] Binder Ya I, Gebel' I D, Nefedov A I, Bakulin V N and Shapiro V A 1999 *Measurement Techniques* **42** 755-760
[4] Zakharov O V and Brzhozovskii B M 2006 *Measurement Techniques* **49** 1094-1097
[5] Sysoev Yu S 1995 *Measurement Techniques* **38** 1108-1114
[6] Witzke F W 1969 *American Machinist* **113** 103-106
[7] Muzakkir S M, Hirani H and Thakre G D 2013 *Tribology Transactions* **56** 1060-1068
[8] Spragg R C 1967 *Proceedings of the Institution of Mechanical Engineers* **182** 397-405
[9] Dovica M, Buša J, Palenčár R, Ďuriš S, Šooš L, Vrba I, Kelemenová T and Škovránek T 2013 *Measurement Technique* **56** 1021-1025
[10] Nikolskii A A and Korolev V V 2011 *Measurement Techniques* **54** 640-649
[11] Rossi A, Antonetti M, Barloscio M and Lanzetta M 2011 *Measurement* **44** 1243-1252

IOP Conf. Series: Materials Science and Engineering **93** (2015) 012025 doi:10.1088/1757-899X/93/1/012025

Lamb wave ultrasonic evaluation of welded AA2024 specimens at tensile static and fatigue testing

M V Burkov[1,2,a], A V Byakov[1,2], R T Shah[1], P S Lyubutin[1,2] and S V Panin[1,2,b]
[1] Lenina ave 30, National Research Tomsk Polytechnic University, Tomsk, Russia
[2] Akademicheskii ave 2/4, ISPMS SB RAS, Tomsk, Russia

E-mail: [a]burkovispms@mail.ru, [b]svp@ispms.tsc.ru

Abstract The paper deals with the investigation of Lamb waves ultrasonic testing technique applied for evaluation of different stress-strain and damaged state of aluminum specimens at static and fatigue loading in order to develop a Structural Health Monitoring (SHM) approach. The experimental results of tensile testing of AA2024T3 specimens with welded joints are presented. Piezoelectric transducers used as actuators and sensors were adhesively bonded to the specimen's surface using two component epoxy. The set of static and cyclic tensile tests with two frequencies of acoustic testing (50 kHz and 335 kHz) were performed. The recorded signals were processed to calculate the maximum envelope in order to evaluate the changes of the stress-strain state of the specimen and its microstructure during static tension. The registered data are analyzed and discussed in terms of signal attenuation due to the formation of fatigue defects during cyclic loading. Understanding the relations between acoustic signal features and fatigue damages will provide us the ability to determine the damage state of the structure and its residual lifetime in order to design a robust SHM system.

1. Introduction

Non-destructive testing (NDT) plays a significant role in development of next generation industry from small watch to huge spacecraft. The most widespread NDT practice of different structures that is used today utilizes defined time intervals of inspection, but the statistics of testing shows that the damages occur in small amount of tested objects while time and funds for diagnostics of the rest were wasted. However the designers cannot expand the inspection intervals because the structures where the damages have been already formed can experience the catastrophic failure which is inappropriate for different industries: aerospace, petrochemical, etc.

To ensure the safe operation of the structure (for example, the passenger aircraft) and simultaneously reduce financial losses during NDT (due to the downtime of the aircraft) scientific and engineering NDT community few years ago focused on research and development of the Structural Health Monitoring (SHM) systems [1,2]. SHM is a non-destructive in-situ sensing and evaluation method that uses a variety of sensors attached to, or embedded in, a structure to monitor the structural response, to analyze the structural characteristics for the purpose of estimating the severity of damage/deterioration and evaluating the consequences thereof on the structure in terms of load bearing capacity and residual lifetime. Various types of data (either continuously or periodically) from the sensors are collected, analyzed and stored for future analysis and reference. The data can be used to identify damage at its onset, to assess strength and integrity, and therefore the performance and safety of the structure as well as to expand the inspection intervals if the system doesn't register the significant changes exceeding the defined threshold.

There are different SHM principles proposed by different research groups, e.g. strain sensing using optical fiber [3], damage location and identification via the network of ultrasonic transducers embedded in the structure [4] or operational load monitoring using different physical principles and gages [5,6]. All SHM approaches require complex algorithms and software for data processing and decision-making [7] which are based on joint consideration of deformation mechanics of different materials, computer modelling of gages response as well as of the results of experimental verification of SHM technique on different specimens (from small used for mechanical properties evaluation to large structurally similar assemblies). The paper presents the experimental results of Lamb waves ultrasonic evaluation of AA2024T3 specimens with welded joints tested with static and fatigue tension.

2. Materials and methods

The investigation of proposed ultrasonic technique was performed during static and cyclic uniaxial tensile testing of the AA2024T3 specimens with welded joints. The argon arc welding was used for joining process after which the specimens were treated with water quenching at 435°C to reduce residual stress and improve the strength properties of welded joint.

Figure 1 presents the drawing of the dogbone-shaped specimen used for testing. The static tensile testing was performed on electromechanical testing machine Instron 5582 with the load rate of 0.3 mm/min. Cyclic loading with the load ratio R = 0.1, 10 Hz loading frequency and P_{max} defined as $0.4\sigma_b$ was performed using servohydraulic testing machine BiSS UTM 150kN. The piezoelectric transducers AW1E12G-190EFL1Z by Audiowell Corp were used as actuators and receivers. These piezoelectric transducers (discs with diameter of 9 mm and thickness of 0.19 mm on steel substrate) were glued on the specimen's surface using 3M Scotch-Weld DP490 epoxy adhesive. The 1st piezoelectric transducer (PZT) was used as an actuator. The 2nd (to characterize the changes outside the highly stressed gage length of the specimen), the 3rd and the 4th (to evaluate the changes in the gage length) were used as receivers. The ultrasonic signals were generated using arbitrary waveform generator AWG-4105 with amplitude of 10 V and frequency in the range from 10 to 500 kHz. The 5-cycle sine modulated by Hanning-window was used as a testing signal.

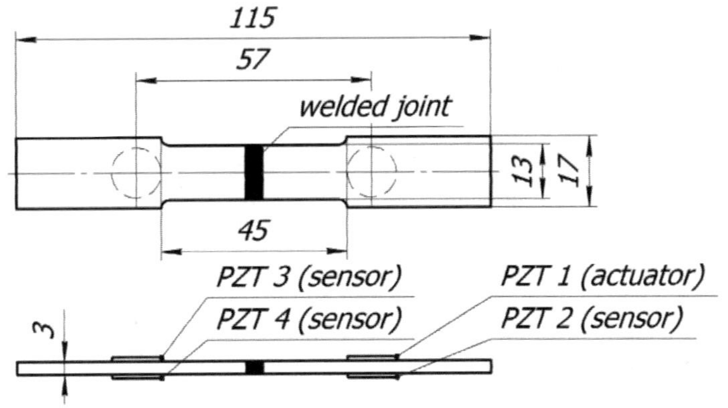

Figure 1. The drawing of the specimen, the position of the PZTs is schematically described.

The signals were recorded using USB oscilloscope Handyscope HS4 with sampling rate of 5 MHz and 12 bit resolution and filtered using band-pass 10-800 kHz filtering. The signal acquisition from sensor PZTs was triggered by the reference signal. The recorded acoustic signals were averaged 100 times and processed to calculate maximum envelope in order to characterize the state of the specimen at different static loads or fatigue lifetime. The calculation of the maximum envelope of received ultrasonic signal was carried out using Hilbert transform procedure in the frequency domain [8].

There are two experiments performed for the uniaxial static tension. In the first the continuous loading was used and the signals were recorded when the loading was stopped – the specimen was fixed in grips and subjected to tensile loads. In the second the loading was performed in the step mode and two signal sets were recorded: in loaded condition and when the load was withdrawn. The main goal of the static tests is to investigate the response of the actuator-receiver pairs during loading and to find the dependence of the sensed signal on the different stress-strain or damaged state. The cyclic loading experiment was carried out in the following manner: after a defined number of cycles the cyclic loading was stopped and the P_{min} was set to obtain the ultrasonic testing data. Thus the sensed signal amplitude dependence on different lifetime was recorded.

3. Results of static testing and discussion

Our previous work [9] was related to the investigation of the Lamb waves testing applied to unnotched AA2024T3 specimens tested with static and fatigue tension. Herein the same testing technique was applied and the same two testing frequencies (50 and 335 kHz) were used. The results of the present paper are quite similar to those obtained earlier.

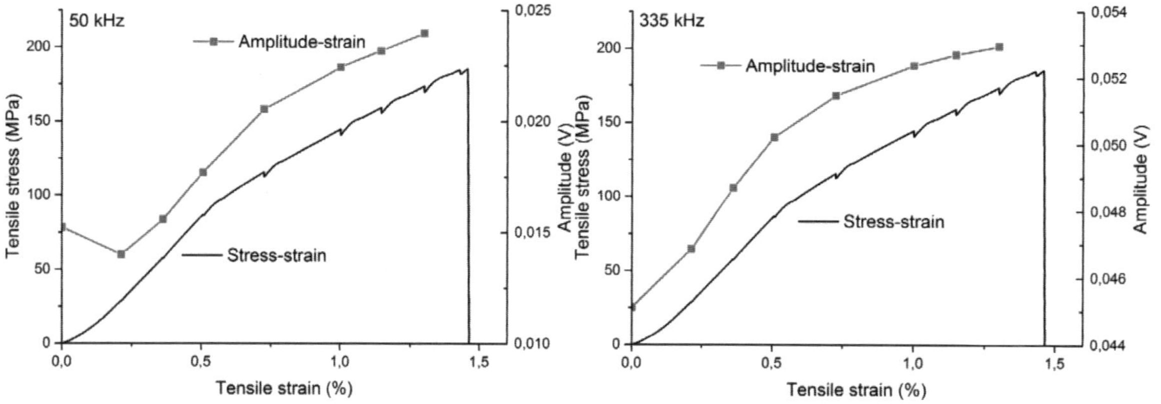

a b

Figure 2. Combined stress-strain and amplitude-strain graphs for the continuous tension test of the welded AA2024T3 specimen.

Figure 2 represents two combined graphs of the stress-strain curve of the AA2024T3 specimen with welded joint loaded continuously until fracture and the amplitude of the signal sensed by the 3rd PZT located on the same side with the generator along the gage section of the specimen. The signal sensed by the 4th PZT is almost the same with the 3rd except the reverse sign of the instantaneous sensed signal magnitude due to the propagation of A_0 Lamb wave mode. There are 7 points of defined load values where the loading was stopped and the ultrasound testing was performed (these points are easily recognized as the teeth on the stress-strain curve due to the stress relaxation during data acquisition). The first graph (a) corresponds to the frequency of 50 kHz; the second (b) to the 335 kHz. It is seen that amplitude of both high and low frequency signals increases during loading of the specimen. This result is similar to one obtained in the previous our work that the ultrasound used is sensitive to the different stress-strain state. After the second experiment two sets of data were obtained because loading was performed in step mode with data acquisition at both loaded and unloaded conditions. The different testing procedure with the stress relaxation occurring during unloading of the specimen results in different propagation of ultrasound during whole experiment. Figure 3 presents two curves: while the specimen was in loaded (red square) and unloaded (black triangle) conditions. Graph (a) corresponds to the low frequency signal and (b) to the high frequency.

It is seen that in contrast to the first experiment the amplitude (loaded state – red square plot) for both high and low frequency testing decreases during extension of the specimen. The amplitude for the low frequency signal starts to decrease from the initial of loading with constant slope till the 1.1 % of tensile strain. Then short nonlinear stage is observed and finally there are three last points with constant amplitude value captured till the specimen fracture. The amplitude of the high frequency signal propagated along the gage length stays constant till 0.8 % of strain and then it starts to decrease nonlinearly up to fracture. The correspondent curves of amplitude recorded in the unloaded condition are plotted on the same graph (black triangle) and they have the same shape compare to the plots recorded in the loaded state. However there are small differences:

- the amplitude of high frequency signal start to decrease after the strain reaches higher values (~1.3 %);
- low frequency signal has lower amplitude throughout all the experiment and its amplitude increases slightly near fracture.

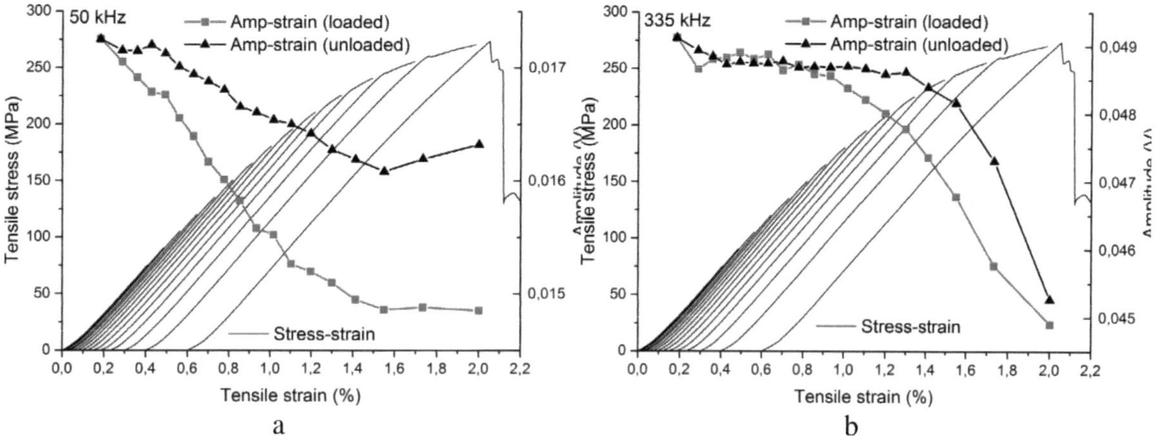

Figure 3. Combined stress-strain and amplitude-strain graphs for the step-mode tension test of the welded AA2024T3 specimen.

By analyzing the data obtained in our previous work [9] it can be noted that both high and low frequency signal propagation behavior during step mode loading of welded AA2024T3 specimen obtained in the present paper were the same compared to the unnotched specimen tested before. Low frequency signal amplitude decreases from the initial of loading and stays constant up to fracture while high frequency sensed signal decreases rapidly after the 2/3 of tensile strain.

4. Results of cyclic testing and discussion

Based on the results obtained during static tensile experiments the parameters for cyclic investigation were chosen. The average ultimate tensile stress for three specimens was obtained (210 MPa) and the maximum stress for cyclic was estimated as $0.4\sigma_u$=82 MPa. Thus three data sets for specimens with lifetime of 9000 cycles, 33500 cycles and 84000 cycles were captured. Such huge scatter of the lifetime is the result of poor quality of welded joints but it was not a problem for the present research where the ultrasonic testing technique was investigated. Moreover the presence of large welding defects such as pores or inclusions can give rise to the formation of large crack during cyclic loading in order to assess the Lamb wave testing technique sensitivity. The testing and data acquisition technique in this section are similar to ones used in static testing: there are two frequencies of testing 50 and 335 kHz, the amplitude of sensed signal corresponds to the 3rd transducer located on the same side with generator along the gage section of the specimen.

Figure 4 presents two graphs of low (a) and high (b) frequency signal amplitude registered during cyclic loading of the specimen fractured after only 9000 cycles. Because the data acquisition interval was set to 2000 cycles there are only four points were captured. Low frequency signal amplitude behaves uncertainly while the high frequency amplitude remains constant at first three points and then large decrease is registered. Low amount of data capturing points doesn't allow us to make any precise analysis of the propagated signal behavior so let's consider the second specimen with 33500 cycles of lifetime (Figure 5).

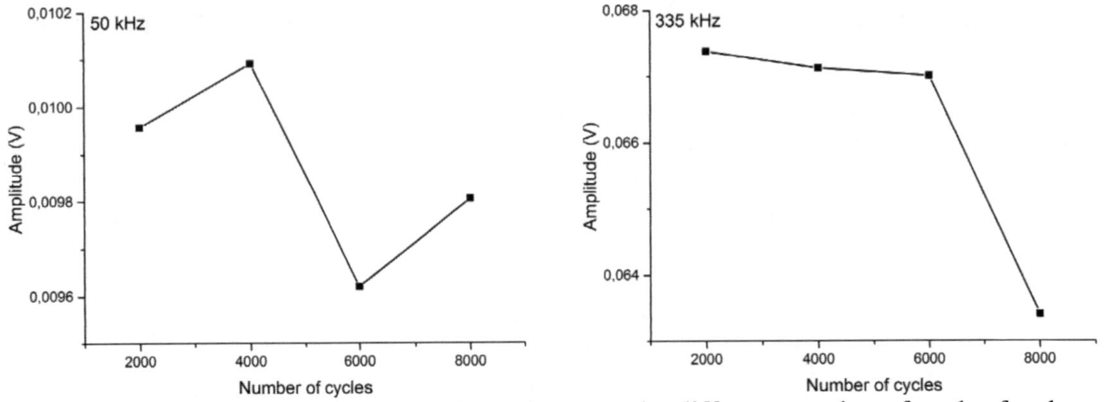

Figure 4. Sensed signal amplitude dependence on the different number of cycles for the specimen with lifetime of 9000 cycles.

It is seen that low frequency sensed signal graph doesn't have any precise trend during cyclic loading. The plot has large hops and drops of amplitude magnitude while the average value stays constant. In contrast to the low frequency behavior the high frequency signal amplitude decreases monotonically from the initial of the experiment till fatigue failure of the specimen.

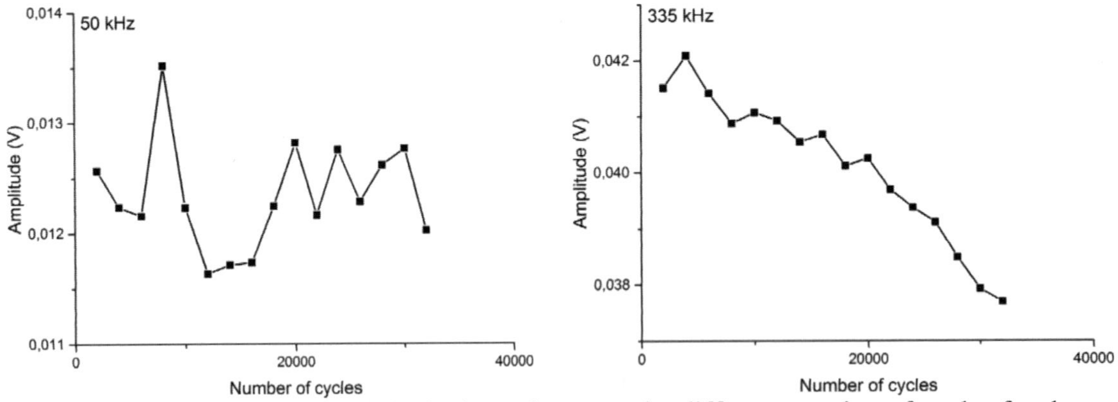

Figure 5. Sensed signal amplitude dependence on the different number of cycles for the specimen with lifetime of 33500 cycles.

Figure 6 presents the graphs of amplitude for the specimen with lifetime of 84000 cycles. The amplitude value for the low frequency testing stays near constant during entire experiment and at the last point near fracture the large jump is registered. This result differs with the explanation according to which the amplitude should decrease due to the higher signal attenuation owing to the micro damaging of the structure or the large crack formation. The second graph for high frequency confirms the assumption of higher attenuation with large drop of the sensed signal amplitude registered at the last point before fatigue failure. Also it should be noted that the specimen withstand more than 1500 cycles of load after the last data acquisition point.

5. Conclusion

The analysis of the obtained results allows us to conclude that the ultrasonic testing technique is sensitive to the different stress-strain state which appears in the increase of sensed signal amplitude during continuous tension test. This increment for both frequencies can be explained like a string being subjected to tension conducts elastic waves with lower attenuation. During the step mode tensile test where due to the stress relaxation after unloading the internal structure of the specimen exhibits significant changes. So based on the results of the present test the following can be concluded: the amplitude of both high and low frequencies decreases during extension of the specimen that is related most likely to the formation of the macro-defects in the welded joint that result in higher attenuation of the sensed signal.

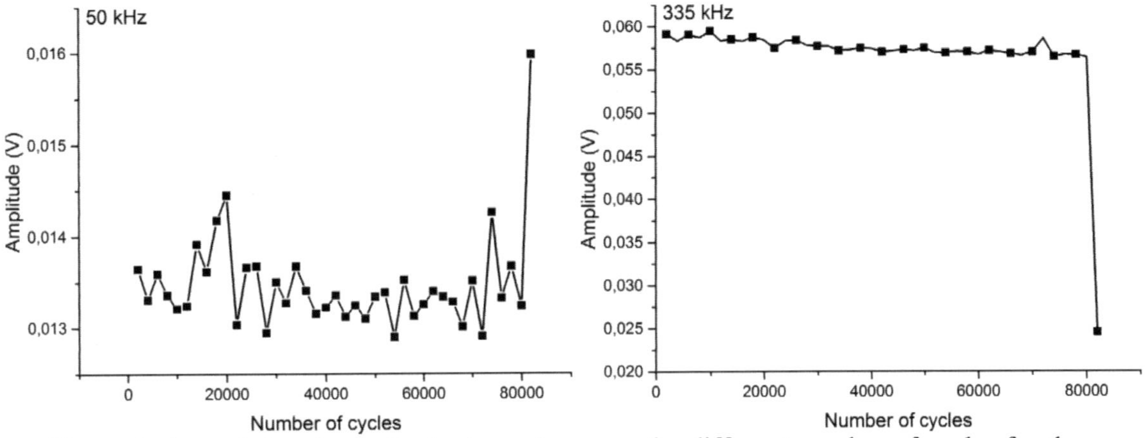

Figure 6. Sensed signal amplitude dependence on the different number of cycles for the specimen with lifetime of 84000 cycles.

The analysis of the fatigue tension test allows us to conclude that during cyclic loading the structure of the tested specimen exhibits the changes which influence the signal propagation. While the behavior of the graphs of low frequency signals amplitude is uncertain and cannot be interpreted in a precise manner the high frequency amplitude graphs behavior for the specimens with different lifetimes is similar to each other: the amplitude value decreases during whole cyclic loading. Thus we can conclude that only high frequency testing provides the results that can be interpreted accurately in order to design a SHM technique. The next our investigation will be related to the application of the present ultrasonic technique for evaluation of aluminum specimen during high-cycle fatigue (up to 1M cycles).

Acknowledgements
The work is supported by the RFBR grants № 13-07-00009_a and № 14-08-31747 mol_a.

References

[1] P J Schubel, R J Crossley, E K G Boateng and J R Hutchinson 2013 *Renew. Energ.* **51** 113-123
[2] W Liu, B Tang and Y Jiang 2010 *Renew. Energ.* **35** 1414–1418
[3] S Minakuchi, Y Okabe, T Mizutani and N Takeda 2009 *Smart Mater. Struct.* **18** 9
[4] Q Lei, Y Shenfang, W Qiang, S Yajie and Y Weiwei 2009 *Chinese J. Aeronautics* **22** 505-512
[5] P Wang, T Takagi, T Takeno and H Miki 2013 *Sensors and Actuators* **198** 46- 60
[6] S V Panin, M V Burkov, P S Lyubutin, Yu A Altukhov and I V Shakirov 2014 *Engineering Fracture Mechanics* **129** 45–53.
[7] Ch E Katsikeros and G N Labeas 2009 *Mechanical Systems and Signal Processing* **23** 372–383
[8] S L Hahn 1996 *Hilbert Transforms in Signal Processing* (Norwood: Artech House)
[9] A V Byakov, A V Eremin, M V Burkov, R T Shah, P S Lyubutin and S V Panin *Applied Mechanics and Materials* – in press.

Deformation characteristics of the near-surface layers of zirconia ceramics implanted with aluminum ions

S A Ghyngazov[1], I P Vasiliev[1], T S Frangulyan[1] and A V Chernyavski[1]
[1]Tomsk, 634050, Russia, National Research, Tomsk Polytechnic University

E-mail: ghyngazov@tpu.ru

Abstract. The effect of ion treatment on the phase composition and mechanical properties of the near-surface layers of zirconium ceramic composition 97 ZrO_2-3Y_2O_3 (mol%) was studied. Irradiation of the samples was carried out by accelerated ions of aluminum with using vacuum-arc source Mevva 5-Ru. Ion beam had the following parameters: the energy of the accelerated ions E = 78 keV, the pulse current density J_i = 4mA / cm^2, current pulse duration equal τ= 250 mcs, pulse repetition frequency f = 5 Hz. Exposure doses (fluence) were 10^{16} и 10^{17} ion/cm^2. The depth distribution implanted ions was studied by SIMS method. It is shown that the maximum projected range of the implanted ions is equal to 250 nm. Near-surface layers were investigated by X-ray diffraction (XRD) at fixed glancing incidence angle. It is shown that implantation of aluminum ions into the ceramics does not lead to a change in the phase composition of the near-surface layer. The influence of implanted ions on mechanical properties of ceramic near-surface layers was studied by the method of dynamic nanoindentation using small loads on the indenter P=300 mN. It is shown that in ion-implanted ceramic layer the processes of material recovery in the deformed region in the unloading mode proceeds with higher efficiency as compared with the initial material state. The deformation characteristics of samples before and after ion treatment have been determined from interpretation of the resulting P–h curves within the loading and unloading sections by the technique proposed by Oliver and Pharr. It was found that implantation of aluminum ions in the near-surface layer of zirconia ceramics increases nanohardness and reduces the Young's modulus.

1. Introduction

It is well known that the surface layer has a great influence on the physical and mechanical properties of solids. Surface modification of materials appears as the power and widely methodology, as it allows combining different properties of material bulk and material surface, hence designing of new materials characterized by unique properties. One of the most promising modern directions in modification of near-surface properties of materials is the treatment of accelerated ion beams.

Modification of the near-surface region of materials by use of energetic ion beams has been investigated extensively in recent years [1-6]. The energetic ion comes to rest by displacing atoms from their normal lattice sites by atom collisions thus producing a large number of radiation defects of different nature and complexity. The ion implantation allows one to introduce any element into the near-surface region of solids in a controlled and reproducible manner that is independent of most equilibrium constraints.

This may change not only the elemental composition of the surface, but as result atom rearrangement and accumulation of implanted ions new chemical compounds may be formed.

IOP Conf. Series: Materials Science and Engineering **93** (2015) 012026 doi:10.1088/1757-899X/93/1/012026

As a result structural-phase state of the material surface layers changes dramatically. Both radiation damage and implanted species themselves can introduce very large compressive stresses into near – surface layer and these can have marked effects on deformation of the surface layers, as well as the conditions of nucleation and propagation of cracks therein. All previously mentioned structure changes induced by the ion bombarding of materials must to lead to modification of near-surface layers properties.

Ion radiation treatment has been widely used to modify the mechanical properties of the surface of metals and alloys [1-3]. Recently new areas of application of ion implantation emerge, among them implantation into various types of ceramics [3-6]. Zirconium oxide is the most widely used oxide ceramic for functional applications and in particular as structural ceramics in mechanical engineering. The full exploitation of this material requires a fine modification of its properties. The use of ion beams represents one effective approach to achieve this goal. The interaction processes of ion beam with ceramics yet are understood not well. They are different from those observed in metals and alloys. Alterations to the structure and properties of ceramics are complex due to the range of bonding types encountered and the necessity for maintaining local charge balance.

As rule the state of implanted ions, their the reaction with host atoms, the phase evolution and the properties of implanted ceramics cannot be predicted accurately. Structural-phase state and properties of the modified surface layer ceramics depend on the conditions of ion treatment (nature of the implanted ions, ion beam power, radiation dose). By varying of these parameters we can form near-surface layers in ceramics with the most diverse properties. Our studies have shown [7] that processing of zirconium ceramics by powerful pulsed beam of accelerated carbon ions causes a decrease in hardness and increases plasticity of its of surface layers.

In the present work the effect of irradiation with accelerated aluminum ions on mechanical properties of the surface layers of zirconium ceramics is studied.

2. Experimental technique

The object of this study was (mol %) 97ZrO2-3Y2O3 zirconia ceramics sintered from ultrafine powders synthesized by plasma-chemical method. Sintering was conducted in air in a resistance furnace at T = 1400 ° C for 3 hours. The samples were made in the form of pellets with the diameter of 10 mm and thickness of 3 mm. Their density was equal to ρ= 5.7 g/cm^3, and the porosity was 6.5%. After sintering the samples were polished and then subjected to normalizing annealing at T = 1000 ° C for one hour. Aluminum ions were implanted into the ceramics. For generation of metal ion beam we used ion source Mevva-5 Ru based on vacuum arc discharge [8]. Ion beam had the following parameters: the energy of the accelerated ions E = 78 keV, the pulse current density J_i = 4mA / cm^2, current pulse duration equal τ= 250 mcs, pulse repetition frequency f = 5 Hz. Fluences (f) of ion irradiation were equal to 10^{16} and 10^{17} ion / cm^2. Samples were analyzed using technique of glancing incidence X-ray diffraction (GIXRD). The GIXRD was performed using Cu K_α X-rays. X-ray diffractometer ARL X'tra was used in the experiments. The elemental composition of the implanted layers was studied by SIMS with mass spectrometer PHI6300.

The study of mechanical properties of thin modified near-surface layers of specimens was performed by the method of nanoindentation using small loads on the indenter. It consisted in a continuous hardness indentation, with the load on the indenter varying linearly over time. This was also accompanied by the measurement of indentation depth (h), and load (P). The method makes it possible to measure in one cycle the depth of nonrecovered h_m and recovered (plastic) h_0 indents, Young's modulus E (from the tilt of the linear section of the unloading curve), and material hardness H. By applying this method we can follow the deformation process dynamics in a material microvolume and obtain more complete information on its mechanical characteristics. To this end, in this work we used a CSEM Nano Hardness Tester. Indentation tests were performed using a Vickers pyramid. About 8 – 10 indents were applied to the specimen surface.

The spacing between the dips of the indenter exceeded 20 μm. The indenter load P was 300 mN. The statistical error was within 6 – 10 %.

3.Experimental results

TRIM code and method SIMS are used to estimate the depth profile of the implanted ions in ceramics. The results are shown in Figure 1.

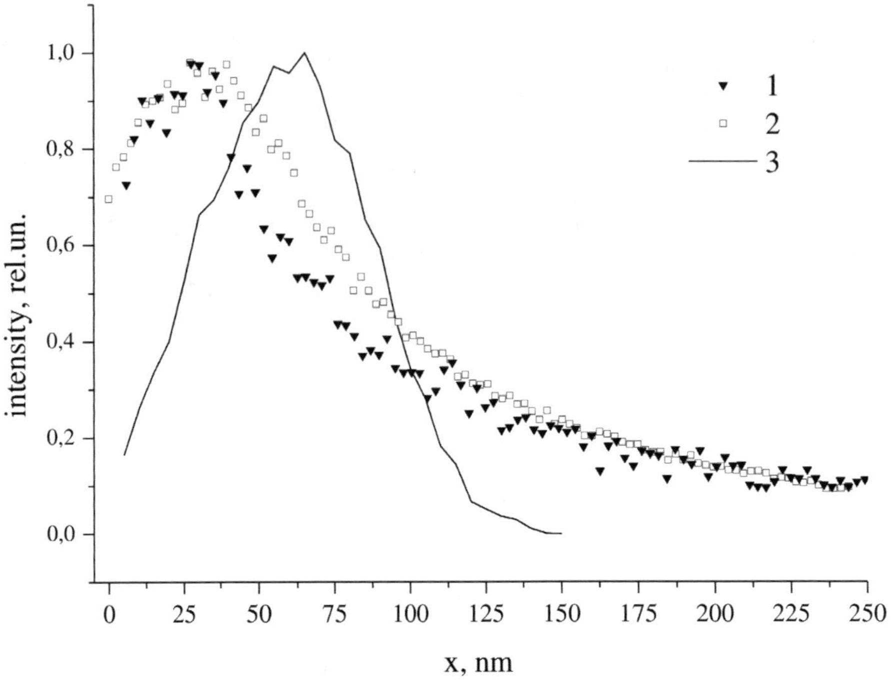

Figure1. Depth profiles of aluminum ions implanted in zirconium ceramics (Curves 1, 2 is experiment, Curve 3 is calculation. 1 – f=10^{16} ion / cm^2, 2- 10^{17}ion / cm^2).

It is seen that maximum of the experimental depth profiles (curves 1 and 2) is shifted towards the surface relative to the curve 3 calculated by TRIM code. Such displacement may be caused by sputtering of ceramic surface under ion bombardment. This virtually shifts implantation profile towards the surface. The research allowed us to determine the thickness of the ceramics implanted layer, which was τ 250 nm.

Effect of ion treatment on the phase composition of the implanted layer ceramics was investigated by X-ray diffraction. Taking x-ray patterns using the geometry of the Bragg-Bretano is incorrect, as according to numerical calculations probing depth is about 6.5 microns. This value is significantly greater than the thickness of the implanted layer ceramics, which, as seen in Figure 1, is about 250 nm. That is why the samples were analyzed using technique of GIXRD. For the case when technique of GIXRD is used, thickness of ceramic layer, which contributes in x-ray diffraction using technique of GIXRD was calculated by the method described in [9]. Calculated values of thickness of layer for different τ are the next: 820 nm for β=1° 2330 nm for β= 3°

It was found that the X-ray diffraction patterns obtained by technique of GIXRD for the unirradiated sample and the samples irradiated by the ion beam with f=10^{16} ion / cm^2 and 10^{17} ion/cm^2 were identical. We can observe diffraction reflections which belong to the tetragonal phase of zirconia dioxide only. Therefore, these modes of ion implantation do not cause any noticeable change in the phase composition of the treated surface.

Figure 2 a, b shows a typical P–h dependences obtained during indentation of ceramic specimen before (Figure. 2 a) and after ion irradiation with f =10^{17}ion/cm^2 (figure. 2 b).

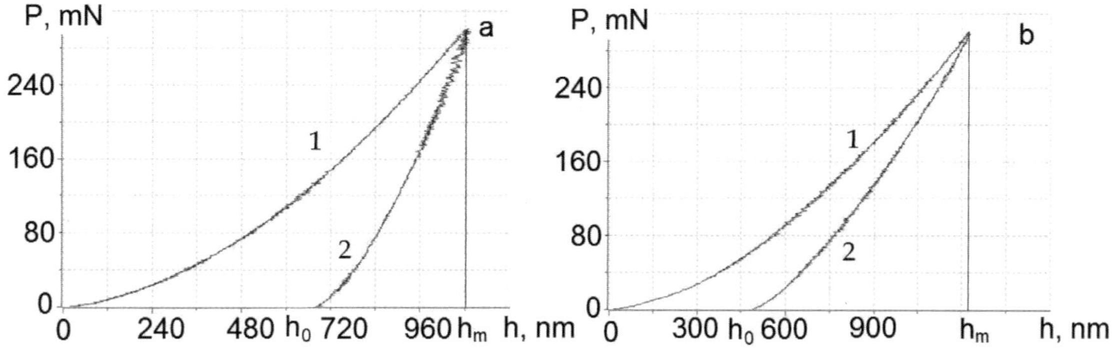

Figure 2. Dependences P(h) of zirconia ceramics: a) non- irradiated , b) irradiated with f =10^{17} ion/cm^2.

The curve consists of two sections. Section 1 corresponds to the process of indenter dipping into the specimen and characterizes material resistance to plastic deformation. Section 2 provides information on deformation behavior of the deformed region in the unloading mode. It is evident that the residual indentation depth h_0 after removal of the indenter is significantly different from its maximum indentation depth h_m. This fact is indicative of an active character of the processes of material recovery in the deformed region in the unloading mode. Their contribution into the changes in indentation size can be estimated using parameter $\alpha = h_m - h_0 / h_m$.

The deformation characteristics were determined from interpretation of the resulting P–h curves within the loading and unloading sections by the technique proposed by Oliver and Pharr [10]. Deformation characteristics of the near-surface layers of zirconia ceramics before and after ion treatment are shown in Table 1.

Table 1 lists the deformation characteristics of the near-surface layers of zirconia ceramics before and after ion treatment.

Table 1. Deformation characteristics of the near-surface layers of zirconia ceramics before and after ion treatment.

	Unirradiated ceramics	Irradiated ceramics	
		f=10^{16} ion/cm^2.	f=10^{17}ion/cm^2.
H, GPa	12.2	15.5	15.2
E, GPa	214	106	97
α	0.36	0.58	0.61
H/E	0.057	0.146	0.156
H^3/E^2, GPa	0.04	0.33	0.37

Analysis of the P(h) curves on Figure 2, and results shown in Table 1, showed that ion treatment significantly stimulates proceeding of process of material recovery in the deformed region in the unloading mode. Table 1 shows that the value α for the irradiated sample almost two times higher than for unirradiated sample. In our opinion this is due to the mechanical stresses that occur in the near-surface layer during implantation of accelerated ions.

Studies have shown that implantation of aluminum ions in the near-surface layer of zirconia ceramics increases nanohardness and reduces the Young's modulus. To evaluate the resistance to elastic deformation of the material and its ability to withstand mechanical stresses without forming is often used value H / E [11].

At the same time the material's resistance to plastic deformation characterizes parameter H^3/E^2 [11]. From the data given in table1 seen that those parameters after ionic treatment of ceramics are substantially increased.

4. Conclusions

The study found:

1. Treatment of zirconia ceramics with accelerated aluminum ions does not change the phase composition of ceramic near-surface layers.
2. Implantation of aluminum ions in the zirconia ceramics significantly stimulates proceeding of process of material recovery in the deformed region in the unloading mode.
3. Implantation of aluminum ions in the near-surface layer of zirconia ceramics increases nanohardness and reduces the Young's modulus.

Acknowledgments

We would like to acknowledge members of Prof. **Okc E.M.** group of Institute of High Current Electronics SB RAS for using their ion source Mevva-5 Ru.

This research is supported by The Ministry of Education and Science of the Russian Federation in part of the "Science" program.

References

[1] Didenko A N, Ligachyov A E and Kurakin I B 1987 Charged particle beam action on a surface of metals and alloys. *Moscow: Energoatomizdat* 186

[2] Anishhik V M and Uglov V V 2003 Modification of tool materials and ion plasma beam *(Minsk:BSU Publisher)* 191 p).

[3] Belyj A V, Kukarenko V A and Lobodaeva O V 1998 Ion-beam processing of metals, alloys and ceramic materialov. *Minsk: Physico-Technical Institute* p 220

[4] Nenadovic T M 2005 *Material in tehnologije* 9-17

[5] Nargue C J 1998 *Material Science and Engineering* A **253** 94-105

[6] Garrido F, Vincent L, Nowicki L, Sattonnay G and Thomé L 2008 *Nuclear Instruments and Methods in Physics Research* Section B **266** 2842–2847

[7] Ghyngazov S A, Vasil'ev I P, Surzhikov A P, Frangulyan T S and Chernyavskii A 2015 *Technical Physics* **60** 128–132.

[8] Nikolaev A G, Okc E M, Yushkov G Yu and Brown I G 2012 *Rev. Sci. Instrum* **83** 02A501

[9] Ivanov A N and Yagodkin Y D 2000 X-ray analysis of the surface layer (review) *Factory Laboratory* **66** 24-34

[10] Oliver W C and Pharr G M 1992 *J. Mater. Res.* **7** 1564-1583.

[11] Leyland A and Matthews A 2000 *Wear* **246** 1-11.

Influence of ion-beam treatment on structure and deformation resistance of 12Cr1MoV steel under static, cyclic and dynamic loading

S V Panin[1,2], I V Vlasov[1,2], P O Maruschak[3] and V P Sergeev[1,2]
[1]Department of Material Science in Mechanical Engineering, Institute of High Technology Physics, Tomsk Polytechnic University, Russia, 634050,Tomsk, Lenin Avenue, 30
[2]Institute of Strength Physics and Materials Sciences SB RAS, Russia, 634021, Tomsk, pr. Akademicheskii 2/4
[3]Ternopil National Ivan Pul'uj Technical University, Ukraine, 46001, Ternopil, Ruska 56

E-mail: svp@ispms.tsc.ru

Abstract. Features of modification of structure and properties of 12Cr1MoV steel subjected to ion-beam irradiation by zirconium ion beam have been investigated with the use of optical and electron microscopy, and microhardness measurement. It was shown that after the treatment the modification occurs across the entire cross-section of specimens with the thickness of 1 mm. Changes in mechanical properties of these specimens under static, cyclic and impact loading were interpreted in terms of identified structure modifications.

1. Introduction

Heat resistant 12Cr1MoV steel is used for the manufacturing of energetic equipment parts operating at temperatures of 570 ... 585 °C under action of (thermal)elastic-plastic loads. One of the main reasons of such elements failure is fatigue fracture due to the thermo-mechanical effect [1, 2]. Ion treatment for a long time has been used for surface layer modification to obtain new structural-phase states there [3]. A new method of treatment by irradiation of the beam of metal ions, providing more deeper structure modification in contrast with traditional ion implantation has been developed in the ISPMS RAS. However, the impact of Zr ions beam is accompanied by heating the surface layer of the steel and formation of new phases of iron with zirconium.

The purpose of the study is to investigate the influence of vacuum-arc ion beam treatment on the structure and mechanical properties of heat resistant 12Cr1MoV steel. For a comprehensive assessment of the effect specimens were tested under static and cyclic tension and impact bending on the machine equipped with a instrumented striker.

2. Materials and research methods

12Cr1MoV steel specimens for testing were cut out of a tube fragment by the wire electroerosion machine, and then subjected to a standard heat treatment: normalization at 960 ... 980°C and subsequent high tempering at 740 ... 760°C [4]. Further, these specimens are referred to as initial state ones.

Tensile tests were carried out for dog-bone shaped specimens with the dimensions of gauge length of 20×5×1 mm. Specimens with dimensions of 55 × 10 × 8 mm were prepared in accordance with Russian Federation standard GOST 9454 to determine the impact toughness. The tests were carried out for 3 temperatures: ambient – 20°C, increased – 375°C and high – 600°C (that is similar to study conducted by the authors earlier [5]).

Flat specimens with the size of 70×10×1 mm were used for the cyclic tensile testing. There is a stress concentrator in the form of a central hole with the diameter of 2 mm. Specimens of the same shape were used for the cyclic alternating bending tests. The following are the basic parameters of the tests: cyclic tension (maximum load – 320 MPa, minimum – 115 MPa, frequency – 15 Hz); cyclic alternating cantilever bending (range – 6 mm, shoulder – 51 mm, frequency – 9.5 Hz). Ion-beam treatment of specimens was performed with the use of high-currency vacuum-arc source of metal ions UVN-0.2 "Quantum" [6]. According to data of pyrometric monitoring during the treatment the surface layer of a specimen experiences a short-term heating up to the temperatures of 600 ... 900 °C.

However, due to the rotation relative to the ion source, the specimens were only for a short time exposure to the effect of the beam that made it possible to avoid continuous heating (taking place at the moment of interaction of specimens with the ion beam) and the consequent loss of strength (tempering). To a certain extent one can say that in addition to the exposure, the specimens were undergoing cyclic thermal impact which would lead to structural changes far beyond the thin surface layer where zirconium ions can penetrate.

Tests on static tension were performed with the help of electromechanical testing machine Instron 5582. Impact tests were conducted on motorized pendulum impact machine Instron 450 MPX equipped with a instrumented striker. Cyclic tension tests were carried out on a servo-hydraulic testing machine Biss UTM 150. Microhardness measurement was performed with the use of PMT-3 microhardness meter with a load onto Vickers pyramid of 0.49 N (50 g).

3. Study of the modified layer microstructure

The microstructure of the specimen without the treatment represents a mixture of ferrite-pearlite grains (figure. 2, b). After the ion beam treatment not only a thin surface layer was modified, whose formation is associated with the possible penetration of zirconium ions, but also with noticeable changes in the structure of the material core. The modified surface layer to a depth of 130 μm when etching has a distinct white color and is characterized by grains of substantially large size (up to several tens of microns, figure 2, c). Its interface with the underlying substrate layers (core) is not clearly defined.

The reason for the grain size growth from 29 ± 3 μm in the as-supplied state specimens to 48 ± 6 μm in the surface layer of this steel can be assumed due to the heating to high temperatures (up to 900°C). On the other hand, keeping the original grain structure of the core can be explained by periodic cooling of the specimens due to their rotation relative to the ion beam (source). Moreover, thermal cycling during the treatment came to decrease in the grain size from 29 ± 3 μm to $d \leq 20$ μm. Analysis of microimages showed that this structure corresponds to the ferrite-sorbite mixture.

It is shown that the microhardness of specimens without the treatment is equal to 1.67 ± 0.1 GPa while the value after the treatment varies considerably at different distances from the surface (figure 1, a). In the measurement of the microhardness through the cross section (on the lateral face) there was revealed its reduction to a depth of 70 ... 100 μm from the irradiated surface and is equal to 1.45 ± 0.04 GPa. At the same time, at a depth more than 100 μm microhardness has increased to about 2.16 ± 0.1 GPa and remains at approximately constant level that exceeds one of specimens in as supplied state ($H_\mu \sim 1.7$ GPa). Thus, there is a softening of the surface layer to a depth of about 100 ... 150 μm in the specimen after the treatment, while it was deeper microhardness contrary increased by ~ 22 %.

4. Results of tests

4.1. Tests on static tensile. The presence of yield tooth on the loading diagram is characteristic for material without the treatment that is typical for deformation of low carbon steels, figure 1, b. Yield point for such specimens makes $\sigma_{0.2}$ = 387 ± 23 MPa, tensile strength σ_U = 495 ± 35 MPa, and value of elongation at failure ε = 20 ± 3 % which is close to one for the standard data of the steel [5]. After the irradiation the tensile strength value σ_U = 570 ± 17 MPa and value of elongation at failure ε = 16 ± 0.7 %; in doing so there is no formation of the yield plateau in the specimens after the treatment. Thus, after the irradiation the tensile strength was increased by $\Delta\sigma$=76 MPa (15 %), and values of relative elongation decreased by $\Delta\varepsilon$ = 4 % (19 %).

4.2. Cyclic testing. Data of the cyclic tensile tests showed that the number of cycles prior to failure of specimens after the Zr^+ ion beam treatment was increased twice, as well as the time before the crack nucleation (also, ~ 2 times), figure 1, c. For specimens after the treatment the characteristic feature is formation of small microcracks on a surface, while the strain relief is formed in a much less degree. Formation of obvious sliding traces within individual grains is evident in specimens without the treatment.

According to the test results, the fatigue life of specimens under the cyclic alternating bending increased twice due to the treatment by Zr^+ ion beam, figure 1, d. The strain induced relief of these specimen looks smoother. Fatigue failure is accompanied by a significantly lower crack opening in comparison with the non-treated specimens.

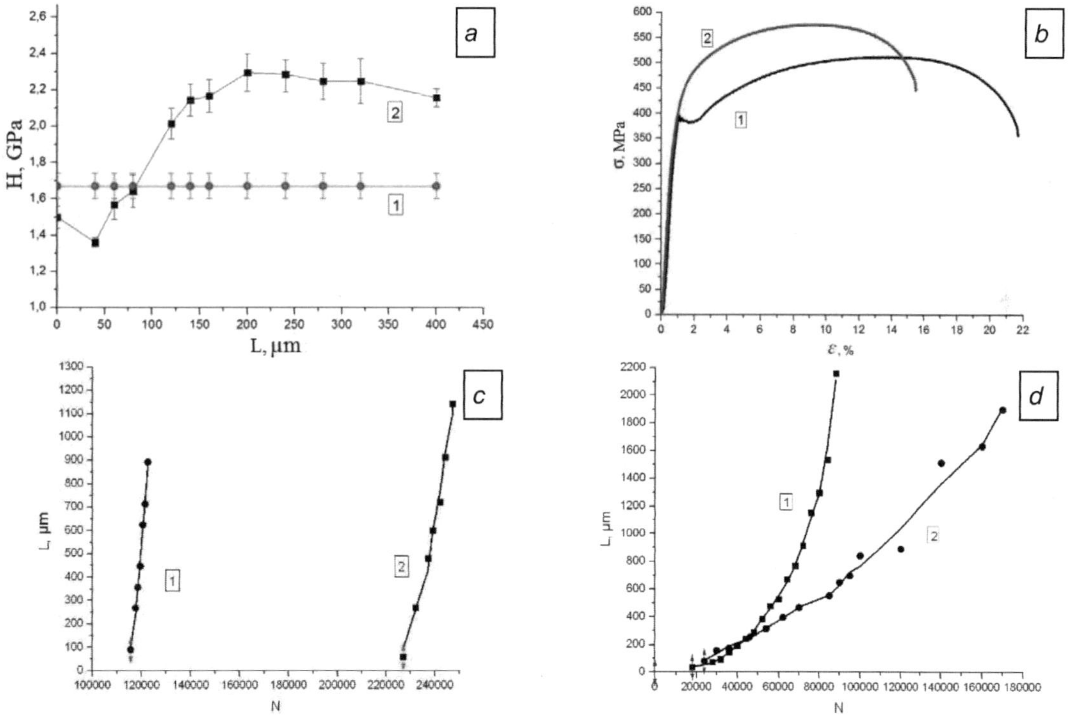

Figure 1. a) Changes in microhardness as a function of the distance from irradiated surface; b) stress-strain curve for tensile testing of 12Cr1MoV steel specimens of dog-bone shape; dependence of crack length vs the number of cycles; c) cyclic tension; d) alternating bending; 1) without the treatment; 2) after the irradiation.

4.3. The impact test. Investigation of the microstructure for all types of specimens was carried out. It was found that with increasing the temperature the grain size of the specimen in initial state does not change. At the same time, after the treatment the grain size in the specimen core is not substantially changed, while it is increased in the surface layer. The depth of the modified surface layer makes 120 μm. The average grain size of the modified surface layer is 27 ± 3 μm.

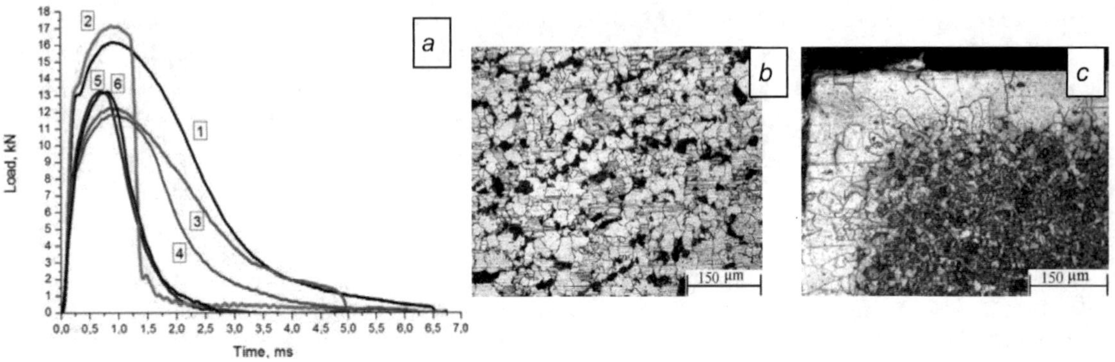

Figure 2. a) diagram of impact loading; 1), 3), 5) specimen without the treatment, 2), 4), 6) after the treatment; Testing temperature – 20ºC (1, 2); 375ºC (3, 4); 600ºC (5, 6); optical micrographs of 12Cr1MoV steel specimen cross sections in as supplied state (a) and after the ion-arc treatment by the Zr^{+} beam (b).

Microhardness measurement on the irradiated surface and through the cross section of the treated specimens was performed. For all specimens it was revealed reduction in microhardness values in surface layer at a depth of about 200 μm which agrees well with metallography data. However, at the depth of 200 ... 1000 μm strain hardening due to cyclic thermal influence during the treatment is registered. Impact test of specimens were carried out at the temperatures of 20°C, 375°C and 600°C. It was shown that the ion-beam treatment results in a substantial decrease in toughness mostly at room temperature.

Investigations of impact toughness of the material in an initial state and after treatment were fulfilled. Recorded diagrams of specimen impact fracture at different temperatures are presented in figure. 2, a. Impact toughness values (KCV) were recalculated on the base of instrumented striker data. Test results showed that the impact toughness of the specimens after the surface treatment was reduced by 5 ... 42 %. The maximum decrease in toughness was observed at 20°C. This is due to softening of surface and subsurface layers, and hence a decrease in energy intensity of crack origin under dynamic loading [7]. Simultaneously, ductilityof the specimen core is reduced, that is proved by the microhardness measurement data.

5. Conclusion

It was revealed the treatment by the zirconium ion beam irradiation of 12Cr1MoV steel specimens with the thickness of 1 mm the structure modification occurs through the entire cross section. There is decrease in microhardness values in the surface layer at the depth up to 100 μm, while in the core it is by 23 % higher than that in the non-treated specimens. Hardening of the core influence the results of the static tension tests: an increase in tensile strength by 15 % and reduction of elongation by 19% takes place. Under fatigue tests by the cyclic tension scheme it was revealed that modification of the structure during the treatment gives rise to delay in the development of localized plastic deformation prior to the origin and growth of the main fatigue crack that is manifested primarily in reducing of the mean value of shear strain intensity. As a result, the increase of the fatigue life time of the modified 12Cr1MoV steel specimen can make 2 times.

Under impact loading of the ion-beam treated specimens, the impact toughness at ambient temperature is decreased by 42 %. Heating up to 375 °C and 600 °C give rise to some structural changes that is accompanied nearly equal value of KCV for non-treated and irradiated specimens.

Offered method of the treatment cannot be characterized as a way of surface modification since locally arising high temperatures lead to a change in the structure through the entire cross section. The observed effects of the mechanical properties changing under static and cyclic loading should be interpreted with consideration of a loaded specimen as a multilevel system: surface layer alloyed with zirconium (a few μm) – thermally "weakened" layer (up to 100 ... 130 μm) - hardened one due to cyclical short-term thermal effect of substrate material. In so doing this multi-level system becomes the most effective to resist crack nucleation and growth under cyclic loading.

References

[1] Smirnov A N 2004 The microstructure and phase composition of steel 12Cr1MoV after pro longed use *Bulletin of Kuzbass State Technical University* **2** 39 67-72
[2] Mikova K, Bagherifard S, Bokuvka O, Guagliano M and Trško L 2013 Fatigue behavior of X70 microalloyed steel after severe shot peening *International Journal of Fatigue* **55** 33 pp 33-42
[3] Nastasi M and Mayer JW 2006 *Ion implantation and synthesis of materials* (Germany, Springer) pp 77–90
[4] Koloskov M M, Kashircky Yu V et.al. 2003 *Database of steels and alloys* 2nd iss ed Zubchenko A S (Moscow: M.: Mashinostroenie) p 784
[5] Yasnii P V, Maruschak P O, Panin S V, Bischak R T, Vukherer T, Ovechkin B B and Panin V E 2010 The effect of temperature on the nature of the impact fracture ferrite-pearlite steel 25Cr1Mo1V *Physical mesomech.* **13** 4 pp 73-84
[6] Sergeev V P 2006 *Modification of surface layers of high-strength steels by composite ion beams. Deposition of nanostructured coatings* ed Panin V E (Novosibirsk: Publishing House of SB RAS) chapter 10 pp 363-403
[7] Samotugin S S, Leschinsky L K and Soyanik N Kh 1996 *Structure and character of fracture of welded joints, weld and hardened materials* (Mariupol: Publ. of PSTU) p 179

IOP Conf. Series: Materials Science and Engineering **93** (2015) 012028 doi:10.1088/1757-899X/93/1/012028

A search for the quantum-chemical methods of germanium-oxygen geometric structure calculation

M Yu Gavalyan[1], V V Turovtsev[1,2] and I A Kaplunov[1]
[1]Tver State University, Zhelyabova str., 33, Tver, 170100, Russia
[2]Tver State Medical University, Sovetskaya str., 4, Tver, 170100, Russia

e-mail: mamikonghavalyan@mail.ru

Abstract. The methods of density functional theory (DFT in Kohn-Shem formalism) currently play the role of main instruments of quantum chemistry. The examination of the oxygen atoms behavior in germanium requires preliminary calibration of the functionals for less resource-intensive problems. With this end calibration of the germanium exchange and correlational functionals was made for the finite calculation of such parameters as oscillation frequencies, dipole moment, full electron energy. A correlation was found between the calculated values of Ge-O bond lengths and harmonic oscillation frequency. The functionals for further study of the oxygen behaviour in crystalline germanium were determined.

1. Introduction

The first step of quantum-mechanical calculations of physico-chemical properties of a compound always starts from the optimization of its equilibrium structure. By now a rather large number of models and methods exist in the disposal of quantum chemistry. This variety is explained by the fact that one or another approximation is introduced during the derivation of equations thus narrowing the sphere of applicability of the specified model. As a result the studies of new properties or classes of substances were accompanied by the development of modified methods.

Crystalline germanium is an inherent part of a large number of semiconductor instruments, units and elements of infrared optics and photonics. As this takes place the creation of modern element base will be impossible without the use of single crystals having low and ultralow concentration of impurities. For example, crystals for detectors of ionizing irradiation require electroactive impurity concentration below 10^{10} cm^{-3}. In addition germanium always contain background impurities, such as oxygen with a concentration of up to 10^{17} cm^{-3}), carbon, nitrogen, etc. Oxygen impurity results in a formation of dislocations, microdefects and thermal donors, which in its turn have an influence on the lifetime of nonequilibrium charge carriers [1, 2].

Calculations of the structure and properties of bulk crystalline bodies are based on the selection of the optimal method/basis combination [3] reproducing the geometry of the object under study with the required accuracy: the bond length inaccuracies should not exceed 0.01 Å, and those of valent and dihedral angles should be less than 5 and 10 degrees, respectively. Large errors in the equilibrium geometry result, as a rule, in divergence of such parameters as oscillation frequency, dipole moment and full electron energy, which have an effect on the overall values of the entropy, enthalpy, free Gibbs energy, etc.

IOP Conf. Series: Materials Science and Engineering **93** (2015) 012028 doi:10.1088/1757-899X/93/1/012028

2. Method of calculation

The methods based on the Hartree-Fock model (MP*n*, QCI, CC), and those of the density functional (DFT in Kohn-Shem formalism) (DFT в формализме Кона-Шема) at the moment play the role of main quantum chemistry instruments. The former put into the foreground more productive computer systems. By this reason the DFT model is most frequently used in quantum-mechanical theoretical studies, while post-Hartree-Fock calculations are exploited as exemplary ones when the experimental data are lacking. However, as mentioned above, the adopted approximations resulted in an appearance of a large number of exchange and correlational functional which may be used to compile a huge number of calculation model combinations. Hence a preliminary calibration of the functionals at less resource-intensive tasks is required for the investigation of oxygen atom behavior in germanium.

Germanium oxide GeO is the most simple composition for making an estimation of the geometric parameters. Its structure has been studied in detail, so at the beginning we reviewed the literature data (Table 1) which is then followed by the calculation of *r*(Ge-O) bond lengths (Table 2) with the aid of program GAUSSIAN 03 [4] in the basis 6-311+G(3df, 3pd) with spherical functions 6d 10f.

Table 1. Bond lengths of *r*(Ge-O) in gaseous germanium oxide. Δ – is the calculation error.

Method/basis	R(GeO), Å	Δ, Å	Method/basis	R(GeO), Å	Δ, Å
B3LYP/SVP	1,632 [5]	**0,007**	B97/6-311+G(2df)	1,601 [10]	-0,024
B3LYP/6-31G(d)	1,647 [6]	0,022	MB97/6-311+G(2df)	1,601 [10]	-0,024
B3LYP/cc-pVQZ	1,634 [7]	**0,009**	HCTH/6-311+G(2df)	1,637 [10]	0,012
B3LYP/SVP	1,632 [8]	**0,007**	TPSS/cc-pVQZ	1,643 [7]	0,018
B3LYP/SVP+f	1,624 [8]	-0,001	LSDA/ECP	1,563 [11]	-0,062
B3LYP/DZP	1,636 [8]	0,011	LDA/ECP	1,593 [12]	-0,032
B3LYP/DZP+f	1,628 [8]	**0,003**	LDA/6–311++G(3df,3pd)	1,627 [12]	**0,002**
B3LYP/6-311G**	1,630 [8]	**0,005**	BLYP/SVP	1,655 [8]	0,030
B3LYP/6-311G	1,648 [9]	0,023	BLYP/SVP+f	1,648 [8]	0,023
B3LYP/6-311+G(2df, 2pd)	1,633 [9]	**0,008**	BLYP/DZP	1,660 [8]	0,035
BPW91/6-31G(d)	1,664 [6]	0,039	BLYP/DZP+f	1,652 [8]	0,027
B3PW91/6-31G(d)	1,642 [6]	0,017	BLYP/6-311G**	1,653 [8]	0,028
LHF/6-311+G(2df)	1,591 [10]	-0,034	BLYP/cc-pVQZ	1,657 [7]	0,032
B972/6-311+G(2df)	1,622 [10]	**-0,003**	BLYP/6-31G(d)	1,671 [6]	0,046
MB97-2/6-311+G(2df)	1,622 [10]	**-0,003**	Phenomenological calculations	1,714 [13]	0,089
Experiment	1,625	–	Experiment [14]	1,625	–

The consideration of Table 1 shows that the required accuracy may be achieved by the use of functionals MB97-2, B972, B3LYP and LDA. The DFT provides a possibility of obtaining the required geometrical parameters, which give an opportunity to calculate other physico-chemical parameters of germanium.

The spectrum of gaseous germaniun oxide is characterized by a large number of lines merging into wide bands with the increase of the temperature. The complexity of the spectrum is related to the existence of several isotopes of both elements comprising natural samples of GeO. Nevertheless combination of $^{74}Ge^{16}O$ is used in most calculations of the properties. Starting from the performed analysis given above we calculated the harmonic frequency (ω_{harm}, Table 3) for the chosen set of isotopes and compared it with the ω_{harm} estimated by the experimental spectrum [15]. The divergence in the calculated length resulted in a divergence in ω_{harm} (Table 3). The values most close to the experimental ω_{harm} ones were given by functionals XAlpha, VSXC and SVWN.

Table 2. *r*(Ge-O) bond lengths in gaseous germanium oxide found in the basis of 6-311+G(3df, 3pd) 6d 10f. Δ is the error of the calculation.

Method	R(GeO), Å	Δ, Å	Method	R(GeO), Å	Δ, Å
SVWN	1,624	**0,000**	B972	1,623	**-0,001**
VSXC	1,638	0,013	B3LYP	1,631	**0,007**
XAlpha	1,628	**0,003**	B3PW91	1,626	**0,001**
HFS	1,645	0,020	MPW1PW91	1,621	**-0,003**
HCTH	1,636	0,012	PBE1PBE	1,622	**-0,003**
B3P86	1,624	**-0,001**	BHandH	1,596	-0,029
HFB	1,677	0,052	BHandHLYP	1,608	-0,016
B1B95	1,620	**-0,005**	M06	1,623	**-0,001**
B1LYP	1,629	**0,004**	B2PLYP	1,648	0,023
BLYP	1,655	0,030	mPW2PLYP	1,641	0,017
B98	1,627	**0,003**	MP2(FC)	1,662	0,037
B971	1,630	**0,005**	MP2(FULL)	1,644	0,020
experiment	1,625		BPW91	1,646	0,021

Table 3. Theoretical harmonic frequencies (wave numbers) of $^{74}Ge^{16}O$ (ω_{harm}), cm^{-1}.

Method	ω_{harm}	Метод	ω_{harm}	Method	ω_{harm}	Method	ω_{harm}
HFB	893	VSXC	978	B972	1015	BHandHLYP	1071
BLYP	937	XAlpha	983	B98	1017	BHandH	1098
HFS	954	SVWN	996	B3P86	1017	UMP2-FC	947
B2PLYP	954	B3LYP	999	M06	1018	MP2(FULL)	954
BPW91	956	B1LYP	1009	PBE1PBE	1025		
HCTH	962	B971	1012	MPW1PW91	1026		
mPW2PLYP	974	B3PW91	1013	B1B95	1028	ω_{harm} [15]	985,5

The calculated harmonic frequencies rarely coincide with the experiment, especially for motions having large amplitudes. As a rule, the deviations from the experiment larger than 15 cm^{-1} results in calculation errors in thermodynamical properties reaching a maximum in the high-temperature region. In this view a large part of ω_{harm} (Table 3) needs to be corrected by scaling. In terms of performed calculations it becomes possible to introduce scaling coefficients for each functional and further apply them for the determination of germanium oxide properties. Our next step in the study of oxygen and germanium interaction will be the determination of GeO thermodynamical functions in the temperature interval of 298-1500 K.

3. Conclusion

Analysis of the Tables 1 - 3 enables to recommend the use of the following functionals MB97-2, B972, B3LYP and LDA(SVWN), XAlpha. B3P86, B1B95, B1LYP, B98, B971, B3PW91,

IOP Conf. Series: Materials Science and Engineering **93** (2015) 012028 doi:10.1088/1757-899X/93/1/012028

MPW1PW91, PBE1PBE, M06 for the study of oxygen behavior in crystalline germanium.

Acknowledgments
The work was performed within the framework of the realization of the Federal Target Program «Investigations and elaboration in the priority directions of scientific–technological development of the complex of Russia for 2014 – 2020 years», agreement 14.577.21.0004.

References

[1] Clauws P 1996 *Materials science and Engineering* B **36** 213
[2] Podkopaev O I and Shimansky A F 2013 *Germanium single crystal growth with low density of dislocations and impuities* (Krasnoyarsk: Sib. Feder. Univ.) p 104
[3] Hehre W J, Radom L, Schleyer Pv R and Pople J A 1986 *Ab Initio Molecular Orbital Theory* (New York: John Willey & Sons)
[4] Frisch M J, Trucks G W, Schlegel H B et al 2007 *Gaussian 03 (Revision E 0.1 SMP)* (Pittsburgh, PA: Gaussian Inc.)
[5] Friesen M, Junker M and Schnoeckel H 2000 *J. Chem. Phys.* **112** 1782
[6] Igel-Mann G, Stoll H and Preuss H 1988 *Mol. Phys.* **65** 1329
[7] Cohen A, Mori-Sanchez P and Yang W 2007 *J. Chem. Phys.* **127** 034101
[8] Zumbusch A and Schnoeckel H 1998 *J. Chem. Phys.* **108** 8092
[9] Johnson J R T and Panas I 1999 *Chem. Phys.* **249** 273
[10] Teale A M and Tozer D J 2005 *J. Chem. Phys.* **122** 034101
[11] Ozaki K 2004 Condensed Matter and Materials Physics. *Physical Review* B **69** 195113
[12] Hartwigsen C and Goedecker S, Hutter J 1998 *Phys. Rev.* B **58** 3641
[13] Chakraborty G 2010 *Eur. Phys. J.* D **59** 183
[14] Appelblad O, Fredin S and Lagerqvist A 1983 *Physica Scripta*, **28** 160
[15] Thompson G. A., Maki A. G. and Weber A. 1986 *J. Mol. Spectr*, **116** 136.

IOP Conf. Series: Materials Science and Engineering **93** (2015) 012029 doi:10.1088/1757-899X/93/1/012029

Sorption ability of the composite sorbent for water treatment from radioactive elements

A V Gonets[1], M P Chubik[1], M V Chubik[1] and A N Tretyakov[1]
[1] Tomsk, 634050, Russia, National Research Tomsk Polytechnic University

E-mail: moskalenkoav@tpu.ru

Abstract. The goal of research is to develop the composite sorbent with application of various metal oxides nanoforms and nonpathogenic mold fungi mycelium modified by these nanoforms. This article describes the producing method of the composite sorbent and the research results of the sorbent sorption ability while the sorption process conditions are changed.

1. Introduction

Nuclear energy makes possible to avoid an energy crisis by using the energy released at certain atomic nuclei reactions. However, with the development of nuclear energy and its related industries, environmental pollution by radioactive ions has become a serious environmental issue, which carries a potential threat to human health as the long-lived radionuclides do not decompose. Instead, they tend to accumulate, can be easily transported over long distances and often included in the biological cycle of matter. Up to 80% of anthropogenic radioactive pollution arrives in the Earth aquatic environment, turning it into the most powerful depot of radionuclides.

The major problem in the technologies development for the radioactive substances disposal from the environment is the invention of materials that can irretrievably, effectively and abundantly absorb radioactive ions from water.

In the last years in many countries research on new class sorbent development is actively progressing. These sorbents should constitute of biological substances or include them as the main element – biosorbents. For example, they can be produced from microbe mass or fungi which are microbiological industry wastes. Apart from that, application of different metal oxides nanoforms as a sorbent is viewed as promising.

The prerequisites for metal oxides nanoforms using as component of composite sorbent are hypotheses on fact that - the metal nanoparticles can be used as matrices for immobilization of plutonium, technetium, uranium and transuranic elements due to the ability to absorb radioactive ions while the deformation process of the nanomaterial takes place [1]. As a result, the absorbed radionuclides are permanently enclosed in the sorbent structure [2].

The prerequisites for nonpathogenic mold fungi mycelium using as component of composite sorbent are hypotheses that state - the main characteristics of metal nanoparticles deposited on the growing mycelium of fungi are not differed from the nanoparticles properties suspended in the solution. At the same time there was no any preliminary modification of the particles or mold for the adsorption of nanoparticles on the surface of the growing mycelium. While deposition of the nanoparticels on the mold there was no any aggregation with each other [3].

IOP Conf. Series: Materials Science and Engineering **93** (2015) 012029 doi:10.1088/1757-899X/93/1/012029

2. Materials and methods

For the research model uranyl nitrate solutions with various initial concentration of $(UO_2)^{2+}$ were used. As the nanosized components of composite sorbent the titanium dioxide nanotubes, obtained by low-temperature sintering of electroexplosive nanopowders and the nanopowders of cupric oxides, obtained by electrical explosion of cupric wire in air were used. The titanium dioxide nanotubes were visualized using scanning electronic microscope Hitachi S-3400N with the additional attachment for microanalysis which was used for fluorescent-spectral analysis of nanomaterials. The length of the nanotubes was varied from 300 nm to 600 nm, the diameter was about 8 nm and the thickness of the walls about 2.5 nm.

As the biological component for sorbent development the mycelium of non-pathogenic mold fungi *Aspergillus niger* was used due to its natural occurrence and because this filamentous fungi exhibits a high tolerance to environmental conditions [4].

2.1. Cultivation of microorganisms

For the cultivation the inoculating of microorganismsnutrient medium "Nutrient broth dry" (the volume is 100 ml, pH 5) was produced by Pasteur loop aseptically. Cultivation of microorganisms was carried out in 250 ml round-bottomed flask at 37°C in a thermostat until the surface protection film fungi certain size (2-3 mm) will appear.

After that the flasks were placed on a shaker and were incubated under constant stirring with heating (350 rev/min) for 7 days.The growth of microorganisms was recorded visually (appearance of the film, blurred the medium, sediment appearance, pH change). Upon completion of cultivation mycelium of nonpathogenic mold fungi was washed three times by distilled water.

2.2. Deposition of nanomaterials on the mycelium of nonpathogenic mold fungi

Firstly, 100 ml of distilled water and samples of nanomaterials (TiO_2, CuO) were added to separate 250 ml flask. It is known that nanoscale materials form poorly stable suspensions [5]. High ability of nanoparticles to agglomerate in an aqueous medium does not allow reaching the maximum surface and, consequently, the maximum sorption ability of the material. The ultrasonic dispergation was used for agglomerates to breaking it up.

Secondly, into each flask 1 g (wet weight) of washed mycelium was added and then the flasks were closed with gauze-cotton plugs and were stirred in the shaker (350 rev/min) for 24 hours.

2.3. The uranyl ions sorption process

The research of the uranyl ions sorption process was carried out under static conditions at the indoor temperature. In all experiments the volume of the solution was 5 ml and the weight of composite sorbent sample was 1 g. The time of contact of the sorbent with the solution was from 1 to 14 days. Upon completion of the sorption solution was separated from the solids by centrifugation or filtration. The series of experiments were carried out to study the changes in the sorption properties of the composite sorbent depending on the variation of the sample mass of nanomaterial, the pH of the solution, as well as the time and temperature conditions of the sorption process.

Measurements of solution were performed at 130-ionomer and glass electrodes ESL-63-07. The pH measurement error was 0.05. Containing of uranyl ions in solution was estimated by luminescent method on spectrofluorimeter «Fluorat-02 Panorama» using standard routine.

Sorption capacity was evaluated by the degree of adsorption:

$$S = \frac{Cin - Cfin}{Cin} \cdot 100\% \qquad (1)$$

where C_{in} *and* C_{fin} - initial and final uranium concentration, mg/L.

3. Results and discussion

3.1. Change in the pH of the solution

The experiment was conducted at the initial uranium concentration 1.212 mg/L. It was noted that the optimal pH value is from 4 to 5. It can be explained by the fact that in the acidic environment (pH = 2) uranyl ions compete with protons for binding sites on the surface of the sorbent. When pH=7, uranyl ions can contact not only with surface of the sorbent, but with hydroxide ions (Figure 1).

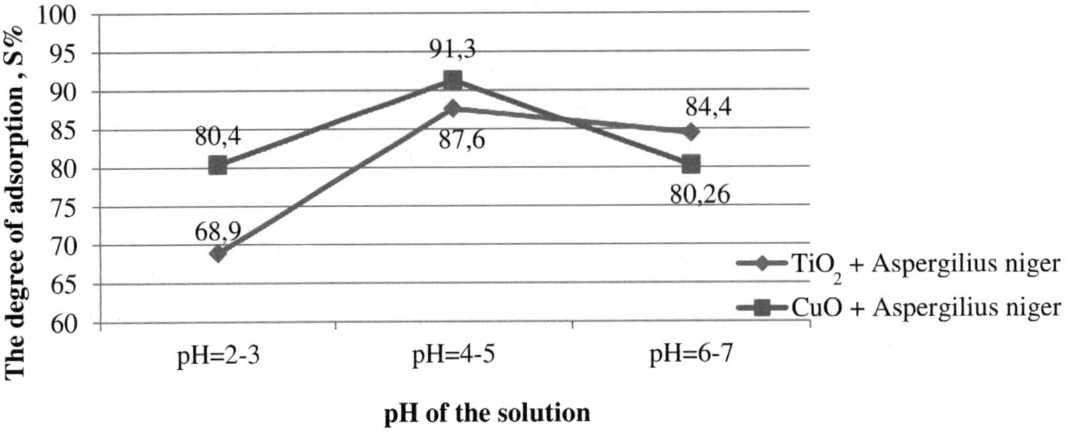

Figure 1. Dependence the degree of adsorption of uranyl ionson pH of the solution.

3.2. Change in the time of the sorption process

The experiment was conducted at the initial uranium concentration 1.17 mg/L. It was established that the maximum value of the degree of adsorption reached after one day and it is practically unchanged for the next two weeks (Figure 2).

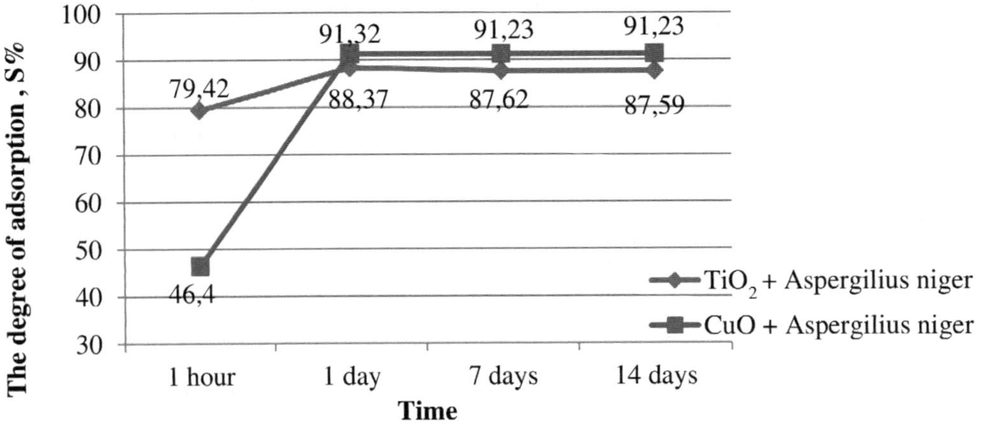

Figure 2. Dependence the degree of adsorption of uranyl ions on time.

3.3. Change in the temperature conditions

The experiment was conducted at the initial uranium concentration 1.17 mg/L. It was noted that the temperature difference has a little influence on the degree of adsorption value (within 10%). It can be explained by the fact that mycelium wall is highly dense, that does not change its structure depending on the temperature, and the degree of adsorption value is practically unchanged (Figure 3).

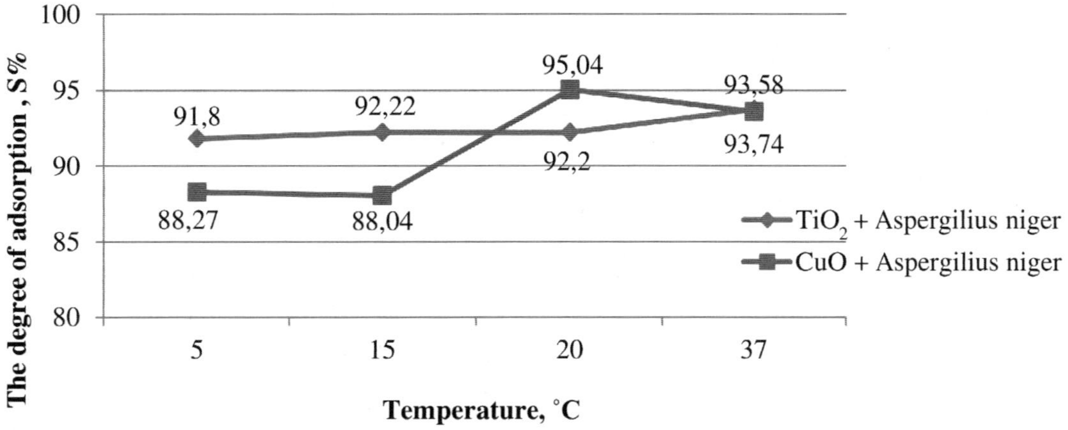

Figure 3. Dependence the degree of adsorption of uranyl ions on temperature conditions.

3.4. Change in the nanomaterial sample mass

The experiment was conducted at the initial uranium concentration 0.4 mg/L. In the study of the influence of nanomaterial sample mass on the absorption of uranyl ions it was founded that the optimal ratio nanomaterial:mycelium was the ratio of 1:100 (Figure 4), as thus binding sites of uranium ions on the surface of the mycelium is available that increases the efficiency of the sorbent.

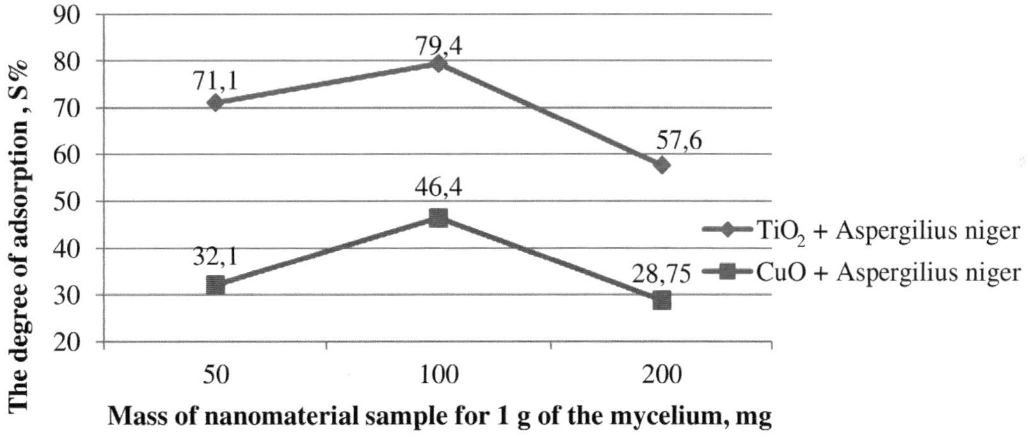

Figure 4. Dependence the degree of adsorption of uranyl ions mass on nanomaterial sample.

These results may indicate that at ratio nanomaterial:mycelium 1:200, the mycelium adsorbs uranium predominantly and the degree of adsorption is 57.6% for titanium dioxide nanotubes and is 28.75% nanopowders of cupric oxides. Raising of sorption with increasing sample mass of nanomaterial (ratio 1:100) leads to increasing the degree of adsorption to maximum value.

With further increasing in shot of nanomaterial (up to 1:50) the sorption degree is reduced to 71.1% for TiO_2 and to 32.1% for CuO. In this case, it can be assumed that the sorption takes place predominantly through the nanomaterial rather than mycelium as they mostly cover the sorption surface of mycelium and thereby reduce its contribution to the sorption process.

Due to visual inspection using a light microscope (40X increase) it was noted that the deposition of nanoparticles on the mycelium surface does not lead to deterioration of the mold growth or a change in its cultural properties. It proves that the formation of the hybrid matrix is not attended by the nanoparticles adhesion. The resulting composite material is a tubular mold mycelium, wrapped in several layers of nanoparticles. Figure 5 shows images of *Aspergillus niger:* before (a) and after (b) cultivation with titanium dioxide nanotubes.

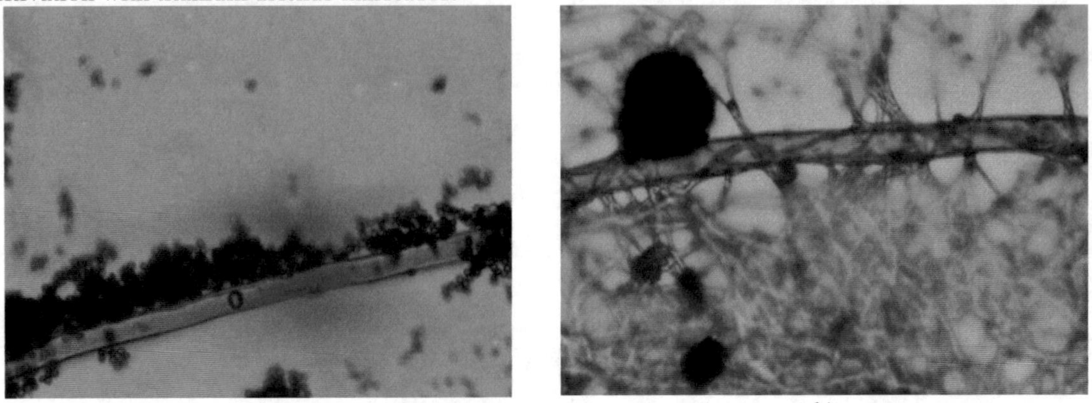

a) b)

Figure 5 Images of *Aspergillus niger*: before (a) and after (b) cultivation with titanium dioxide nanotubes.

Due to separation of the metal nanoparticles, fungal matrixes preserve all the properties of the initial nanoparticles. The absence of aggregation contributes to the significant increasing of the modified mycelium activity in comparison with the nanoparticles suspended in the solution. Also an important advantage of mold matrices using is the fact that the composite sorbent is separated easily from the reaction mixture.

4. Conclusion

Due to application of non-pathogenic fungi as the active matrix for deposition of metal oxides nanoforms on the mycelium of nonpathogenic mold fungi it is assumed to be safe and simplified process for operation and utilization of composite sorbent in comparison with using separate nanosorbents. The optimal conditions to provide high sorption ability of the composite nanobiosorbent for water treatment from radioactive elements were founded.

Acknowledgments

The presented work and the current paper are funded within the framework of realization of Strategic Programme on National Research Tomsk Polytechnic University Competitiveness Enhancement in the Group of Top Level World Class Research and Academic Institutions.

References

[1] Krivovichev S V, Kahlenberg V, Kaindl R, Mersdorf E, Tananaev I G and Myasoedov B F 2005 Nanoscale tubules in uranyl selenates *Angew. Chem. Int. Ed.* **44(7)** 1134-6.

[2] Dong J Y, Zhan F Z and Huai Y Z 2008 Titanate Nanofibers as Intelligent Absorbents for the Removal of Radioactive Ions from Water *Adv. Mater.* **20(14)** 2777-81.

[3] Bigall N C, Reitzig M, Naumann W, Simon P, van Pée K H and Eychmüller A 2008 Fungal Templates for Noble-Metal Nanoparticles and Their Application in Catalysis *Angew. Chem. Int. Ed.* **47(41)** 7876-9.

[4] Schuster E, Dunn-Coleman N, Frisvad J and van Dijck P 2002 On the safety of Aspergillusniger *Appl. Microbiol. Biotechnol.* **59** 426-35.

[5] Suetin M V, Fedorov A Yu, Vahrushev A V and Vahrushev A A RF Patent No. 2301771 Byull. Izobret., No. 18 (2007).

IOP Conf. Series: Materials Science and Engineering **93** (2015) 012030 doi:10.1088/1757-899X/93/1/012030

Ceramics based on titanium nitride and silicon nitride sintered by SPS-method

A A Sivkov[1], D Yu Gerasimov[1] and A A Evdokimov[1]
[1]Russia, Tomsk, Lenin ave, 30, National Research Tomsk Polytechnic University

e-mail: kraamis@gmail.com

Abstract. The dependences of the microstructure and physical and mechanical properties of ceramic mixtures Si_3N_4/TiN in the full range of mass ratios of the components. Was also investigated directly, and the process of sintering occurring during a physical or chemical processes, in particular, has been obtained and the hardness of the material density on the ratio of the conductive titanium nitride phase and a silicon nitride insulating phase with values above and below the percolation threshold. Also obtained was pure ceramics based on titanium nitride with high physical-mechanical characteristics (H = 21.5 GPa).

1. Introduction

Today, there is a high demand for metal-cutting tools, created from light ceramic materials with relatively high hardness [1]. These materials are in demand in various industries. Oxygen-free ceramic based on silicon nitride is quite popular due to its high chemical resistance, high temperature resistance and dielectric properties [2, 3]. However, this ceramics has a number of disadvantages, in particular, silicon nitride is an insulator and therefore it is unacceptable for spark plasma sintering (SPS). In fact, for pure silicon nitride, this process is similar to the hot isostatic pressing. This problem can be solved by various ways, in particular, by doping the silicon nitride with small additions of oxides (aluminum oxide Al_2O_3 and yttrium oxide Y_2O_3) in an amount of 10-12% [4-8]. However, the disadvantage of this method is the clogging of the ceramics sample with silica and aluminum. This can lead to decreasing in physical and mechanical properties of the sample due to the appearance of local dense centers.

TiN is often used as the doping agent in the preparation of metal-cutting samples [4-6]. TiN is selected because of its high electrical conductivity, which allows increasing the sinterability of Si_3N_4, without a strong fall in strength characteristics [1, 7, 8]. Also, TiN has a high melting point, good hardness, chemical, tribological and corrosion resistance.

SPS is a relatively new method for consolidating nanocomposites in high-speed process, and this method is highly effective for obtaining nanocomposites in a short time. In previously published literature it was observed that nanostructured ceramics with high physical-mechanical characteristics based on Si_3N_4/TiN can be obtained after a short preparation of raw powder with following sintering [9, 10].

The aim of this work is to consider the use of titanium nitride simultaneously in two roles: as a doping factor at low levels, and as a component of the composition at a concentration above the percolation threshold. Spark plasma sintering method is chosen because of its advantages over the two-step methods. In particular, using this technology large recrystallizing ceramics grains can be avoided both with maintaining nanostructures.

2. Experimental

Pure silicon nitride nanopowders [10-25 nm, SSA=60m^2/g, $\alpha/(\alpha+\beta)$=50%, X-ray amorphous phase] and titanium nitride produced by using the system based on coaxial magneto plasma accelerator [11-17] [TiN - 90%, αTi - 10%] were used for sintering.

Five types of powder blanks were prepared and their types are indicated in Table 1.

Table 1. The content of titanium nitride and silicon in experiments.

Experiment	1	2	3	4	5
TiN / Si_3N_4, %	100 / 0	75 / 25	50 / 50	25 / 75	0 / 100

The required weight of the powder was weighed using high-precision electronic scales, to the nearest hundredth of a gram. Further the mixed powder compositions were subjected to thorough mixing, activating, and deagglomerating in planetary mill for one hour. The process was carried out with using steel grinding balls.

Significant increase in bulk density from 0.06 g/cm^3 to 0.48 g/cm^3 for pure Si_3N_4, and from 0.18 g/cm^3 to 0, 6 g/cm^3 for pure TiN is attributed to important results of such preparation.

Sintering was performed by SPS bot in helium at 1 atmosphere pressure and under vacuum at 1300°C. Other sintering conditions were the same and as follows: the heat rate V_T = 100°C/min, the pressing force P=60 MPa (~ 20 kN) and the exposure absence with a constant temperature. The material was placed into a graphite mold with an inside diameter of 20 mm.

After sintering the samples were cleaned, ground and polished using special machine. The sample hardness was studied by microhardness measuring using a Vickers indenter at a load of 1 kg for 10 seconds. Determination of hardness was performed by the method of unrestored print. The microstructure was investigated by scanning electron microscopy on a polished surface of the untreated and cleaved sample. Phase composition was determined by X-ray diffraction.

3. Results and discussion

Table 2 shows the results of XRD analysis for powder samples and ceramics materials, obtained in a series of experiments, conducted at the sintering temperature 1300°C. This range has been chosen as a basis, because of the critical temperature for titanium nitride is 1450°C and at this temperature TiN begins to react with the carbon from which the mold is made.

The divergence in XRD analysis data with the original data on the composition (Table 1) can be explained by the following factors: firstly, the grinding time was enough for partial mechanosynthesis of titanium nitride; secondly, the presence of amorphous phase in silicon nitride shifts the equilibrium phases in the direction of titanium nitride. The presence titanium suboxides (sample 1), appearing after sintering, can be explained by oxidation of free titanium either during reacting with oxygen in the chamber or after removal of the sample and cooling it in the open air. Also in experiments 3 and 4, free silicon appeared. This fact can be explained by the growth of silicon grains due to pressure load.

Also in experiments 2, 3 and 4 there is a presence of such titanium silicide as $TiSi_2$ and Ti_5Si_3. The high concentration of silicide in the experiment 2 is explained by both the high content of titanium metal in the sample and the presence of free silicon. When heating the powder sample by pulse current during the SPS the differences in current between the sample and the mold are determined by the conductivity of the powder sample. Due to the small process time (up to 20 minutes) the system does not have time to warm up evenly and there is heterogeneity in the distribution of temperature. It leads to the fact that the formation of titanium silicide is directly determined by the conductivity of the sample. The transition of the silicon nitride from α-phase to β-phase does not occur due to insufficient temperature. There was also no transition from β-phase and γ-phase due to low pressure molding.

Figure 1 shows the curves of the SPS- process, such as temperature, pressure, current and sealing. The temperature curve is removed by the thermometer from the outside of the graphite mold. The constant temperature level at the beginning and end of the process is due to the physical limitations of the pyrometer, the temperature begins to register from the value 467°C. It should also be noted that the temperature inside the sample is above then the outside one, the difference can be up to 100-120°C [18].

Table 2. Results of X-ray diffraction analysis of the powder materials and ceramic samples.

Phase	%	CSR, нм	Phase	%	CSR, нм
	*** powder 1 ***			*** ceramic 1 ***	
cTiN	90	--	cTiN	90	--
αTi	10	15	αTi	5	59.54
----	----	----	Ti_xO_y	5	--
	*** powder 2 ***			*** ceramic 2 ***	
cTiN	73	--	cTiN	69	--
αTi	7	17	$TiSi_2$ / Ti_5Si_3	7 / 17	27 / 63
α/β Si_3N_4	3 / 17	41 / 12	α/β Si_3N_4	3 / 3	23 / 24
	*** powder 3 ***			*** ceramic 3 ***	
cTiN	71	--	cTiN	72	46
αTi	7	12	$TiSi_2$ / Ti_5Si_3	3 / 1	12 / 22
α/β Si_3N_4	6 / 16	62 / 56	α/β Si_3N_4	10 / 10	35 / 48
----	----	----	aSi	3	49
	*** powder 4 ***			*** ceramic 4 ***	
cTiN	38	--	cTiN	47	49
αTi	1	111	$TiSi_2$ / Ti_5Si_3	2 / 2	13 / 34
α/β Si_3N_4	25 / 36	92 / 51	α/β Si_3N_4	26 / 20	43 / 23
----	-----	----	aSi	3	70
	*** powder 5 ***			*** ceramic 5 ***	
α/β Si_3N_4	55 / 45	47 / 53	α/β Si_3N_4	50 / 50	38 / 52

The main curves in the figure are the sealing curves. In every case, the sealing is substantially different. For a pure titanium nitride pre sealing on the step of pressure growth (t_0-t_1) does not occur because the titanium nitride powder has a fairly rigid structure (as shown below) and there is no possibility to further increase in the bulk density before sintering. Further, at step t_1-t_2 when the powder is impacted only with pressure over an extended period of time ~250-270 seconds, the partial seal occurs that is similar to the process of cold isostatic pressing. At this stage, the heating does not occur, because of the sample temperature is below the soft threshold pyrometer (467°C) and the current switching point (600°C). The third stage (t_2-t_3) is characterized by uncontrolled temperature growth due to increasing heating current. During130-150 seconds the temperature rises up to 600°C. After that the speed is normalized and reduced to a predetermined program values. At this point, it is possible to see a small decompression due to thermal expansion exceeding hydraulic compression. Further, at t_3-t_4 there is a controlled heating of the sample at a constant pressing force. Sealing curve knee can be explained due to the fact that the processes of liquid-phase sintering on the grain boundaries begin to prevail over the processes of thermal expansion. It leads to increasing density of the sample at this stage. At the time t_4 the heating current is switched off and the sample begins to cool down both by radiation and by convection processes in the atmosphere of helium. At the time t_5 the compaction pressure reduction begins. There is no matter substantially sealing at this stage because of the lack of sealing factors.

As for the second sample, consisting of equal mass fractions of titanium nitride and silicon nitride, it should be noted that the above processes description is applicable in this case, but with some additives. The partial compaction during stages t_1-t_2 and t_2-t_3 is observed to a much lesser degree than for a pure titanium nitride. This is because of the mixture conductivity is far less than the conductivity of pure titanium nitride, and liquid-phase processes occur at the grain boundaries are much less active.

IOP Conf. Series: Materials Science and Engineering **93** (2015) 012030 doi:10.1088/1757-899X/93/1/012030

However, due to decreasing Joule heating, the nature of seal and decompaction in the area around t_3-t_4 is similar to the sample made of pure titanium nitride with a slightly lower seal. Also in the sample there is the presence of free silicon particles, which rose above the limit of X-rays, as mentioned above. It also fits in a sealing scheme, which depends on the composition of the sample.

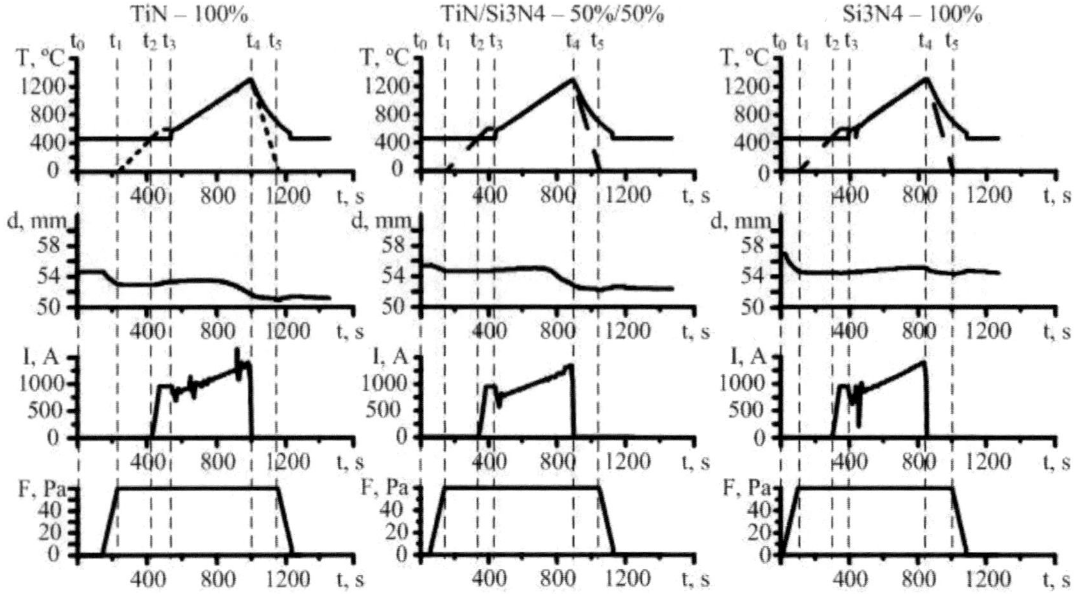

Figure 1. The process parameters for the spark plasma sintering (experiments 1, 3, 5).

For the third sample, consisting of pure silicon nitride, the spark plasma sintering process differs significantly. This is due to the fact that the sample is below the percolation threshold [4], and its conductivity is lower by orders of magnitude than that of the previous samples. In step t_0-t_1 there is a sharp seal due to applying large force to the sample. The titanium nitride matrix is absent in the sample, and the sample has almost no resistance to the primary compression. The sample relatively quick condenses almost to the end value. During several steps from t_1 to t_4 there is some decompaction that can be explained by the fact that there is no liquid phase process. Thus the Joule heating process dominates. The seal during the cooling step is also explained by thermal contraction.

Table 3 and Figure 2 show the mechanical testing data for the samples sintered at 1300°C. The figure 2 also clearly shows two abrupt changes of hardness. These data can be explained by a significant difference between the conductivity of titanium nitride (conductor) and silicon nitride (dielectric). The low hardness at low TiN contents (experiments 4 and 5) can be explained by the fact that the content of the conductive fraction is below the percolation threshold. When the threshold is exceeded, equal to 30%for a given pair of substances [4], both the conductivity and the sinterability sharply increase. The second hardness surge occurs when there is no dielectric fraction (exp. 1).

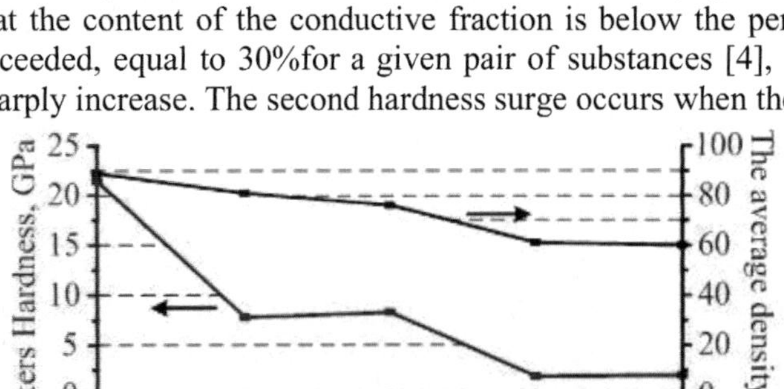

Figure 2. Curves of density and hardness for the ceramic samples, in dependence on the phase ratio of titanium and silicon nitride, sintered by spark plasma sintering in the helium environment at a temperature of 1300°C, with heating speed rate of 100°C/min without time delay, and a compression pressure 60 GPa.

Similar abrupt changes are also clearly visible in the density curve, which can also be explained by a sharp change in conductivity samples.

Figure 3 shows SEM images of the samples sintered at a temperature 1300°C with different ratios of titanium and silicon nitride. Series a (a1-a3) shows feedstock surface and chipping of the ceramics sintered from pure titanium nitride. Series b (b1-b3) is for mix of titanium nitride and silicon nitride in equal amounts. Series c) (c1-c3) is for pure silicon nitride.

Table 3. Data of the density and hardness for the ceramic samples, in dependence on the phase ratio of titanium and silicon nitride, sintered by spark plasma sintering in the helium environment at a temperature of 1300°C, with heating speed rate of 100°C/min without time delay, and a compression pressure 60 GPa.

Experiment	Structure		Ceramic	
	TiN, %	Si_3N_4, %	Density, g/cm^3 (%)	Hardness, GPa
1	100	0	4.84 (89%)	21.5
2	75	25	3.82 (81%)	7.8
3	50	50	3.21 (76%)	8.3
4	25	75	2.30 (61%)	1.9
5	0	100	2.05 (60%)	2.0

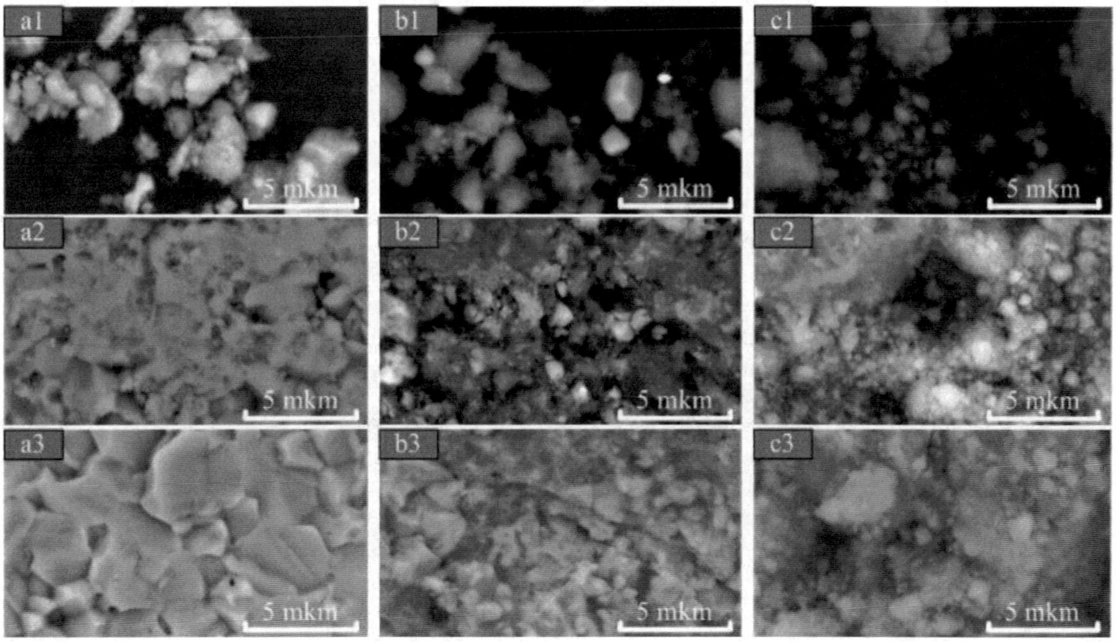

Figure 3. SEM-microscopy data for ceramics sintered by spark plasma sintering at 1300°C. Column A) TiN - 100% (Experiment 1, Table 1), b) TiN/Si_3N_4 - 50%/50% (Experiment 3, Table 1), c) Si_3N_4 - 100% (Experiment 5, Table 1). Rows: 1) ultrafine feedstock after machining, 2) SEM-picture of polished sample surface, 3) SEM-picture of raw chipping of the sample.

In line 1 the powdered materials, treated in a planetary mill directly before sintering, are shown. Titanium nitride particles have strict faceted structure and a darker color. It is seen that the range of particle size distribution is quite wide, from a few microns to less than 100 nm nanoparticles. Silicon nitride has a cotton-like structure due to its high dispersion and availability of silicon in the form of fine amorphous phase. Due to the high dispersion, silicon nitride has a high tendency to agglomerate, gathering in agglomerates with a size of several microns. In consequence of the fact that after activating on the planetary mill, the bulk density dramatically increases, it leads to an increasing the individual agglomerates and expanding the particle size distribution.

In line 2 SEM-images of the polished surface of the sintered samples are shown. Here it is possible to see the grain structure with a clear division into components. Moreover, according to data of sintering for all samples, there is no significant particle growth of both titanium nitride and silicon nitride due to insufficiency of this temperature for the grain growth of both titanium nitride and silicon nitride [4]. It can be seen in more details on the raw chipping of the samples (line 3). Also the wide range of size distribution is more a plus than a minus, because the smaller particles fill the pores between large ones, thereby achieving high packing density of particles.

Conclusion

In this work, the study is carried out to obtain ceramics based on titanium nitride and silicon nitride by spark plasma sintering. There is information about the processes occurring during sintering, in particular, about the formation of by-products from the impurities in the starting components. Also there is information about the dependence of the density and hardness on the ratio of the components and their electrical conductivity. Also the pure ceramics based on titanium nitride with high physical-mechanical characteristics (H = 21.5 GPa) is obtained.

Acknowledgment

This work was supported by the Russian Science Foundation grant 15-19-00049

References

[1] Lee C, Lu H, Wang C, Nayak P and Huang J 2010 *Journal of alloys and compounds* **508** 2
[2] Gogotsi Y 1994 *Journal of Materials Science* **29** 2541
[3] Huang J, Chen S and Lee M 1994 *Journal of Materials Research* **9** 2349
[4] Guo Z, Blugan G, Kirchner R, Reece M, Graule T and Kuebler J 2007 *Ceramics International* **33** 1223
[5] Chunyan T, Hai J and Ning L 2011 *International Journal of Refractory Metals and Hard Materials* **29** 14
[6] Liu C and Huang J 2003 *Ceramics International* **29** 679
[7] Yoshimura M, Komura O and Yamakawa A 2001 *Scripta Materialia* **44** 1517
[8] Gao L, Li J, Kusunose T and Niihara K 2004 *Journal of the European Ceramic Society* **24** 381
[9] Tokita M 2004 *Journal of Material Science* **5** 78.
[10] Ayas E, Kara A and Kara F 2008 *Journal of the Ceramic Society of Japan* **116** 812
[11] Evdokimov A, Sivkov A, Gerasimov D, Saigash A and Khasanov A. 2013 *Russian Physics Journal* **55** 983
[12] Sivkov A, Gerasimov D and Evdokimov A 2014 *Instruments and Experimental Techniques* **57** 222
[13] Sivkov A, Gerasimov D, Saigash and Evdokimov A 2012 *Russian Physics Journal* **54** 1160
[14] Sivkov A, Gerasimov D, Saigash A and Evdokimov A 2012 *Russian Electrical Engineering* **83** 39
[15] Pak A, Sivkov A, Shanenkov I, Rahmatullin I and Shatrova K 2015 *Int. J. Refract. Met. Hard Mater.* **48** 51
[16] Shanenkov I I, Pak A Ya, Sivkov A A and Shanenkova Yu L 2014 *MATEC Web of Conferences* **19** 01030
[17] Sivkov A, Saygash A, Kolganova J and Shanenkov I 2014 *IOP Conf. Series Mater. Sci. Eng.* **66** 012048
[18] Wan J, Duan R and Mukherjee A. 2005 *Scripta Materialia* **53** 663.

IOP Conf. Series: Materials Science and Engineering **93** (2015) 012031 doi:10.1088/1757-899X/93/1/012031

Design features of matrix nanoscale pointed graphene/SiC field emission cathodes

E Volkov[1], I Jityaev[2] and A Kolomiitsev[2]
[1] Southern Laser Innovation Technological Center, Taganrog, 347923 Russia
[2] Institute of Nanotechnologies, Electronics, and Electronic Equipment Engineering, Southern Federal University, Taganrog, 347922, Russia

E-mail: jityaev.igor@gmail.com

Abstract. The article describes the design features of the matrix nanoscale pointed field emission cathodes based on silicon carbide with a graphene film. The influence of the distance between electrodes, the distance between cathodes, the cathode height at the fixed values of the rounding-off radius of the top, the potential difference and the half-angle of the opening of cone forming the cathode is researched. The effective parameters of the matrix pointed field emission structure with nanoscale cathodes that provides homogeneity of the electrical field are determined.

1. Introduction

The interest to graphene electronics is growing every year [1-3]. Exceptional geometrical, electrical, mechanical and thermal properties of graphene expand the scope of its application [4-7]. One of the research areas of graphene is the field emission micro- and nanoelectronics [8, 9]. It is expected that the use of field emission devices will advance terahertz electronics.

The disadvantage of field emission cathode is the instability of field emission current due to the destruction of the emissive surface in strong electric fields. Ion bombardment, adsorption and desorption of residual gas molecules, local overheating can be reasons of destruction [10-12]. Field emission cathodes based on carbon materials are more resistant to destructive factors [13]. Silicon carbide has high thermal conductivity and is stable in a wide range of temperatures and radiation [14]. The method of thermal decomposition of silicon carbide in a vacuum permits to obtain graphene on the entire surface of the wafer. In this case, graphene is a natural continuation of the silicon carbide substrate [15]. As a result, the use of field emission cathodes based on silicon carbide with a graphene film will make it possible to create devices for operation in extreme, inaccessible to the solid-state electronics conditions.

Optimization of field emission cathodes structure is also important in engineering development along with a choice of material. There are pointed, edge, film and other constructions of field emission cathodes. Pointed cathodes are the most promising because they have the greatest curvature of the surface, which permits to obtain the necessary high values of the electric field strength for the start of the emission. The decrease of geometric dimensions of field emission cell to nanometer size contributes to lower operating voltage up to 2-3 orders of magnitude compared to modern characteristics of field emission devices [8, 16, 17].

The matrix of field emission cathodes is applied to increase the values of the field emission current. The value of field emission current is determined by the number of single cathodes in the matrix.

There is a limit on the maximum number of individual field emission cathodes per unit area related to reduction of electric field strength under the screening effect. The gain factor tends to 1 at extremely short distances between the cathodes in the matrix. Increasing the distance between the cathodes contributes to the rise of gain factor up to a value corresponding to the single cathode. Currently, the influence of screening effect in the matrix with nanoscale pointed field emission cathodes has not been studied sufficiently. The purpose of the work is the research of design and identification of the effective geometrical parameters of the matrix pointed nanoscale field emission cathodes based on silicon carbide with a graphene film on the surface.

2. Simulation

Simulation of the electric field strength distribution in the interelectrode space at different values of the cathode height (h), the distance between cathode (L) and the interelectrode distance (d) by the finite elements method with adaptive meshing algorithm was carried out to determine the effective parameters of matrix nanoscale pointed field emission cathodes based on silicon carbide with a film of graphene [18]. The choice of the design features of individual field emission cathodes was based on the theoretical and experimental studies [8, 19]. The general view of a matrix nanoscale pointed field emission cathode is shown in Figure 1, where K – the cathode, A – the anode, α - the half-angle of the opening of cone forming the cathode.

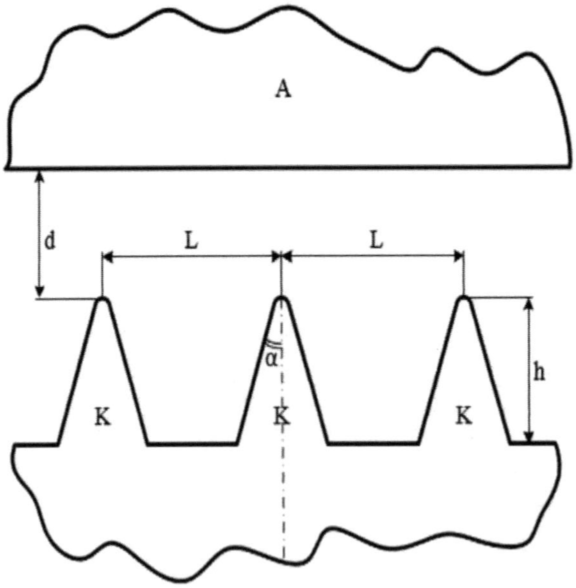

Figure 1. General view of a matrix pointed field emission cathode.

Table 1. Simulation parameters.

Parameter	Symbol	Value
Rounding-off radius of the field emission cathode top	r	10 nm
Half-angle of the opening of cone forming the cathode	α	10°
Potential difference	U	4 V
Interelectrode distance	d	10 - 50 nm
Height of the field emission cathode	h	50 nm - 3 μm
Distance between the field emission cathode top	L	50 - 500 nm

The values of the parameters which were used for the simulation of the electric field distribution in interelectrode gap are shown in Table 1. Wide range of the interelectrode distance, the distance between the field emission cathodes top and cathode height permitted to determine the effective parameters of the matrix field emission cell.

The main results of the simulation in the form of the electric field distribution in the interelectrode space with equipotential lines are presented in Figure 2.

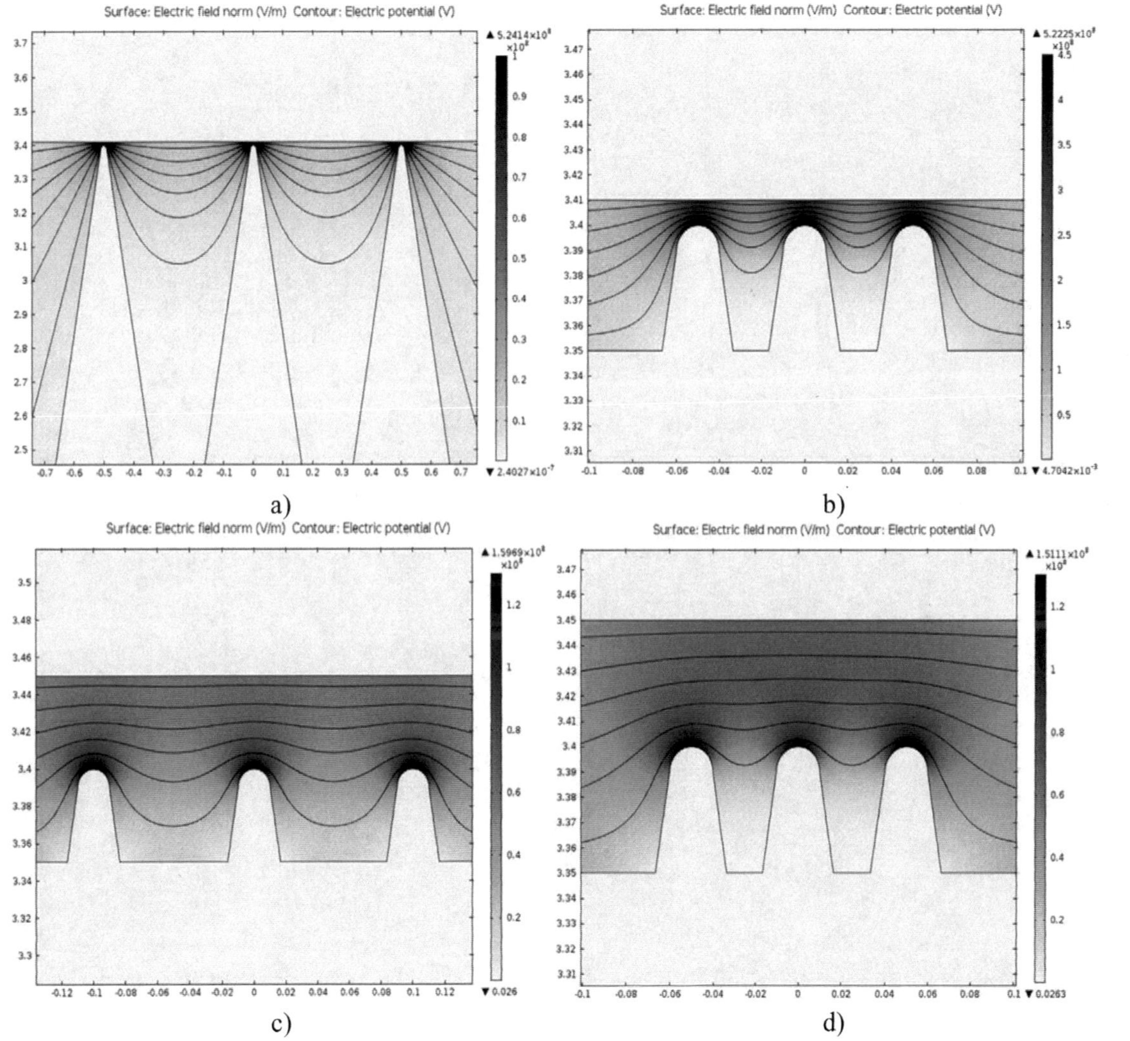

Figure 2. The simulation results of the matrix pointed field emission cathode at the parameters: a) h = 3 μm, r = 10 nm, d = 10 nm, L = 500 nm; b) h = 50 nm, r = 10 nm, d = 10 nm, L = 50 nm; c) h = 50 nm, r = 10 nm, d = 50 nm, L = 100 nm; d) h = 50 nm, r = 10 nm, d = 50 nm, L = 50 nm.

3. Results and Discussion

The dependences of the electric field strength on the x coordinate (along the cathodes) at fixed values of the potential difference, rounding-off radius of the field emission cathode top, the half-angle of the opening of cone forming the cathode and variables the parameters of the interelectrode distance, field emission cathode height and the distance between the cathodes tops were determined with the simulation results. The plots of the distribution of electric field strength along the matrix field emission cathodes permitted to visually estimate the impact of screening effect or its absence (Figure 3). Figures 2 and 3 shows only the results that have practical importance and permit to draw conclusions about the impact of the specific parameters of the matrix field-emission cathodes on the electric field distribution. The obtained dependences showed a reduction of the electric field strength in the central cathodes due to the screening effect at high density of cathodes per unit area. This effect is becoming stronger with increasing the distance between electrodes. It was found that the emission from the substrate begins with small values of the field emission cathode height.

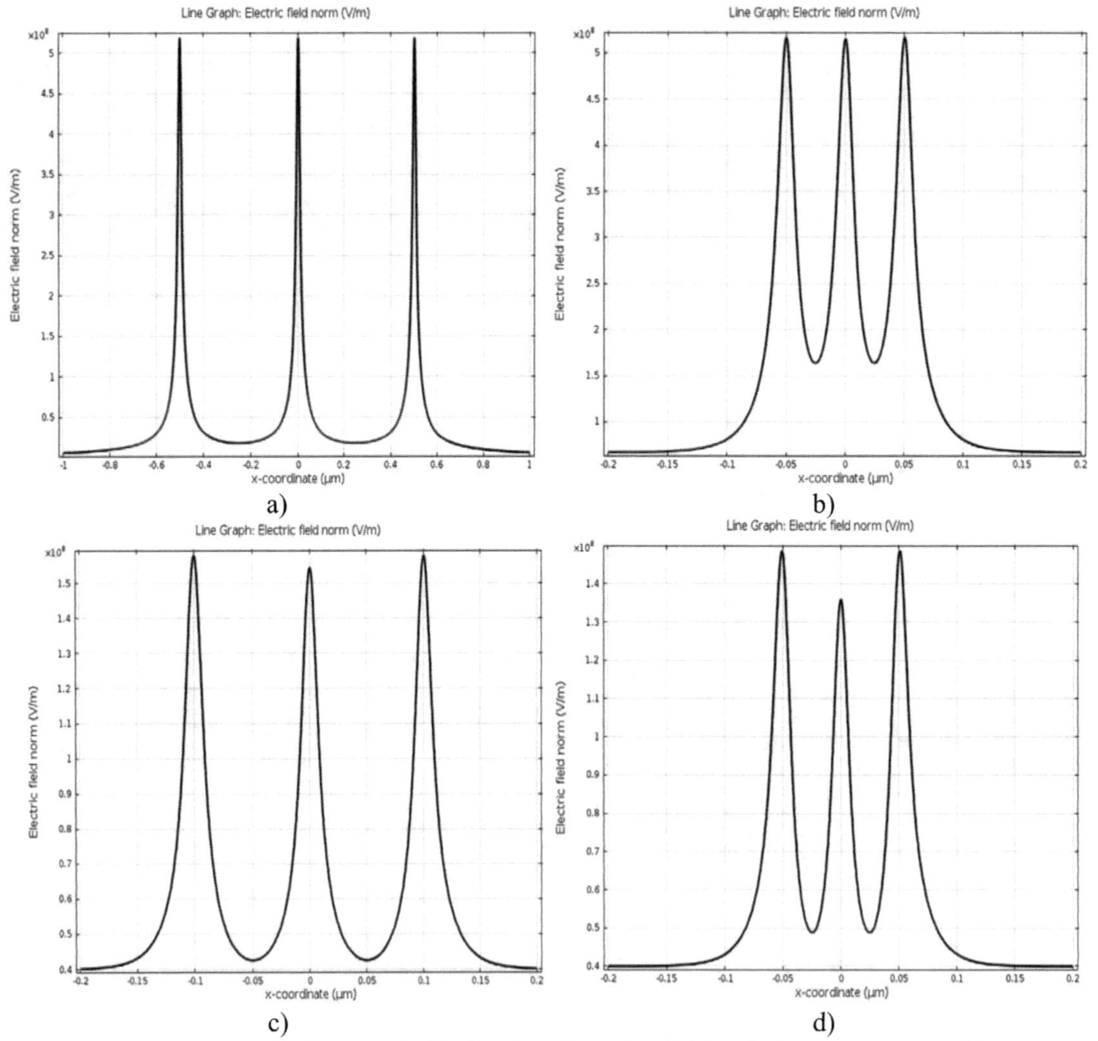

Figure 3. Electric field strength distribution along the field emission cathodes of different configurations:
a) h = 3 µm, r = 10 nm, d = 10 nm, L = 500 nm; b) h = 50 nm, r = 10 nm,d = 10 nm, L = 50 nm;
c) h = 50 nm, r = 10 nm, d = 50 nm, L = 100 nm; d) h = 50 nm, r = 10 nm,d = 50 nm, L = 50 nm.

4. Conclusions

The crucial factor of the number of field emission cathodes per unit area is the distance between their peaks. The simulation of the electric field distribution in the electrode gap of matrix field emission cell with nanosized pointed cathodes was carried out in this work. The simulation results shows that the screening effect in matrix nanoscale pointed field emission cathodes is low when the interelectrode distance is much less than the distance between the cathodes and their height. The appearance of the screening effect is observed when the interelectrode distance, distance between the cathode and their heights are approximately equal. There is no screening effect on conditions that: $d \leq 0{,}5L$ and $d \leq 0{,}5h$.

Acknowledgments
This work was supported by The Ministry of Education and Science of Russian Federation, the State Task in the Sphere of Scientific Activities (project no.16.1154.2014/K). The equipment of the Sharing Center and REC "Nanotechnologies" of Southern Federal University was used for this study.

References
[1] Novoselov K S, Geim A K, Morozov S V, Jiang D, Katsnelson M I, Grigorieva I V, Dubonos S V and Firsov A A 2005 *Nature* **438** 197
[2] Arefinia Z and Asgari A 2015 *Solar Energy Materials and Solar Cells* **137** 146
[3] Yue W, Jiang Sh, Xu Sh, Ma Y and Ba Ch 2015 *Sensors and Actuators B: Chemical* **214** 204
[4] Lee C, Wei X, Kysar J W and Hone J 2008 *Science* **321** 385
[5] Berger C, Song Z, Li X, Wu X, Brown N, Naud C, Mayou D, Li T, Hass J, Marchenkov A N, Conrad E H, First P N and de Heer W A 2006 *Science* **12** 1191
[6] Ghosh S, Calizo I, Teweldebrhan D, Pokatilov E P, Nika D L, Balandin A A, Bao W, Miao F. and Lau C N 2008 *Applied Physics Letters* **92** 151911
[7] Das Sarma S, Adam Sh, Hwang E H, and Rossi E 2011 *Reviews of Modern Physics* **83** 407
[8] Konakova R V, Okhrimenko O B, Svetlichnyi A M, Ageev O A, Volkov E Yu, Kolomiytsev A S, Jityaev I L and Spiridonov O B 2015 *Semiconductors* **49** 1278
[9] Okhrimenko O B, Konakova R V, Svetlichnyi A M, Spiridonov O B and Volkov E Yu 2012 *Nanosystems, Nanomaterials, Nanotechnologies* **10** 335
[10] Dyke W P, Charbonnier F M, Strayer R W, Floyd R L, Barbour J P and Trolan J K 1960 *Journal of Applied Physics* **31** 790
[11] Todokoro H, Saitou N and Yamamoto Sh 1982 *Japanese Journal of Applied Physics* **21** 1513
[12] Cui Y, Zhang X, Lei W, Di Y, Wang J and Yang X 2009 *Vacuum Electronics Conference, 2009. IVEC '09. IEEE International* 269
[13] Bormashov V S, Baturin A S, Nikolskiy K N, Tchesov R G and Sheshin E P 2007 *Surface and Interface Analysis* **39** 155
[14] Lebedev A A, Kozlovski V V, Strokan N B, Davydov D V, Ivanov A M, Strel'chuk A M and Yakimova R. 2002 *Semiconductors* **36** 1270
[15] Konakova R V, Kolomys O F, Okhrimenko O B, Strelchuk V V, Volkov E Yu, Grigoriev M N, Svetlichnyi A M and Spiridonov O B 2013 *Semiconductors* **47** 812
[16] Chen L, He H, Yu H, Cao Y, Lei D, Menggen Q, Wu Ch and Hu L 2014 *Journal of Alloys and Compounds* **610** 659
[17] Deng J H, Wang F J, Cheng L, Yu B, Li G Zh, Hou X G, Li D J and Cheng G A 2014 *Materials Letters* **124** 15
[18] Reddy J N 2005 *An Introduction to The Finite Element Method (Third Edition)* (New York: McGraw-Hill)
[19] Svetlichnyi A M, Ageev O A, Volkov E Yu, Jityaev I L and Dem'yanenko M V 2015 *Applied Mechanics and Materials* **752-753** 163

IOP Conf. Series: Materials Science and Engineering **93** (2015) 012032 doi:10.1088/1757-899X/93/1/012032

Combustion and agglomeration of aluminized high-energy compositions

A G Korotkikh[1,2], V A Arkhipov[2], O G Glotov[3] and K V Slyusarskiy[1]
[1] Institute of Power Engineering, Tomsk Polytechnic University, 634050, Tomsk, 30 Lenin Av., Russia
[2] Research Institute of Applied Mathematics and Mechanics, Tomsk State University, 634050, Tomsk, 36 Lenin Av., Russia
[3] Institute of Chemical Kinetics and Combustion of SB RAS, 630090, Novosibirsk, 3 Institutskaya Str., Russia

E-mail: korotkikh@tpu.ru

Abstract. The results of combustion study for high-energy compositions (HECs) based on ammonium perchlorate (AP), butadiene rubber and ultrafine powder (UFP) aluminum Alex, and agglomeration of metal particles on the burning surface and composition of condensed combustion products (CCPs) are presented. It was found that partial replacement 2 wt. % of Alex by iron UFP in HEC increases the burning rate 1.3–1.4 times at the range of nitrogen pressure 2.0–7.5 MPa and reduces the mean diameter of CCPs particles d_{43} from 37.4 μm to 33.5 μm at pressure ~4 MPa. Upon partial replacement 2 wt. % of Alex by boron UFP in HEC the recoil force of gasification products outflow from burning surface is increased by 9 % and the burning rate of HEC does not change in the above pressure range, while the mean diameter of CCPs particles is reduced to 32.6 μm at $p \sim 4$ MPa.

1. Introduction

The modern high-energy compositions (HECs) contain as the oxidizer crystals of ammonium perchlorate (AP), ammonium nitrate (AN) and nitramines (RDX, HMX, CL-20) [1–2], combustible-binder – inert or energetic polymer rubber, and metal fuel – aluminum powder with typical content up to 22 wt. % [3–5]. These systems are used as the energy source of solid rocket motors and gas generators for various purposes. Adjustment of the HEC burning rate is mainly achieved by introduction in propellant formulation the catalysts, through partial or complete replacement of AP and AN by nitramines, by changes of the excess oxidizer factor, as well as the changing of the particle size of oxidizer and metal powders.

As the burning catalysts of HECs metal oxides: MnO_2, Ni_2O_3, Cr_2O_3, MgO, Fe_2O_3, Co_2O_3, CuO, SiO_2, copper, iron, zinc, cadmium and magnesium salts of chromic and metachromic acids, complex cyanides of copper, iron and nickel are used [6–9]. The most common and versatile in this group of catalysts are the systems containing copper, chromium, boron and iron [10–12].

This paper presents the experimental results on combustion processes of HECs based on aluminum ultrafine powder (UFP) Alex, containing additives of iron and amorphous boron UFPs and the methods for determine of parameters: the recoil force of gasification products outflow, the burning rate of samples, and the quantitative and phase compositions, the size distribution of particles of condensed combustion products (CCPs).

2. Experimental

2.1 The HEC samples

We studied the samples of HEC on the basis of bidispersed AP (fraction less than 50 μm and 160–315 μm in the ratio 40/60), inert combustible-binder (19.7 wt. %) – butadiene rubber plasticized by transformer oil, and aluminum UFP Alex (15.7 wt. %) obtained in argon using electric explosion of conductors with a shelf life of more than two years. In the second and third compositions of HECs 2 wt. % of the Alex UFP was partially replaced by 2 wt. % catalyst additives – UFPs of iron and amorphous boron. According to the measurements on the BET analyzer Nova 2200e in nitrogen the specific surface area of Alex amounts to 7.04 m^2/g, iron – 1.08 m^2/g, amorphous boron – 8.63 m^2/g. The studied cylindrical samples of HECs in diameter 10 mm and height 30 mm produced in the laboratory by extrusion pressing with the subsequent curing. The density of solid samples depending on the component composition was in an interval of 1.53–1.59 g/cm^3.

2.2 Combustion of HECs

The study of effect of iron and amorphous boron UFPs additives on the combustion process of HECs was carried in blow-through sampling bomb in nitrogen at different pressures [14]. We used the HEC samples in diameter 10 mm and height 25–30 mm in experiments with sampling of condensed combustion products (CCP). The lateral surface of samples was inhibited by heat resistant rubber «solpren» (a styrene/butadiene copolymer). During combustion of solpren the carbonaceous particles were formed, the contents of which in CCPs was not exceed 1 wt. %.

The burning rate of HECs was evaluated by known length and the burning time of samples, defined by a signal from a pressure sensor in bomb during the combustion. During combustion of HEC sample the gaseous and condensed combustion products passed through the metal sieve mesh packet and analytical aerosol filter which are captured solid particles of different sizes. Sampled CCP particles were divided onto four fractions – less than 80 μm, 80–160 μm, 160–315 μm and greater than 315 μm, and then subjected to a visual inspection, particle size and phase analyses.

The value of recoil force of gasification products outflow from the end surface of HEC sample during the heating of reaction layer, ignition and combustion of HECs was determined using recoil force transducer [13].

3. Results and discussion

3.1 The recoil force

The values of the time of gasification and ignition, the recoil force of gasification products outflow for the solid propellants samples under study in atmospheric conditions were determined in air under normal conditions. Processing of signal recording from the recoil force transducer we get the value of recoil force F outflow of gasification products from the burning surface of HEC samples. Examples of records are shown in Figure 1.

Partial replacement of Alex by iron UFP in HEC leads to the increase of the recoil force of gasification products outflow from burning surface 1.3 times in the period of stationary combustion of HECs. For the partial replacement of Alex by boron UFP in the HEC composition the recoil force of gasification products outflow increases 1.1 times.

3.2 The burning rate

The values of the burning rate at different nitrogen pressures were determined in the sampling bomb (Figure 2). The values of the HECs burning rate and the relative mass of CCPs m_{ccp} equal to the ratio of the CCPs mass M_{ccp} to the mass of HEC sample M_{samp} at pressure ~ 4 MPa are presented in Table 1.

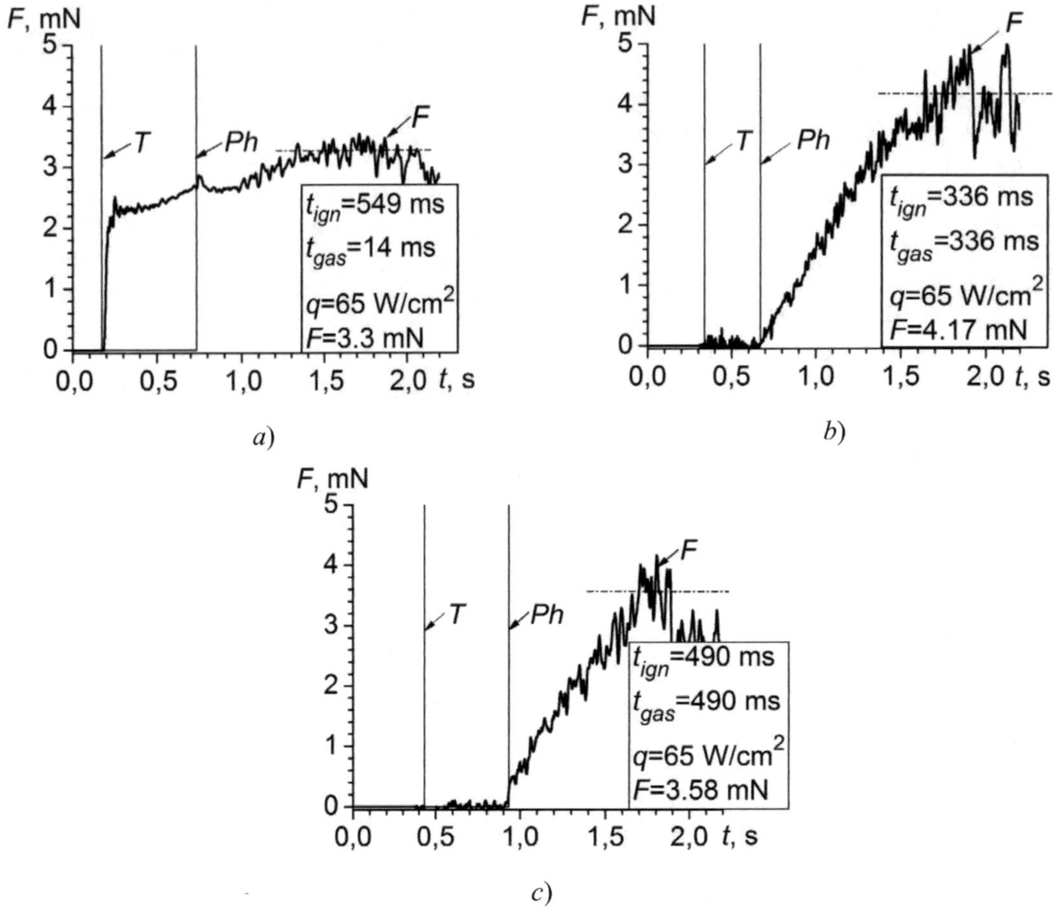

Figure 1. The signals of thermocouple (*T*), photodiode (*Ph*) and the recoil force transducer (*F*) that have been recorded over time interval embracing the heating, ignition and combustion processes for HECs samples with Alex (*a*), Alex+Fe (*b*) и Alex+B (*c*).

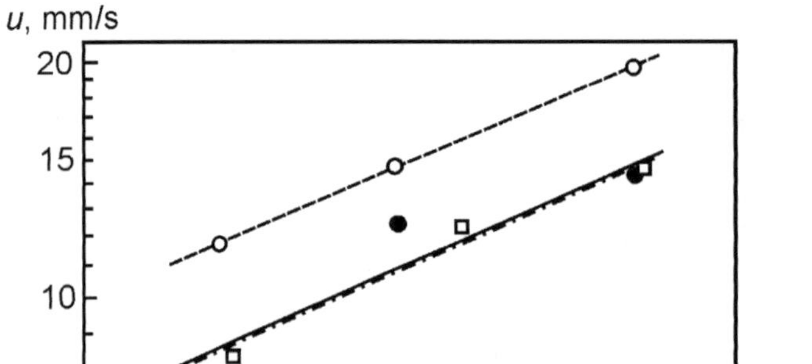

Figure 2. The burning rate of aluminized HECs from pressure.

Table 1. The burning rate of HECs and relative mass of CCPs at $p \sim 4$ MPa.

HEC sample	u (mm/s)	$m_{ccp} = M_{ccp}/M_{samp}$
1, Alex	12.4 ± 0.2	0.30 ± 0.04
2, Alex+Fe	14.7 ± 0.2	0.29 ± 0.03
3, Alex + B	12.3 ± 0.3	0.29 ± 0.04

Partial replacement of Alex by iron UFP in HEC composition slightly decreases the exponent of burning rate law from 0.45 to 0.43 and the increase of the burning rate from 7.5 to 11.7 mm/s at pressure 2.2 MPa and from 14.3 to 19.7 mm/s at pressure 7.5 MPa. The mass of sampled CCPs of tested HECs practically does not change at nitrogen pressure ~ 4 MPa.

For partial replacement of Alex by boron UFP in HEC the burning rate and relative mass of CCPs at pressure ~ 4 MPa are virtually unchanged. The burning rates of HECs with Alex+B are equal 8.4 mm/s at pressure 2.3 MPa and 14.6 mm/s at pressure 7.7 MPa.

Obtained data of the HECs burning rate correlates well with the data of recoil force of gasification products outflow from burning surface of samples (see Figure 1).

3.3 Compositions of CCPs

Sampled CCPs at nitrogen pressure ~4 MPa were subjected to particle size and X-ray diffraction (XRD) analyses. The mass-size distribution function of CCPs particles measured with use of Analysette 22 particle size analyzer in distilled water under the ultrasound action are presented in Figure 3.

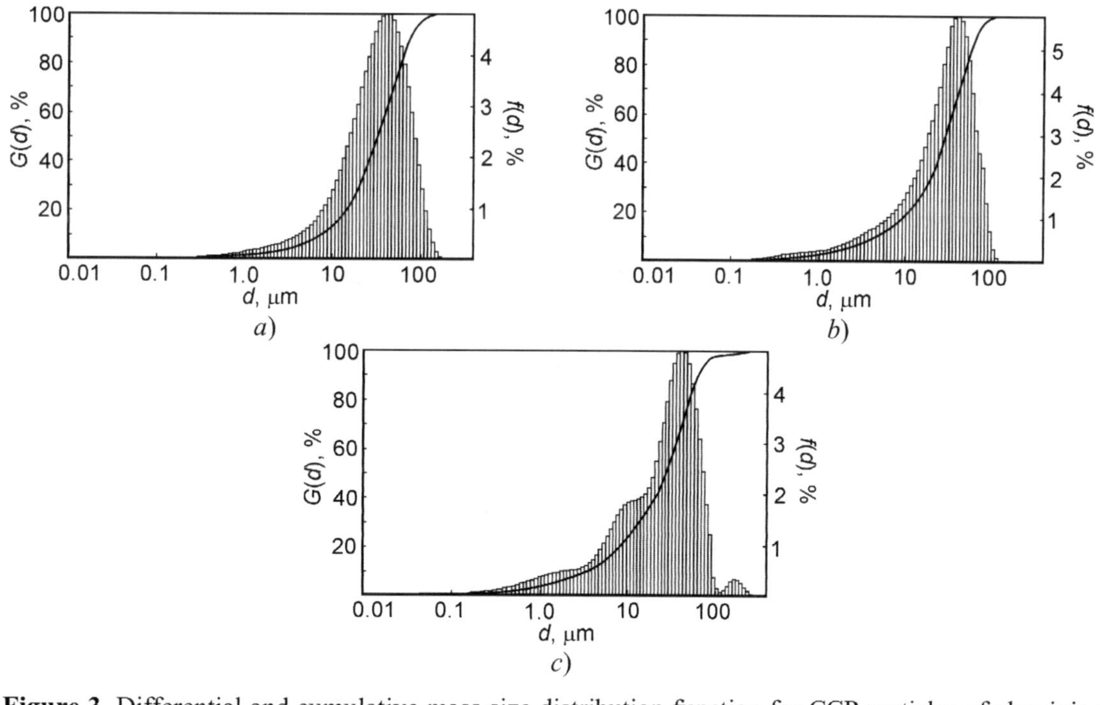

Figure 3. Differential and cumulative mass size distribution function for CCP particles of aluminized HECs with Alex (*a*), Alex+Fe (*b*) and Alex+B (*c*).

Particle size measurements showed that the mass fraction of CCPs particle fraction less than 10 μm amounted 0.16, 0.18 and 0.27, for HECs with Alex, Alex+Fe and Alex+B, respectively, the particle fraction of CCPs less than 45 μm – 0.67, 0.71, 0.74, the particle fraction of CCPs less than 100 μm – 0.97, 0.99 and 0.98.

Partial replacement of Alex by iron UFP in the HEC composition decreases the mean particle diameter d_{43} from 37.4 to 33.5 μm. Partial replacement of Alex by boron UFP in HEC leads to the decrease of the CCPs d_{43} to 32.6 μm.

XRD analysis of CCPs performed using the Shimadzu XRD 6000 diffractometer showed that the content of amorphous phase in the sampled CCPs in the case of propellant with Alex is equal 18 wt. %, in the case of propellant with Alex+Fe – 22 wt. % and in the case of propellant with Alex+B – 26 wt. %. The phase composition of CCPs, sampled at nitrogen pressure ~ 4 MPa, is presented in Table 2.

Table 2. The phase composition (without the amorphous phase) of CCPs of aluminized HECs.

HEC sample	Content of phases in CCPs (wt. %)				
	α-Al_2O_3	θ-Al_2O_3	γ-Al_2O_3	AlN	C_3N_4
1, Alex	7.1	66.7	2.0	7.1	17.1
2, Alex+Fe	9.4	54.7	1.6	8.3	26.0
3, Alex+B	43.3	–	–	–	56.7

It was found that partial replacement Alex by Fe UFP in HEC increases the content of alumina α-Al_2O_3 (corundum) by 2.3 wt. %, aluminum nitride AlN – by 1.2 wt. %, carbon nitride C_3N_4 – by 8.9 wt. % and reduces the content of alumina θ-Al_2O_3 (monoclinic) by 12 wt. % in composition of sampled CCPs due to possible the catalytic effect, which reduces the temperature of the beginning of AP high temperature decomposition in the reaction layer of HEC, and the increase of the temperature in gas phase zone of chemical reactions, and also the exothermic reaction of iron and aluminum particles on the reaction layer near the surface of HEC sample.

Partial replacement of Alex by boron UFP in HEC increases the content of alumina α-Al_2O_3 by 6.1 times and carbon nitride C_3N_4 by 3.3 times in composition of sampled CCPs due to the possible increase of the heat release during combustion of boron UFP and, respectively, the temperature in the gas phase zone of chemical reactions.

4. Conclusions

The effects of additives of iron and amorphous boron UFPs in high-energy compositions on the basis of AP, butadiene rubber and 15.7 wt. % of aluminum Alex UFP on combustion characteristics were studied. Burning rate law and condensed combustion products were studied with use of a special designed blow-through sampling bomb at nitrogen pressure from 2.0 to 7.5 MPa.

It was found that partial replacement 2 wt. % of Alex by iron UFP in HEC increases of the recoil force of gasification products outflow from burning surface 1.3 times in the period of stationary combustion of samples and increases the burning rate 1.3–1.4 times at the range of nitrogen pressure 2.0–7.5 MPa and reduces of the mean diameter d_{43} of CCP particles from 37.4 μm to 33.5 μm at nitrogen pressure ~4 MPa. In the sampled combustion products at ~4 MPa the content of aluminum oxide α-Al_2O_3 is increased by 2.3 wt. %, aluminum nitride AlN – by 1.2 wt. % and carbon nitride C_3N_4 – by 8.9 wt. % due to possible the catalytic effect, which reduces the temperature of the beginning of AP high temperature decomposition in the reaction layer of HEC, and interaction of thermite mixture of aluminum and iron particles on the surface of the reaction layer of sample, and also the increase of the temperature in the gas phase zone of chemical reactions.

Partial replacement 2 wt. % of Alex by amorphous boron UFP in HEC increase of the recoil force of gasification products outflow from the burning surface 1.1 times. In this the burning rate of HEC does not change at the range of pressure 2.0–7.5 MPa, the mean diameter d_{43} of CCPs particles is reduced to 32.6 μm at pressure ~4 MPa, and the content of alumina α-Al_2O_3 and carbon nitride C_3N_4 are increased 6.1 times and 3.3 times in CCPs due to the possible increase of the heat release during combustion of boron UFP and the temperature in the condensed and gas phase zone of chemical reactions.

Acknowledgement
This work was partially supported by the Russian Foundation for Basic Research (project No. 15-33-50066 «Research of ignition and combustion processes of high-energy nanocomposites based on ammonium nitrate» and project No. 15-03-04321 «Ignition and combustion of metal fuel particles based on aluminum and boron»).

References

[1] Beckstead M W, Puduppakkama K, Thakreb P and Yang V 2007 *Progress in Energy and Combustion Science* **33** 497–551.
[2] Takahashi K, Oide Sh and Kuwahara T 2013 *Propellants, Explosives, Pyrotechnics* **38** 555-562.
[3] Sossi A, Duranti E, Manzoni M, Paravan C, DeLuca L T, Vorozhtsov A B, Lerner M I, Rodkevich N G, Gromov A A and Savin E N 2013 *Combustion Science and Technology* **185** 17–36.
[4] Arkhipov V A and Korotkikh A G 2012 *Combustion and Flame* **159** 409–415.
[5] Arkhipov V A, Bondarchuk S S, Korotkikh A G, Kuznetsov V T, Gromov A A, Volkov S A and Revyagin L N 2012 *Combustion, Explosion, and Shock Waves* **48** 625–635.
[6] Arkhipov V A, Bondarchuk S S and Korotkikh A G 2010 *Combustion, Explosion, and Shock Waves* **46** 570–577.
[7] Shioya S, Kohga M and Naya T 2014 *Combustion and Flame* **161** 620–630.
[8] Ishitha K and Ramakrishna P A 2014 *Combustion and Flame* **161** 2717–2728.
[9] Farley C W, Pantoya M L and Levitas V I 2014 *Combustion and Flame* **161** 1131–1134.
[10] Berner M K, Talawar M B and Zarko V E 2013 *Combustion, Explosion, and Shock Waves* **49** 625–647.
[11] McDonald B A, Rice J R and Kirkham M W 2014 *Combustion and Flame* **161** 363–369.
[12] Zarko V E and Glotov O G 2013 *Science and Technology of Energetic Materials* **74** 139–143.
[13] Arkhipov V A, Kiskin A B, Zarko V E and Korotkikh A G 2014 *Combustion, Explosion, and Shock Waves* **50** 622–624.
[14] Glotov O G, Zyryanov V Ya 1995 *Combustion, Explosion and Shock Waves* **31** 72–78.

IOP Conf. Series: Materials Science and Engineering **93** (2015) 012033 doi:10.1088/1757-899X/93/1/012033

Orientation dependence of the shape memory effect and superelasticity in ferromagnetic $Co_{49}Ni_{21}Ga_{30}$ single crystals with γ'-phase particles

I V Kuksgauzen[1], I V Kireeva[1], Y I Chumlyakov[1] and H Maier[2]
[1]National research Tomsk State University, Tomsk, Russia
[2]Institut für Werkstoffkunde, Leibniz Universität Hannover, Garbsen, Germany

E-mail: irbas@sibmail.com

Abstract. This paper reports on the orientation dependence of the shape memory effect and superelasticity in [001] and [$\bar{1}$23] single-phase and aged at 623K, 1 hour single crystals of ferromagnetic $Co_{49}Ni_{21}Ga_{30}$ (at.%) alloy with B2-$L1_0$ martensitic transformation. It was demonstrated that in single-phase crystals the values of reversible strain and the values of thermal and stress hysteresis depend on the crystal orientation. Precipitation of γ'-phase particles reduces the value of the shape memory effect and superelasticity and reduces their orientation dependence, increases the thermal and stress hysteresis in comparison with single-phase crystals.

1. Introduction

Currently, the ordered CoNiGa alloys with B2-$L1_0$ thermoelastic martensitic transformation (MT) (B2 – ordered phase based on volume-centered cubic lattice, $L1_0$ – tetragonal martensite based on a face-centered tetragonal lattice) are promising ferromagnetic alloys with high strength and good ductility [1-8]. In CoNiGa crystals, firstly, a high stress level of the high-temperature phase can be achieved by choosing the orientation. It is known [9] that in alloys with B2 structure the dislocation slip takes place along the <100> direction on {110}, {100} slip planes. Under tension/compression deformation the [001] crystals are characterized by high stresses level of B2-phase because of the Schmid factor for operating slip systems is zero in contrast to other orientations, in which Schmid factors for these systems are high. Therefore crystals oriented along the [001] direction should show a wider temperature interval of stress-induced MT compared with other crystal orientations. Secondly, a high stress level of high-temperature phase in CoNiGa alloy can be achieved by precipitation of γ'-phase particles (γ'-phase – an ordered face-centered cubic with $L1_2$ structure) [1-4, 10]. Precipitation of γ'-phase particles will allow to expand the temperature interval of superelasticity (SE) not only in the [001]-oriented crystals, but also in other orientations, result from increased stress levels of high-temperature phase and suppression of local plastic flow processes during the formation of martensite crystals under stress and to receive alloys with high-temperature of SE, a manifestation which will not depend on the crystal orientation. To date there have been no systematic investigations of the simultaneous influence of the crystal orientation and dispersed γ'-phase particles on the functional properties of single-crystal of $Co_{49}Ni_{21}Ga_{30}$ alloys. Therefore, in this paper presented the results of investigations on the effect of crystal orientation and γ'-phase particles on the value of shape memory effect (SME), ε_{SME}, and SE, ε_{SE}, thermal ΔT^{σ} and stress $\Delta\sigma$ hysteresis under stress, the temperature interval of SE, ΔT_{SE}, in [001]- and [$\bar{1}$23]-oriented single crystals of $Co_{49}Ni_{21}Ga_{30}$ (at.%) alloy at compression. This choice of orientation was defined as follows: firstly, the crystals are characterized by different values of the lattice deformation: in [001]-oriented single crystals under compression ε_0=4.5%, and [$\bar{1}$23]-oriented single crystals – ε_0=3.2%, for this reason, in the work is supposed to receive orientation dependence of the value of reversible strain on the experiments of SME and SE. Secondly, in the [001]-oriented single crystals the contribution of detwinning strain of $L1_0$-martensite, ε_{detw}, to the total lattice deformation at B2-$L1_0$ MT is zero, and in the [$\bar{1}$23]-oriented single crystals $L1_0$-martensite detwinning takes place and the associated strain of ε_{detw}=0.8% [6] . This choice of orientations will allow not only to trace the influence of orientation on the functional properties due to different values of ε_0, but also reveal the influence of $L1_0$-martensite detwinning on energy dissipation processes.

2. Materials and methods

Single cryslals of $Co_{49}Ni_{21}Ga_{30}$ (at.%) alloy were grown by the Bridgman technique in inert gas atmosphere. Samples for compression tests were in parallelepiped form in size of 3x3x6 mm^3 with the compression axis oriented along the [001] and [$\bar{1}$23] directions. For single-phase state the samples were kept at T=1430 K for 30 minutes in quartz tubes in inert gas atmosphere followed by water quenching. To precipitate the nanometric dispersed γ'-phase particles selected low temperature aging at 623 K for 1 hour. After this aging the γ'-phase particles are spherical of size d=5 nm, and the volume fraction of f=15% [1]. Shape memory effect was measured using a specially designed installation when cooling/heating at different levels of external stresses. SE effect of single crystals were examined by the Instron 5969 at the temperature interval from 273 K to 623 K under compression. MT temperatures in free state were determined by differential scanning calorimetry (DSC). In the single-phase crystals peaks of direct and reverse transformation are observed by the DSC method and B2-$L1_0$ MT is characterized by low thermal hysteresis: $\Delta T=A_f-M_s$=296–272=24K (M_s – start temperature of the forward MT on cooling; A_f – the finish temperature of the reverse MT on heating). At precipitation of nanometric γ'-phase particles after aging at T=623 K, 1h the MT temperatures shifted to lower temperatures: M_s=165K, A_f=274K, and the thermal hysteresis $\Delta T=A_f-M_s$=109K increased in 4.5 times in comparison with single-phase crystals without γ'-phase particles.

3. Results and discussion

Figures 1 and 2 presents the results of a study of SME recorded during cooling/heating experiments under different external stresses from 2.5 MPa to 350 MPa for crystals oriented along the [001] and [$\bar{1}$23] directions in a single-phase state and after aging at 623K, 1 h. Such experiments allow us to determine the B2-$L1_0$ MT temperatures, the value of thermal hysteresis, ΔT^{σ}, and transformation strain, ε_{SME}, depending on external applied stresses.

It is seen that in single-phase crystals oriented along the [001] and [$\bar{1}$23] directions one-stage B2-$L1_0$ MT is realized, which is fully reversible by heating and thus SME is observed. The value of SME in single-phase crystals depends on the crystal orientation. Thus, in crystals oriented along the [001] direction, at σ_{ext}=2.5 MPa the $\varepsilon_{SME}^{[001]}$=3.5%, and this value is close to theoretically calculated value of the lattice deformation ε_0=4.5% for a given crystal orientation at B2-$L1_0$ MT [6]. Consequently, even at minimal external stresses of σ_{ext}=2.5 MPa in single-phase crystals oriented along the [001] direction, occurs the destruction of the self-accommodated $L1_0$-martensite microstructure and growth oriented twinned $L1_0$-martensite [4]. In [001]-oriented single crystals at σ_{ext}=20 MPa reaches a maximum reversible strain $\varepsilon_{SME}^{[001]}$=4.2%, which is equal to the value of ε_0=4.5% for a given orientation of the crystals at B2-$L1_0$ MT [6]. With increasing σ_{ext}>60 MPa the value of $\varepsilon_{SME}^{[001]}$ decreases slightly. In crystals oriented along the [$\bar{1}$23] direction at σ_{ext}=2.5 MPa $\varepsilon_{SME}^{[\bar{1}23]}$=2.5% and this value is close to theoretically calculated value of the lattice deformation ε_0=3.2% for the [$\bar{1}$23] crystals at B2-$L1_0$ MT as well [6]. This means that in [$\bar{1}$23] crystals at the minimum stress at σ_{ext}=2.5 MPa self-accommodated $L1_0$-martensite structure is destroyed and there is a growth oriented $L1_0$-martensite containing no twins as the value of ε_{SME}=2.5% at σ_{ext}=2.5 MPa exceeds value of $\varepsilon_0=\varepsilon_{CVP}$=2.4% (the contribution of detwinning strain of L1_0-martensite in $[\bar{1}23]$ crystals at B2-L1_0-MT is 0.8% [6]). When σ_{ext}>20 MPa the ε_{SME}=2.9% reaches a maximum value close to the value of ε_0 for a given orientation of the crystals at B2-L1_0 MT [6] and with the increasing of σ_{ext} the ε_{SME} remains practically constant. In both crystal orientations with minimal external compressive stresses the start temperature of the forward B2-L1_0 MT on cooling under stress is identical and M_s^σ=263 K. This temperature is close to M_s, that determined by DSC for single-phase crystals in the free state. However, the value of the thermal hysteresis $\Delta T^\sigma = A_f^\sigma - M_s^\sigma$, characterizing the energy dissipation at MT under stress, in [001]-oriented single crystals has smaller value than in $[\bar{1}23]$-oriented single crystals, and in stress interval of σ_{ext}=2.5÷40 MPa decreases from ΔT^σ=27K to ΔT^σ=15K and then remains constant, and in $[\bar{1}23]$-oriented single crystals, on the contrary, increases from ΔT^σ=42K to ΔT^σ=55K (figure 1). In $[\bar{1}23]$-oriented crystals with increasing applied external stresses the position of habit plane of L1_0-martensite under detwinning is changed relatively non-detwinning state. This leads to additional internal stresses and energy dissipation under unloading and it associated with the increase of ΔT^σ in the $[\bar{1}23]$ crystals. In [001]-oriented single crystals, where the detwinning strain of L1_0-martensite suppressed because of equality to zero Schmid factors, habit plane of L1_0-martensite does not change its position [11], energy dissipation when unloading there is no and ΔT^σ in [001]-oriented single crystals does not increase.

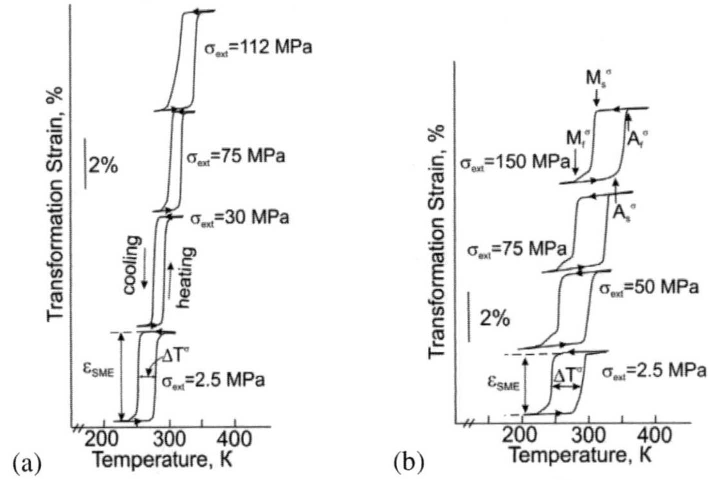

Figure 1. «σ-T» curves for the [001] (a) and $[\bar{1}23]$ (b) single-phase crystals of $Co_{49}Ni_{21}Ga_{30}$ alloy

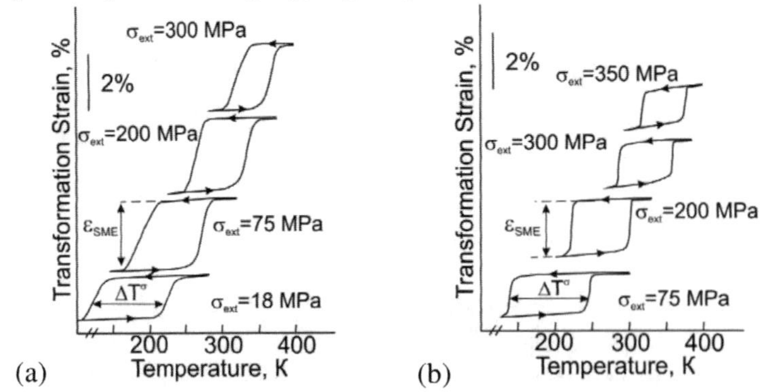

Figure 2. «σ-T» curves for the [001] (a) and $[\bar{1}23]$ (b) single crystals of $Co_{49}Ni_{21}Ga_{30}$ alloy aged at T=623K for 1 h.

With increasing applied stress levels σ_{ext} in [001]- and [$\bar{1}$23]-oriented single crystals observed the temperature M_s^σ rise. With increasing M_s^σ the stresses $\sigma_{ext}=\sigma_{cr}$ lay down on the linear relationship (figure. 4), which is described by the Clausius-Clapeyron relation [11]:

$$\frac{d\sigma_{cr}}{dT}=-\frac{\Delta H\rho}{\varepsilon_0 T_0}=-\frac{\Delta S\rho}{\varepsilon_0}, \quad (1)$$

here ΔH and ΔS – respectively enthalpy and entropy change at B2-$L1_0$ MT; ε_0 – lattice deformation which depends on the crystal orientation; T_0 – temperature of chemical phase equilibrium which may be calculated as $T_0=1/2(M_s+A_f)$ and ρ – mass density.

From figure 3 shows that in the single-phase [001] and [$\bar{1}$23] crystals value of $\alpha=d\sigma_{cr}/dT$ depends on the crystal orientation: $\alpha^{[001]}=1.74$ MPa/K, $\alpha^{[\bar{1}23]}=2.95$ MPa/K. Orientation dependence of $\alpha=d\sigma_{cr}/dT$ is due to the orientation dependence of the $\varepsilon_{SME}=\varepsilon_0$. From (1) that smaller value of $\alpha=d\sigma_{cr}/dT$ should corresponds to larger value of ε_0 and, in contrast, the smaller value of ε_0 should corresponds to larger value $\alpha=d\sigma_{cr}/dT$. This is confirmed experimentally. Thus, in the [001]-oriented single crystal smaller value of $\alpha^{[001]}=1.74$ MPa/K corresponds to a larger value of $\varepsilon_{SME}=4.2\%$, and in the [$\bar{1}$23]-oriented single crystals, in contrast, a larger value of $\alpha^{[\bar{1}23]}=2.95$ MPa/K corresponds to a smaller value of $\varepsilon_{SME}=2.9\%$. At that $\alpha^{[\bar{1}23]}/\alpha^{[001]}=1.7$ ratio and $\varepsilon_{SME}^{[001]}/\varepsilon_{SME}^{[\bar{1}23]}=1.5$ ratio are close to each other in accordance with equation (1).

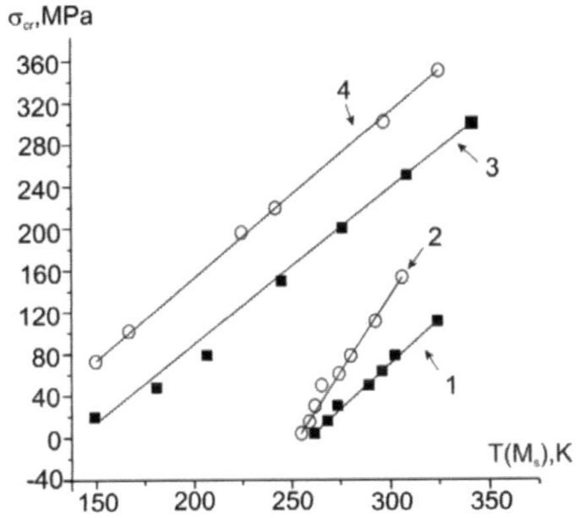

Figure 3. Temperature dependence of stresses σ_{cr} for single-phase (1, 2) and aged at T=623K, 1 h (3, 4) [001]- (1, 3) and [$\bar{1}$23] - (2, 4) oriented single crystals of $Co_{49}Ni_{21}Ga_{30}$ alloy.

In addition, using the relation (1) can calculate theoretical values of $\alpha=d\sigma_{cr}/dT$ and compare them with the experimental values of $\alpha=d\sigma_{cr}/dT$. Entropy change ΔS, obtained in experiments by DSC on $Co_{49}Ni_{21}Ga_{30}$ single crystals equal 10.0 J/kg K, and the mass density ρ of the $Co_{49}Ni_{21}Ga_{30}$ crystals is equal 8470 kg/m^3 was directly obtained by the measurement of the single crystal [2]. Using the data of ΔS, ρ and experimental values of ε_{SME} for the appropriate orientation were obtained theoretical values of $\alpha=d\sigma_{cr}/dT$: $\alpha^{[001]}=2.02$ MPa/K, a $\alpha^{[\bar{1}23]}=2.9$ MPa/K and these values are close to the experimental obtained values of α for the corresponding orientation.

At precipitation of γ'-phase particles in crystals oriented along the [001] and [$\bar{1}$23] directions does not change the nature of the MT and one-stage B2-$L1_0$ MT is observed as well as in single-phase crystals, which completely reversible after heating. Consequently, SME is realized. As can be seen from figure 2 at the precipitation of γ'-phase particles the SME under stress at σ_{ext} exceeds of σ_{ext} for single-phase crystals without the particles and the value of SME becomes smaller. Thus, in the [001]-oriented single crystal of $\varepsilon_{SME}=1.5\%$ is observed when $\sigma_{ext}=18$ MPa, and in [$\bar{1}$23]-oriented sungle crystals $\varepsilon_{SME}=1.6\%$ at $\sigma_{ext}=75$ MPa. When $\sigma_{ext}=100$ MPa values of SME are close: $\varepsilon_{SME}[001]=2.2\%\approx\varepsilon_{SME}[\bar{1}23]=1.9\%$ and with increasing of σ_{ext} values of SME vary slightly (Figure 3, b, curves). Therefore, the dispersed γ'-phase particles creates significant resistance to the movement of interphase and twin boundaries. Decrease of ε_{SME} at precipitation of γ'-phase particles may be due to several factors: 1) a decrease in the volume fraction of the matrix, which undergoes MT with precipitation of γ'-phase particles, which do not undergo the MT and 2) difficulties of detwinning deformation of $L1_0$-martensite [12].

At precipitation of γ'-phase particles after aging at T=623K, 1 h in both orientations there is a decline of M_s^σ temperature on ~110K as compared to single-phase state (Figure 1, 2). That also correlated with the data obtained by DSC [1]. At that the value of ΔT^σ is larger than in single-phase crystals and with growth of σ_{ext} in both orientations the value of ΔT^σ is decreased. For example, in [001]-oriented single crystals with γ'-phase particles at minimal σ_{ext}=18 MPa the ΔT^σ=100K and with increasing stress up to σ_{ext}=300 MPa the ΔT^σ is reduced to 50 K. In the $[\bar{1}23]$ - oriented single crystal with γ'-phase particles at minimum σ_{ext}=75MPa value of ΔT^σ=110K with increasing of σ_{ext} to 350 MPa the ΔT^σ is reduced by 2 times. As well as in single-phase state in the crystals with particles $\Delta T^\sigma[\bar{1}23]>\Delta T^\sigma[001]$ (figure 2). The physical reason of the increase of ΔT^σ in crystals with particles compared with single-phase crystals without the particles is associated with resistance of particle to movement interphase and twins boundaries, with interaction of martensitic variants (interaction of martensite arising near the particles with stress-induced martensite) and with increasing density of twins martensite as has previously been shown in [1]. When the stress increases selection variant of stress-induced martensite is advantageous and the interaction between martensite variants makes a smaller contribution to the energy dissipation and this may be due to a decrease of ΔT^σ with growth of external applied stress [4].

With an increase in the level of external applied stress σ_{ext} in [001]- and $[\bar{1}23]$ - oriented single crystals with γ'-phase particles as well as single-phase crystals the M_s^σ temperature rise is observed and with increasing of M_s^σ the stresses of $\sigma_{ext}=\sigma_{cr}$ are falling on linear dependence. As seen from figure 4 the value of $\alpha=d\sigma_{cr}/dT$ in the crystals with γ'-phase particles becomes smaller than a single-phase state for corresponding orientation and weakly dependent on orientation: $\alpha^{[001]}$=1.49 MPa/K и $\alpha^{[\bar{1}23]}$ =1.55 MPa/K. According to relation (1), when the value of SME decreases the values of $\alpha=d\sigma_{cr}/dT$ must increase and, therefore, must be larger than in single-phase crystal for corresponding orientation. This contradiction of the experimental data with the relation (1) is due to the dependence of the ε_{SME} and ΔT^σ and, accordingly, energy dissipation on the level of externally applied stresses, that by us it was shown in [2]. The weakening of the orientation dependence of $\alpha=d\sigma_{cr}/dT$ in crystals with the γ'-phase particles is due to degeneration orientation dependence of ε_{SME} value and the ratio of $\varepsilon_{SME}^{[001]}/\varepsilon_{SME}^{[\bar{1}23]}$=1.1 is equal to the ratio of $\alpha^{[\bar{1}23]}/\alpha^{[001]}$=1.0.

Experimental studies of the superelastisity in single crystals of $Co_{49}Ni_{21}Ga_{30}$ alloy showed that in [001]-oriented single-phase crystals the SE is observed a wide temperature interval from A_f=283K to 573K, ΔT_{SE}=290K that is associated with the development of reversible stress-induced B2-$L1_0$ MT. Therefore, the studied alloys exhibit high-temperature SE. In $[\bar{1}23]$ -oriented single-phase crystals SE is observed in the temperature interval of T=318÷473K and ΔT_{SE}=155K. Precipitation of γ'-phase particles leads to an expansion of temperature interval of SE in both orientations: in [001]-oriented single crystals the SE observed from T=273K to T=623K and $\Delta T_{SE}[001]$=350K and in $[\bar{1}23]$ – from T=298K to T=573K and $\Delta T_{SE}[\bar{1}23]$=275K that exceeds the ΔT_{SE} in single-phase crystals for the corresponding orientations. In the crystals oriented along the $[\bar{1}23]$ direction SE also has a place at T=573K. Consequently, the precipitation of γ'-phase particles provides a large temperature interval of SE including at high temperature at T>573K, not only in [001]- orientation, but also in other orientations due to increasing the stresses level of high-temperature phase and suppressing local plastic flow processes during stress-induced martensite.

Figure 4 shows the «σ-ε» curves at T=323K for single-phase and aged at T=623K, 1 h [001]- and $[\bar{1}23]$ - oriented single crystals at compression. It is seen that the γ'-phase particles influence the development of stress-induced MT. In single-phase crystals value of SE, ε_{SE}, and value of stress hysteresis, Δσ, which is defined as the difference stresses required for direct and reverse MT during deformation at half of SE loop, depend on the crystal orientation: in [001]-oriented single crystals at T=323K ε_{SE}=4.5%, Δσ=25 MPa, and in $[\bar{1}23]$-oriented single crystals at T=323K ε_{SE}=3.2%,

$\Delta\sigma$=73MPa. Precipitation of γ'-phase particles leads to a decrease of ε_{SE}, increase of $\Delta\sigma$ and weakening their orientation dependence: in [001]-oriented single crystals at T=323K ε_{SE}=2.2%, $\Delta\sigma$=90MPa, and in [$\bar{1}$23]-oriented single crystals ε_{SE}=2.1% and $\Delta\sigma$=95MPa. It should be noted that the value of SE equals to the value of SME for corresponding crystals orientation both in single-phase state and with the γ'-phase particles.

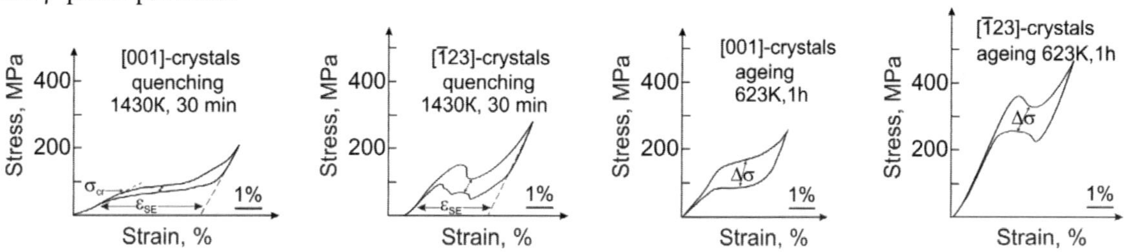

Figure 4. «σ-ε» curves at T=323K for the [001]- and [$\bar{1}$23]- oriented single-phase and aged at T=623K, 1h single crystals of $Co_{49}Ni_{21}Ga_{30}$ alloy.

4. Summary

It was established experimentally that in single-phase $Co_{49}Ni_{21}Ga_{30}$ (at. %) crystals the values of SME and SE, the temperature interval of SE, the values of thermal and stress hysteresis depend on the crystal orientation. In [001]-oriented single-phase crystals the value of SME and SE has a maximum value equal to 4-4.5%. SE is observed over a wide temperature interval ΔT_{SE}=290K and high temperature at T=573K and characterized of narrow stress hysteresis.

Precipitation of γ'-phase particles leads to a decrease values of SME and SE, an increase of the temperature interval of SE, of the values of stress and thermal hysteresis under stress and a weakening them depending on the crystal orientation. It is shown that in crystals with the γ'-phase particles due to changes in the level of external applied stresses can control the value of thermal hysteresis from 100-110K up to 50 K.

It is shown that an increase in strength properties of high-temperature phase due to the γ'-phase precipitate can be obtained the SE over wide temperature interval and at high temperatures not only in the crystal of [001] orientation but in [$\bar{1}$23] -oriented single crystals.

Acknowledgements

This study was supported by The Tomsk State University Academic D.I. Mendeleev Fund Program in 2015 and by Russian Fund for Fundamental Research No 13-08-98036 и 14-08-91334.

References

[1] Kireeva I V, Pons J, Picornell C, Chumlyakov Yu I, Cesari E and Kretinina I V 2013 *Intermetallics* **35** 60–66
[2] Kireeva I V, Picornell C, Pons J, Kretinina I V, Chumlyakov Yu I and Cesari E 2014 *Acta Mat.* **68** 127–139
[3] Chumlyakov Yu I, Kireeva I V, Panchenko E Y, et al. 2012 *Russ. Phys. J.* **8** 973–950
[4] Chumlyakov Yu I, Kireeva I V, Kretinina I V, et al. 2013 *Russ. Phys. J.* **11** 1290–1295
[5] Kireeva I V, Pobedennaya Z V, Chumlyakov Yu I, et al. 2009 *Tech. Phys. Lett.* **35** 186–189
[6] Dadda J, Maier H J, Niklasch D, Karaman I, Karaca H E and Chumlyakov Y I 2008 *Metall. Trans. A* **39** 2026–2039
[7] Dogan E, Karaman I, Chymlyakov Y I and Luo Z P 2011 *Acta Mater.* **59** 1168–1183
[8] Chumlyakov Yu I, Panchenko E Y, Kireeva I V, et al. 2008 *Mater. Sci. Eng. A.* **481** 95–100
[9] Myazaki S, Kimura S, Otsuka K and Suzuki Y 1984 *Scripta Metall.* **18** 883–888
[10] Liu J, Xie H, Huo Y, Zheng H and Li J 2006 *J. Alloy Compd.* **420** 145–157
[11] Otsuka K and Wayman C M 1998 *Cambridge University Press* 284
[12] Hamilton R F, Sehitoglu H, Chumlyakov Yu I and Maier H J 2004 *Acta Mater.* **52** 3383–3402

Shape Memory Effect and Superelasticity in [001] Single Crystals of FeNiCoAlNb(B) Alloys

O A Kuts, M Yu Panchenko, I V Kireeva and Yu I Chumlyakov
National Research Tomsk State University, Siberian Physical-Technical Institute, Tomsk, 634050, Russia

E-mail: bolga@sibmail.com

Abstract. In given paper presents data research of influence of boron on the functional properties - the shape memory effect and superelasticity in the [001] single crystals FeNiCoAlNb(B) alloys aged at 973 K for 5 hours. On the [001] single crystals FeNiCoAlNbB at aging at T = 973 K for 5 hour, it is shown, that boron leads to decrease the start M_s temperature of γ-α' martensitic transformation on cooling, to the development of γ-α' stress-induced martensitic transformation at higher stress at one test temperature and to increase of thermal ΔT and stress Δσ hysteresis is compared to [001] crystals without boron.

1. Introduction

In recent years, interest to the ferromagnetic iron-based alloys with thermoelastic γ-α' martensitic transformations (MT) have been increasing. Firstly, iron-based alloys are more durable, have low cost, better ductility and weldability as compared to known TiNi alloys [1]. Secondly, a large shape change can obtain not only under the influence of temperature, the stress, but also the magnetic field [2]. It is known [3-14], that in order to γ-α'-MT in FeNiCoTi, FeNiCoAlTa, FeNiCoAlTi iron-based alloys was thermoelastic the precipitation of disperse particles of ordered γ'-phase is necessary. In the polycrystals at precipitation of γ'-phase particles in grain body occur precipitation of β-phase on the grain boundaries, which leads to brittle fracture and does not allow to get the shape memory effect (SME) and superelasticity (SE). In the polycrystals FeNiCoAlTa, FeNiCoAlTi alloys to suppress the precipitation of β-phase additionally alloyed with boron [5, 6]. In the present work, was set a problem to investigate the effect of boron on the value of the SME, SE, the value of stress hysteresis Δσ in single crystals of Fe-28Ni-17Co-11.5Al-2.5Nb(0.05 B) (at.%) alloys at aging at T = 973 K for 5 hours at tensile deformation. The single crystals of these alloys in the absence of grain boundaries in a pure form will allow to reveal a difference in development of stress-induced γ-α'-MT in crystals on iron-based alloys with boron and without boron. For study the γ-α' stress-induced martensitic transformation were selected single crystals oriented along the [001] direction, because in given orientation the theoretical value the strain lattice ε_0 = 8.7 % has a maximum value for γ-α'-MT at tensile deformation [5].

2. Materials and discussion of the results

Single crystals of Fe-28Ni-17Co-11.5Al-2.5Nb(0.05 B) (at.%) alloy were grown by the Bridgman's method in inert gas atmosphere. Single crystal was confirmed by the X-ray method and optical microscopy of the etched crystal surface. Test samples were cut on electro-discharge machine in dog-bone-shaped with dimensions working part 1.5×12×3 mm. The single crystals after growth were homogenized at T = 1550 K 6 hours in an inert gas atmosphere followed by the quenching in water. The aging of single crystals FeNiCoAlNb(B) alloy was performed at a temperature T = 973 K for 5 hours in a helium atmosphere. Mechanical properties of crystals tested on Instron testing machine in a temperature range from 77 to 525 K with a strain rate of $4\cdot10^{-4}$ s^{-1}. The shape memory effect was determined on a specially designed installation at cooling/heating under different levels of external stresses. The study of crystal surfaces after removal of the load was conducted by the digital optical microscope VHX-2000.

Investigation of temperature dependence of electrical resistance ρ(T) for the single crystal FeNiCoAlNb(B) alloys after quenched and aging at T = 973 K, 5 hours is showed that both in the quenched state and after aging at 973 K for 5 hours on ρ(T) curve the specific stage for development of MT in these alloys is not observed and ρ is independent of temperature [4, 10]. For determination the possibility of development of reversible stress-induced γ-α'-MT in single crystals of FeNiCoAlNb(B) alloys the investigated of the temperature dependence of the critical stresses σ_{cr}, the SME at cooling/heating under a constant load and the SE.

The figure 1 shows the results of studies of the temperature dependence of the critical stresses σ_{cr} at tension for [001]-oriented single crystals of FeNiCoAlNb(B) alloys after quenching and aging at T = 973 K.

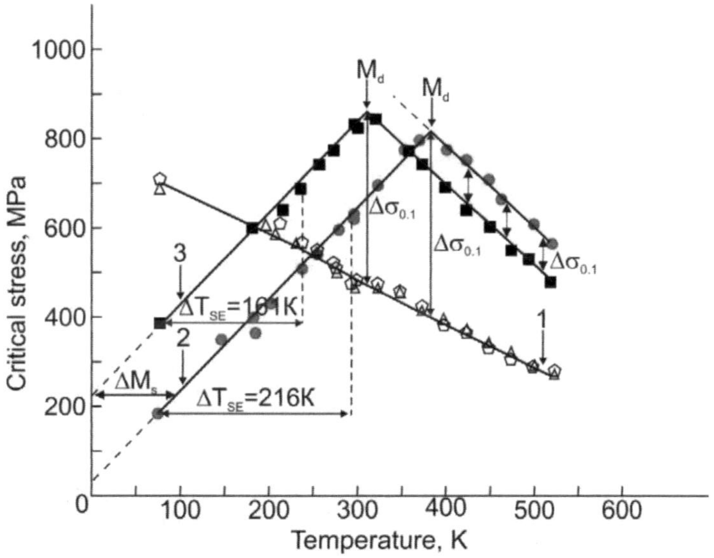

Figure 1. Temperature dependence of critical stresses σ_{cr} for [001] single crystals FeNiCoAlNb(B) alloys at tensile deformation: curve 1 – single-phase state after quenching; the curve 2 – the FeNiCoAlNb crystals, aged at T = 973 K, 5 hours; curve 3 – FeNiCoAlNbB crystals, aged at T = 973 K, 5 hours.

It is seen that in the single-phase state after quenching the $\sigma_{cr}(T)$ curve has a normal temperature dependence characteristic for single- and polycrystals of FCC-alloys – with increasing test temperature the σ_{cr} was decreased (figure 1, curve 1) [15]. It is qualitative confirmation that stress-induced γ-α' martensitic transformation in single-phase crystals in the temperature range T = 77 – 550 K does not develop at stresses near the yield stress, which also agrees with the experimental data obtained at the study of temperature dependence of ρ(T). Under alloying with boron up to 0.05 at.% the σ_{cr} in single crystals of FeNiCoAlNbB alloy are equal σ_{cr} in single crystals of FeNiCoAlNb alloy (figure 1, curve 1). Therefore, at the concentration of boron equal to 0.05 atm. % the hardening effect of it is zero. The temperature dependence $\sigma_{cr}(T)$ at aging at T = 973 K, 5 hours differs from $\sigma_{cr}(T)$ for a single-phase state and has the kind characteristic for alloys which undergoes stress-induced MT [7], namely, on the temperature dependence $\sigma_{cr}(T)$ in single crystals FeNiCoAlNb(B) alloy in the T = 77 – 550 K temperature range two stages is observed. In the first stage at 77 K < T < M_d (M_d is the temperature at which the yield stress level of high-temperature phase is equal to the critical stress level for stress-induced martensite) σ_{cr} increases with increasing temperature, which is described by the Clausius-Clapeyron relation [7]:

$$\frac{d\sigma_{cr}}{dT} = -\frac{\Delta H}{\varepsilon_0 T_0}. \quad (1)$$

Where ΔH – enthalpy change at γ-α'-MT; ε_0 – transformation strain which depends of crystal orientation; T_0 – chemical phase-equilibrium temperature. The second stage is characterized by a normal temperature dependence $\sigma_{cr}(T)$, which is usually observed under the slip deformation in FCC-crystals.

Analysis of the data presented in fig. 1 shows firstly that upon alloying boron $\sigma_{cr}(T)$ dependence of the parallel shifted to the area of the lower temperatures. The $\sigma_{cr}(T)$ curves which are extrapolated on σ_{cr} = 0 show that the values of M_s in crystals with a boron and without boron lie less than the temperature of liquid nitrogen and in crystals with boron the M_s and M_d temperatures are respectively on 90 K and 80 K lower than in crystals without boron. Consequently, the boron leads to stabilization of high-temperature phase. The value of $\alpha = d\sigma_{cr}/dT$ in crystals with boron and without boron prove to be same and equal $\alpha_{Nb} = \alpha_{NbB}$ = 2.05 MPa/K.

Secondly, at alloying by boron the stress of high-temperature phase at T > M_d less on 50 MPa than in crystals without boron. Since the solid-solution hardening by boron at low concentrations in single crystals FeNiCoAlNb(B) is equal to zero (figure 1, curve 1), the hardening of high-temperature phases of crystals with boron and without boron at aging T = 973K 5 hours can determine in pure form at the test temperature, which above M_d temperature, for example, 400 K, by the ratio [16, 17]:

In crystals without boron

$$\Delta\sigma_{cr} = \sigma_{cr}(Nb) - \sigma_{cr}(WQ) \quad (2a)$$

In crystals with boron

$$\Delta\sigma_{cr} = \sigma_{cr}(NbB) - \sigma_{cr}(WQ). \quad (2b)$$

Where $\sigma_{cr}(WQ)$ – stress of high-temperature phase of crystals without boron and with boron after quenching while $\sigma_{cr}(Nb)$ and $\sigma_{cr}(NbB)$ the stress in crystals without boron and with boron after aging at 973 K 5 h, respectively. Hardening of the high-temperature phase in crystals with boron and without boron after aging at 973 K for 5 hour at test temperature T = 400 K are equal, respectively, 380 and 415 MPa and in crystals without boron above on 35 MPa. Investigation of the crystals structure with boron and without boron after aging at T = 973 K, 10 hours shows that in crystals with boron the size and the volume fraction of γ'-phase particles is less than in crystals with boron. Therefore, boron slows the aging process and at one aging at T = 973 K, 5 hours crystals are characterized by different structure (size and volume fraction of particles in crystals with boron and boron are different). And because level of the strength properties of high-temperature phase determined by the size and volume fraction of particles, according to the relation [16, 17]:

$$\sigma_{cr} = 3 \cdot G \cdot E^{3/2} \cdot (f \cdot r / b)^{1/2}, \quad (3)$$

here G – the shear modulus of the high-temperature phase, E = Δa/a – parameter misfit of lattice of high-temperature the phase and the particles ($\Delta a = a_m - a_{par}$), f – volume fraction of particle, r – radius of the particle, b – a module of the Burgers vector of a slip dislocation, that in the crystals without boron the level of stress to must be larger than in crystals with the boron as observed experimentally.

It has been experimentally established that after aging at T = 973 K, 5 hours in crystals FeNiCoAlNb(B) alloys the γ-α'-MT is thermoelastic and SME and SE are observed.

Figure 2 presents the results of the research of the SME in the experiment during cooling/heating under external tensile stresses for the [001]-crystals FeNiCoAlNb(B) alloys, aged at 973 K, 5 hour. From the figure 2 it can be seen that in crystals FeNiCoAlNb(B) under load is realized one-stage γ-α'-

MT. In crystals FeNiCoAlNb γ-α'-MT is reversible, thus SME is observed. At the external stresses σ = 400 MPa the maximum value of reversible strain ε_{SME} = 2.5 % is observed, which is 3 times less than the theoretically calculated values of the lattice strain ε_0 = 8.7 % for the [001]-crystals at the γ-α'-MT [5].

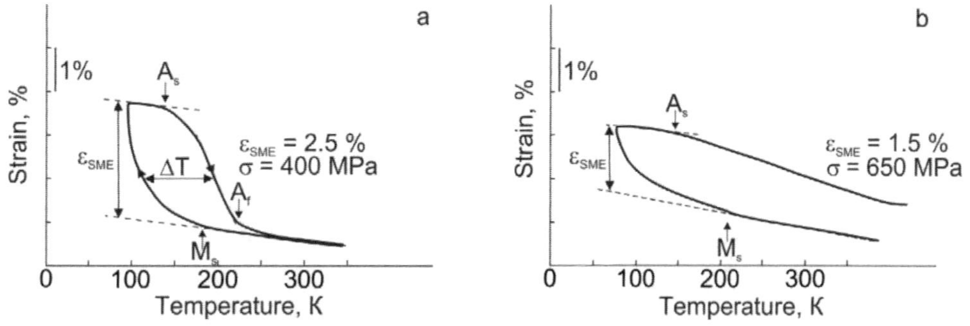

Figure 2. The shape memory effect ε_{SME} at constant external tensile stresses in the [001]-oriented single crystals aged at 973 K 5 hour: a) FeNiCoAlNb; b) FeNiCoAlNbB.

This is due to the fact that the crystal at stress σ = 400 MPa is destroyed before it reaches the theoretical value ε_0. The value of thermal hysteresis under load ΔT = 75 K (fig. 2a). In FeNiCoAlNbB crystals the ε_{SME} = 1.5 % is observed when the external stresses σ = 600 MPa and in this case γ-α'-MT is irreversible. The irreversible strain is disappears when the load is removed with followed the keeping the sample at room temperature, sample does not restore its original dimensions and the residual martensite on the surface is observed (fig. 3). Given that the form of temperature hysteresis loop not closed, that finish temperature of reverse MT A_f under load σ during heating is not achieved and can assume that the hysteresis in single crystals FeNiCoAlNbB alloy ΔT > 200 K.

Figure 3. The residual α'-martensite on the surface of [001]-oriented FeNiCoAlNbB crystals aged at 973 K for 5 hour at the load is removed σ = 650 MPa.

It has been experimentally established that at aging 973 K 5 hour in single crystals FeNiCoAlNb SE is observed from T = 77 K to 293 K, and its temperature range is equals ΔT_{SE} = 216 K, and in single crystals FeNiCoAlNbB SE is observed from T = 77 K to 238 K and ΔT_{SE} = 161 K (Fig. 1).

Figure 4 shows the results of studies SE in the [001]-single crystals FeNiCoAlNb(B) alloys aged at 973 K for 5 hours at tensile deformation, depending on a given strain value in cycle at T = 77 K.

Visible that in [001] crystals FeNiCoAlNb(B) the value of SE ε_{SE} = 5 % and it is limited due to brittle fracture of crystals at strain ε_{SE} > 5 %. The value of stress hysteresis $\Delta\sigma$, which is defined as the difference of stress required for the forward and reverse MT on half of deformation cycle, in the crystal FeNiCoAlNb equals 50 MPa and in crystals FeNiCoAlNb (B) is 4 times more. It should also be noted that in crystals without boron the γ-α'-MT starts at the stress σ_{cr} = 180 MPa and «σ-ε» curve is characterized by two stages: the first stage with the strain-hardening coefficient $\Theta = d\sigma/d\varepsilon$ close to zero with ε = 3.5 % and the second stage with ε > 3.5 % with high Θ = 1330 MPa. In crystals with a boron the γ-α'-MT develops at σ_{cr} = 380 MPa, which is 2 times higher than in the crystals without boron and on «σ-ε» curve the one stage with Θ = 3600 MPa is observed.

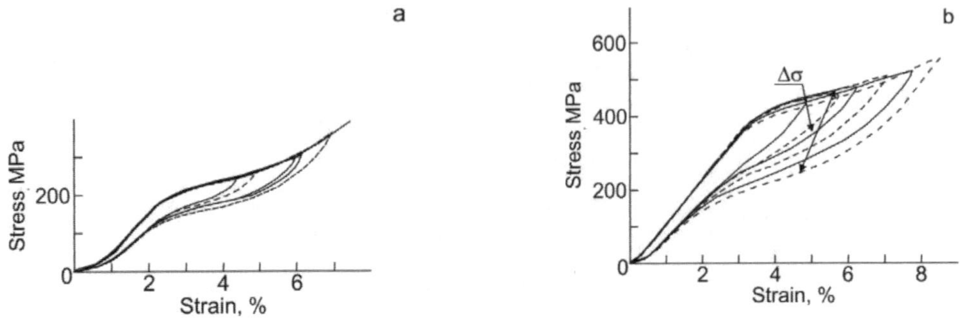

Figure 4. «σ-ε» curves for [001]-oriented single crystals of alloy aged at 973 K for 5 hours under tension: a) FeNiCoAlNb and b) FeNiCoAlNbB; test temperature T = 77 K.

Thus, it has been experimentally established that in the [001] single crystals of FeNiCoAlNb(B) aged at 973 K for 5 hours the thermoelastic stress-induced γ-α' MT with the SME and SE is observed, and their value, temperature range of SE ΔT_{SE}, the value of stress hysteresis $\Delta\sigma$ in single crystals are found to depend on the presence of boron in crystals of FeNiCoAlNb alloy with the same aging time T = 973 K 5 hour. From [11, 13] it follows that the level of stress of high-temperature phase determines the value ΔT_{SE} and $\Delta\sigma$: the high strength properties of high-temperature phase the more T_{SE} and the less $\Delta\sigma$. In crystals FeNiCoAlNb aged at 973 K for 5 hours the level of the strength properties of high-temperature phase is higher than it in crystals FeNiCoAlNbB (figure 1), in consequence the temperature range of SE is greater, and $\Delta\sigma$ is less. In FeNiCoAlNbB crystals the boron leads to a significant decrease of M_s compared to crystals without boron and at temperature T = 77 K the γ-α'-MT is developing at the stress which is 2 times less stress at T = M_d. This leads to the fact that the stress-induced γ-α' martensitic transformation is followed by processes of local plastic current and this is cause an increase $\Delta\sigma$ in crystal with boron compared to crystals without boron.

3. Conclusion

1. It has been established experimentally that in the single crystals of Fe-28%Ni-17%Co-11.5%Al-2.5%Nb (at.%) with boron and without boron aged at 973 K for 5 hours the γ-α' thermoelastic martensitic transformation with the shape memory effect and superelasticity at tensile deformation is develops.

2. It was shown that the boron in single-crystal Fe-28%Ni-17%Co-11.5%Al-2.5%Nb alloy leads to reduction the temperature M_s, to reduction the temperature range of superelasticity ΔT_{SE} and to increase the value of stress hysteresis $\Delta\sigma$ compared to the crystal without boron under the same conditions of aging (temperature and time aging).

3. It has been found out that the temperature dependence for the start of a stress-induced γ-α'-martensitic transformation in tension in the [001] single crystals of Fe-28%Ni-17%Co-11.5%Al-2.5%Nb alloy aged at 973 K for 5 hours with boron and without boron is described by the Clausius-Clapeyron equation. It is shown that boron is leaded to a parallel shift of the curves $\sigma_{cr}(T)$ in area the low-temperature of test without changing the value of $\alpha = d\sigma_{0.1}(T)/dT$, which is associated with decreasing temperature M_s.

Acknowledgements
This work was financially supported by RNF grants 14-29-00012.

References

[1] Chumlyakov Y I, Kireeva I V, Panchenko E Y and etc. 2003 *Rus. phys. J.* **46** 811– 823
[2] Sehitoglu H, Efstathiou C, Maier H J and Chumlyakov Y 2005 *Journal of Magnetism and Magnetic Materials*. **292** 89–99
[3] Kokorin V V 1987 Martesitic transformations in nonhomogeneous solid solutions *Naukova Dumka Kiev*
[4] Kokorin V V and Gun'ko L P 1995 *Metal Physics and the latest technology* **17** 30– 35
[5] Tanaka Y, Himuro Y, Kainuma R and et al. 2010 *J Science* **327** 1488–1490
[6] Omori T, Abe S, Tanaka Y, Lee G Y, Ishida K and Kainuma R 2013 *Scripta Mater* **69** 812–815
[7] Otsuka K and Wayman C M 1998 Shape memory materials *Cambridge University Press* 284
[8] Kroos P, Niendorf T, Karaman I, Chumlyakov Y I and Maier H J 2012 *Functional Materials Letters* **5** 1250045-1–1250045-1
[9] Ma J, Hombuckle B C, Karaman I, Thompson G B, Luo Z P and Chumlyakov Y I 2013 *Acta Mat.* **61** 3445–3455
[10] Chumlyakov Y I, Kireeva I V, Kretinina I V, Kaynih K S, Kuts O A, Kirillov V A, Karaman I and Maier H 2013 *Rus. phys. J.* **56** 66–74
[11] Chumlyakov Y, Kireeva I, Panchenko E, Timofeeva E, Kretinina I, Kuts O, Karaman I and Maier H 2014 *Advanced Materials Research* **1013** 15–22
[12] Chumlyakov Yu I, Kireeva I V, Poklonov V V, Pobedennaya Z V. and Karaman I 2014 *Tech. Phys. Letters* **40** 747–750
[13] Kretinina I V, Kuts O A, Kuksgauzen D A, Poklonov V V, Pobedennaya Z V and Platonova Yu N 2015 *IOP Conf. Series: Materials Science and Engineering* **71** 1–6
[14] Chymlyakov Yu I, Kireeva I V, Kuts O A and Kuksgauzen D A 2015 *Rus. phys. J.* **57** 1328–1335
[15] Chumlyakov Y I, Kireeva I V, Korotaev A D, Litvinova E I and Zuev Y L 1996 *Rus. phys. J.* **39** 5–33
[16] Nembach E 1997 Particle strengthening of metals and alloys *John Wiley & Sons, Inc.*
[17] Gerold V and Haberkorn H 1966 *Phys. Status Solidi.* **16** 675–684

Influence of mechanical milling conditions on the dispersity of lithium ferrite

S A Lamonova, E N Lysenko and A V Malyshev
National Research Tomsk Polytechnic University, Tomsk, Russia

E-mail: sal17@tpu.ru

Abstract. The effect of mechanical milling in a planetary ball mills on dispersity of the synthesized lithium ferrite powder was investigated by laser diffraction technique. The mechanical milling of powder was carried out by two planetary ball mills: SPEX 8000M and Fritsch Pulverisette 5 with different time period. The lithium ferrite powder was milled for 30, 60 and 120 minutes in the first ball mill. The powder milled for 60 minutes in the second ball mill was performed. The results showed that an increase in the milling time slightly decreases the average particle size of the ferrite powder in case both ball mills. However, only milling in a SPEX planetary mill increases homogeneity of the ferrite powder, while the milling in Fritsch causes intensive processes of particle aggregation.

1. Introduction

Lithium ferrites have currently of great interest to researchers due to their technological applications as a low cost magnetic material for microwave techniques [1-5]. As was shown in [6-7], lithium ferrite is considered a promising cathode material due to its low environmental impact and lower cost compared to other commonly used materials such as Ni and Co. Also, $LiFe_5O_8$ can be used as an element of gas sensors because of high temperature and chemical stabilities [8].

The perfection of traditional methods of synthesis and the elaboration of new methods for lithium ferrites were implemented in order to improve the quality of production. In most practical applications, it is required to produce nanostructured $LiFe_5O_8$ of homogeneous composition [9-12].

Pre-milling operations in a planetary ball mill have become of great importance for production of ferrite materials by ceramic technology. This type of processing enables homogenization, milling and mechanical activation of the initial powder reactants. This leads to the stimulation of physico-chemical processes in powder materials [13-14].

In [15-19], it was shown that mechanical activation of the initial reactants allows great increase in its reactivity and enables to obtain lithium ferrites at significantly lower temperatures if compared to with those obtained by the conventional method.

In [20], inclusion of mechanical activation into the technological cycle enables to the produce of lithium ferrite ceramic with a higher density.

In the research, we studied the effect of the duration of mechanical activation in planetary ball mills of different types on the dispersion of synthesized lithium ferrite powders.

2. Experimental

Lithium ferrite samples were prepared by standard technology. Iron oxide and lithium carbonate were used as the initial reagents to produce lithium ferrite. The ratio of the initial components in the reaction mixture was calculated by the equation: $Li_2CO_3+5Fe_2O_3 \rightarrow 2LiFe_5O_8+CO_2\uparrow$ Before weighing, initial powders were dried for three hours at a temperature of 200°C in a drying oven. Synthesis of lithium ferrite was conducted at 800°C for 120 minutes in "HimLabo" standard laboratory oven with programmable thermostat "VARTA". The parameter of heating and cooling was ~ 10°C/min.

The obtained powder was divided into two parts. The powder from the first part (sample A) was milled dry in SPEX 8000M planetary ball mill using tungsten carbide balls. The powder was consistently milled for 30, 60 and 120 minutes.

The milling of lithium ferrite from the second part (sample B) was performed for 60 minutes using Fritsch Pulverisette 5 ball mill with tungsten carbide balls.

For samples A and B, the particle size was analyzed by laser diffraction method using Fritsch Analysette 22 MicroTec Plus analyzer. Additionally for sample A, micrographs of non-milled and milled powders were obtained with Supra 50 VP scanning electron microscope (SEM).

3. Results and discussion

Figure 1 shows the SEM micrographs for a non-milled sample A (figure 1a), milled lithium ferrite 30 min (figure 1b), 60 min (figure 1c), and 120 min (figure 1d).

Figure 1. SEM micrographs for non-milled sample (a), milled for 30 min (b), 60 min (c), and 120 min (d).

The average particle size was 0.59 μm for a non-milled sample. After milling, the average particle sizes were 0.57 μm in milling for 30 min, 0.54 μm in milling for 60 min, 0.42 μm in milling for 120 min.

As can be seen in figure 1, mechanical activation of lithium ferrite powder leads to a slight reduction in particle size and higher powder homogenization. Thus, to obtain a homogeneous composition of the lithium ferrite powder, the time of mechanical treatment should be at least 2 hours.

The characteristics of particle size distributions by laser diffraction method are summarized in table 1. For the milled powders, a narrow monomodal particle size distribution are observed as a result of higher homogeneity of the ferrite powder. The results showed that as the duration of milling increases, the average particle size decreases.

Table 1. Particle size distribution for sample A.

Sample	D_{10} (μm)	D_{50} (μm)	D_{90} (μm)
Initial sample	1.28±0.02	1.95±0.02	3.00±0.10
Milled for 30 min	1.13±0.09	1.80±0.06	2.88±0.04
Milled for 60 min	1.00±0.01	1.54±0.05	2.38±0.18
Milled for 120 min	1.90±0.05	1.32±0.02	1.91±0.03

The differential and integral particle size distribution for sample B is constructed and compared with the samples milled in a ball mill SPEX 8000M (figure 2).

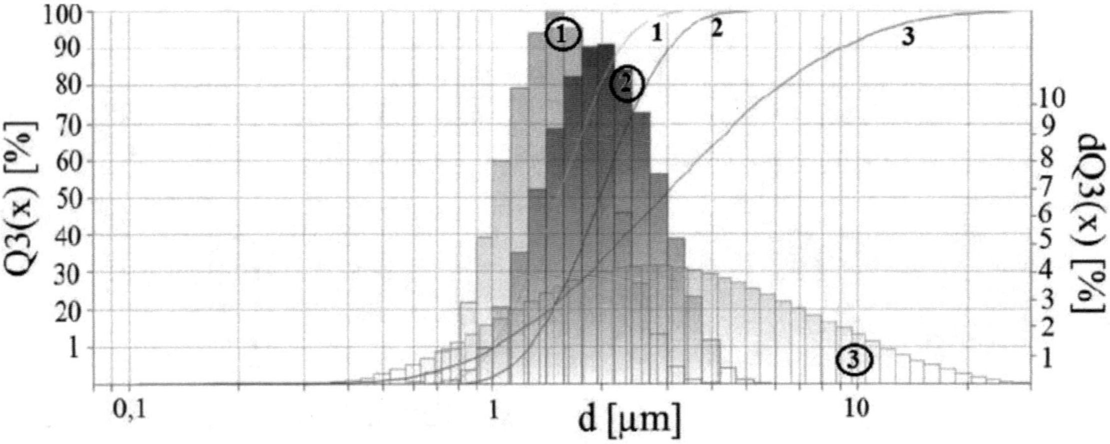

Figure 2. Differential and integral particle size distribution for samples B milled for 60 min: 1- initial sample, 2- milling in SPEX 8000M ball mill; 3- milling in Fritsch Pulverisette 5 ball mill.

The characteristics of the particle size distributions are shown in table 2. The date indicate, enhances both of the process and more intensive process of particle aggregation occur after powder milling in Fritsch Pulverisette 5 ball mill.

Table 2. Particle size distribution for sample B.

Sample	D_{10} (μm)	D_{50} (μm)	D_{90} (μm)
Initial sample	1.28±0.02	1.95±0.02	3.00±0.10
After milling in Fritsch Pulverisette 5	1.13±0.09	1.80±0.06	2.88±0.04
After milling in SPEX 8000M	1.00±0.01	1.54±0.05	2.38±0.18

4. Conclusion

The results showed that mechanical activation of lithium ferrite powder in air at room temperature causes slight reduction of the particle size and increases the homogeneity of the ferrite powder.

In this case, the optimal conditions to obtain a homogeneous composition of the lithium ferrite powder are:

- mechanical activation in SPEX 8000M planetary ball mill;
- at least 2 hour of milling.

The obtained powders can be used for further sintering of ceramic from lithium ferrite.

Acknowledgements
This work was financially supported by The Ministry of Education and Science of the Russian Federation in part of the science activity program.

Reference

[1] White G O and Patton C E, 1978 *J. Magn. Magn. Mater.* **9** 299.
[2] Baba P D, Argentina G M, Courtney W E, Dionne G F and Temme D H 1972 Fabrication and properties of microwave lithium ferrites *IEEE Trans. Magn.* **8** 83.
[3] Smit J and Wijn H P J 1959 *Ferrites. Physical properties of ferrimagnetic oxides in relation to their technical applications* (Philips Technical Library: Eindhoven) p 299.
[4] Surzhikov A P, Lysenko E N and Malyshev A V 2014 *Rus. Physic. J.* **57** 621.
[5] Green J J and Van Hook H J 1977 *IEEE Trans. Microwave Theory Tech.* **25** 155.
[6] Tabuchi M, Ado K, Sakaebe H, Masquelier C, Kageyama H and Nakamura O 1995 Preparation of $AFeO_2$ (A = Li, Na) by hydrothermal method. *J. Solid State Ion.* **79** 220.
[7] Wang X, Gao L, Li L, Zheng H, Zhang Z, Yu W and Qian Y 2005 *Nanotechnology* **16** 2677.
[8] Rezlescu N, Doroftei C, Rezlescu E and Popa P D 2008 Lithium ferrite for gas sensing applications *J. Sens. Actuators.* **133** 420.
[9] Surzhikov A P, Pritulov A M, Lysenko E N, Sokolovsky A N, Vlasov V A and Vasendina E A 2012 Dependence of lithium–zinc ferrospinel phase composition on the duration of synthesis in an accelerated electron beam *J. Therm. Anal. Calorim.* **110** 733.
[10] Humaira A and Asghari M 2013 Microwave Magnetic and Absorption Properties of $Li_{0.5}Mn_{x/2}Zn_{0.75-x/2}Fe_2O_4$ Soft Nano Ferrites Prepared by Sol-Gel Auto Combustion Method, *J. Electronic Materials Letters.* **9** 641.
[11] Reddy P V B, Ramesh B and Reddy Ch G 2010 Electrical conductivity and dielectric properties of zinc substituted lithium ferrites prepared by sol-gel method, *J. Physica* B. **405** 1852.
[12] Surzhikov A P, Lysenko E N, Vlasov V A, Malyshev A V and Nikolaev E V 2013 Investigation of the process of ferrite formation in the Li_2CO_3–ZnO–Fe_2O_3 system under high-energy actions *Russian Physics J.* **56** 681.
[13] Zyraynov V V 2008 Mehanohimicheskiy sintez slozhnih oxidov, *J. Uspehi himii.* **2** 107.
[14] Giri A K 1997 Nanocrystalline materials prepared through crystallization by ball milling, *J. Adv. Mater.* **9** 163.
[15] Berbennia V, Marinia A, Matteazzib P, Riccerib R andWelhamc N J 2003 Solid-state formation of lithium ferrites from mechanically activated Li_2CO_3–Fe_2O_3 mixtures *J. Eur. Ceram. Soc.* **23** 527.
[16] Surzhikov A P, Lysenko E N, Malyshev A V, Vasiljeva O G and Pritulov A M 2012 Influence of mechanical activation of initial reagents on synthesis of lithium ferrite, *Rus. Physic. J.* **6** 672.
[17] Kavanlooee M, Hashemi B and Maleki-Ghaleh H 2012 Effect of annealing on phase evolution, microstructure, and magnetic properties of nanocrystalline ball-milled LiZnTi ferrite, *J. Electron. Mater.* **41** 3082.
[18] Widatallah H M, Johnson C and Berry F G 2002 The influence of ball milling subcequent calcination on the formation of $LiFeO_2$, *J. Mater. Sci.* **37** 4621
[19] Gee S H, Hong Y K, Park M H, Erickson D W and Lamb P J Synthesis of nanosized (Li0.5xFe0.5xZn1-x)Fe_2O_4 particles and magnetic properties, *J. Appl. Phys.* **91** 7586.
[20] Surzhikov A P, Frangulyan T S, Ghyngazov S A, Lisenko E N and Galtseva O V 2006 Physics of magnetic phenomena: Investigation of electroconductivity of lithium pentaferrite, *Rus. Physic. J.* **49, 5** 506.

IOP Conf. Series: Materials Science and Engineering **93** (2015) 012036 doi:10.1088/1757-899X/93/1/012036

Structure and Properties of Ti-Nb-C Coatings Obtained by Non-vacuum Electron Beam Cladding

O G Lenivtseva[1], I A Polyakov[1], D V Lazurenko[1] and V S Lozhkin[1]
[1] Material Science in Mechanical Engineering Department, Novosibirsk State Technical University, 20 Prospekt K. Marksa, Novosibirsk, 630073, Russia

E-mail: lenivtseva_olga@mail.ru

Abstract. In this study the structure and properties of surface-alloyed cp-titanium layers obtained by non-vacuum electron beam cladding of niobium carbide powders were analyzed. A thickness of coatings fabricated by single-layer cladding was 1.3 mm. Cladding of the second layer led to an increase in the thickness by 0.8 mm. It was found that titanium carbide particles of different morphology acted as strengthening structural elements. The X-ray diffraction (XRD) analysis revealed the presence of α-Ti (α′-Ti), β-Ti, and TiC in the cladded layer. The results of the energy dispersive X-ray (EDX) analysis indicated the presence of Nb in the titanium matrix as well as in the carbide phase. However, such phases as NbC and (Nb, Ti)C were not identified by the XRD analysis. Transmission electron microscopy (TEM) revealed zones containing an increased amount of Nb. The structure of these zones was represented by the β-Ti and ω-Ti precipitation. An average microhardness value of cladded layers was approximately 330 HV.

1. Introduction

A number of machine components and structural elements, which have critical applications, cannot be produced without using titanium and its alloys. These materials possess unique mechanical properties including a high specific strength, a low density and a high corrosion resistance [1, 2]. Titanium alloys are most widely used in aircraft, shipbuilding, space technology, medicine, chemical engineering and sports equipment manufacturing.

However, these materials have also a number of disadvantages, such as high cost and high activity at enhanced temperatures which lead to gas saturation. Moreover, low tribological properties can be classified as serious drawbacks which limit a wide application of titanium alloys. Titanium has a tendency to frictional seizure [1-3].

One of the efficient solutions of the aforementioned problem is based on the formation of high strength and wear resistant coatings on the titanium workpieces [4-9]. Among the variety of methods of surface hardening developed by now only a limited number can be recommended for treatment of titanium alloys.

Laser cladding is considered to be a one of the most widely applied and advanced methods of surface treatment [7-9]. A limitation of this technology is a high reflectivity of metals and consequently a low efficiency of laser radiation.

Electron beam treatment can be suggested as a method alternative to laser cladding [3-6, 10]. In contrast to a laser beam, an electron jet, being a volume energy source, penetrates into the depth of a treated material.

Industrials accelerators such as ELV-6 were developed in the Budker Institute of Nuclear Physics (Russia, Novosibirsk) to generate relativistic electrons [11]. These accelerators are equipped with a unique device providing the injection of electrons in the air atmosphere. Interaction of a beam with a treated material induces accelerated heating of a surface layer to temperatures higher than the melting point of this material. A molten pool, in which modifying components are dissolved, is formed in these conditions.

The technology based on non-vacuum electron beam treatment allows manufacturing large-scale parts with a high degree of precision. Their dimensions are limited only by the table geometry and a shop floor area. Using an ELV-6 industrial accelerator allows treating almost any metallic materials as well as carbides, borides, nitrides and other compounds [3-6, 10-13]. Operational experience of this equipment showed that the parts made of titanium alloys could be surface–hardened to a high depth [4-5, 10].

In this paper the structure and properties of coatings obtained by non-vacuum electron beam cladding of niobium carbide on titanium substrates were investigated.

2. Materials and methods

In this study VT1-0 cp-titanium plates with a thickness of 10 mm were used as substrates. A powder mixture consisted of 50 wt. % of niobium carbide and 50 wt. % of calcium fluoride was used as a surfacing material. CaF_2 provided protection of a molten pool from atmospheric oxygen. Technological experiments were carried out in the Budker Institute of Nuclear Physics (SB RAS).

The scheme of the cladding process is presented in [4]. A surfacing material in an amount of 10 g / 1 cm^2 was uniformly distributed over a surface of the titanium workpiece and processed by an electron beam. A workpiece, mounted on a table, moved in the longitudinal direction with a speed of 25 mm /sec. In order to increase a volume fraction of a hardening phase in a coating, double-layer electron beam cladding of powder mixtures was employed. Cladding of the second layer was preceded by removal of a slag layer from the workpiece surface. Then the alloying mixture was applied to a surface and a treatment was repeated. The second layer cladding was accompanied by increasing the current (from 31 mA to 38 mA). Treatment of a wide area equal to the width of the workpiece (50 mm) was carried in a scanning mode with a scanning frequency of 50 Hz. The distance from the outlet to the sample was 90 mm. The values of specific surface energy introduced in the material during cladding of the first and second layers were 3.17 kJ/cm^2 and 4.26 kJ/cm^2 respectively.

Investigations of the structure of materials in the magnification range from x 25 to x 1500 were carried out using the *Carl Zeiss Axio Observer A1m* optical microscope. The structure of cladded layers was revealed by sample etching with Kroll's solution. Particularities of structural transformations were investigated using the *Carl Zeiss EVO*50 *XVP* scanning electron microscope (SEM).

A fine structure of coatings was studied by transmission electron microscopy (TEM) method using a *FEI Tecnai G*2 20 *TWIN* microscope at an accelerating voltage of 200 kV. TEM samples were prepared using the Gatan *Dimple Grinder* 656 equipment and the *Gatan PIPS* 659 precision ion polishing system.

A phase composition of caddied layers was investigated using the *ARL X'TRA* θ-θ diffractometer. A copper tube (*Cu Kα*) was used as a source of X-ray radiation. The recording of diffraction patterns was carried out in a step mode.

The microhardness of materials was estimated using the *Wolpert Group* 402 *MVD* semi-automatic microharness tester. The load on a diamond pyramid was 0.98 N.

3. Results and discussion

Samples under investigation with one and two cladded layers were identified as "1NbC" and "2NbC" respectively. An X-ray diffraction (XRD) pattern of the niobium carbide powder used for cladding is shown in Figure 1 (red line). The results of the phase analysis of cladded layers are presented in Figure 1 (black line). The main phases formed during cladding of both one and two layers of niobium carbide were α-Ti (α'-*Ti*), β-Ti and titanium carbide. The analysis of the XRD patterns revealed the absence of peaks corresponding to niobium carbide or (*Nb*, *Ti*)*C* complex carbide [15].

Figure 2 shows the distribution of chemical elements along the line obtained by the energy dispersive X-ray (EDX) analysis. It is clearly seen in the picture that the concentration of niobium in the carbide phase was lower than in the matrix. Analysis of the elemental composition of the carbide phase and the titanium matrix revealed a 2-fold rise in the niobium concentration in the titanium matrix together with a rise in the niobium quantitative content from about 8 wt. % to 13.5 wt. % as a consequence of a layer amount increase.

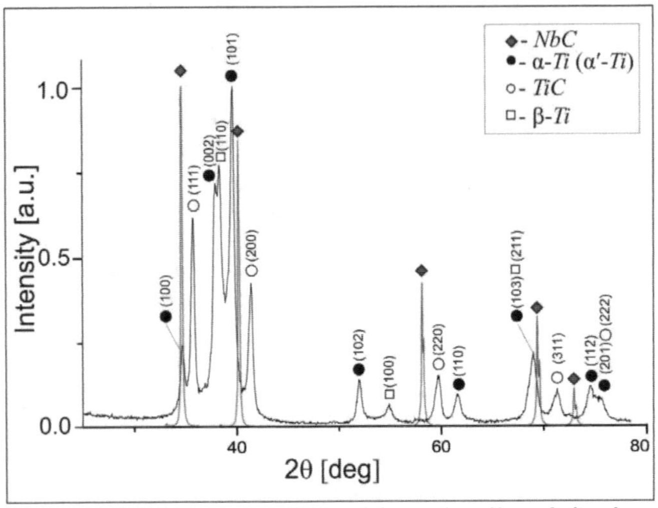

Figure 1. XRD patterns obtained from the alloyed titanium layer: red line – niobium carbide powder; black line – a sample fabricated by one-layer cladding of niobium carbide.

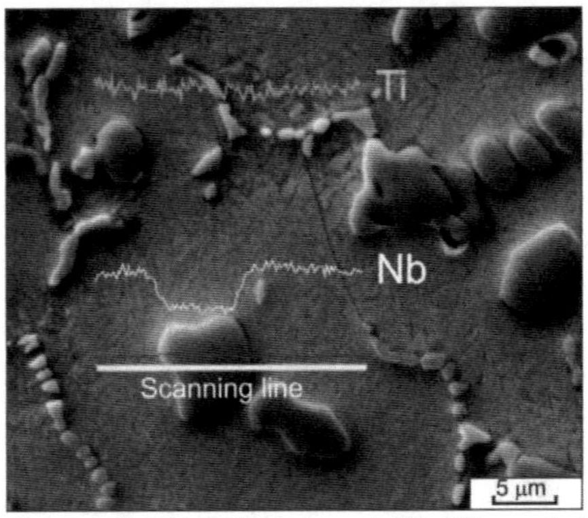

Figure 2. Distribution of Ti, Nb and C along the scanning line.

Figure 3 shows the structure of the coating. The material can be nominally divided in several zones: a cladded layer (1-3), a heat affected zone (4) and a base material (5).

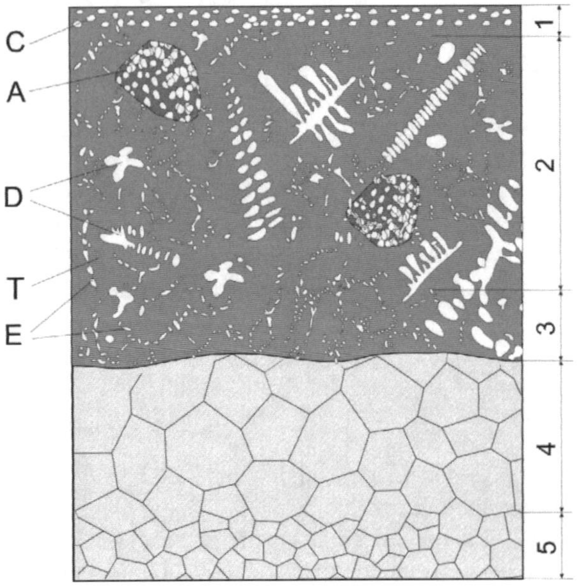

Figure 3. Scheme of the structure of the titanium workpiece obtained by non-vacuum electron beam cladding of niobium carbide: 1 – an area containing fine close-packed TiC particles ; 2 – an area containing TiC particles of different morphology; 3 - an area containing eutectic TiC particles; 4 – a heat affected zone; 5 – the initial structure of titanium; C - fine close-packed TiC particles; T – a lamellar structure of the titanium matrix; A – TiC aggregations; E – fine TiC particles found in the TiC-Ti eutectic; D – dendritic TiC particles.

The cladded layer is an area of the particular interest. During electron beam treatment niobium carbide particles are dissolved in the melted material and the diffusion of Nb and C occurs. As a result, niobium partially dissolved in the alloy contributes to the stabilization of the titanium β-phase at room temperature. On the other hand, some of the niobium partially replaces titanium atoms in the carbide phase. As a consequence of a short time of the material being in a melted state, an inhomogeneous distribution of chemical elements across the depth of the cladded layers occurs.

Weakly etched areas were observed in the structure of the cladded layers (Figure 4, a). This phenomenon was generally observed in the areas characterized by the carbide particles accumulation (Figure 4 a, b). Cladding of the second layer resulted in increasing the structural uniformity of the material. Aggregations of carbide particles were formed in the layer with a thickness of 800 – 1000 μm.

A duplex microstructure of the titanium surface layer was revealed by SEM (Figure 4, d). The doping level determined the structure of local zones of the caddied layer: some areas of the coating were represented by a β-Ti-based solid solution with nanosized precipitations of ω-Ti (Figure 4, b; Figure 5 a, b); the others were characterized by the presence of a α- and β-Ti mixture (Figure 4, d).

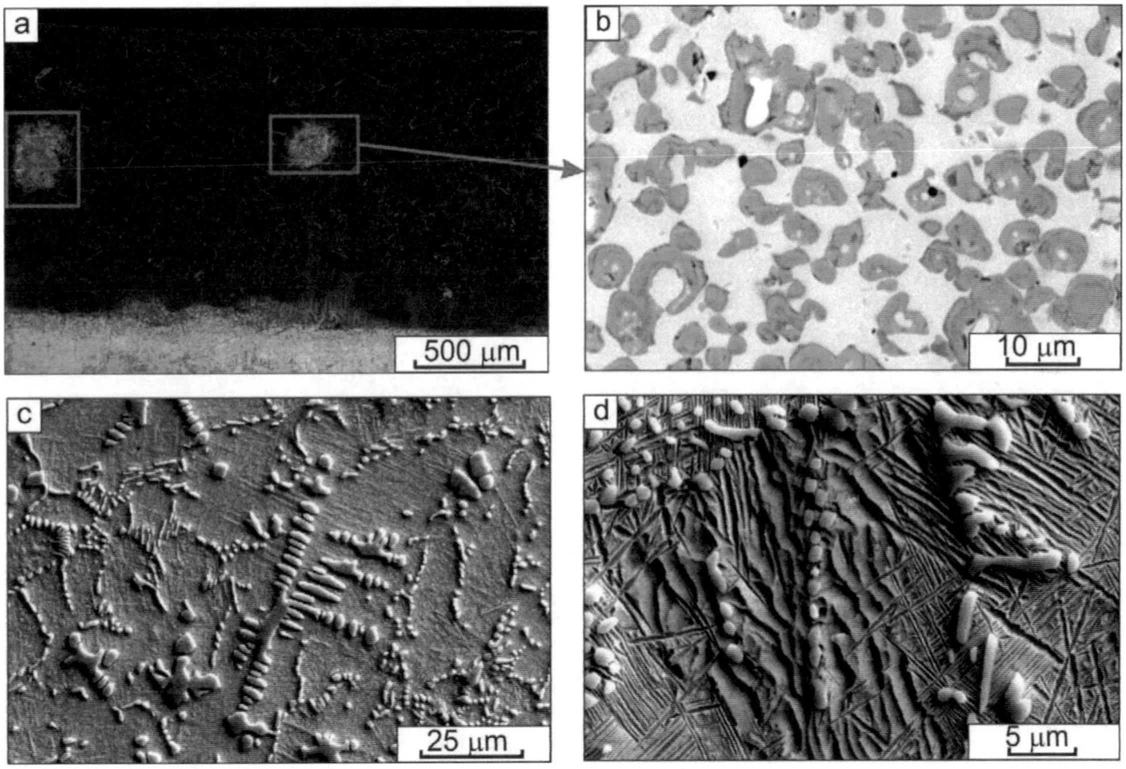

Figure 4. SEM microphographs of the structure of titanium surface layers after non-vacuum electron beam cladding of niobium carbide powders.

The presence of martensitic structure attributes was revealed in some zones (Figure 5, c). Taking into account a high Nb concentration in these areas and a high cooling velocity, it can be supposed that martencitic crystals were represented by the α″-phase with a composite orthorhombic lattice. The formation of the α″-*Ti* martensitic phase in similar conditions was also observed in [14].

Titanium carbides alloyed with niobium had a different morphology (see Figure 3). In the 1st zone (near the surface of the cladded layer) the carbide phase was represented by round particles Dendritic crystals with long one-fold axes and short secondary branches as well as fine particles distributed along the grain boundaries and within the grains can be seen in the direction from the surface to the center of the cladded layer (Figure 4).

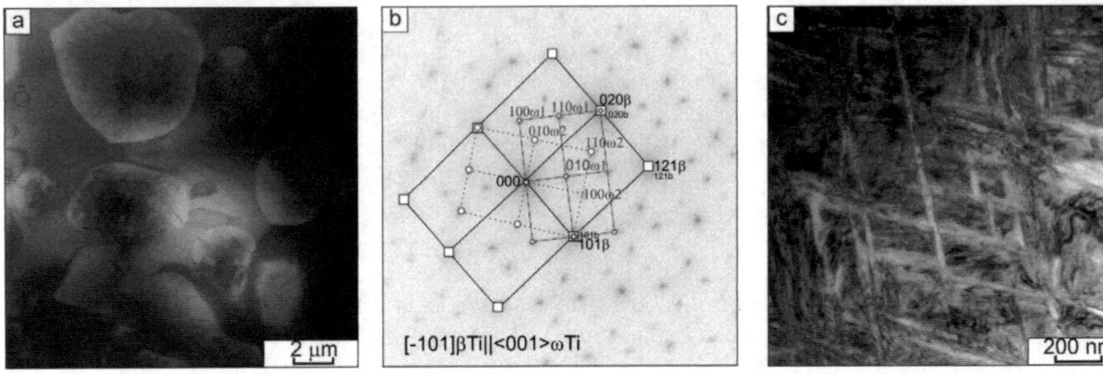

Figure 5. Bright-field microphotographs of the coating structure and the microdiffraction pattern: a – the aggregation zone of carbide particles; b –the electron diffraction pattern obtained from the areas in Figure 4, a and corresponding to β- and ω-Ti; c –the martensitic structure of titanium in the cladded layer.

Analysis of the relationship between microhardness and the depth of layers cladded on "1*NbC*" and "2*NbC*" samples (Figure 6) showed that the maximum microhardness level (480 HV) was observed in the double-layer coating. The material microhardness decreased to 330 HV at a depth of 0.4 mm. An increase in the microhardness level occurred at the depth of 0.9 mm due to the accumulation of carbide particles (Figure 4 a, b) and the formation of ω-Ti. An average microhardness level of the modified layer formed on a "1*NbC*" sample was 330 HV.

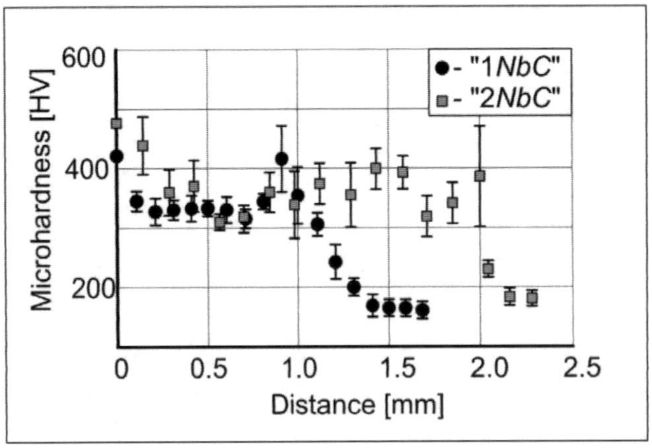

Figure 6. Distribution of microhardness in the depth of the cladded layer obtained by non-vacuum electron beam cladding of the niobium carbide powder.

4. Conclusions

Non-vacuum electron beam cladding of niobium carbide powders allows efficient hardening of surface layers of titanium workpieces to a depth of approximately 2.1 mm. The main phases formed in the cladded layers were hexagonal titanium, cubic titanium and titanium carbide. The presence of niobium carbide (NbC) and complex carbide ((Nb, Ti)C) was not observed by TEM and XRD. It can therefore be said that niobium is contained in the titanium matrix in the form of a solid solution as well as in the carbide phase. High velocities of the coating formation resulted in the areas enriched with niobium where the presence of ω-Ti was observed. An average microhardness value of the cladded layer was approximately 330 HV.

Acknowledgements
The reported study was partially supported by RFBR, research project No. 15-38-20776 mol_a_ved.

References

[1] Leyens C and Peters M 2003 *Titanium and Titanium Alloys: Fundamentals and Applications* (New York: Wiley-VCH Verlag GmbH & Co. KGaA)
[2] Lütjering G and Williams J C 2007 *Titanium (Engineering Materials and Processes)* (Berlin: Springer)
[3] Oh J, Lee S and Golkovski M G 2001 *Metal. Mater. Trans.* **32** 2995–3005
[4] Lenivtseva O, Golovin E, Samoylenko V, Mul D and Golovin D 2014 *Advanced Materials Research* **1040** 784-9
[5] Lenivtseva O G, Bataev I A, Belousova N S, et al 2013 *Obrabotka metallov (tekhnologiya, oborudovanie, instrumenty)* **61** 49-57
[6] Golkovski M G, Bataev I A, Bataev A A, et al *Mater. Sci. Eng., A.* **578** 310-7
[7] Saleh A F, Abboud J H and Benyounis K Y 2010 *Optics and Lasers in Engineering* **48** 257-67
[8] Courant B, Hantzpergue J J and Benayoun S 1999 *Wear* **236** 39–46
[9] Li J, Yu Z and Wang H 2011 *Thin Solid Films* **519** 4804-8
[10] Bataev I A, Bataev A A, Golkovski M G, et al 2013 *Appl. Surf. Sci.* **284** 472– 81
[11] Fadeev S N, Golkovski M G, Korchagin A I, et al 2000 *Radiation Physics and Chemistry* **57** 653-5
[12] Bataev I A, Golkovski M G, Bataev A A, et al 2014 *Surf. Coat. Technol.* **242** 164–9
[13] Bataev I A, Bataev A A, Golkovski M G, et al 2012 *Surf. Coat. Technol.* **207** 245-53
[14] Popov A A, Illarionov A G, Rossina N G and Grib S V 2013 *Metallovedenie i termoobrabotka splavov titana. Struktura i svojstva* : *uchebnoe posobie* (Ekaterinburg: URFU)
[15] Lenivtseva O G 2014 *Poverhnostnoe uprochnenie titanovyh splavov karbidnymi chasticami s ispol'zovaniem tehnologii vnevakuumnoj jelektronno-luchevoj naplavki* (Novosibirsk: NSTU)

Application of spark plasma sintering for fabricating Nd-Fe-B composite

A A Sivkov [1], A S Ivashutenko [1] and A A Lomakina[1]
[1]Russia, Tomsk, Lenin ave, 30, National Research Tomsk Polytechnic University

E-mail: svechkanevaaa@mail.ru

Abstract. Constant magnets are applied in such fields as electric equipment and electric generators with fixed rotor. Rare earth metal neodymium is well known as promising material. Production of magnets by sintering three elements (neodymium, iron and boron) is one the most promising methods. But there are difficulties in choosing the right temperature for sintering and further processing. Structure and properties of the product, consisted of rare earth metals, was analyzed. X-ray analysis of the resulting product and the finished constant magnet was performed. Vickers microhardness was obtained.

1. Introduction

The permanent magnets (PM) are widely used in various fields of science and technology, such as radio and electrical engineering, electronics, devices for recording information, microelectronics, etc. The advantages of the permanent magnets are as follows: long life, material savings, and reliability. The main aim of modern technology is to decrease sizes at the same time with the saving high operating characteristics. This requirement is also attributed to the PM. Now the most promising materials are magnets based on cobalt, neodymium and samarium. PM based on rare earth elements have a hexagonal structure, strong magnetic anisotropy and high Curie temperature.

In 1984 the compound of neodymium (Nd) and iron (Fe) was firstly mentioned [1]. The most successful materials in this group are based on cobalt and neodymium magnets. The main advantage of the neodymium-iron-boron magnets in comparison with the samarium-cobalt alloy is the price and the scale of production. Also such class of rare-earth compounds shows record values of residual magnetic induction B_r = 1.1-1.2 T and coercive force H_c = 950-1750 kA/m. Papers [2-4] show the possibility of obtaining a composite material Nd-Fe-B, having practical advantages, such as thermal stability, high corrosion resistance and fracture toughness. The mentioned above high values of magnetic characteristics can be explained by the fine structure of materials and they are produced with using high-intensity actions. Spark plasma sintering (SPS) can be attributed to these kinds of effects. One of the most important benefits of SPS-method is its high sintering speed, which can effectively restrain the grain growth and ensure the formation of a dense fine grain structure in a short time. This is the purpose of this paper.

2. Experimental

For sintering industrially produced nanopowders were used as precursors. Figure 1 shows scanning electron microscopy (SEM) images and X-ray diffraction patterns of powders: Nd (figure 1, a), Fe (figure 1, b) and B (figure 1, c). These precursors were used to obtain a mixture which was sintered by SPS method.

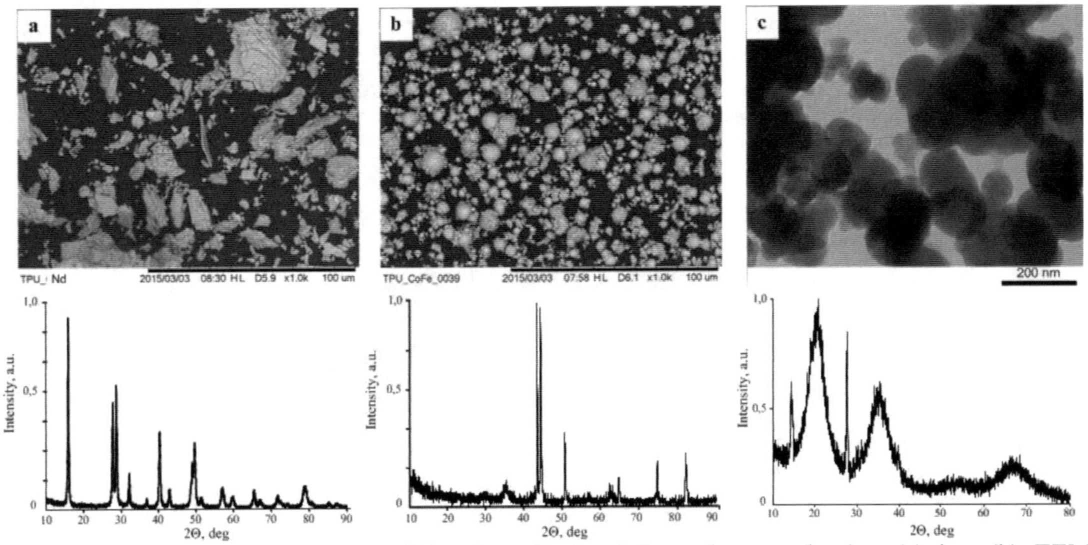

Figure 1. SEM-images and X-ray diffraction of industrial powders: neodymium (a), iron (b); TEM-image and X-ray pattern of boron (c).

The percentage of precursors in the mixture was as follows: Nd - 33 mass.%, Fe - 65 wt.%, B - 2 wt.%. The resulting mixture was mixed for 2 hours in a planetary mill SPEX Sample Prep. Mixer / Mill 8,000 M. After that the X-ray analysis for the powder mixture was carried out using diffractometer XRD Shimadzu 7000S. The XRD pattern is shown in Figure 2.

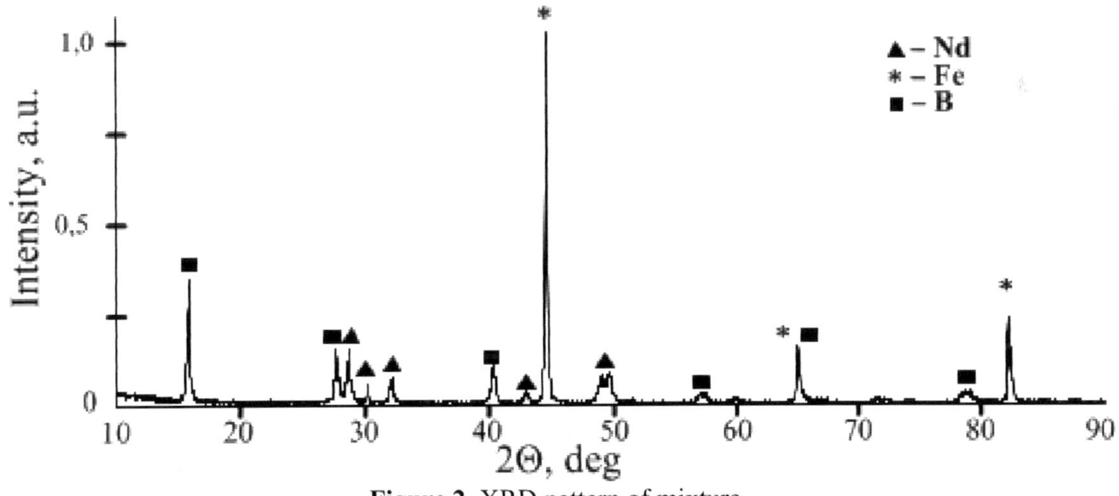

Figure 2. XRD pattern of mixture.

The finished mixture was sintered using the setup SPS10-4 (Advanced Technology, USA). The sintering was carried out in vacuum at 700 °C. The sintering mode was as follows: heating rate V_t= 100 °C/min, the pressing force 50 MPa, holding at the sintering temperature for 10 minutes, the current flow mode with a pulse repetition rate 25/5. The mixture material was molded in a graphite die with an inside diameter of 20 mm. The temperature monitoring was carried out using the K-type thermocouple, placed in the closed hole of the lower punch in 2.5 mm from the powder. The pressure was applied at the beginning of the sintering process and was removed after complete cooling of the sample.

Figure 3 shows the diagram of the geometric dimensions of the sample during sintering. The onset temperature of the sintering process (shrinkage) was 600 °C. In the time interval t_1-t_2 the sintering process went more intensively due to increasing temperature. In the case of isothermal hold at 700 °C the intensity of the shrinkage decreases.

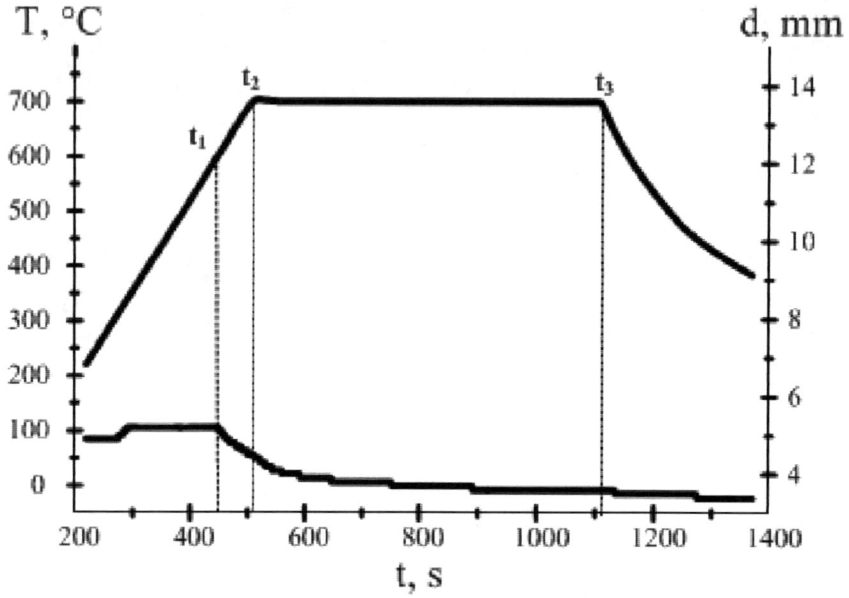

Figure 3. Diagram of the sintering process and the shrinkage of the sample during SPS-sintering.

After sintering, the sample was subjected to grinding. The hardness of the samples was studied using microhardness Isoscan OD HV2 Vickers. The microstructure of the polished surface and the chipping were examined using a scanning electron microscope HITACHI TM3000. The phase composition was determined using X-ray diffraction XRD Shimadzu 7000S (CuKα-radiation).

3. Results and discussion

The results of qualitative X-ray analysis were obtained using PDF2+. The figure 4 shows the phase analysis confirms the presence of such main PM phases as Nd, Fe and B.

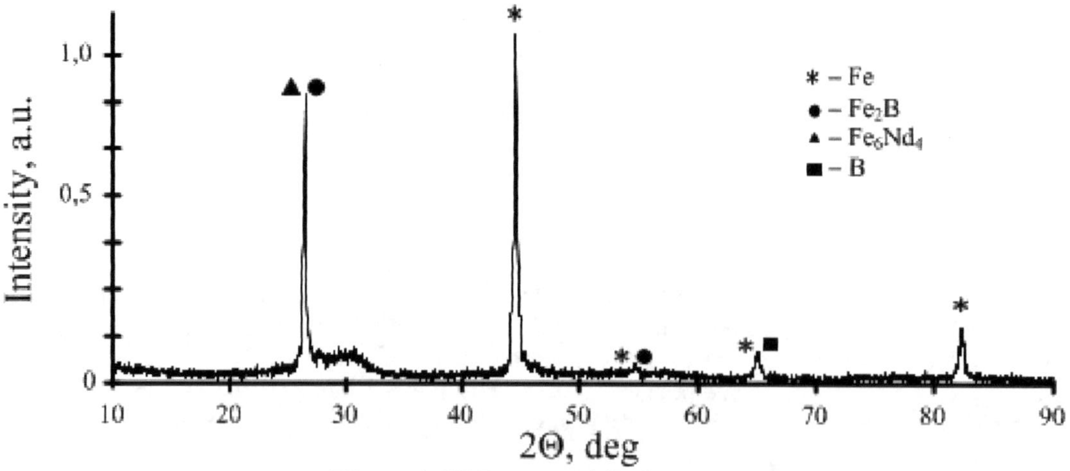

Figure 4. XRD pattern of Nd-Fe-B.

Data of the Vickers hardness and fracture toughness were obtained in the [5]. The hardness measurement was carried out using Vickers indenter according to the method of unrestored print. The average value of the microhardness was 142.06 HV. The maximum value of known hardness for permanent Nd-Fe-B magnets is 600HV. Also the average diagonal indentation equal to 0.1144 mm was measured.

Figure 5 shows the SEM-images of the surface of the sintered sample. The investigation of the sample surface showed the presence of sintered objects (Figure 5, b). In particular, there is an object not subjected to sintering and connections to other parts of the sample (see Figure 5, c and d).

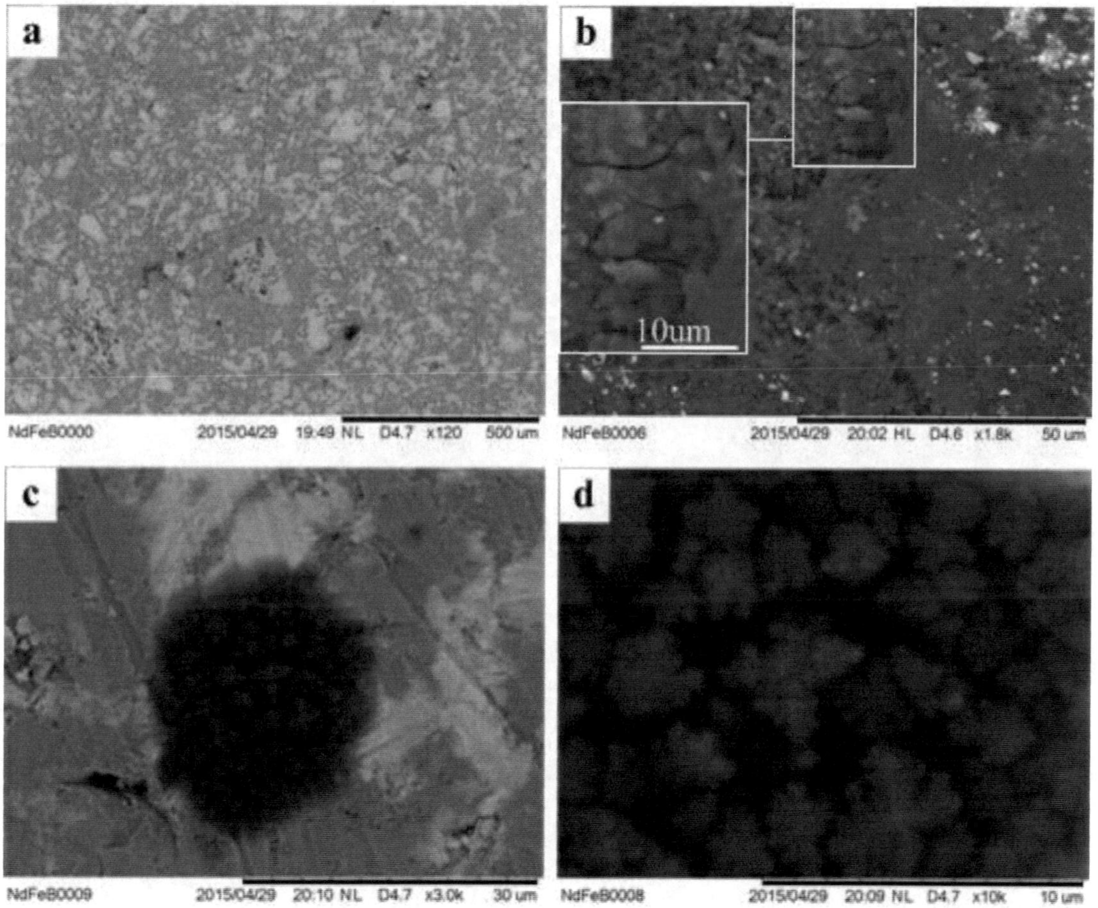

Figure 5. SEM-images of the sintered surface of the Nd-Fe-B sample.

In order to study the internal structure of the material the chipping was made. Figure 6 shows the SEM-images of chipping. An analysis of images allows us to conclude that the magnet material has a residual porosity. The density of the sample, which is equal to 6.32 g/cm^3, (theoretical density is 7.4 g/cm^3) was measured by hydrostatic method. Therefore, to obtain a dense structure it is necessary to increase the sintering temperature, and, possibly, the time of isothermal hold. The similar structure of the material is presented in [6,7].

To study the magnetic properties of the sample the magnetometer H-04 was used, taking the measurements in fields up to 30 kOe [8]. Figure 7,a shows the field dependence of the specific magnetization σ(H). The value of the specific magnetization is determined by the tangent line and is equal to 156 Gs·cm^3/g (dashed line in Fig. 9), which is in good agreement with published data (paper

[9] shows the value of specific magnetization of a permanent magnet $Nd_2Fe_{14}B$ with a value of more than 150 ° Gs·cm^3/g).

Figure 6. SEM-images of the chipping Nd-Fe-B.

The sample presented in our work has the effective anisotropy field at the level of 3,6 kOe (Figure 8). This value is less than that was previously described (6,4 kOe [2]). This difference can be explained by the residual porosity in our sample.

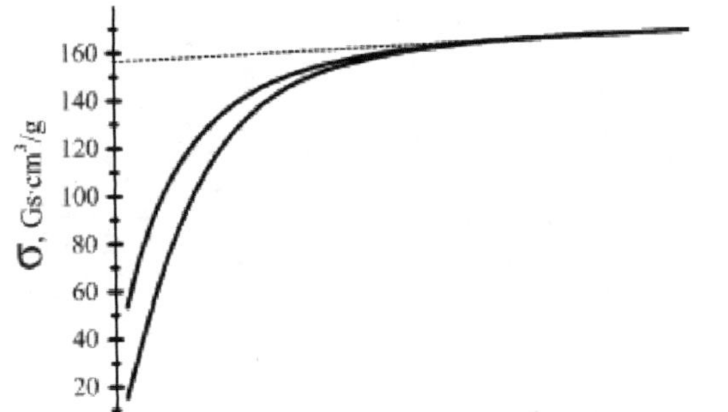

Figure 7. The curve of specific magnetization of SPS sample Nd-Fe-B

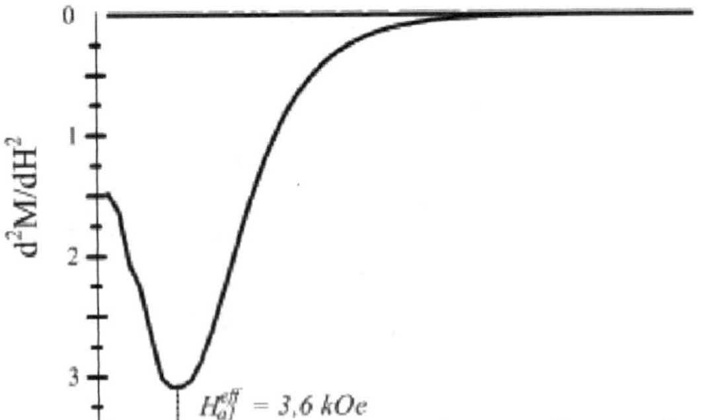

Figure 8. The curve of effective anisotropy field of SPS sample Nd-Fe-B

4. Conclusion

Thus the SPS-sintering method provides the synthesis of Nd-Fe-B permanent magnets in a forward and short process cycle (20 min.). This method allows synthesizing magnetic material, which combines high technology, the stability of the magnetic characteristics, relatively low cost. The experimental results provide further refinement of sintering parameters of magnetic material to improve its magnetic characteristics.

References

[1] Sagawa M, Fujimura S, Togawa N, Yamamoto H and Matsuura Y 1984 *J. Appl. Phys.* **55** 2083–2087

[2] Yue M, Tian M, Zhang J X, Zhang D T, Niu P L and Yang F 2006 *Mater. Sci. Eng., B* **131(3)** 18-21

[3] Song J, Yue M, Zuo J, Zhang Z, Liu W, Zhang D, Zhang J, Guo Zi and Li W 2013 *Journal of Rare Earths* **31(7)** 674-678

[4] Li X, Li L, Hu K, Chen Z, Qu S and Yang C 2014 *Transactions Nonferrous Metals Society of China* **24** 3142-3151

[5] Hu Z H, Chu L H, Li J and Liu Y 2012 *J. Magn. Magn. Mater.* **324(2)** 101–104

[6] Hu Z H, Qu H, Zhao J., Luo C., Li J and Liu Y 2012 *Journal of Rare Earths* **30(11)** 1112-1115

[7] Hu Z H, Li J, Chu L H and Liu Y 2011 *J. Magn. Magn. Mater.* **323(1)** 104–107

[8] Sivkov A A , Ivashutenko A S, Lomakina A A and Shanenkov I I 2015 *Appl. Mech. Mat.* **756** 325-328

[9] Neznakhin D S, Bolyachkin A S, Volegov A S, Markin P E, Andreev S V and Kudrevatykh N V 2015 *Magn. Magn. Mater.* **377** 477–479

Multifunctional composite material based on carbon-filled polyurethane

T Malinovskaya[1], S Melentyev[2] and S Pavlov[2]
[1]National Research Tomsk State University, Innovation and Technology Center, Tomsk, 634050, Russia
[2]Tomsk State University of Architecture and Building, Applied Mechanics and Material Sciences Department, Tomsk, 634003, Russia

E-mail: sergey.melentev.88@mail.ru

Abstract. The research paper deals with the performance of composite resistive material heating coatings based on the polyurethane binder, filled with colloidal-graphite preparation C-1, which can be used in structures of electric heaters. Frequency dependences of transmission and reflection coefficients, dielectric permeability of composite materials with the various content of carbon fillers (technical carbon, graphite) in polyurethane varnish in ranges of frequencies 26–40 GHz and 110–260 GHz are experimentally investigated.

1. Introduction

Among composite materials developed nowadays of greatest interest are multifunctional composites used in various fields. Polymer-based composite materials with various fillers belong to these materials [1–3]. They are widely used in building, machine-building, oil refining, and light industries. These mostly include composite resistive materials (CRM), used as coatings for heating elements of technological equipment designs for industry and construction, such as heating press plates and ovens, wall plating and oil baths, tanks, pipelines, electric, and metal shields thermosetting formwork [4, 5].

The research analysis of CRM and their use as coatings for heating elements of technological equipment showed that some problems are still unresolved due to the lack of such heating coatings characteristics as stability and performance of electrical resistance in a long time and wide temperature ranges, high adhesive strength to metal surfaces, uniform temperature distribution on the surface and others. Different research organizations around the world try to find solution of this problem, including: Research Institute of chemical fiber and composite materials, Scientific and Production Center «Carbon fibers and composites», PERI, MITSUBISHI and others. Currently, when creating a CRM as binders such heat-resistant polymers as fluoroplastic, polyamide, polypropylene and others are widely used. However, these materials obtain a certain number of characteristics which complicate their use in CRM. They are low adhesion, a significant dimensional instability, the allocation of large amounts of volatile substances in polymerization. As a conductive component the dispersed metal or carbon fillers, where some nanoscale forms of carbon such as carbon black, carbon fibers, carbon nanotubes, graphite, graphene, being particularly attractive for the creation of CRM and rather advantageous with lightness and chemical inertness.

The use of a polyurethane CRM binder is promising as it has good adhesion to almost all existing materials, the ability to provide a solid contact with the conductive polymer molecules filler particles. This can lead to a unimodal distribution of particles in the mixture and, consequently, improve the conductive properties of CRM with minimum heat deflection temperature. In recent years, there was a number of scientific publications describing advances in obtaining and using conductive composites based on polyurethane filled with carbon black, carbon fibers, carbon nanotubes, graphite, graphene [6, 7]. However, these studies lack the information on the use of polyurethane filled with carbon in CRM and study of composites performance on its basis, which can be used as a heating coating in the construction of electric heaters technological equipment.

In addition, it is well known that composites with carbon fillers are used as screens reflecting or absorbing electromagnetic radiation to protect biological objects against the harmful effect of microwave radiation and radio equipment against internal or external noise. Despite of many developments in this direction, search is continued of new materials having small weight, relatively small thickness, and well developed production technology.

The purpose of this work is to study of CRM performance based on carbon filled polyurethane and radioabsorbing and radioreflection ability of composite materials containing polyurethane varnish and dispersed carbon fillers in the form of graphite and carbon black.

2. Experimental procedure

In this research paper flat samples of composite materials at the thickness of 0.3 mm were used, comprising of binder and carbon fillers. As the binder two-component polyurethane varnish VM 700 GLOSS is used. These materials have high indicators of chemical firmness, durability, and elasticity, shock firmness as well as high adhesion to metals and functionality in the conditions of high humidity in a wide temperature interval from minus 60 °C to 130 °C without essential deterioration of mechanical properties. On the basis of conductive fillers properties analysis for introducing into a polyurethane binder colloidal-graphite preparation C-1, graphite element GE-3 and channel balck K-163 were chosen. These carbonaceous materials are chemically inactive with respect to oxygen and to most of the known binders in wide temperature range, which in its turn, according to minimizing the contact resistance between the particles, has a number of advantages compared with metals.

Coverings have been made under the following scheme: dispersive mixing of polyurethane varnish and carbon fillers during 2 hours; placing composition on a surface of a metallic substrate; drying of carbon-filled polyurethane coverings during one day at room temperature of 23 °C and its polymerisation at 120 °C during 2 hours [8].

For reliable assessment of specific volume resistance, hardness and adhesive strength CRM, experimental studies were performed using certified methods to certified equipment. Specific volume resistance was measured using two-probe method in accordance with GOST RF 20214-74 «Electrical conductive plastics. Test method for determination of specific volume electrical resistance at dc voltage». Hardness was determined in accordance with GOST RF 9450-76 «Measurements microhardness by diamond instruments indentation» hardness tester on Nano Hardness Tester company CSEM Instruments. The study was carried out on the adhesive strength measurement setup Micro Scratch Tester company CSEM Instruments in accordance with ASTM D 2197 «Standard Test Method for Adhesion of Organic Coatings by Scrape Adhesion». Frequency dependences of electromagnetic response properties (coefficients of reflection (*R*) and transmission (*T*)), and dielectric permeability ($\varepsilon^*=\varepsilon'-\varepsilon''$) of polymer composite films were investigated at a radio spectroscope constructed on the basis of the vector analyzer of chains E8363B of firm Agilent Technologies in frequency range 26–40 GHz or Mach–Zehnder interferometer based on backward wave oscillator for frequency range 110–260 GHz.

3. Results and discussion

The results obtained in experimental studies indicate that using VM 700 GLOSS as the binder polyurethane varnish and conductive filler C-1 (24 wt %) with a thickness CRM allows (301 ± 4) μm to obtain coatings with specific volume resistance (1.59 ± 0.06) Ω·cm, hardness (407 ± 9) MPa, adhesive strength to the metal (28.12 ± 0.29) N and the stability of these parameters at the maximum operating temperature (130 ± 0.7) °C during 8000 hours [8]. The resulting figures are optimal for use of CRM as heating surfaces in the construction of electric heaters technological equipment.

The technological process of CRM formation as a heating coating on the metal surfaces of the process equipment structural elements includes a series of sequential operations. The initial step is to prepare the metal surface of the structural element, which consists in making holes for electrical contacts (copper bars), cleaning and degreasing surfaces, insulation contact surfaces fasteners, electrical contact with the metal surface. Next, the application is carried out on the metal surface of the dielectric coating thickness of 100 μm of polyurethane varnish VM 700 GLOSS by spraying, followed by drying at 23 °C within 60 min. It is necessary for "joining" dielectric coating with CRM by formation of intermolecular hydrogen bonds, increasing the strength of the connection layers thick film electric heater. The dielectric layer thickness of 100 μm prevents electrical breakdown between the CRM and the metal surface due to excellent dielectric properties. Electrical contacts are secured to the CRM by bolting solid. The next stage is the optimal dosage component percentage. Further carbon filled polyurethane polymer composition as a low viscosity slurry was subjected to dispersive mixing within 120 min in a ball mill for 1 SHLM-fracture aggregates C-1 and its uniform distribution in the polyurethane binder. After that, the layering thickness of 300 μm CRM is carried out on the dielectric surface of the structural element with contacts by spraying. Heat treatment of the intermediate layers CRM 100 μm and 200 μm is carried out at a temperature of 50 °C during 30 min to remove the solvent and partial polymerization of the composition. Full cure and stabilizing heat treatment is carried out at CRM 120 °C during 120 min. If necessary, CRM is covered applied dielectric coating of polyurethane varnish thickness of 100 μm by spraying to prevent electrical shock.

The process CRM formation on metal surfaces of technological equipment allows to create heaters, which are characterized by simplicity of installation and operation, reliability, absence of external influences on CRM, providing a uniform distribution over the underlying surface of the structure.

To study the performance of the heating coatings for thermo shield casing (Figure 1a) and press plate (Figure 1b), by the method described above some electric heaters of CRM were made. Heater, disposed on the outer side of the shuttering panel and the press plate is formed as a thick CRM (301 ± 4) μm and electrodes, connected to the coating and attached to the formwork panel and the press plate. The insulating layers on both sides of electric heater are formed as coatings, one of which is a dielectric, and the second – thermal insulation, which reliably insulates heater, preventing heat loss during heating up and operation. The selected design scheme of the heating panel formwork and thermo heating press plate (shield plate – heater – insulating layer) ensures the absence of external influences on CRM and its durability and meets all the safety requirements of the work undertaken. The process of forming electric heater provides a choice of its configuration for specific tasks.

The uniformity of the press heating plates temperature field is an important prerequisite for producing high-quality material or product by pressing heated. Tolerance temperature heating plates for the manufacture or processing of most materials is ± 5 °C. Using CRM electric heater design to achieve uniform temperature field over the entire area of the heating shield thermosetting formwork (Figure 2) and the surface of the heating plate press. Temperature 80 °C, and its differential ± 0.5 °C at the inner side of the heating panel (Figure 2b), in contact with the concrete, corresponds to the allowable temperature deviation in the thermosetting heating concrete formwork (± 3 °C), as well as heat treatment materials under pressure in the heating press platens.

It was also found out that the maximum operating temperature (130 ± 0.7) °C and specific volume resistance (1.59 ± 0.06) Ω·cm electric heater thermo heating panels formwork and heating press platens during 2000 hours retain their values, indicating CRM performance. Repeated experiments have not led to a change in the values of these parameters and demonstrated good stability in electric heater with a lifetime of not less than 8000 hours.

Frost testing of the electric heater revealed that after 150 cycles of its haeting to a temperature of (130 ± 0.7) °C and subsequent cooling to –(40 ± 0.2) °C, mechanical electrical properties and CRM performance do not change.

IOP Conf. Series: Materials Science and Engineering **93** (2015) 012038 doi:10.1088/1757-899X/93/1/012038

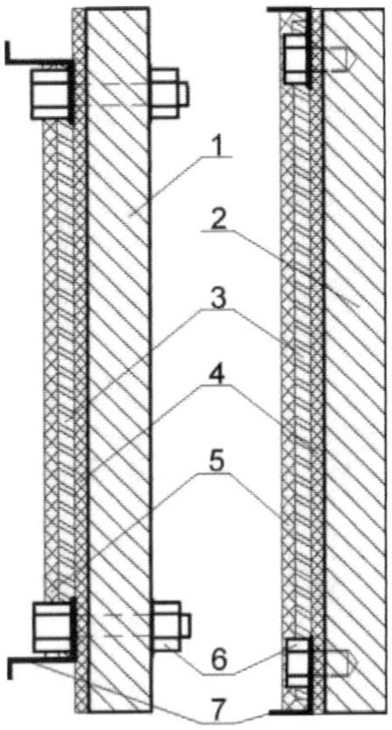

a b

Figure 1. Design of heating thermo shield casing *(a)* and the heating plate press *(b)* in electric heaters of CRM: shield (1); plate (2); CRM (3); dielectric coating (4); thermal insulation coating (5); fasteners (6); electrodes (7).

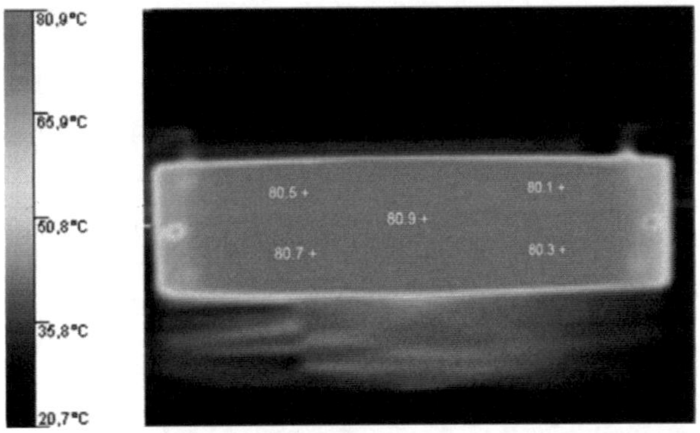

a

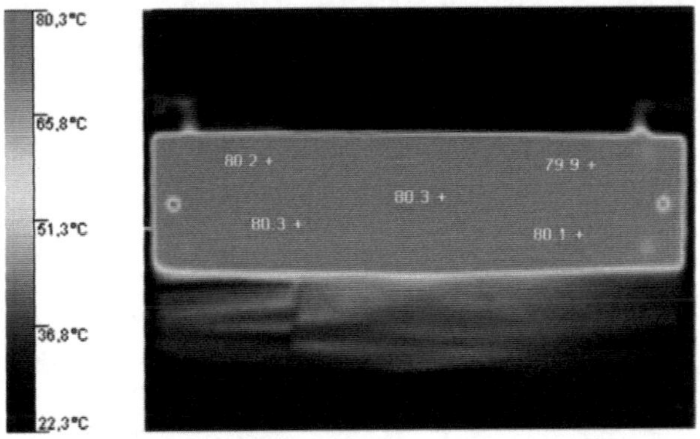

b

Figure 2. Thermogram of heating thermo shield casing with CRM:
a) electric heater; *b)* heating the inside of the shield.

3.1. Electromagnetic response and dielectric permittivity

Measurements of the parameters of electromagnetic response allowed us to retrieve the complex dielectric permittivity for the frequency band at which measurement was performed. Values of the parameter ε^*, being the numerical characteristic of polarizability of materials in an electric field, allow the electromagnetic characteristics of specimens at preset frequency, thickness, and geometric shape to be modeled. Our investigations demonstrated that with increasing mass content of the filler, the real (ε') and imaginary (ε'') components of the dielectric permittivity increased practically without frequency dispersion. Table 1 presents the results of investigations of electromagnetic radiation interaction with the developed composite material based on the polyurethane depending on the nature and content of the carbon fillers.

Table 1. Electrophysical characteristics of composites with C-1, GE-3, and K-163 fillers.

Frequency, GHz	Content of the filler in the binder, wt %	Reflection coefficient, dB	Transmission coefficient, dB	ε', rel. units	ε'', rel. units
		Channel black K-163			
140	1	6.1	0.1	3.4	0.12
240	1	7.5	1.5	3.5	0.12
140	2	5.5	1.7	3.8	0.17
240	2	6.8	1.5	3.8	0.17
140	5	4.2	3.1	5.7	0.07
240	5	2.4	2.6	5.8	0.53
140	10	0.5	5.2	10.3	2.9
240	10	0.5	7.7		
		Colloidal graphite preparation C-1			
140	1	6.1	2.0	3.6	0.34
240	1	13	1.0	3.4	0.11
140	2	7.2	2.0	5.2	0.11
240	2	7.1	3.0	5.0	0.16
140	5	5.2	4.8	9.5	0.95
240	5	5.1	5.3	9.4	0.92
140	10	0.5	10.5	20.1	3.1
240	10	0.5	11.3	21.3	2.7
		Elemental graphite GE-3			
140	1	4.8	2.1	5.1	0.5
240	1	5.1	4.0	5.1	0.5
140	2	10.1	2.1	7.1	0.9
240	2	6.3	7.7	7.1	1.2
140	5	7.5	10.2	17.0	3.67
240	5	7.6	20.1	14.2	6.1
140	10	Near 0	23.0		
240	10	Near 0	Exceeding 40		

The least value of the real dielectric permittivity component for concentration exceeding 1 % was recorded for the composite based on channel black K-163. This can be explained by two ways. First, particles of this material are the least ones, and the surface unstructured layer starts to produce considerable effect on the effective dielectric permittivity of the composite as a whole. Second, particles of channel black, as demonstrated above, have spherical shapes, whereas particles of other carbon structures are flat. The spherical particles have the depolarization ratio equal to 1/3 in all directions, whereas the flat particles arrange in parallel in the process of composite preparation, that is, the structured material is obtained whose effective dielectric permittivity is higher.

It was demonstrated that composites in which channel black K-163, elemental graphite GE-3, or colloidal graphite preparation C-1 were used as fillers actively interacted with electromagnetic radiation in the range 26–260 GHz. With filler concentration exceeding 10 wt %, they can be used as reflecting coatings, and with filler concentrations up to 5 wt %, they can be used as coatings reducing the transmitted signal level. The most efficient were composite with GE-3 filler that reduced the level of the transmitted electromagnetic radiation signal by 7 dB in a wide frequency range and up to 20 dB in individual spectral regions when the composite thickness was 0.4 mm. The least values of the dielectric permittivity had composites with K-163 filler, which demonstrated the spherical shape of their particles (the highest depolarization ratio) and the least particle sizes (greater influence of the surface layer). Composite material with C-1 filler can also be used as a protective coating, because they transmit less than half of incident radiation. The efficiency of the examined composites can be increased by increasing the thickness of specimens up to 1–2 mm that is characteristic for modern wideband coatings.

4. Conclusions

Thus, the technological process of forming composite resistive material on the metal surfaces of structural elements technological equipment is developed, allowing to create electric heaters, which are characterized by simplicity of installation and operation, reliability, service life more than 8000 hours, the stability of the electrical resistance in a long time and a wide temperature range, high adhesive strength to metal surfaces, a uniform temperature distribution over the entire surface. Conducted CRM performance tests on the basis of carbon filled polyurethane in the construction of electric heaters heating panels thermosetting formwork, heating plate presses, oil baths for shrink fitting details are conducted, which confirm the effectiveness of its use in mechanical engineering.

Our investigations have demonstrated that the functionality of the developed CRM with carbon fillers can be expanded by their application as radio absorbing and radio reflecting coatings. It was demonstrated that the composites with the active phase from channel black, colloidal graphite preparation, and elemental graphite interacted effectively with electromagnetic radiation of teraherz range: with filler concentrations exceeding 10 % they could be used as reflecting screens, and with filler concentrations up to 5 %, they could be used as devices reducing the transmitted signal level.

References

[1] Sengupta R, Bhattacharya M, Bandyopadhyay S and Bhowmick A 2011 *Prog. in Polym. Sci.* **36** 638–670.
[2] Krupa I, Prokes J, Krivka I and Spitalsky Z 2011 *Handb. of Multiphase Polym. Syst.* **1** 425–477.
[3] Zucolotto V, Avlyanov J and Mattoso L 2004 *Polym. Compos.* **25** 617–621.
[4] Bokobza L 2013 *Rubber Chem. and Technol.* **86** 423–448.
[5] Brigandi P J, Cogen J M and Pearson R A 2014 *Polym. Eng. And Sci.* **54** 1–16.
[6] Mendes R, Claro-Neto S and Cavalheiro E 2002 *Talanta* **57** 909–917.
[7] Petit L, Guiffard B, Seveyrat L and Guyomar D 2008 Sens. and Actuators A. **148** 105–110.
[8] Malinovskaya T D and Melentyev S V 2013 *Russ. Phys. J.* **56** 970–972.

On the impact of the plasma jet energy on the product of plasmadynamic synthesis in the Si-C system

D Nikitin[1] and A Sivkov[1]
[1]National Research Tomsk Polytechnic University,
634050, Tomsk, Lenin Avenue, 30

E-mail: dima_n@sibmail.com

Abstract. Silicon carbide (SiC) nanoparticles can be used for ceramics reinforcement, creation of nanostructured ceramics, microelectromechanical systems. The paper presents the results of plasmadynamic synthesis of silicon carbide nanopowders. This method was realized by the synthesis in an electrodischarge plasma jet generated by a high-current pulsed coaxial magnetoplasma accelerator. Powdered carbon and silicon were used as precursors for the reaction. Four experiments with different energy levels (from 10.0 to 30.0 kJ) were carried out. The synthesized products were analysed by several modern techniques including X-ray diffractometry, scanning and transmission electron microscopy. According to analysis results all the products mainly composed of cubic silicon carbide (b-SiC) with a small amount of unreacted precursors. Silicon carbide particles have a clear crystal structure, a triangular shape and sizes to a few hundred nanometers. Comparison of the results of experiments with different energy levels made it possible to draw conclusions on ways to control product phase composition and dispersion. The silicon carbide content and particles sizes increase with increasing the energy level.

1. Introduction

Silicon carbide is the only stable compound in the Si-C system known about hundred years. This material has a combination of different physical and chemical properties as high hardness and mechanical strength, low density, excellent thermal, corrosion radiation resistance, high thermal conductivity [1], [2], [3], [4], [5]. So silicon carbide ceramics is one of the most important non-oxide ceramic materials and it is a very attractive material for a wide range of industrial applications. SiC is a wide band gap semiconductor and it is applied for creation of electronic systems for harsh environments [6], [7], [8]. Nanomaterials are known to have unique and higher properties than conventional materials [9]. Various silicon carbide nanostructures are widely used as ceramics reinforcement, for creation of nanostructured ceramics, micro- and nanoelectromechanical systems (MEMS and NEMS) [10], [11], [12].

The main way of silicon carbide production is the Acheson's process developed in the late of 19th century and still used [13]. It is the carbothermal reduction of silica by coke at 2200–2500 °C. Despite the industrial volumes of the obtained product the Acheson's method has obvious disadvantages as large grain size, long process time and contamination of the product. So this method is not suitable for nanodispersed SiC powders production. There are many successful attempts to synthesize a silicon carbide nanopowder including carbothermal synthesis [14], self-propagating high-temperature process [15], microwave heating method [16], sol-gel process [17], laser synthesis [18] and planetary ball milling [19]. The above synthesis ways have such disadvantages as the high cost, time and energy expenses and small product volumes. Finding a way of simple and cheap method for the silicon carbide synthesis became the urgent task.

Plasmodynamic method of the nanopowders synthesis is based on the chemical reaction of silicon and carbon in a hypersonic electrodischarge plasma jet. A plasma jet is generated by a high-current pulsed coaxial magnetoplasma accelerator (CMPA). Plasmodynamic method was reported to be used for obtaining nanopowders of materials based on tungsten, silicon and carbon [20], [21], [22].The possibility of silicon carbide synthesis has been shown [22].The present article is devoted to the study of the influence of the energy level on the product characteristics.

2. Experimental Conditions

The design of the used experimental installation is shown in Figure 1. The installation consists of a CMPA, a capacitive energy storage and a working chamber. The CMPA includes the central electrode with the graphite insert and the insulator and the graphite accelerator channel. A mixture of amorphous carbon black and micron crystalline silicon powder with the Si:C ratio 3:1 was used as precursors. The averaged sizes of silicon and carbon particles were ~50 μm and ~30 nm, respectively. The precursors mass was ~ 1.0 g. The mixture was placed into the plasma forming zone of accelerating channel. The Si-C mixture also provided the initiation of an arc discharge between the central electrode and the graphite accelerating channel after circuit locking. The CMPA was supplied by the capacitive energy storage with a capacity C = 6 mF. An accumulated energy Wc was changed by varying of the charging voltage Uc from 2.0 kV to 3.5 kV. Four experiments with different voltages were conducted.

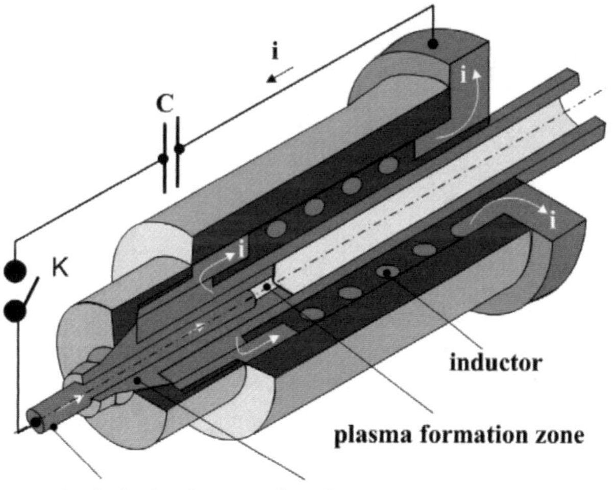

Figure 1. Schematic diagram of the coaxial magnetoplasma accelerator.

Typical oscillograms of working current and voltage on electrodes were obtained by a Tektronix TDS1012 oscilloscope. Average values of energy parameters (operating current Im, the voltage on electrodes Um, discharge power Pm and energy W) are shown in Table 1. Plasma shots were produced in a sealed volume of the cylindrical reactor chamber filled with argon at normal pressure and room temperature. The plasma jet impacted on the copper barrier installed at a distance of 23 mm from acceleration channel. The powder synthesized product was collected from the wall of the reactor chamber after precipitation of suspended particles.

Synthesized in the experiments powder products were analyzed without any preparation by such modern methods: X-ray diffraction (XRD) using a Shimadzu XRD 6000 diffractometer (CuKα-radiation, λ = 0.15406 nm), transmission electron microscopy (TEM) using Philips CM 12 microscope. Qualitative X-ray analysis was made by PowderCell 2.4 software package using a database PDF4+.

Table 1. The average size of nanoparticles and aggregates in suspensions.

Energy parameters	Experiments			
	1	2	3	4
U_c [kV]	2.0	2.5	3.0	3.5
I_m [kA]	65	76	98	111
U_m [kV]	1.0	1.1	1.3	1.8
P_m [MWt]	65	76	123	204
W_c [kJ]	12.0	18.8	27.0	36.8
W [kJ]	10.5	13.0	19.0	29.7

3. Results and discussion

The XRD-studies of synthesized powders are presented in the form of X-ray diffraction patterns in Figure 2. The results of structural and phase analysis are shown in Table 2. The character of XRD-patterns and the set of different coherent reflections show the practical absence of amorphous component in the synthesis product and demonstrate the presence of several crystalline phases in all cases. According to computer calculations by PowderCell 2.4 software the powder product of plasmodynamic synthesis in all cases consists of four crystalline phases with the following crystallographic parameters of structural models: cubic silicon carbide β-SiC, space group SPGR – F-43m {216}; cubic silicon cSi, SPGR – F-43/d-32/m {227}; graphite gC, SPGR – P6-3mc {186}; carbon onions structure C-Onions, SPGR – P6-3mc {186}. The most intense peaks belong to respective planes (111), (200), (220), (311), (222) of cubic silicon carbide. Results of the comprehensive analysis of the X-ray diffraction pattern of the synthesized products (including phases contents and coherent scattering areas) are presented in Table 2.

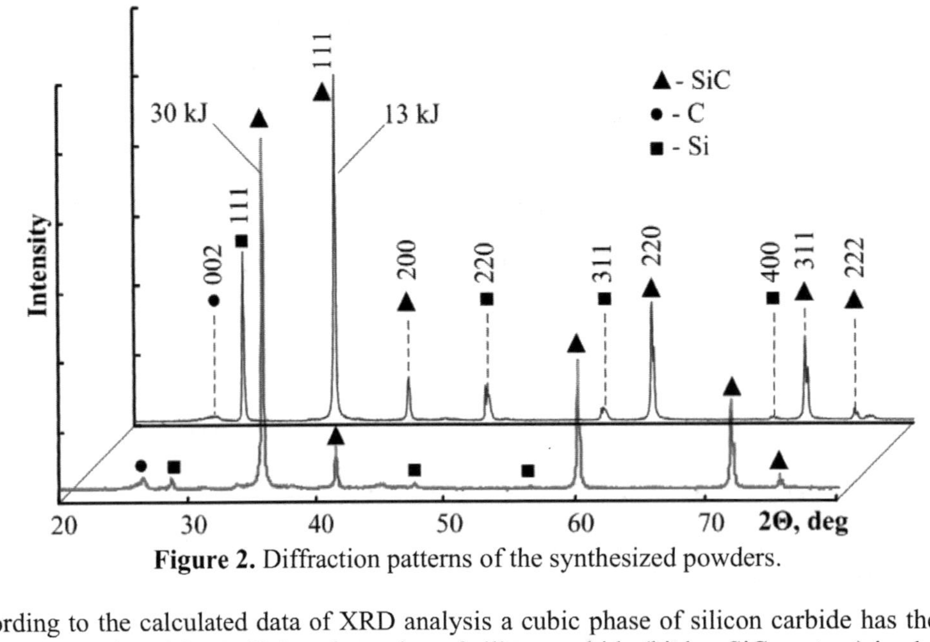

Figure 2. Diffraction patterns of the synthesized powders.

According to the calculated data of XRD analysis a cubic phase of silicon carbide has the highest content in the product. More efficient formation of silicon carbide (higher SiC content) is observed at higher values of discharge energy W. The carbon content do not decreases in the range of discharge energy from W = 13.5 to W = 29.7 kJ due to increase of the graphite eroding by the central electrode. The growth of SiC particle sizes is observed with increasing energy in the entire range. The growth of content and particle sizes of silicon carbide can be explained by more fully ionization precursors due to the higher pressure and temperature parameters (pT-parameters). Dependences of the content ω in the product of silicon carbide, silicon and carbon on discharge energy W are shown in Figure 3.

Table 2. The average size of nanoparticles and aggregates in suspensions.

Phase *W* [kJ]		SiC	Si	C
1. 10.5	Content [%]	60.0	24.0	16.0
	CSA [nm]	35	20	10
2. 13.0	Content [%]	73.0	19.0	8.0
	CSA, [nm]	70	80	20
3. 19.0	Content [%]	80.0	13.0	7.0
	CSA, [nm]	75	65	35
4. 29.7	Content [%]	90.0	3.5	6.5
	CSA, [nm]	90	40	20

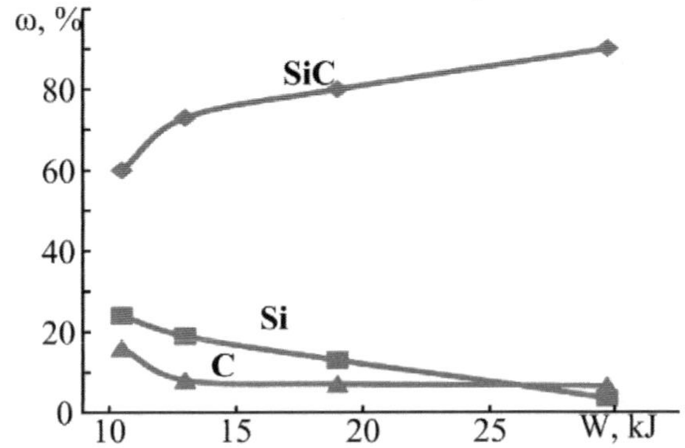

Figure 3. Dependences of the content ω in the product of silicon carbide, silicon and carbon on discharge energy W.

Results of the analysis of synthesis product by transmission electron microscopy confirm and complement the XRD data. Bright-field TEM-image of a powder produced at the maximum energy W = 30 kJ is shown in Figure 4. According to the TEM-image the synthesized product consists of two different fractions of particles. The first fraction (1) is represented by particles with close to "ideal" crystallographic design in the shape of triangles with truncated vertices and hexagons of sizes to 400 nm. Such particle shapes is typical for silicon carbide crystals [23]. The selected area electron diffraction (SAED) is presented in Figure 5. The most striking and the main diffraction reflexes relate to reflecting planes of cubic silicon carbide. There are weak-intensive reflections related to the carbon phase.

Dark-field TEM-image of a powder produced at the maximum energy W = 30 kJ is presented in Figure 5. These pictures were obtained at the diffracted light beam on the packets plane SiC (111). The glow of reflecting planes of particles (1) can be seen. It proves the particles of this fraction relate to silicon carbide phase.

Fraction 2 represents particles of size to a few tens of nanometers in the form of amorphous structures and crystalline objects of uncertain morphology. They can be considered as belonging to the phases of the unreacted precursors of silicon and carbon [21].

4. Conclusions

Silicon carbide (SiC) nanopowders were produced by plasmadynamic synthesis. This method was realized by the synthesis in an electrodischarge plasma jet generated by a high-current pulsed coaxial magnetoplasma accelerator. Four experiments with different energy levels (from 10.0 to 30.0 kJ) were carried out. The synthesized products were analyzed by several modern techniques including X-ray diffractometry, scanning and transmission electron microscopy. According to analysis results all the products mainly composed of cubic silicon carbide (β-SiC) with a small amount of unreacted precursors. Silicon carbide particles have a clear crystal structure, a triangular shape and sizes to a few hundred nanometers. Comparison of the results of experiments with different energy levels made it possible to draw conclusions on ways to control product phase composition and dispersion. The silicon carbide content and particles sizes increase with increasing the energy level.

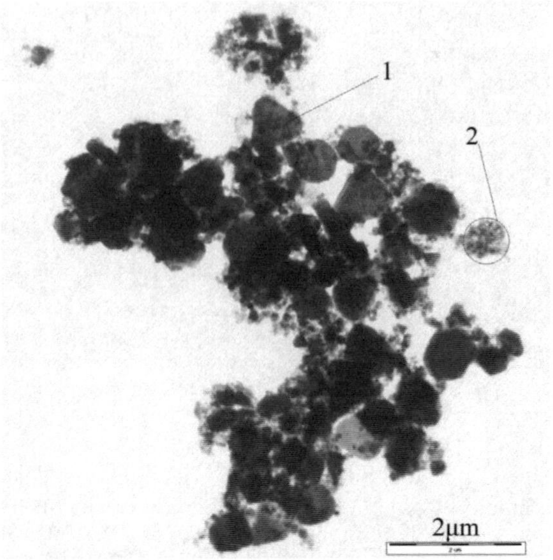

Figure 4. Bright-field TEM-images of the plasmadynamic synthesis product (W = 29.7 kJ).

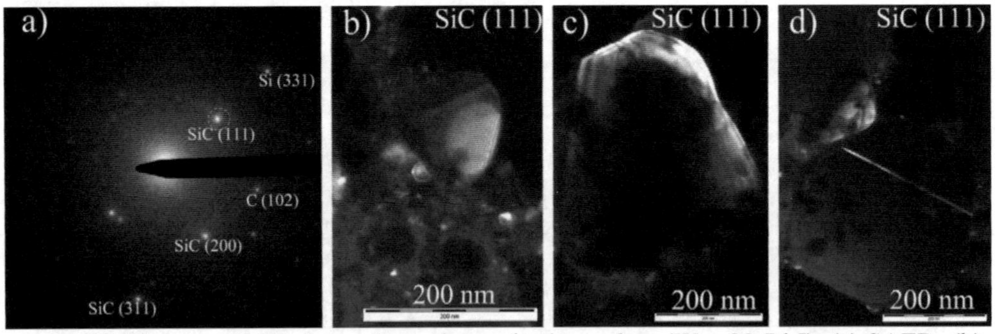

Figure 5. TEM-images of the plasmadynamic synthesis product (W = 29.7 kJ): (a) SAED, (b), (c), (d) dark-field images

Acknowledgments
This work was supported by the Russian Science Foundation (project no. 15-19-00049).

References

[1] Harris G L 1995 *Properties of Silicon Carbide* (London: NSPEC)
[2] Gerhardt R 2011 *Properties and Applications of Silicon Carbide* (Rijeka: InTech)
[3] Sarin V K 2014 *Comprehensive Hard Materials. Volume 2, Ceramics* (Oxford: Elsevier Ltd.)
[4] Watari K 2001 *J. Ceram. Soc. Japan.* **109** S7-S16
[5] Wu R, Zhou K, Yue C Y, Wei J and Pan Y 2015 *Prog. Mater Sci.* **72** 1-60
[6] Chin H S, Cheong K Y and Ismail A B 2010 *Metall. Mater. Trans. B.* **41** **4** 824-832
[7] Willander M, Friesel M, Wahab Q and Straumal B 2006 *J. Mater. Sci. - Mater. Electron.* **17** 1-25
[8] Wang Z, Shi X, Tolbert L M, Wang F F, Liang Z, Costinett D and Blalock B J 2015 *IEEE T. Power Electr.* **30** **3** 1432-1445

[9] Cao G and Wang Y 2011 *Nanostructures and Nanomaterials: Synthesis, Properties, and Applications*, (Singapore: World Scientific Publishing Co. Pte. Ltd.)
[10] El-Daly A A, Abdelhameed M, Hashish M and Daoush W M 2013 *Mater. Sci. Eng. A.* **559** 384-393
[11] Filipescu M, Stokker-Cheregi F, Colceag D, Nedelcea A, Birjega R, Nistor L C and Dinescu M 2013 *Appl. Surf. Sci.* **278** 96-100
[12] Maboudian R, Carraro C, Senesky D G and Roper CS 2013 *J. Vac. Sci. Technol. A.* **31** 1-18
[13] Acheson G 1893 *Production of artificial crystalline carbonaceous material* U S Patent 492 767.
[14] Rajarao R, Ferreira R, Sadi S H F, Khanna R and Sahajwalla V 2014 *Mater. Lett.* **120** 65-68.
[15] Huczko A, Kurcz M, Dabrowska A, Baranowski P, Bhattarai A and Gierlotka S 2014 *J. Cryst. Growth* **401** 469-473.
[16] Van Laar J H, Slabber J F M, Meyer J P, van der Walt I J, Puts G J and Crouse P L 2015 *Ceram. Int.*, **41 3 B** 4326-4333.
[17] Omidi Z, Ghasemi A and Bakhshi S R 2015 *Ceram. Int.* **41, 4** 5779-5784.
[18] D'Amato R, Falconieri M, Gagliardi S, Popovici E, Serra E, Terranova G and Borsella E 2013 *J. Anal. Appl. Pyrolysis* **104** 461-469.
[19] Abderrazak H and Abdellaoui M 2008 *Mater. Lett.* **62, 23** 3839-3841.
[20] Pak A, Sivkov A, Shanenkov I, Rahmatullin I and Shatrova K 2015 *Int. J. Refract. Met. Hard Mater.* **48** 51-55.
[21] Sivkov A A, Nikitin D S, Pak A Y, Rakhmatullin I A 2013 *J. Superhard Mater.* **35, 3** 137-142.
[22] Sivkov A A, Pak A Y, Nikitin D S, Rakhmatullin I A, Shanenkov I I 2013 *Nanotech. in Rus.* **8** 489-494.
[23] Feng A and Munir Z A 1995 *Metall. Mater. Trans. B.* **26 3** 587-593.

Activated sintering of ceramics on the basis of Al_2O_3

S V Matrenin[1], B B Ovechkin[1], B S Zenin[1] and R V Tayukin[1]
[1] Tomsk Polytechnic University, 30 Lenin Avenue, Tomsk, Russia, 634050

E-mail: msv@tpu.ru

Abstract. Ceramic sintering on the basis of Al_2O_3 activated by nanopowder additives Al_2O_3 and Al has been studied. The given paper shows that adding nanopowder Al_2O_3 in the amount of up to 20 wt. % into the coarse powder α-Al_2O_3 activates the sintering process and, as a result, leads to the increase in density and microhardness of sintered alumina ceramics. The study has revealed a significant effect of alumina ceramic density growth which is due to introducing the submicron powder TiO_2 to the initial blends composition.

1. Introduction

The most widespread methods of obtaining strong alumina ceramics are powder techniques, i.e. various types of pressing and sintering modified as applied to ceramics. Nevertheless, the complexity and poor performance by hot and hot isostatic pressing techniques [1, 2] that make it possible to obtain materials with maximum high structural characteristics [3] handicap a wide practical application of strong alumina ceramics. As a rule, a relatively simple technique of uniaxial pressing followed by free sintering does not allow obtaining ceramics with a high level of mechanical properties [4]. Therefore, the problem of activating processes of alumina ceramic consolidation is of great practical importance.

This work aims at investigating the activating methods of ceramic sintering based on Al_2O_3 through powder mechanical processing in a planetary mill adding nanopowders (NP) Al, Al_2O_3 and submicron powder TiO_2 into a blend and applying the technology of spark plasma sintering (SPS).

2. Experiment

Manufactured oxide nanocrystalline powders (NP) Al_2O_3, Al_2O_3 - ZrO_2 - Y_2O_3, obtained by plasma-chemical synthesis were used. Besides plasma-chemical NP, coarse pure aluminum oxide powder, technical alumina powder and white electrocorundum were applied in the work. The electro-explosive NP Al was used as an activating additive.

Particle size analysis of the coarse pure Al_2O_3 powder and technical alumina was conducted with the help of the analyzer A 20. Powder test portions with a weight of 50 gr were taken; vibration motor frequency was 70 Hz with sizing time 10 min. After sizing the fractions were weighed and the percentage of each fraction was calculated. Also bulk density, tapping density, flowability of all the powders under study were determined.

Oxide powders were annealed in air in a high-temperature resistance furnace at 1450°C for an hour to transform γ-Al_2O_3 into α-Al_2O_3.

To improve the processing characteristics and to increase activity, the annealed powders were processed in the high-energy planetary ball mill for 20 minutes at the rotation frequency of the grinding vessels 30 Hz. Zirconia balls were used as grinding bodies.

In order to determine the activating effect of machine processing on the consolidation process, white electrocorundum was machined at various modes (the rotation frequency was *f* – 20 and 30 Hz with processing time τ – 10, 20, 30 and 40 minutes at each frequency).

On being treated, the powder blends were sieved through the sieve No. 0045 for 10 minutes on the vibration motor C.1 to obtain fractions < 45 μm and plasticized by carboxymethylcellulose (CMC) water solution in an amount of 5 wt. % CMC – 95 % of the powder.

After granulation and drying, the plasticized powders were molded by uniaxial pressing method in a steel mold, extrusion pressure was 400 MPa. The obtained compacts had the form of cylinders with a diameter of 10±0.01 mm and a height of 5±0.01 mm.

Compact sintering was carried out in a high-temperature resistance furnace according to the following mode: heating rate was 10 deg/min, isothermal holding temperature was 1600°C with time 1 h, cooled with the furnace. The processed unplasticized plasma-chemical NP Al_2O_3 was consolidated by means of SPS method in the unit SPS-515S (Sumitomo).

The specimens sintered by SPS method had the form of cylinders with a diameter of 15,0±0,1 mm and a height of 2,0±0,2 mm.

The density of the sintered specimens ρ was determined by hydrostatic weighing method in 96% ethanol (ρ_{et}=0,807 g/cm^3 at 20°C) with an accuracy of ±0,001 g. Also, the relative density of specimens Θ was calculated from:

$$\theta = \frac{\rho}{\rho_t} \cdot 100\,\%,$$

where ρ_t is the theoretical density of ceramics.

The analysis of processing characteristics according to the abovementioned methods has shown that oxide nanopowders have a very low bulk density and flowability is substantially absent. The coarse powders Al_2O_3 possess a satisfactory level of processing characteristics. The data are presented in table 1.

Table 1. Processing characteristics of initial powders.

No.	Powder composition	Bulk density, (g/cm^3)	Tapping density, (g/cm^3)	Flowability, (g/s)
1	NP Al_2O_3	0.07	0.06	-
2	NP 80% Al_2O_3-19% ZrO_2-1% Y_2O_3	0.14	0.14	-
3	Coarse pure Al_2O_3	1.47	1.65	0.4
4	Technical alumina	1.50	1.66	0.4

The electro-explosive NP Al has bulk density equal to 0,2 g/cm^3 and it does not possess flowability.

Thus, the analysis of plasma-chemical NP Al_2O_3 properties allows making a conclusion that it is practically impossible to apply them in their initial condition as ready-to-use process feedstock.

table 2 shows the results of particle-size analysis of the initial coarse oxide powders. It indicates that the powders have a similar fractional particles distribution.

Table 2. Grain-size composition of coarse powders.

Particles fraction	Fraction composition X, %	
	Coarse pure Al_2O_3	Technical alumina
+025	0	0
-025+020	1.0	3.3
-020+016	6.1	4.2
-016+0125	9.6	10.1
-0125+008	15.3	13.8
-008+0063	24.0	23.8
-0063+0045	23.9	24.6
-0045	20.1	19.8

To improve the processing characteristics of plasma-chemical and coarse powders on the basis of Al_2O_3, they were processed in a planetary mill for 20 minutes at the grinding vessels rotation frequency of 30 Hz. Such a mode of oxide powder processing is optimal [5]. The data are presented in table 3.

Table 3. Processing characteristics of machined powders.

No.	Powders composition	Bulk density, (g/cm^3)	Tapping density, (g/cm^3)	Flowability, (g/s)
1	NP Al_2O_3	0.82	1.23	0.2
2	NP 80% Al_2O_3-19% ZrO_2-1% Y_2O_3	0.95	1.26	0.2
3	Coarse pure Al_2O_3	1.22	1.47	0.4
4	Technical alumina	1.36	1.49	0.4

During the processing, spherical particles of plasma-chemical powders were destroyed and grouped into hard agglomerates. The processing of coarse powder Al_2O_3 significantly increased the composition of fine fractions (-0063) up to 60%.

It has been found that the most effective mode for processing white electrocorundum powder in the planetary ball mill is as follows: the rotation frequency of the grinding vessels should be f – 30 Hz with the processing time τ – 40 min.

When powders are processed according to this mode, coarse fractions substantially disappear (+010), the number of medium fractions reduces (-010+008) and fine fraction outcome significantly increases (-0063) to more than 75% (Table 4). The bulk density of processed white electrocorundum powder was found to be 1.23 g/cm^3.

Table 4. Grain-size composition of white electrocorundum powder (grinding vessels rotation frequency f was 30 Hz).

Particles fraction	Processing time, min			
	10	20	30	40
	Fraction composition X, %			
+020	0.5	0.1	0	0
-020+014	4.4	2.5	2.1	0
-014+010	2.2	2.1	2.0	1.4
-010+008	18.5	14.4	9.7	10.1
-008+0063	8.8	13.3	12.6	13.3
-0063+0045	53.7	51.4	48.8	47.7
-0045	11.8	16.2	24.8	27.5

Figure 1 shows the diagrams of extruding powders without activating additives which were processed in a planetary mill and plasticized. These studies were conducted to determine the optimal extrusion pressure. It can be observed that when the pressure exceeds 400 MPa, the increase in pressed density is quite insignificant. Thus, the optimal extrusion pressures of the studied powders are within a range of 300 to 400 MPa.

Coarse pure Al_2O_3 powder was excluded from further experiments since the properties of the alumina ceramics sintered from this powder was analyzed by the authors in the work [6].

The processed powder compositions 1,2 and 4 (Table 3) and white electrocorundum powder were sieved into franction–0063, mixed with activating additives in the planetary mill. The mixtures were plasticized in accordance to the above method.

The compositions of the obtained powder mixtures are presented in table 5.

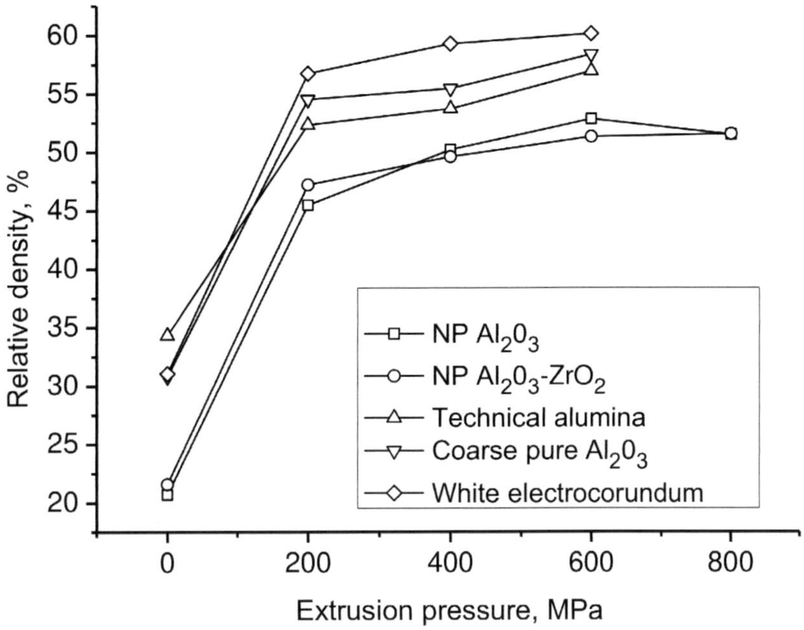

Figure 1. The diagrams of extruding the processed oxide powders (the grinding vessels rotation frequency is *f* – 30 Hz with the processing time τ – 40 min).

Table 5. Powder mixture compositions.

No.	Composition, wt. %					
	NP Al_2O_3	NP 80% Al_2O_3-19% ZrO_2-1% Y_2O_3	Technical alumina	White electrocorundum	NP Al	TiO_2
1	100	0	0	0	0	0
2	95	0	0	0	5	0
3	98,5	0	0	0	0	1,5
4	0	100	0	0	0	0
5	0	95	0	0	5	0
6	0	0	100	0	0	0
7	5	0	95	0	0	0
8	10	0	90	0	0	0
9	20	0	80	0	0	0
10	0	0	95	0	5	0
11	0	0	98,5	0	0	1,5
12	0	0	0	100	0	0
13	10	0	0	90	0	0
14	0	0	0	95	5	0
15	0	0	0	98,5	0	1,5

The prepared mixtures were used to obtain a compact sintered according to the above modes.

Figure 2 shows the results of measuring the density ratio of the sintered alumina ceramics. Adding NP Al_2O_3 into technical alumina powder increased the density of sintered ceramics. This activating effect is due to the expanding area for particle contacts which results from adding NP Al_2O_3. The sintering activation mechanism is due to the increased structural and surface activity of NP Al_2O_3, which is determined by the crystal structure faultiness, the size and shapes of particles. The most considerable growth in density was observed in the ceramics containing NP additive Al_2O_3 in an amount of 5 to 20 wt. %.

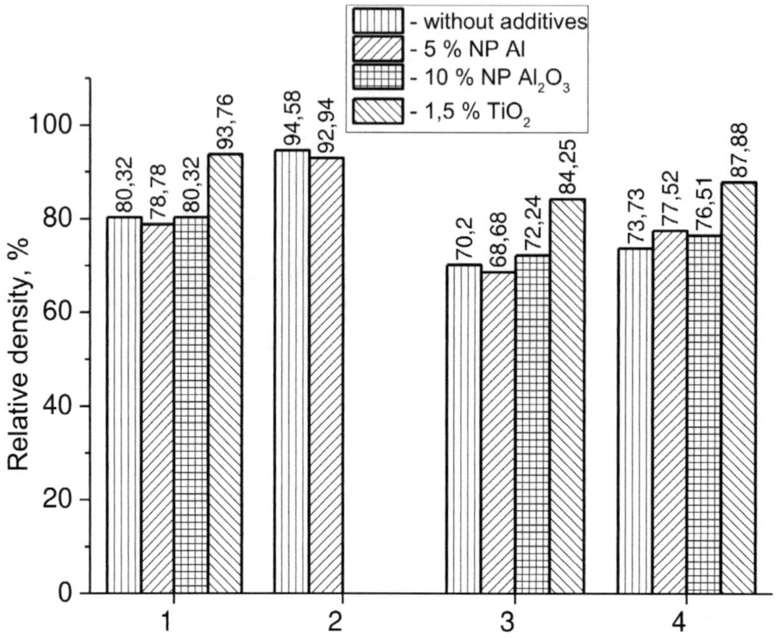

Figure 2. The dependence of ceramic density ratio on the additives amount and chemical composition: 1, 2, 3, 4 – ceramics sintered from NP Al_2O_3, NP 80% Al_2O_3-19% ZrO_2-1% Y_2O_3, technical alumina and white electrocorundum powders respectively (the first and the third columns are equivalent for the composition 1; NP Al_2O_3 and the powder TiO_2 were not added into the composition 2).

Additional contribution to activating sintering of NP Al_2O_3 owing to adding NP Al is the subject of practical interest. The reduction in the density of sintered ceramics was observed when nanodispersed aluminum was added to NP Al_2O_3. This is due to the increase in the sintered ceramic porosity caused by the oxidation of additive NP Al up to α-Al_2O_3 within the sintering process. The oxidation was accompanied by a significant decrease in the specific volume of introduced additive caused by a significant difference in density of Al (2.7 g/cm^3) and α-Al_2O_3 (3.96 g/cm^3) which was observed in a number of experiments. Thereby, the main benifit of NP as a sintering activator – its ability to form a great number of particle contacts at its very small amount in the sintering compact – turned to be unimplemented.

3. Summary

Ceramic sintering on the basis of Al_2O_3 activated by nanopowder additives of Al_2O_3 and Al has been studied. It was shown that adding nanopowder Al_2O_3 up to 20 wt. % into the coarse powder α-Al_2O_3 activates alumina ceramics sintering process. Consequently, there is an increase in density and microhardness of the sintered alumina ceramics. The research has indicated a significant effect of alumina ceramic density growth resulted from introducing the submicron powder TiO_2 in the amount of 1,5 wt. % into the initial blends composition.

IOP Conf. Series: Materials Science and Engineering **93** (2015) 012040 doi:10.1088/1757-899X/93/1/012040

References

[1] Kiparisov S S and Padalko O V 1988 Equipment of Power Metallurgy Enterprises (Moscow: Metallurgy)
[2] Hunold K and Kempten 1985 *Interceram.* **39(2)** 40-43
[3] Lukin E S, Makarov N A and others 2003 *Glass and Ceramics* **9** 32–34
[4] Smirnov A I 1990 Structural Ceramics *Itogi nauki i tekhniki* (All-Russian Institute of Scientific and Technical Information Series Powder Metallurgy) 64–106
[5] Matrenin S V, Belokrylova A O and Ovechkin B B 2012 *Izvesiya vuzov. Physics* **55(5/2)** 205-208
[6] Matrenin S V, Ilyin A P, Tolbanova L O and Zolotareva E V 2010 *Bulletin of the Tomsk Polytechnic University* **317(3)** 24-28

Effect of one variant of Ti_3Ni_4 particles on stress-induced martensitic transformations in <111>-oriented $Ti_{49.2}Ni_{50.8}$ single crystals

K S Osipovich, N G Vetoshkina, E Yu Panchenko and Yu I Chumlyakov
National Research Tomsk State University, 36, Lenina ave., Tomsk, 634050, Russia

E-mail: osipovich_k@mail.ru, vetoshkina23011991@mail.ru, panchenko@mail.tsu.ru, chum@phys.tsu.ru

Abstract. In the present study the effects of stress-assisted aging of the $Ti_{49.2}Ni_{50.8}$ single crystals oriented along $[\bar{1}11]$ direction on the stress-induced B2-R-B19′ thermoelastic martensitic transformations and superelasticity are investigated. It is experimentally established that aging at 823 K for 1h under compression stress of 150 MPa along $[\bar{1}11]$ direction leads to the precipitation of one crystallographic variant of Ti_3Ni_4 particles of 350(±30) nm in size. Precipitation the single variant of Ti_3Ni_4 particle results in an appearance of homogeneous long-range internal stress field $|\langle\sigma_G\rangle| \approx 65$ MPa, that defines the main features of stress-induced B2-R-B19′ transformation and determines the increase in the characteristic temperatures of martensitic transformation and the existence of two-way shape memory effect.

1. Introduction

It is known [1-3], that in the widely used Ni-rich alloys of TiNi ($C_{Ni} \geq 50.6$ at. %) the control of the temperatures of thermoelastic B2-R-B19′ martensitic transformation, strength properties of high temperature B2-phase, shape memory effect and superelasticity is achieved by the precipitation of the Ti_3Ni_4 disperse particles from 10 up to 500 nm in size during the ageing in the temperatures interval from 623 to 823 K. The Ti_3Ni_4 disperse particles have a lenticular form with a habit plane of $\{111\}_{B2}$ and do not undergo martensitic transformation [1]. Thus so-called nanocomposites are formed in which the matrix undergoes martensitic transformation, but dispersed particles deform only elastically.

Stress-free ageing results in the precipitation of four crystallographic equivalent variants of particles with four habit plane of $\{111\}_{B2}$ [1-3].The crystallographic variants of Ti_3Ni_4 particles are equivalent in stress-free aged crystals. By contrast, at the aging under uniaxial tensile or compressive stress the crystallographic variants of Ti_3Ni_4 particles possess different elastic energy. Therefore during stress-assisted aging particle variations with low elastic energy should precipitate, and the growth of high energy particle variations will be suppressed. Ideally, only one particle variant oriented in a certain way to the axis of the applied stress should form in stress-assisted single crystals oriented along the $[\bar{1}11]_{B2}$ direction [4]. At the ageing under tensile stress of $[\bar{1}11]$-oriented single crystal the habit plane of Ti_3Ni_4 particles tends to be parallel to the tensile axis and under a compression stress it tends to be perpendicular to the compressive axis. Up to the present the theoretical energy-based approaches for TiNi single crystals and the experimental confirmations of the oriented growth of Ti_3Ni_4 particle at the ageing under the action of external stresses from 50 to 150 MPa by the transmission electron microscopy (TEM) method were carried out [4,5]. The long-range internal stress fields $\langle\sigma_G\rangle$ in the aged under stress single crystals are created by combining of local stress fields from oriented formation of coherent and semicoherent Ti_3Ni_4 particles. Such long-range internal stress fields result in the oriented growth of R - and B19'-martensite during stress-free cooling of TiNi crystals and, accordingly, in the manifestation of two-way shape memory effect [1, 5].

It was previously shown [6] that in the aged at 673 K for 1 h under compressive/tensile stresses single crystals of $Ti_{49.2}Ni_{50.8}$ alloy the internal stress field $<\sigma_G>$ depending on their orientation relative to the external stress determines the value of stress hysteresis $\Delta\sigma$, contributes to the increase in the temperature range of the superelasticity and leads to the appearance of elastic twinning of R-phase with the reversible strain up to $\varepsilon_{rev}\sim1.4$ % [6]. However, systematic studies of the effects of the number of crystallographic variants of Ti_3Ni_4 particles with different size and volume fraction on the thermoelastic martensitic transformations, mechanism of the shape memory effect and superelasticity in TiNi single crystals have not yet been carried out.

Thus, the objective of the present work is to study the dependence of the stress-induced thermoelastic B2-R-B19' martensitic transformations, superelasticity on the number of crystallographic variants of Ti_3Ni_4 particles in the $[\bar{1}11]$-oriented $Ti_{49.2}Ni_{50.8}$ single crystals aged at 823 K for 1 h in free-stress state and under a compressive stress of 150 MPa. Such researches are attractive to further development of the theory of thermoelastic martensitic transformations in structural non-uniform materials, as well as to practical use for selection of a structural state at elaboration of the TiNi shape memory alloys with specified properties.

2. Materials and methods

Single crystals from ingots of TiNi with a nominal composition of $Ti_{49.2}Ni_{50.8}$ (in at. %) were grown using the Bridgman technique in an inert gas atmosphere. From the large single crystals, dog-bone shaped samples with 16 mm gauge length and a 6 mm x 3 mm cross section were electro-discharge machined with the loading direction oriented along $[\bar{1}11]_{B2}$ in austenite. The sample orientation was verified with a DRON-3 X-ray diffractometer.

Before tests, the specimens were annealed at 1273 K for 1 h and quenched in water. Aging at 823 K, 1 h under compression stress $|\sigma|$=150 MPa is carried out in vacuum and stress-free aged samples were used as a reference condition. Stress to a sample was put after heating in a free condition up to 473 K and sample was unstressed after its cooling to 473 K to avoid stress-induced martensitic transformation during aging. After the aging process was completed, tensile specimens with 16 mm gauge length and 1.5 mm x 3 mm cross section were cut.

The microstructures were examined with a Philips CM 12 transmission electron microscope (TEM) operated at a nominal accelerating voltage of 120 kV.

The isothermal loading/unloading test was performed in an Instron 5969 universal testing machine, the strain rate was $1 \times 10^{-3}\ 1\cdot s^{-1}$. Isostress thermal-cycling tests were carried out using a miniature load frame capable of +/- 0.1 kN. The heating/cooling from 77K to 373 K was achieved by flow of cold nitrogen through a contact heating-cooling system, the heating/cooling rate was 10 $K\cdot min^{-1}$. For all cycles, stresses were first applied after an initial heating to well above A_f.

3. Results and discussion

The TEM bright field image shows that four crystallographic variants of Ti_3Ni_4 particles of 350(±30) nm in size are formed in stress-free aged crystals (figure 1(a)). In contrast, in stress-assisted aged $[\bar{1}11]$-oriented $Ti_{49.2}Ni_{50.8}$ single crystals only one variant of Ti_3Ni_4 particles with the normal to the particle habit plane parallel to the $[\bar{1}11]$ compressive axis are precipitated (figure 1(b)).

The research results of the effects of the number of particle crystallographic variants on the stress-induced thermoelastic B2-R-B19' martensitic transformation and superelasticity are presented on figures 2-4. It was experimentally established that the characteristic temperatures of martensitic transformation (M_s and M_f are the start and finish temperatures of the forward R-B19' transformation during cooling, A_s and A_f are the start and finish temperatures of the reverse B19'-B2 transformation upon heating) and the critical stresses of the stress-induced martensitic transformation are determined by the number of crystallographic variants of the disperse Ti_3Ni_4 particles.

The stress-strain responses $\sigma(\varepsilon)$ of stress-assisted aged crystals with one particle variant demonstrate two critical stresses (figure 2). The first critical stress σ_{cr1} is observed within temperature range from M_s up to T_R ($T_R \sim 325$ K is the start temperature of the B2-R martensitic transformation during cooling) and it is associated with the reorientation of R-martensite crystals under applied stress.

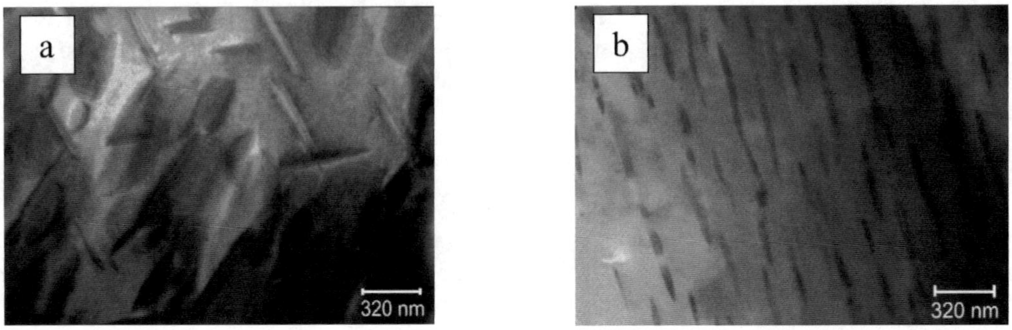

Figure 1. Microstructure of a $Ti_{49.2}Ni_{50.8}$ single crystal aged at 823 K for 1 h in stress-free state (a) and under compression stress of 150 MPa along $[\bar{1}11]$ direction (b).

The critical stress σ_{cr1} weakly decreases with increase in the test temperature. In the investigated stress-assisted aged at 823 K crystals the given strain about 0.5% arising from the reorientation of the R-phase at the applied stress $\sigma < \sigma_{cr2}$ is irreversible after unloading in contrast to aged at 673 K under compressive stress $Ti_{49.2}Ni_{50.8}$ crystals with one variant of the particle of 30-40 nm in size [6]

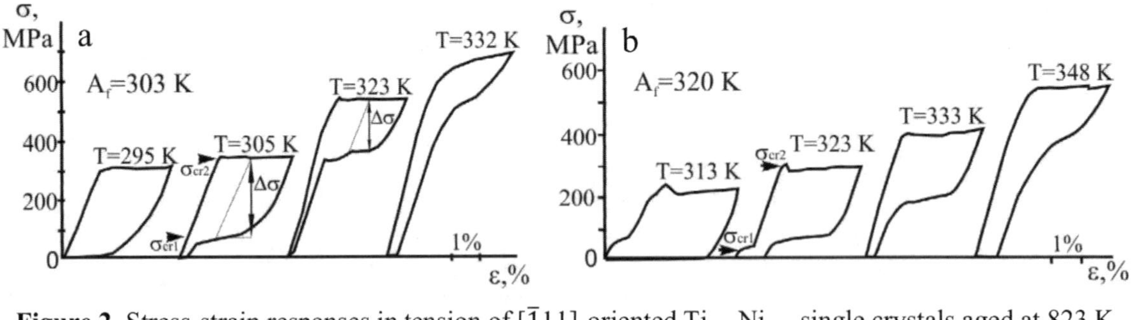

Figure 2. Stress-strain responses in tension of $[\bar{1}11]$-oriented $Ti_{49.2}Ni_{50.8}$ single crystals aged at 823 K for 1 h in stress-free state (a) and under compression stress of 150 MPa (b).

In stress-free aged at 823 K crystals the first critical stress σ_{cr1} within the temperature range $M_s<T<T_R$ is not clearly evident (figure 2) that was also observed for stress-free aged at 673 K single crystals [6]. The second critical stress σ_{cr}^2 is associated with stress-induced R-B19' martensitic transformation and it linearly increases at $T>M_s$ with the growth of the test temperature in accordance with the Clausius-Clapeyron relationship [1]

$$\frac{d\sigma}{dT} = -\frac{\Delta S^{A-M}}{\varepsilon_{tr}^{A-M}}, \qquad (1)$$

where $\Delta S^{A-M} < 0$ is the change of entropy associated with the forward transformation, ε_{tr} is the transformation strain.

Figure 3 shows the temperature dependences of the second critical stress $\sigma_{cr2}(T)$ in stress-free and stress-assisted aged at 823 K for 1 h $Ti_{49.2}Ni_{50.8}$ single crystals. These $\sigma_{cr2}(T)$-dependence can be divided into three temperature ranges, which are characteristic for alloys with a thermoelastic martensitic transformation. In the first range at $T \leq M_s$, the stress σ_{cr2} equals the critical stress for reorientation and detwinning of the B19′ martensite crystals. The aged under stress crystals have the lower level of the critical stress σ_{cr2} for reorientation and detwinning of martensite crystals in this temperature range as compared to the stress-free aged crystals (figure 3). The minimum values of σ_{cr2} are observed at $T = M_s$. In the second range $M_s < T < M_d$, the initial crystal structure is the B2-phase, and the onset of deformation is associated with the stress-induced martensitic transformation to B19′-martensite.The critical stress σ_{cr2} linear increases up to the M_d temperature. Experimentally the M_d point is defined as the temperature at which the maximum in the $\sigma_{cr2}(T)$-dependence is observed. At this point the critical stresses for formation of stress-induced martensite and for plastic flow of austenite are equal to each other. The third range $T > M_d$ is associated with the plastic flow of the B2-phase.

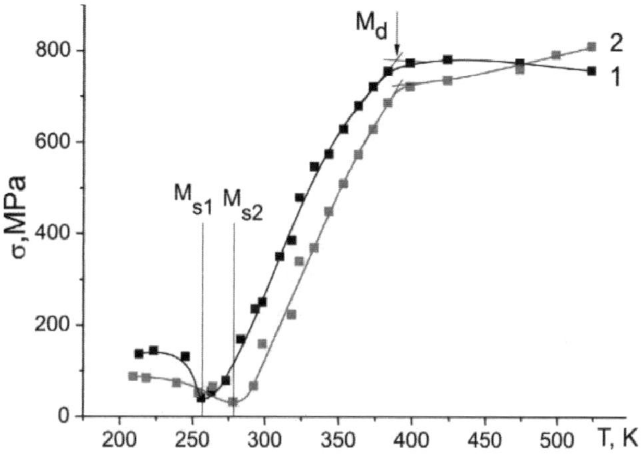

Figure 3. Temperature dependence of the critical stress level σ_{cr2} of the $[\bar{1}11]$-oriented $Ti_{49.2}Ni_{50.8}$ single crystals in tension: for aged at 823 K for 1 h crystals in stress-free state (curve 1) and under compression stress of 150 MPa (curve 2).

Figure 3 shows that for stress-assisted aged crystal the start temperature of R-B19' transformation M_{S2} is higher by 18 K than temperature M_{S1} for stress-free aged crystals with four particle variants. Such increase in the start transformation temperature M_{S2} in stress-assisted aged crystal caused by the action of the internal long-range stress field $<\sigma_G>$ which is created by combining of local stress fields from one variant of particles. In the stress-free aged crystals the long-range stress field is missing because the local stress fields from the neighboring particle variants compensate for each other. The Clausius-Clapeyron equation (1) provides to estimate the long-range internal stress field in the stress-assisted aged crystals [7]:

$$|<\Delta\sigma_G>|=\frac{\Delta S}{\varepsilon_0}\Delta M_s. \quad (2)$$

Using the experimental values of $\left|\frac{\Delta S}{\varepsilon_0}\right| = \frac{d\sigma}{dT} = 3.6\ MPa \cdot K^{-1}$ and equation (2) we determined that the long-range internal stress field $|<\sigma_G>|$ is equal to 65 MPa in the aged under stress $[\bar{1}11]$-oriented $Ti_{49.2}Ni_{50.8}$ single crystals. This internal stress field $|<\sigma_G>|$ is conducive to start of martensitic transformation at higher temperatures $\Delta M_s \approx 18$ K as compared to $[\bar{1}11]$-oriented stress-free aged crystals.

In the temperature range of stress-induced martensitic transformation ($A_f < T < M_d$) the $[\bar{1}11]$-oriented single crystals aged at 823 K for 1 h in stress-free and stress-assisted states demonstrate the superelasticity. The temperature range of the superelasticity ΔT_{SE} is defined as a temperature range in which the given strain by stress-induced martensitic transformation is complete reversibility at unloading. Regardless of the number of crystallographic variants of the disperse Ti_3Ni_4 particles ΔT_{SE}=30 K in [$\bar{1}$11]-oriented aged $Ti_{49.2}Ni_{50.8}$ single crystals. However, in crystals with one variant of particles the temperature interval of superelasticity ΔT_{SE} is shifted towards to higher temperatures due to increase in A_f as compared to stress-free aged crystals with four particle variants (figure 2).

The stress hysteresis $\Delta\sigma$ which was determined as the stress difference between the forward and the reverse stress-induced transformation at the middle of the superelasticity curve characterizes the dissipation energy at stress-induced martensitic transformation. A reduction of the stress hysteresis $\Delta\sigma$ with increase in test temperature is observed in investigated crystals. A similar reduction in $\Delta\sigma(T)$ by a factor of 3-8 is reported for TiNi (C_{Ni}> 50.6 at.%) single crystals that contained Ti_3Ni_4 particles with sizes of d < 40 nm [3].

Figure 4 shows the strain-temperature responses ε(T) recorded during cooling/heating under constant tensile stress of 0.7, 10 and 100 MPa for the stress-free and stress-assisted aged crystals. Stress-assisted aging of the [$\bar{1}$11]-oriented single crystals resultes in appearance of two-way shape memory effect as it is shown in figure 4(b). Upon cooling stress-assisted aging crystal spontaneously expands in the [$\bar{1}$11]-direction at near zero stress. It should be noted that a minimum tensile stress of σ < 1 MPa was required to securely clamp the samples in the setup. Upon heating, the sample dimensions are completely restored and the reversible two-way shape memory effect strain ε_{TWSME} = +2.3(±0.3)% is observed (figure 4(b)). The internal stress field of $|\langle\sigma_G\rangle| \approx 65$ MPa from the oriented growth of dispersed particles in the stress-assisted aged crystals promotes the oriented growth of martensite in cooling, which in turn gives rise to the two-way shape memory effect.

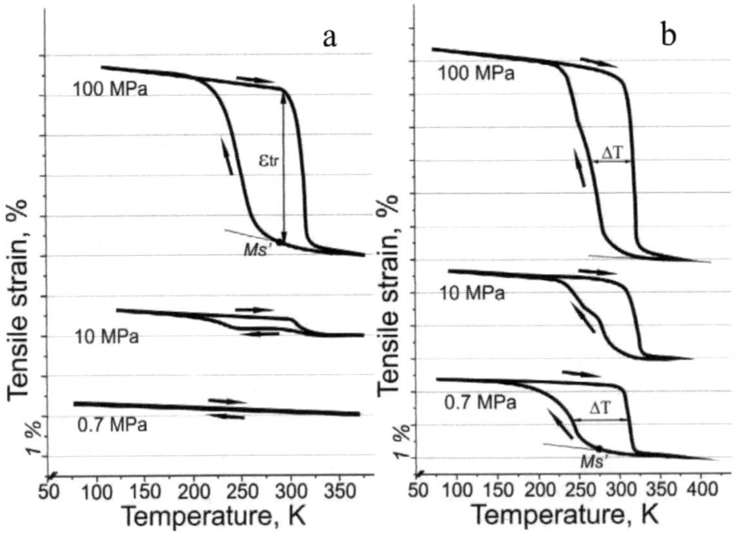

Figure 4. Strain-temperature ε(T) responses of [$\bar{1}$11]-oriented $Ti_{49.2}Ni_{50.8}$ single crystals during cooling/heating under constant tensile stress: for aged at 823 K for 1 h crystals in stress-free state (a) and under compression stress of 150 MPa (b).

In contrast, the stress-free aged crystals do not demonstrate the two-way shape memory effect (figure 4(a)). So, in these crystals during the cooling process a self-accommodated microstructure of martensite is formed in the stress range from 0 to 8 MPa. Under larger tensile stresses the oriented martensite is formed, its volume fraction and the reversible transformation strain increase with stress magnitude.

The martensite start temperature (M_s') increases linearly with the stress value according by Clausius-Clapeyron relationship (1) for both aged crystals. As marked for the ε(T)-respons in figure 4 transformation strains ε_{tr} and thermal hysteresis ΔT were defined as the vertical and horizontal width of the loops, respectively. A decrease in the thermal hysteresis ΔT and growth in ε_{tr} are observed for both crystal, the value of the ε_{tr} at cooling-heating cycle at tensile stress of 100 MPa is, however, larger in the crystals aged under stress (figure 4).

4. Summary

In this study the effects of stress-free and stress-assisted aging at 823 K for 1h on the stress-induced B2-R-B19′ martensitic transformation and superelasticity in $Ti_{49.2}Ni_{50.8}$ (at. %) single crystals are established. It is experimentally shown that aging in a free condition results in precipitation of four variants of Ti_3Ni_4 particles of 350(±30) nm in size oriented along four cristallographically equivalent $\{111\}_{B2}$ planes. Thus, the local stress fields from the neighboring particle variant compensate for each other in stress-free aged crystals. Upon aging under compression stress of 150 MPa along $[\bar{1}11]$ direction only one variant of Ti_3Ni_4 particles grows. The precipitation of one variant of Ti_3Ni_4 particles in the_stress-assisted aged $Ti_{49.2}Ni_{50.8}$ single crystals leads to appearance of homogeneous long-range internal stress field $|\langle\sigma_G\rangle| \approx 65$ MPa, which is created by combining of local stress fields from one variant of particles. The existence of internal long-range stress field in $Ti_{49.2}Ni_{50.8}$ single crystals aged under stress results in, firstly, the increase in the characteristic temperatures of martensitic transformation similar to the effect of external stresses according to the Clausius-Clapeyron equation (1). Secondly, during stress-free cooling in the crystals with one variant of particles R-martensite is formed in accordance with the internal stress field and it is reoriented when an external tensile stress reaches to critical stress level of σ_{cr1} within the temperature range from M_s up to T_R. Thirdly, the growth oriented martensite during stress-free cooling the stress-assisted single crystals with one variant of Ti_3Ni_4 particles is confirmed by observation of the two-way shape memory effect. The sample during stress-free cooling spontaneously undergoes strain of 2.3%, which is fully reversible by heating.

Acknowledgments

The results obtained within the framework of the government contract of the Russian Ministry No. 16.1346.

References

[1] Otsuka K and Ren X 2005 Physical metallurgy of TiNi-based shape memory alloys *Prog in Mater Sci* **50** 511-678

[2] Karaca H E, Kaya I, Tobe H, Basaran B, M. Nagasako M, Kainuma R and Chumlyakov Y 2013 Shape memory behavior of high strength $Ni_{54}Ti_{46}$ alloys *Mater Sci & Eng* **A 580** 66-70

[3] Chumlyakov Y I, Kireeva I V, Panchenko E Y, Timofeeva E E, Kretinina I V and Kuts O A 2015 Physics of Thermoelastic Martensitic Transformation in High-Strength Single Crystals ed N. Resnina and V. Rubanik (Switzerland: *Shape Memory Alloys: Properties, Technologies, Opportunities, Trans Tech Publications Ltd*) 107-174

[4] Chen L Q and Li D Y 1997 Selective variant growth of coherent $Ti_{11}Ni_{14}$ precipitate in a TiNi alloy under applied stresses *Acta Mater* **45** 471-479

[5] Fukuda T, Deguchi A, Kakeshita T and Saburi T 1997 Two-way shape memory properties of a Ni-rish Ti-Ni alloy aged under tensile-stress *Materials Trans JIM* **38** 514-520

[6] Chumlyakov Yu I, Kireeva I V, Karaman I, Panchenko E Yu, Zakharova E G, Tverskov A V, Ovsyanikov A V, Nazarov K M and Kirillov V A 2004 Orientation dependence of Shape memory effect, and Superelasticity in single crystals of CoNiGa, NiMnGa,CoNiAl, FeNiCoTi, TiNi *Russian Phys J.* **47** 893-912

[7] Panchenko E, Chumlyakov Y, Eftifeeva A and Maier H J 2014 Two-way shape memory effect in ferromagnetic $Co_{35}Ni_{35}Al_{30}$ single crystals aged under stress *Scripta Mater* **90-91** 10-13

IOP Conf. Series: Materials Science and Engineering **93** (2015) 012042 doi:10.1088/1757-899X/93/1/012042

Investigation of high-energy external influences on structural heredity of the Ti-Nb alloy

M A Khimich[1], E A Parilov[2,4] , Zh G Kovalevskaya[2,3] and Yu P Sharkeev[2,3]
[1]National Research Tomsk State University, 36, Lenin ave., Tomsk, 634050, Russia
[2]National Research Tomsk Polytechnic Uniersity, 30, Lenin ave., Tomsk, 634050, Russia
[3]Institute of Strength Physics and Material Science SB of RAS, 2/4, Akademichesky ave., Tomsk, 634055, Russia

E-mail: magicinlyfar@mail.ru

Abstract. The effects of high-energy external influences on structural heredity of Ti-Nb alloy is investigated in this paper. By the methods of XRD, SEM, EDX and optical microscopy it was founded that thermal treatment and severe plastic deformation lead to the phase transformations in the alloy, the dendritic segregation occurs and retains in the alloy under external influences.

1. Introduction

The primary dendritic structure forms in the process of the ingots crystallization during production of Ti-Nb alloys. Parameters of this structure depend on conditions of crystallization and bulk of produced ingot [1]. The formation of dendritic structure goes with the formation of chemical heterogeneity of grains which is termed as dendritic segregation. Retaining of dendritic segregation on the next stages of material treatment (structural heredity) can provide heterogeneity of physical and mechanical properties in finished product. It is unacceptable for materials which are applied in the field of medicine [2].

It was explored earlier that the alloy containing 40 mas % niobium is most suitable of all Ti-Nb system alloys for producing of medical materials including bioimplants [3].

This alloy has the lowest elasticity modulus retaining after external influences. This is very important for biocompatibility of implant and bone tissue [2]. Progressive methods of external high-energy influence on material such as electron beam and laser treatment, severe plastic deformation, electron beam and laser treatment can remove structural heredity, to increase strength properties of the alloy and to increase elasticity modulus of finished product insignificantly [4]. The question of structural heredity in the alloy of titanium with 40 mas % niobium after thermal treatment and severe plastic deformation is studied in this paper.

2. Materials and methods

Ti-40 mas % Nb alloy ingots were used as research material. They were produced by the method of electro-arc melting with nonconsumable electrode into water-cooled copper crucible [3]. Ingots after remelting were quenched in two different regimes: heating in argon atmosphere to 1100 °C during one hour and heating to 1000 °C during three hours. Cooling was carried out into water. The specimens were cut from the ingots after quenching. Specimens had parallelepiped form. They were exposed to the SPD different schemes different schemes of SPD. These schemes included forging in cramped conditions and free pressing and rolling [5]. The temperature was decreased during each operation of forging and pressing in the range 500-400 °C. The multistage rolling in grooved rolls at a room temperature was the third stage of SPD. The last one was annealing at 350 °C to remove the residual stresses.

X-ray diffraction analysis was carried out to determine the phase composition. The Bragg-Brentano scheme on Panalytical Empyrean diffractometer (Almelo, Netherlands) in monochromatic CuK_α- and CoK_α-radiations was used. Optical microscope Carl Zeiss Axio Observer (Jena, Germany) and scanning electron microscope Zeiss EVO 50 XVP (Jena, Germany) were used to study the microstructure. Energy dispersive analysis was also carried out on Zeiss EVO 50 XVP.

3. Results and discussion

The primary morphology of a cellular-dendritic structure can be seen on metallographic images of the alloy in the cast state. This morphology is observed in contrast mutually to the secondary grain structure (fig. 1 a).

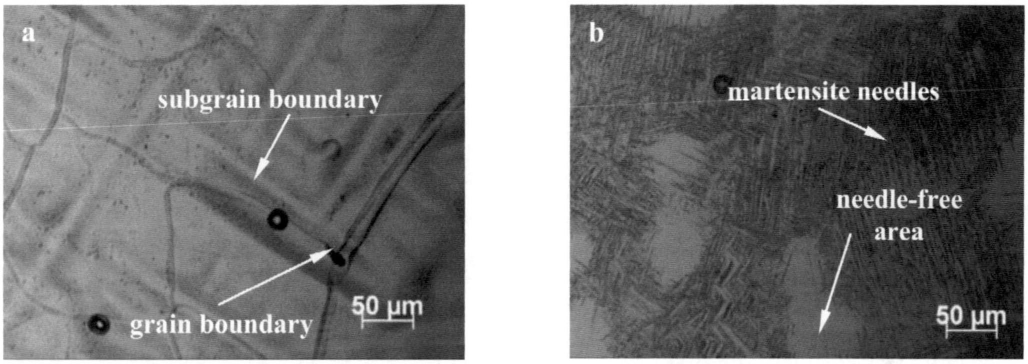

Figure 1. Microstructural images of the specimens in cast state (a) and after quenching (b).

The dendritic structure in the ingot means presence of dendritic segregation in the alloy. The dendritic segregation leads to the heterogeneity of the alloy component's content in different dendrites zones. The results of EDX analysis have shown that niobium concentration in dendrites is more than in interdendritic space. Niobium concentration in dendrites is about 45 mas %. Niobium concentration in interdendritic space is about 39 mas %. Thus, the difference of the niobium concentration in different dendrites zones is 6 %. Presence of dendritic segregation is the feature of the alloys of Ti with β-stabilizing elements [3]. This fact is negative. Following treatment can lead to the formation of different phases in the areas of dendrites with different niobium concentration. Different structures can be formed in the alloy. Usually this problem is solved by many hours alloy annealing [6].

Peaks of α" and β can be seen on X-ray diffraction pattern of the specimen in cast state. Both phases are solid solution of niobium in titanium with the difference at the type of lattice: α" is orthorhombic lattice and β is bcc one. The content of β-phase in the ingot is significantly more than that of α"-phase (fig. 2 a). α"-phase peaks are mild. β-phase is equilibrium and forms as a result of crystallization at equilibrium conditions. α"-phase is non-equilibrium and forms at conditions of quenching from liquid stage. Therefore, process of the ingot's formation occurs at non-equilibrium conditions. The results of investigation of the Ti-Nb alloy ingots with other chemical composition confirm this conclusion [3, 7].

β-phase grains can be seen on metallographic image of the ingot. They have clear and strongly pronounced boundaries. The martensitic structure which is characteristic of non-equilibrium cooling conditions is not observed. Probably, its quantity is so small that the structural non-equilibrium phase is not visualized.

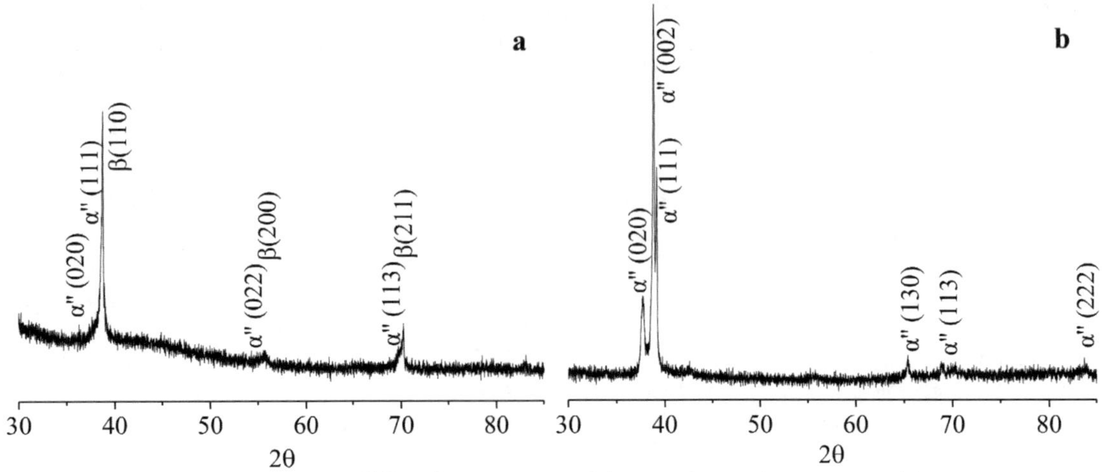

Figure 2. X-ray diffraction patterns of the specimens in cast state (a) and after quenching at 1000 °C (b).

The structure which is characterized of the alloy after non-equilibrium cooling forms in the alloy after quenching. Two types of structure are observed on metallographic images of quenched specimen (fig. 1 a). These two types are areas with needle structure and needle-free areas. Grains of nonequilibrium α"-phase (martensite) have the needle microscope one can see that the primary dendritic structure with some blur contrast retains in the structure. The needle structure forms in the branches of dendrites. Needle-free areas are formed in interdendritic space. Hence, double-phase structure of the alloy after quenching is due to partially retaining of dendritic segregation after heat. The results of EDX analysis shown that niobium concentration in the martensite-free areas is about 44 mas %. The niobium concentration in martensite is equaled to 40 mas %. Difference of Nb concentration after quenching in different dendrites zones is 4%. The variability of the niobium concentration reduce is due to the diffusion processes accompanying keeping of the specimens at a high temperature in the process of quenching. However, complete disappearance of dendritic segregation on standing during three hours was failed. Nonequilibrium conditions of cooling of the alloy led to the formation of martensitic structure in the niobium depleted fields.

Peaks of α"-phase can be seen on X-ray diffraction patterns of quenched specimens. Some peaks which are characterized of α"-phase are absent. The redistribution of peaks intensities is observed. These features of X-ray diffraction pattern indicate the formation of directed textures in the specimen after quenching. These features of structural-phase state is typical as for quenching at 1100 °C during one hour as for quenching at 1000 °C during three hours (fig. 2 b).

However, the values of the alloy's all peaks intensities on X-ray diffraction patterns are higher at three hours that at one hour. X-ray peaks-reflections from the planes of the specimen quenched at 1000 °C are much wider than that of the specimen quenched at 1100 °C. The simultaneous broadening and increase of the values of X-ray peaks intensities could indicate increasing of phases quantity in the specimen. Consequently, increasing the time of quenching leads to the increasing of non-equilibrium α"-phase quantity. Peaks of other phases are not identified at X-ray diffraction pattern of the specimen quenched at 1000 °C (fig. 2 b).

Severe plastic deformation leads to the phase transformation. The complete α" → β phase transformation occurs at SPD as shown by XRD (fig. 3 a). The β-phase peaks are identified on X-ray diffraction pattern of transverse section of the specimen subjected to the three-stage SPD. Broadening of X-ray peaks means reducing of grains size. A small shift of X-ray peaks toward small angles indicates a change of the lattice parameter. The peaks of other phases are not identified on X-ray diffraction pattern. The redistribution of all peaks intensities means the formation of directed textures.

Band contrast created by the directed slip bands can be seen on metallographic image of longitudinal section (fig. 3 b). It seems that presence of dark and light areas caused by partial retaining of dendritic segregation after thermal treatment [3]. However, the results of EDX analysis have shown the following. Concentration of niobium in different areas of the section is not same. The distribution of the areas of enriched and depleted Nb does not coincide with the metallographic contrast. Areas of depleted Nb can have as light as dark contrast. Areas of enriched titanium also are not connected with metallographic contrast. Hence, presence of metallographic contrast is due to the structural transformations which characterize plastic deformation. And it is also due to the formation of crystallographic grain's orientation and formation of dislocation substructure, fragments and subgrains [8].

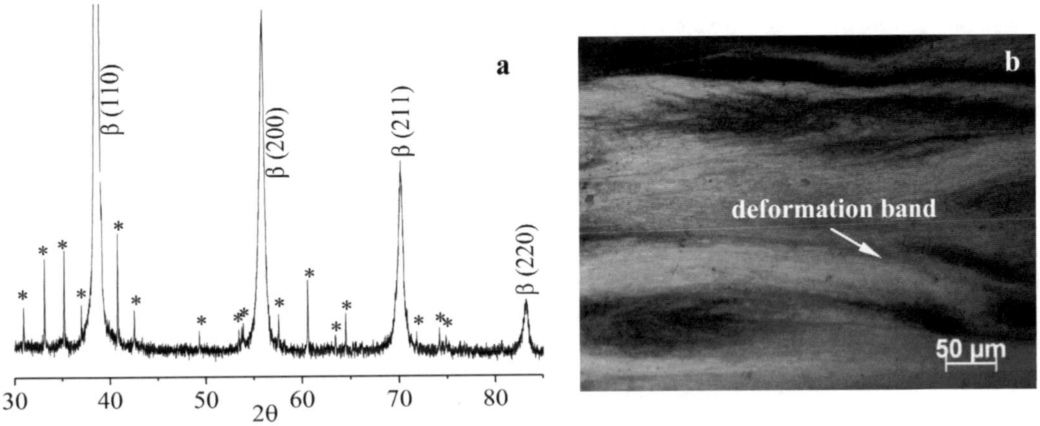

∗ - adhesive composition

Figure 3. X-ray diffraction pattern of transverse section (a) and microstructural image of longitudinal section (b) of the specimens after SPD

4. Conclusions

1. The primary dendritic structure forms in the alloy with 40 mas % Nb during crystallization in the entire volume of the ingot. Polyhedral grains of β-phase are the secondary structure. The dendritic segregation has place in the alloy.
2. The β → α" phase transformation occurs in the Ti-40 mas % Nb alloy as a result of quenching. Retaining of dendritic segregation causes structural heterogeneity of the quenched alloy.
3. The structure consisting of oriented refined grains forms after SPD. The reverse α" → β phase transformation occurs in the process of plastic deformation. The dendritic segregation is not inherited from previous operations. Heterogeneity of elemental composition forms during plastic deformation and is due to the structural transformations.

Acknowledgement
This work is performed under financial support of RSF, project № 15-19-00191.

The authors wish to acknowledge the Leonhard Euler DAAD program for support and personally thank Prof. M. Epple, Prof. C. Mayer, Dr. O. Prymak (University of Duisburg-Essen, Essen), Prof. Q. Zhu and Prof. Z. Sun (General Research Institute for Nonferrous Metals, Beijing) for their assistance and help in research.

References:

[1] Collings E W 1984 *The Physical Metallurgy of Titanium alloys* (Ohio: American Society for

Metals)
[2] Jackson M J and Ahmed W 2007 *Surface Engineered Surgical Tools and Medical Devices* (US : Springer) 533–576
[3] Sharkeev Yu P, Kovalevskaya Zh G, Zhu Q, Khimich M A and Parilov E A 2015 *Advanced Materials Research* **1085** 307–311
[4] Shishkovsky I V 2009 *Laser synthesis of functional mezostructures and bulk products* (Moscow: PHYSMATHLIT)
[5] Sharkeev Yu P, Eroshenko A Yu, Glukhov I A, Zhu Q and Tolmachev A I 2014 *Physical Mesomechanics of Multilevel Systems* **1623** 567–570
[6] Golkovsky M G, Zhuravina T V, Bataev I A, Bataev A A, Veselov S V, Bataev V A and Prikhodko E A 2011 *Advanced Materials Research* **314-316** 23-27
[7] Kovalevskaya Zh G, Khimich M A, Belyakov A V and Shulepov I A 2014 *Advanced Materials Research* **1040** 39-42
[8] Elmay W, Prima F, Gloriant T, Bolle B, Zhong Y, Patoor E and Laheurte P 2013 *Journal of the Mechanical Behavior of Biomedical Materials* **18** 47–56

The influence of hydrogen on the mechanical and functional properties of TiNi and TiNiFe single crystals

Yu N Platonova[1], I V Kireeva[1] and Yu I Chumlyakov[1]
[1]National Research Tomsk State University, Siberian Physical-Technical Institute, Tomsk, 634050, Russia

E-mail: lesoto@sibmail.com

Abstract. The effect of hydrogen on temperature dependence of critical stress $\sigma_{cr}(T)$, the shape memory effect and superelasticity for $[\bar{1}11]$-oriented single crystals of Ti-50.7Ni (at.%) and $[\bar{1}23]$-oriented single crystals of Ti-49%Ni-1%Fe (at.%) alloys at tensile deformation has been studied. It was shown that the change of temperatures of the martensitic transformation, symmetry of thermal hysteresis and level of $\sigma_{cr}(M_s)$ and $\sigma_{cr}(B2)$ stresses are determined by the phase state of the material (martensite or austenite) in the process of hydrogenation.

1. Introduction

Nickel titanium (TiNi) alloys are used in technology and medicine due to their unique functional properties, such as shape memory effect (SME) and superelasticity (SE) [1-2]. It is known, that titanium and nickel are hydride-forming materials. Conducted investigations on single- and polycrystals TiNi have shown [2-5], that hydrogen changes start temperature of the direct martensitic transformation (MT) on cooling M_s, decreases the stability of the crystal lattice of the high-temperature B2- phase, and reduces the plasticity. In addition, the observed effects after hydrogenation in TiNi alloys are dependent on the phase, which saturated with hydrogen. The martensitic phase absorbs hydrogen in greater quantities than the high-temperature phase at same conditions of hydrogenation [6]. However, systematic studies for compare the effects of hydrogen on the temperature MT and, respectively, on the mechanical and functional properties of TiNi-based alloys during hydrogenation of martensitic and austenitic phases was not conducted. Therefore, the purpose of the present work is investigation of the effect of hydrogen on temperature dependence of critical stress, $\sigma_{cr}(T)$, in the temperature range T = 173–473K, SME and SE in single crystals of Ti–50.7%Ni (at.%) and Ti-49%Ni-1%Fe (at.%) alloys at tensile strain in a single-phase state.

The choice of composition of TiNi alloys is due to the following circumstances. Firstly, in Ti-50.7%Ni single crystals observed one-stage B2-B19' MT with start and finish temperatures for the direct on cooling and the reverse on heating MT: M_s = 175K, M_f = 133K, A_s = 192K, A_f = 223K. At room temperature this alloy is in high-temperature B2-phase. This allows us to investigate the effect of hydrogen on the properties of high-temperature phase and MT during hydrogenation of the high-temperature B2-phase. Secondly, Ti-49%Ni-1%Fe single crystals exhibit a two-stage MT through a intermediate R-phase: B2-R-B19' MT [1], with MT temperatures: M_s = 276K, M_f = 260K, A_s = 307K, A_f = 322K. At room temperature this alloy is in the martensitic state. This allows us to investigate the effect of hydrogen on the properties of high-temperature phase and MT when saturated with hydrogen of the martensitic phase. At last, thirdly, the choice Ti-50.7%Ni and Ti-49%Ni-1%Fe alloys will allow to find out the influence of R-phase on B2-B19'- MT after hydrogenation. The $[\bar{1}11]$- and $[\bar{1}23]$-oriented single crystals for studying the influence of hydrogen on the SME and SE were selected due to maximum value of the lattice deformation, ε_0, respectively, ε_0=9.8% and ε_0=6% [7]. The study on single crystals allows excluding inhomogeneity of hydrogen distribution between grain boundaries and grain body.

2. Materials and discussion of the results

Single crystals of Ti-50.7%Ni and Ti-49%Ni-1%Fe(at.%) alloys were grown by the Bridgeman technique in an inert gas atmosphere. After growing samples of single crystals was homogenized at T=1173K in a helium atmosphere for 12 hours, followed by water quenching at room temperature. The crystal orientation and phase analysis of the samples were determined on a diffractometer DRON-3 using Fe kα-radiation. The samples were cut in the form of double pads with 12x2.5x1 mm^3 working part dimensions. Hydrogenation performed electrochemically in solution of 1M H_2SO_4 at a temperature T=295K, and current density j=1400 A/m^2 with using an anode made austenitic stainless steel. MT temperatures were determined from inflections on temperature dependences of electric resistivity ρ(T). The mechanical tensile tests were performed on an Instron-5969 machine at a deformation rate of $4x10^{-4}$ s^{-1}.

Figure 1 shows the temperature dependence of the critical stress $\sigma_{cr}(T)$ for tensile deformation in single crystals measured before and after hydrogenation at room temperature for 2 hours while the current density j = 1400 $A/м^2$.

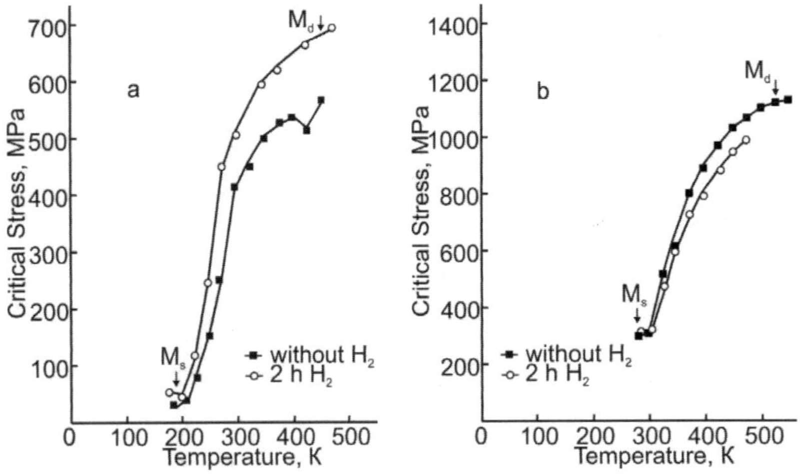

Figure 1. Temperature dependences of critical stress σ_{cr} in $[\bar{1}11]$-oriented single crystals of Ti–50.7Ni alloy (a) and $[\bar{1}23]$-oriented single crystals of Ti-49Ni-1Fe alloy (b) at tensile deformation.

It can be seen, that in both cases the $\sigma_{cr}(T)$ dependence has a staging, analogous to those usually observed in the alloys with stress-induced MT [1]. The minimum stresses on the $\sigma_{cr}(T)$ curve are observed at T=M_s, which coincides with temperature M_s, determined from the ρ(T) curve. The maximum σ_{cr} stresses corresponds to M_d temperature, at which the stress for start of stress-induced MT is equal to the plastic flow stress of the high-temperature B2- phase. The stage at M_s<T<M_d is characterized by an anomalous temperature dependence $\sigma_{cr}(T)$, which is due to the stress-induced nucleation of B19' martensite crystals [1]. The dependence $\sigma_{cr}(T)$ at this stage is described by the Clapeyron-Clausius equation:

$$\frac{d\sigma_{cr}}{dT} = -\frac{\Delta H}{\varepsilon_0 T_0}, \qquad (1)$$

where ΔH is the enthalpy change at B2-B19' МП; ε_0 is the lattice deformation at B2-B19' MT; which depends on the orientation; T_0 is temperature of chemical equilibrium of B2 and B19' phases. At T>M_d on the curve $\sigma_{cr}(T)$ is observed a stage, which is related to the plastic deformation of a high-temperature B2- phase.

When hydrogenation of crystals are at room temperature in a high-temperature phase (Ti-50.7Ni single crystals) for 2 hours and current density j = 1400 A/m^2 the following relationships detected. Firstly, at the hydrogenation the start temperature M_s is increased as compared to the initial state without hydrogen at 25K. The increase of M_s temperature after hydrogenation of single crystals agrees with the data obtained earlier in the hydrogenation of polycrystalline alloys of similar composition under similar conditions of hydrogenation [8]. However, when hydrogenation of single crystals TiNi alloys this composition at a current density of j = 700 A/m^2, on the contrary, it was observed a decrease in M_s temperature [4]. Consequently, the temperature M_s is sensitive to hydrogenation modes. Secondly, the hydrogen increases σ_{cr} at a temperature T=M_s from 25 MPa to 40 MPa. Third, hydrogen increases the value of $\alpha = d\sigma_{cr}/dT$: from 4.26 MPa/K for hydrogen-free crystals to 4.6 MPa/K for hydrogenated crystals. At last, fourth, the hydrogenation at a current density j = 1400 A/m^2 for 2 hours leads to an increase in critical stress of high-temperature B2- phase. The increase stresses during hydrogenation is $\sigma^{H}_{cr}(400) - \sigma_{cr}(400) = 120$ MPa ($\sigma_{cr}(400)$ - critical stress of high-temperature B2- phase at T = 400 K in a state before hydrogenation, $\sigma^{H}_{cr}(400)$ - after hydrogenation). At the same time the change of the phase composition of Ti-50.7Ni single crystals after hydrogenation at a current density of j = 1400 A/m^2 for 2 hours by electron microscopy and X-ray is not detected. It follows that hydrogen is in a state of solid solution at hydrogenation of high-temperature B2- phase at a current density j = 1400 A/m^2 for 2 hours.

Table 1. Mechanical and functional characteristics of $[\bar{1}11]$-oriented single crystals of Ti–50.7Ni alloy and $[\bar{1}23]$-oriented single crystals of Ti-49Ni-1Fe alloy.

	Ti–50.7%Ni		Ti–49%Ni–1%Fe	
	Without H_2	2 h H_2	Without H_2	2 h H_2
Temperature M_s, (K) at σ_{ext} = 0MPa:	175	200	276	280
MT temperatures at σ_{ext}:	σ_{ext} = 75MPa		σ_{ext} = 250MPa	
M_s^σ (K)	214	207	300	310
M_f^σ (K)	178	117	249	257
A_s^σ (K)	235	216	313	311
A_f^σ (K)	272	268	363	365
Interval of direct stress-induced MT (supercooling) $\Delta_1 = M_s - M_f$ (K)	36	90	51	53
Interval of reverse stress-induced MT (superheating) $\Delta_2 = A_f - A_s$ (K)	37	52	50	54
The value of direct thermal hysteresis under stress $\Gamma_1 = A_f - M_s$ (K)	58	61	63	55
The value of reverse thermal hysteresis under stress $\Gamma_2 = A_s - M_f$ (K)	57	99	64	54
$\sigma_{cr}(M_s)/\sigma_{cr}(400)$ (MPa)	25/530	40/650	-	-
$\sigma_{cr}(M_s)/\sigma_{cr}(475)$ (MPa)	-	-	300/1065	315/980
$\alpha = d\sigma_{cr}/dT$ (MPa/K)	4.26	4.6	5.9	5.2

When hydrogenation for 2 hours and the current density j = 1400 A/m^2 of single crystals (Ti-49Ni-1Fe crystals) which are at room temperature in the martensitic state, is detected, firstly, that temperature M_s and critical stress σ_{cr} at T = M_s increase less, than in crystals Ti–50.7%Ni, during hydrogenation under the same conditions, respectively, 4K and 15 MPa. $\alpha = d\sigma_{cr}/dT$ decrease after hydrogenation from 5.9 MPa/K to 5.2 MPa/K. Secondly, at hydrogenation for 2 hours in Ti-49Ni-1Fe crystals observed the decrease of stress of high-temperature B2- phase: $\sigma_{cr}(475) - \sigma^{H}_{cr}(475) = 85$MPa, that is the effect of a softening is observed. Ti-50.7Ni single crystals after hydrogenation of high-temperature phases at room temperature at current density of j = 1400 A/m^2 the softening effect has been observed at higher hydrogenation times [5]. It was associated with the formation of $TiNi_3$ and TiH particles during hydrogenation, as shown by X-ray diffraction phase analysis and transmission electron microscopy [5,9]. Therefore, it is believed that lowering the strength properties of high-temperature phase at hydrogenation of Ti-49Ni-1Fe single crystals in the martensitic state at room temperature occurs due to the change in the phase composition of these crystals at hydrogenation.

Figure 2 presents results of investigation of the SME effect during cooling/heating under tensile stresses in test temperatures interval 77–400K. Such experiments allow determining the influence of external stresses σ_{ext} not only on the MT temperatures, but also on the thermal hysteresis ΔT and the value of transformation strain ε_{tr}. The value of external tensile stress for both alloys was equal to, respectively, the tensile stress at $T = M_s$, obtained from studying the temperature dependence of $\sigma_{cr}(T)$.

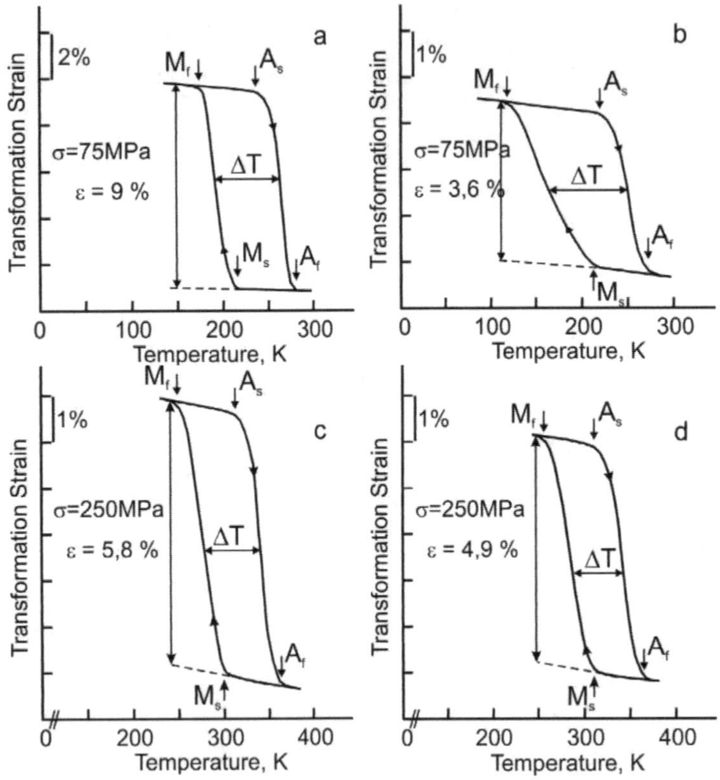

Figure 2. Value of SME at constant tensile stress in $[\bar{1}11]$-oriented single crystals Ti–50.7Ni alloy (a,b) and $[\bar{1}23]$-oriented single crystals Ti-49Ni-1Fe alloy (c,d): (a,c) in the absence of hydrogen; (b,d) upon hydrogenation at T = 295K and current density j = 1400 A/m^2 for 2 hours.

As can be seen for $[\bar{1}11]$-oriented single crystals of Ti-50.7Ni alloy, which are in the B2- phase at room temperature, both hydrogen-free and hydrogenated crystals at σ_{ext} = 75 MPa is realized single-stage B2–B19' MT, that is fully reversible upon heating, and the SME is observed. The value of SME effect ε_{SME} in hydrogen-free $[\bar{1}11]$-oriented single crystals is 9% and close to theoretically calculated value of lattice deformation ε_0 = 9,8% in $[\bar{1}11]$-orientation for B2–B19' MT at tension [7]. In the hydrogenated material at σ_{ext} = 75 MPa the value of ε_{SME} is 2.5 times lower than in the crystals without hydrogen. That is to cause fracture of the self-accommodating B19' -martensite structure in hydrogenated crystals these stresses are insufficient as compared with crystals without hydrogen. Therefore, hydrogen increases the resistance to motion of the intervariant and twin boundaries in martensite as compared with the hydrogen-free state. This agrees with the increasing the value of $\sigma_{cr}(M_s)$. In order to reach the theoretical values of the lattice deformation in the hydrogenated crystals, it is necessary to increase external stresses of B2–B19' MT.

In $[\bar{1}23]$-oriented single crystals of Ti-49Ni-1Fe alloy, which are in the martensitic phase at room temperature, both in hydrogen-free and hydrogenated crystals at σ_{ext} = 250 MPa is realized single-stage B2–B19' MT, that is fully reversible upon heating, and the SME is observed. B2-R MT appears at much lower stresses, and in this work is not considered. In hydrogen-free $[\bar{1}23]$-oriented single crystals of Ti-49Ni-1Fe alloy at σ_{ext} = 250 MPa the value transformation strain ε_{SME} = 5,8% and is close to the theoretically calculated value of the lattice deformation ε_0 = 6% in $[\bar{1}23]$-orientation for B2–B19' MT at tension [7], which is a qualitative confirmation of stress-induced B2–B19' MT in Ti-49Ni-1Fe crystals at σ_{ext} = 250 MPa. After hydrogenation the value of deformation strain decreases at σ_{ext} = 250 MPa and becomes equal to 4.9%.

The table 1 shows the mechanical and functional characteristics of the B2–B19' MT for Ti-50.7Ni and Ti-49Ni-1Fe crystals, obtained in experimental studies of SME under stress. It is visible, that in hydrogen-free Ti-50.7Ni crystals the values of supercooling Δ_1=M_f–M_s=36K and superheating Δ_2=A_s-A_f=37K and the value of thermal hysteresis Γ_1= A_f–M_s=58K and Γ_2= A_s–M_f=57K are equal to each other. As a result of thermal hysteresis loop is symmetric: $\Delta_1/\Delta_2 = 0.97$, $\Gamma_1/\Gamma_2 = 1.02$. Hydrogenation increases Δ_1, Δ_2, Γ_1 and Γ_2 as compared to the state without hydrogen (Table 1, Figure 2.b) due to the influence of the M_f, temperature which is shifted to lower temperatures in 61K. As a result, for hydrogenated of Ti-50.7Ni crystals the thermal hysteresis loop has an asymmetrical form: $\Gamma_1 < \Gamma_2$ (Γ_1 = 61 K, Γ_2 = 99 K).

In $[\bar{1}23]$-oriented single crystals of Ti-49Ni-1Fe alloy thermal hysteresis both in the initial state and after hydrogenation is symmetric: in hydrogen-free crystals – Δ_1/Δ_2 =1.02, Γ_1/Γ_2 =0.98, after hydrogenation – 0.98 and 1.02, respectively. It should be noted, that in hydrogenated Ti-49Ni-1Fe crystals observed narrowing trend of thermal hysteresis by increasing the M_s and M_f temperatures under stress, respectively, on 10K and 8K, compared with crystals without hydrogen.

Analysis of the data shows, that the characteristics of the B2-B19' MT depend on the condition of the crystal during hydrogenation. Thus, if the crystal during hydrogenation at room temperature is in the high-temperature B2- phase, the hydrogenation accompanied by a change of the form of thermal hysteresis loops from symmetric in hydrogen-free state to asymmetric after hydrogenation and a value of Γ_2 increased. If the crystal during hydrogenation at room temperature is in the martensitic phase, form of thermal hysteresis loop after hydrogenation remains symmetrical, as in the state without hydrogen and thermal hysteresis on the contrary decreases. And this is determined by the presence of R- phase in Ti-49Ni-1Fe crystals and its absence in Ti-50.7Ni crystals. R- phase apparently slows down the effect of hydrogen on the characteristics of the MT as a result of the weak impact of hydrogen on the R- phase.

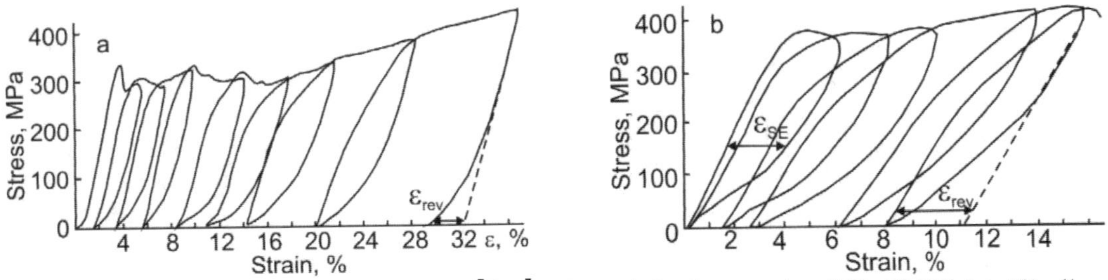

Figure 3. Stress-strain curve at tension for $[\bar{1}11]$-oriented single crystals of Ti–50.7Ni (at. %) alloy in hydrogen-free state (a) and after hydrogenation (b); test temperature 275K, A_f = 223K.

In $[\bar{1}11]$-oriented hydrogen-free crystals of Ti-50.7Ni alloy the SE is not observed (Figure 3.a). At a preset strain of ε_{exp} = 2.5% in the first cycle at test temperature T = 275K, a perfect loop is not formed, and reversible deformation ε_{rev} = 1%. At ε_{exp} = 35% the ε_{rev} increases to 2.5%, and at a further increase to ε_{exp} > 35% the crystal is not destroyed. After hydrogenation for 2 h., when the hydrogen in the crystal is a solid solution the SE equal to 2.5% is observed at T = 275K (Figure 3.b). With increasing ε_{exp} to 12.5% ε_{rev} becomes 3%. After ε_{exp} > 12.5% hydrogenated crystals are destroyed.

In $[\bar{1}23]$-oriented hydrogen-free crystals of Ti-49Ni-1Fe alloy the SE was not detected. At a preset strain of ε_{exp} =3 % in the first cycle at test temperature T = 345K the reversible deformation ε_{rev} = 1 %.

At ε_{exp} = 10 % the ε_{rev} becomes equal 2.5%, and a further increase in ε_{exp} > 10% the crystal is destroyed. After hydrogenation at ε_{exp} = 2.5% at T = 345K reversible strain ε_{rev} = 1.6%. At ε_{exp} > 2.5% crystals are destroyed. This sharp decline of plasticity in TiNi crystals associated with the advent of titanium hydrides [5].

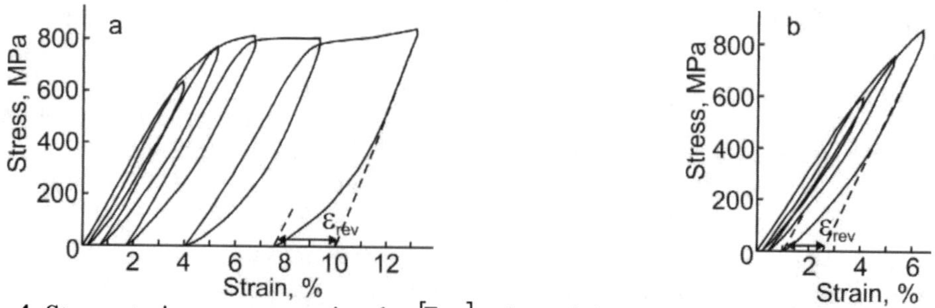

Figure 4. Stress-strain curve at tension for $[\bar{1}23]$-oriented single crystals of Ti–49Ni–1Fe alloy: (a) in hydrogen-free state, (b) after hydrogenation; test temperature 345K, A_f = 322K.

3. Conclusion

It was established experimentally, that the mechanical and functional properties of single crystals of TiNi alloys depend on the phase state (martensite or high-temperature phase), at which electrochemical hydrogenation in 1M solution of H_2SO_4 at T = 295K and current density j = 1400 A/m^2 for 2 hours occurs.

The hydrogenation of $[\bar{1}11]$-oriented single crystals of Ti-50.7%Ni (at. %) alloy, which are in a high-temperature B2- phase at room temperature, increases the start temperature of the MT upon cooling M_s, increases a stress σ_{cr} in the point M_s and in the high-temperature B2- phase, increases of the thermal hysteresis, leads to asymmetrical loop form of thermal hysteresis, to decrease of the plasticity and to appearance of SE equal 2.5% compared with hydrogen-free crystals.

The hydrogenation of $[\bar{1}23]$-oriented single crystals of Ti-49%Ni-1%Fe (at. %) alloy, which are in martensitic phase at room temperature, leads to an increase of temperature M_s and stress $\sigma_{cr}(M_s)$, to decrease of the thermal hysteresis while maintaining its symmetry, to decrease a stress of the high-temperature B2- phase, and the plasticity compared with hydrogen-free crystals.

This difference in behavior TiNi crystals due, apparently, to the difference in the hydrogenation of high-temperature phase and martensite.

Acknowledgements

The results of the work are obtained in part by the Ministry of Education and Science of the Russian Federation, project no. 1346.

References

[1] Otsuka K and Wayman C M 1998 Shape memory materials *Cambridge University Press* 284
[2] Spivak L V 2008 *Phys. UPS* **178** 897–922
[3] Pelton B L, Slater and Pelton A 1997 *Proceedings of the 2nd Int'l Conference on Shape Memory and Superelastic Technologies SMST* 395–400
[4] Chumlyakov Yu I, Kireeva I V, Panchenko E Yu, Kirillov V A, Timofeeva E E, Kretinina I V, Danilson Yu N, Karaman I, Maier H and Cesari E 2011 *Rus. phys. J.* **54** 937–950
[5] Kireeva I V, Chumlyakov Yu I and Platonova Yu N 2015 *Tech. Phys. Letters* **41** 284–287
[6] Yokoyama K, Tomita M and Sakai J 2009 *Acta Mater.* **57** 1875–1885
[7] Sehitoglu H, Hamilton R, Canadic D, Zhang X Y, Karaman I, Chumlyakov Yu I and Maier H J 2003 *Metall. And Mat. Trans. A* **34** 5–13
[8] Stepanov I A, Flomenblit Yu M and Zaymovsky V A 1983 *PMM* **55** 612–614
[9] Pelton T and Chen G W 2003 *ASM Mater. & Proc. for Medical Devices Conference* 285–289

Shape memory effect and superelasticity in single crystals of iron-based alloys

Z V Pobedennaya[1], K A Reunova[2], I V Kireeva[1] and Y I Chumlyakov[1]
[1] National Research Tomsk State University, Siberian Physical-Technical Institute, Tomsk, 634050, Russia
[2] National Research Tomsk State University, Department of Metal Physics, Tomsk, 634050, Russia

E-mail: pobedennaya_zina@mail.ru

Abstract. It was demonstrated that in single crystals of disordered iron-based (Fe-28%Ni-17%Co-11.5%Al-2.5%X (at. %) (X = Ti, Ta, Ti + Ta) alloys, which undergoing γ-α' thermoelastic martensitic transformation (γ – fcc – face-centered cubic lattice, α' – bct – body-centered tetragonal lattice), the precipitation of ordered γ'-phase particles at aged at 973 K for 0.5 h and 7 h results in the shape memory effect and superelasticity. It was experimentally discovered, that the both mechanical and functional properties – the shape memory effect and superelasticity – can be controlled by changing the chemical composition, the size and the volume fraction of dispersed particles of γ'-phase.

1. Introduction

It is well known that the shape memory effect (SME) and superelasticity (SE) related to reversible thermoelastic martensitic transformations (MT). In turn, reversible thermoelastic MT at cooling/heating without stress and under stress usually occur in the alloys with atomic-ordered structure [1, 2]. And the alloys with disordered structure as a rule show non-thermoelastic MT. In disordered iron-based alloys such as FeMnSi, FeNiCoTi, FeMnNiAl, FePt, FeNiCoAlTa thermoelastic MT result from the precipitation of dispersed γ'-phase particles with $L1_2$ structure, which themselves not be transformed [3-5]. These materials, as it is demonstrated in several studies, possess recoverable transformation strain up to 2-13 %, and therefore have a great interest from the point of view of their practical use, primarily due to the low cost of the material and its treatment as compared with NiTi. The analysis of data, contained in [4, 6, 7], shows that not only the precipitation of ordered γ'-phase but also its chemical composition resulting in thermoelastic γ-α' MT in iron-based alloys by controlling the stability and strength of the phase at γ-α' MT.

By now there are no studies on how the chemical composition, size and volume fraction of dispersed γ'-phase particles influence on mechanical and functional properties of disordered iron-based single crystals. Therefore this work contains the investigation results of the influence of the chemical composition, size and volume fraction of dispersed γ'-phase particles on mechanical and functional properties (SME and SE) in new ferromagnetic single crystals alloys such as Fe-28%Ni-17%Co-11.5%Al-2.5%X (at. %) (X = Ti, Ta, Ti + Ta). In single crystals FeNiCoAlX (X = Ti, Ta, Ti + Ta) alloys the change of chemical composition of γ'-phase particles was achieved by the change of the fifth element, while maintaining of atomic concentration of alloy. The chemical elements Ti and Ta were selected for alloying first of all because Ti is lighter than Ta and has a larger diffusion mobility that leads to a more uniform distribution of the Ti atoms in the melting of the alloy and crystal growth than Ta does. Thus, it can lead to increased plasticity of crystals with Ti as compared with Ta [8]. Secondly, Ta is heavier than Ti, and therefore, can more effectively increase the particle strength and the stress level in the high temperature phase. Thirdly, in complex-alloyed crystals when the fifth element is defined as a combination of the elements X = Ti + Ta, the particle strength may depend on the aging time at the same aging temperature and this can lead to a difference in the stress level in the high-temperature phase and the stress-induced γ-α' MT [9-11]. γ'-phase particles have a chemical composition $(FeNiCo)_3(AlTi)$, $(FeNiCo)_3(AlTa)$ and $(FeNiCo)_3(AlTiTa)$, respectively, for FeNiCoAlTi, FeNiCoAlTa and FeNiCoAlTiTa [4, 10]. For this work the crystals with [001] orientation were selected, because the maximum lattice strain ε_0 = 8.7 % can be achieved only in this orientation under tension for γ-α' MT [4].

2. Materials and discussion of the results

Single crystals of iron-alloys Fe-28%Ni-17%Co-11.5%Al-2.5%X (at. %) (X = Ti, Ta, Ti + Ta) were grown using the Bridgeman technique in inert gas atmosphere. The single crystals after growth was homogenized at T = 1550 K for 6 h under an inert gas atmosphere, followed by water quenching. The orientation of the crystals was determined by diffractometer DRON-3. Samples for tension along the [001] orientation were cut on electro-discharge machine. SME was measured using a specially designed device when cooling/heating at different levels of external stresses. Mechanical properties of the crystals were examined by the Instron at the temperature interval from 77 K to 525 K at a deformation rate of $4\cdot10^{-4}$ c^{-1}. The samples were aged at T = 973 K for 0.5 h (FeNiCoAlTi – crystal 1, FeNiCoAlTa – crystal 2, FeNiCoAlTiTa – crystal 3) and for 7 h (FeNiCoAlTi – crystal 4, FeNiCoAlTa – crystal 5, FeNiCoAlTiTa – crystal 6) in inert gas atmosphere, followed by water quenching. The temperatures of γ-α' MT were determined from change of the temperature dependence of the electrical resistivity ρ(T). In crystals 1-4 the γ-α' MT temperatures are lower than the temperature of liquid nitrogen. In crystals 5 the start temperature of γ-α' MT under cooling M_s = 125 K, and the end temperature of reverse transformation under heating A_f = 200 K, in crystals 6 – M_s = 195 K and A_f = 220 K.

Figure 1 shows the results of a study of value of SME at cooling/heating under constant tensile stress for all the investigated crystals. It is evident that in crystals one-step γ-α' MT occurs, which is irreversible in crystals 1-3 aged at 973 K for 0.5 h (Figure 1, a-c) and reversible in crystals 4-6 aged at 973 K for 7 h (Figure 1, d-f).

In crystals 1-3 with applied maximum external stress the maximum value of transformation strain, ε_{SME}, does not achieve the theoretical values of lattice deformation ε_0 = 8.7 % for tension in [001]-orientation crystals with the γ-α' MT [11]. Thus, in crystals 2 the ε_{SME} = 3 % at σ = 575 MPa, and in crystals 1 and 3 the ε_{SME} = 2.4 %, respectively at σ = 600 MPa and 450 MPa. The theoretical values are not achieved due to the crystals destruction under external stress, respectively, σ > 600 MPa in crystals 1, σ > 575 MPa in crystals 2, σ > 450 MPa in crystals 3. And as in crystals 1-3 when heated up to 350-400 K the temperature A_f under stress is not achieved due to the limitation of device of heating to high temperatures, it can be assumed that γ-α' MT in crystals 1-3 aged at 973 K for 0.5 h is characterized by a wide thermal hysteresis under stress ΔT > 150 K. In crystals 4, aged at 973 K for 7 hours, the maximum value of the transformation strain – ε_{SME} = 7.8 % at σ = 375 MPa, and it is close to the theoretical value of the lattice deformation ε_0 = 8.7 % at tension of [001]-orientation crystals with the γ-α' MT [11]. In crystals 5 and 6, aged at 973 K for 7 hours, the maximum value of ε_{SME} does not achieve theoretical values of lattice deformation ε_0: in crystals 5 – ε_{SME} = 4.2 % at σ = 125 MPa; in crystals 6 – ε_{SME} = 2.4 % at σ = 250 MPa. By increasing the external applied stresses crystals are destructed: crystals 4 at σ > 375 MPa, crystals 5 at σ > 125 MPa, crystals 6 at σ > 250 MPa. Thermal hysteresis ΔT under stress for crystals 4, 5 and 6 is respectively 75 K, 60 K and 90 K. It should be noted that when increasing aging time at a given temperature of aging from 0.5 hour to 7 hours, the externally applied stress to achieve maximum transformation strain decreases, that is associated with an increase in M_s temperature by increasing the aging time [12, 13].

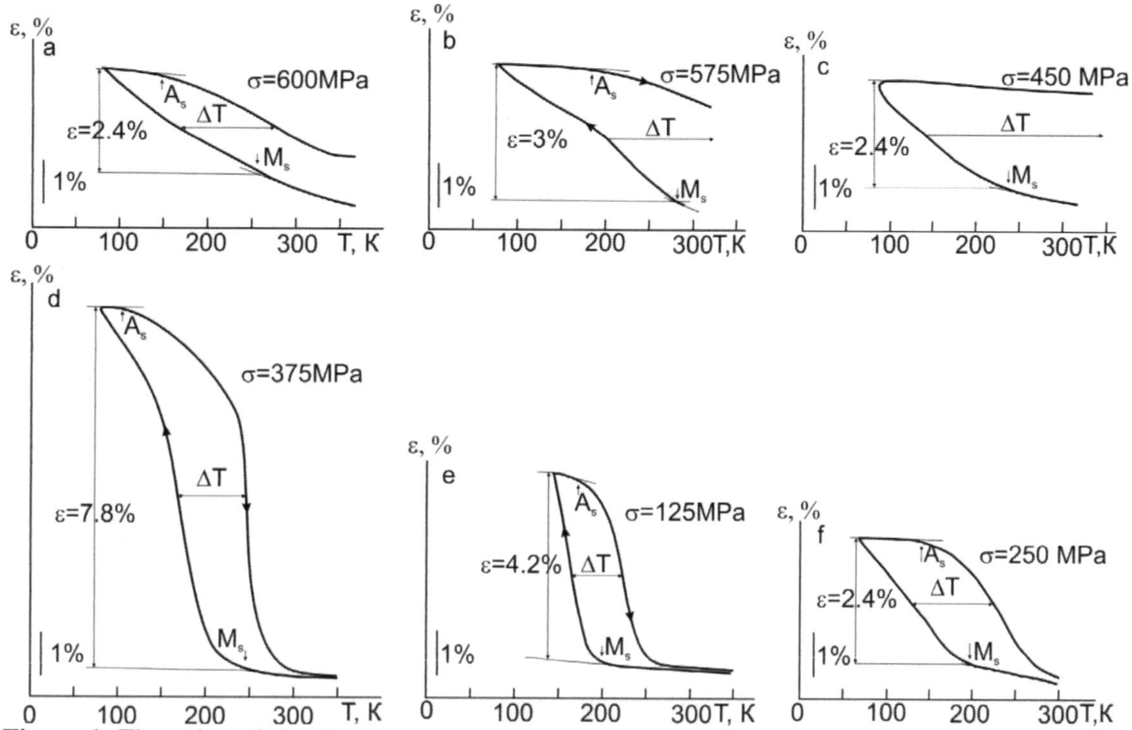

Figure 1. The value of the shape memory effect under constant tensile stress for the [001]-oriented single crystals, aged at T = 973 K, 0.5 hour: a – FeNiCoAlTi, b – FeNiCoAlTa, c – FeNiCoAlTiTa; 7 hours: d – FeNiCoAlTi, e – FeNiCoAlTa, f – FeNiCoAlTiTa.

Figure 2 shows the results of research of SE in depending on the aging time, the test temperature, the value of applied strain in the cycle. In crystals 1-3 according to dependence ρ(T) the γ-α' temperatures are lower than the temperature of liquid nitrogen. At T = 77 K, as can be seen from Figure 2, the SE is observed only in the crystals 2, aged at 973 K for 0.5 h. At the maximum applied strain ε_{ap} = 14.3 % the reversible strain ε_{SE} = 10.8 %, and in size it exceeds value of the lattice deformation ε_0 = 8.7 % at tensile for crystals oriented along the [001] direction with γ-α' MT [11]. Loop SE is characterized by extensive stress hysteresis Δσ = 290 MPa which is defined as the difference stresses required for direct and reverse martensitic transformation during deformation at half the deformation in cycle. In crystals 1 and 3 at T = 77 K the perfect SE loop is not observed. At the total applied strain 3.4 % and 7.9 %, respectively, in crystals 1 and 3 reversible strain is equal 1 %. This difference in the temperature for appearance of SE at T = 77 K is connected with the difference in temperature M_s, and in the crystals 2 M_s is much lower than in the crystals 1 and 3, apparently.

In crystals 4-6, aged at 973 K, 7 hours, the temperature A_f of γ-α' MT on the data of depending ρ(T) is higher than the temperature of liquid nitrogen and at T > 77 K in these crystals perfect SE loop is observed. In crystals 4 maximum value of SE ε_{SE} = 6 % (Figure 2, d), and in crystals 5 and 6 value of SE ε_{SE} = 4.3-4.5 % (Figure 2, e, f). The value of stress hysteresis Δσ in the crystals 4-6 is the same and equal to 170 MPa (Figure 2, d-f).

Thus, it is experimentally established that in the [001] single crystals of FeNiCoAlX (X = Ti, Ta, Ti + Ta) alloys thermoelastic γ-α' MT with SME and SE is implemented. Values of ε_{SME}, ε_{SE}, Δσ depend not only on the size and volume fraction of γ'-phase particles, but also on their composition, and in FeNiCoAlTiTa crystals value of SME and SE is less than in FeNiCoAlTi crystals.

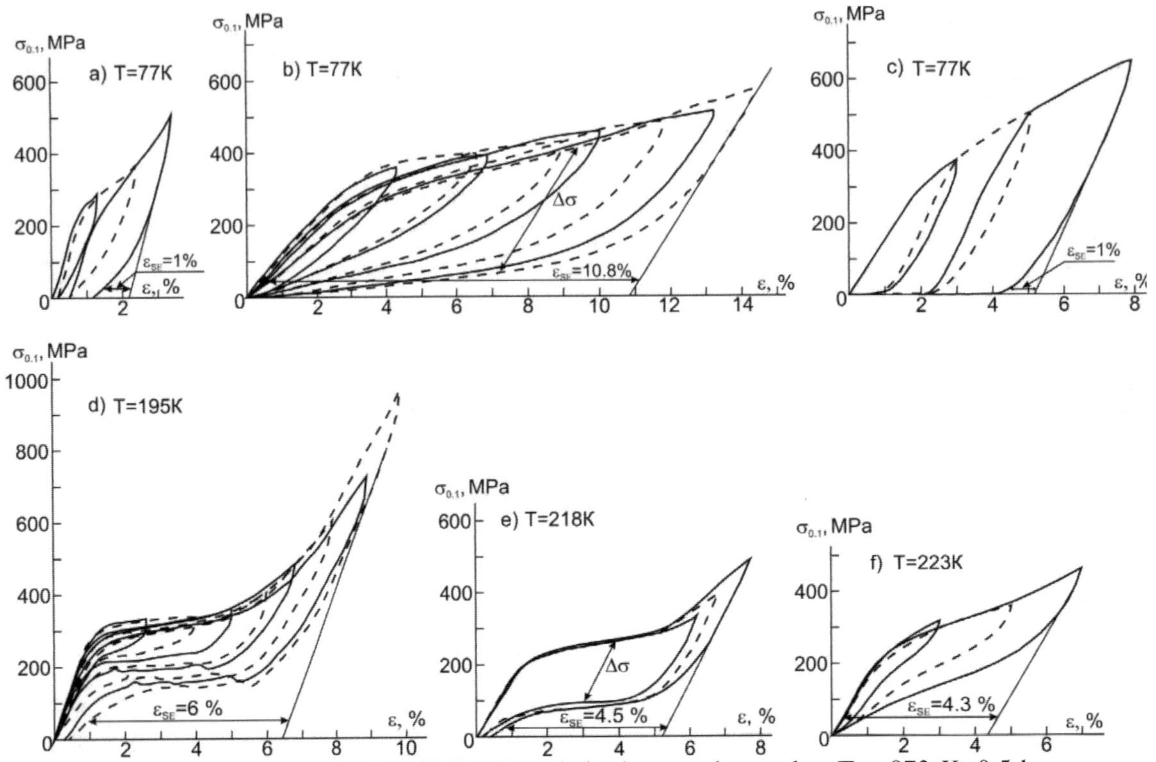

Figure 2. The superelasticity in the [001]-oriented single crystals, aged at T = 973 K, 0.5 hour: a – FeNiCoAlTi, b – FeNiCoAlTa, c – FeNiCoAlTiTa; 7 hours: d – FeNiCoAlTi, e – FeNiCoAlTa, f – FeNiCoAlTiTa.

Diffusion mobility of elements Ti and Ta is different, and the coefficients of diffusion of these elements in iron are in the following sequence $D_{Ti} > D_{Ta}$. Different diffusion mobility of elements Ti and Ta can has impact the processes of formation of the particles in FeNiCoAlTiTa alloy, on their atomic structure at the same aging conditions (temperature and aging time) compared with FeNiCoAlTi and FeNiCoAlTa alloys, and this may affect in the level of strength properties of high-temperature phase, and respectively, on the development of thermoelastic stress-induced γ-α' MT [11].

Electron microscopy studies of the structure of crystals after aging at 973 K for 0.5 h and 7 h show that upon aging for 0.5 hour and 7 hours in crystals FeNiCoAlX (X = Ti, Ta, Ti + Ta) there is precipitation of γ'-phase particles with atomic ordered by type $L1_2$ structure and size of 3-10 nm. The volume fraction of the particles is 15-20 %. Effect of change of the chemical composition on the structure of γ'-phase particles at aging is not established yet and requires further investigation.

In Figure 3 the temperature dependence of axial stresses $\sigma_{0.1}$ for [001]-oriented single crystals FeNiCoAlX (X = Ti, Ta, Ti + Ta), aged at 973 K for 0.5 hour and 7 hours in the temperature range of T = 77-523 K under tensile deformation is presented.

It is clear that for all the crystals FeNiCoAlX (X = Ti, Ta, Ti + Ta), aged at 973 K for 0.5 hour and 7 hours, the dependence $\sigma_{0.1}(T)$ has a characteristic view for alloys experiencing stress-induced MT, and is observed two stages. In the first stage in the temperature range 77 K < T < M_d with increasing test temperature the stresses of $\sigma_{0.1}$ is increased, which is described by the Clausius-Clapeyron relation:

$$\frac{d\sigma_{0.1}}{dT} = -\frac{\Delta H}{\varepsilon_0 T_0} = -\frac{\Delta S}{\varepsilon_0}, \tag{1}$$

here ΔH and ΔS – respectively enthalpy and entropy change at γ-α' MT; ε_0 – transformation strain; T_0 – temperature of chemical equilibrium of γ- and α'-phases.

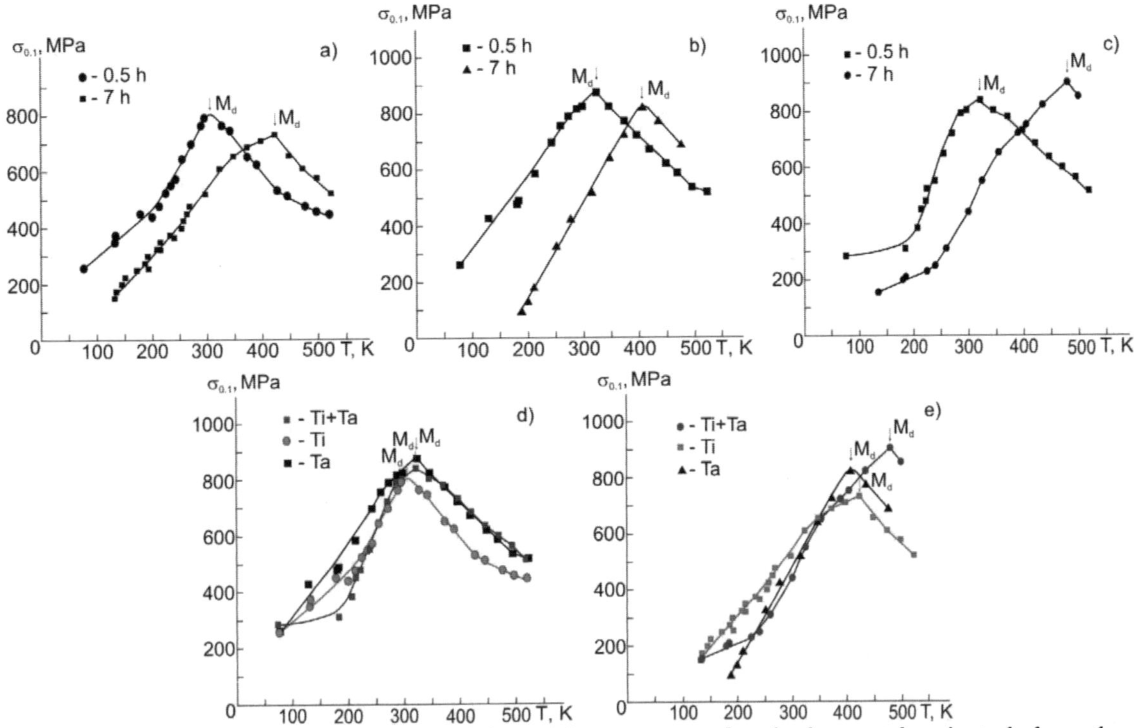

Figure 3. The temperature dependence of the axial stresses $\sigma_{0.1}$ for single crystals oriented along the [001] direction, at tensile deformation: a) FeNiCoAlTi, b) FeNiCoAlTa, c) FeNiCoAlTiTa, d) aging at 973 K, 0.5 hour, e) the aging at 973 K, 7 hours.

The second stage at T > M_d associated with plastic deformation of high-temperature phase and it is characterized by the normal temperature dependence of the yield stress is typical for fcc materials [14]. Data analysis of $\sigma_{0.1}(T)$ in Figure 3 shows that, firstly, in all crystals FeNiCoAlX (X = Ti, Ta, Ti + Ta) with increasing aging time parallel shift of $\sigma_{0.1}(T)$ curves observed and, the less the aging time the lower direct temperature of γ-α' MT. The value of $\alpha = d\sigma/dT$ at this weakly dependent on aging time and is in the range 2.5-3.2 MPa/K (Figure 3, a-c). Secondly, at same time of the aging, stress level of the high temperature phase at T = M_d depends on the chemical composition of the γ'-phase particles. Thus, upon aging at 973 K for 0.5 hour $\sigma_{0.1}^{Ti}(M_d) < \sigma_{0.1}^{Ti+Ta}(M_d) < \sigma_{0.1}^{Ta}(M_d)$, and at aging at 973 K 7 hours $\sigma_{0.1}^{Ti}(M_d) < \sigma_{0.1}^{Ta}(M_d) < \sigma_{0.1}^{Ti+Ta}(M_d)$ (Figure 3, d, e). Therefore, in complex-alloyed crystals FeNiCoAlTiTa, level of strength properties of high-temperature phase determined by the chemical composition of γ'-phase particles, which is determined by the aging time. Due to the different diffusion mobility of Ti and Ta ($D_{Ti} > D_{Ta}$) γ'-phase particle in complex-alloyed crystals FeNiCoAlTiTa can has different concentration of Ta and respectively characterized by different strength, which has an impact on the level strength properties of high-temperature phase [9]. For a small aging time – 0.5 hour, when the γ'-phase particles are less durable, stresses level of high-temperature phase at T = M_d in crystals with Ti and Ti + Ta is close, and the γ-α' MT is characterized by close regularities: large thermal and stress hysteresis under stress and lack of SE at T = 77 K as compared with crystals Ta. At aging during 7 hours the γ'-phase particles in Ti + Ta crystals the richer Ta than at aging for 0.5 hour, and so should be stronger. This is qualitatively confirmed by that in FeNiCoAlTiTa crystals, aged at 973 K for 7 hours, the stresses level of high-temperature phase is greater, and energy dissipation at γ-α' MT, which is determined by the value of stress hysteresis Δσ, and the value of transformation strain in the experiments studying of SME and SE is smaller than in the crystals of Ti and Ta.

3. Conclusion

In the [001]-oriented single crystals of disordered iron-based Fe-28%Ni-17%Co-11.5%Al-2.5%X (at. %) (X = Ti, Ta, Ti + Ta) alloys it is shown that the precipitation of ordered γ'-phase particles during the aging at T = 973 K for 0.5 h and 7 h leads to the development of thermoelastic γ-α' MT with SME and SE. It is experimentally discovered that variation of the chemical composition, size and volume fraction of the dispersed γ'-phase particles allows to control both mechanical and functional properties. It is shown that the maximum values of reversible strain in experiments on research of the SE are equal to 10.8 % observed in single crystals of Fe-28%Ni-17%Co-11.5%Al-2.5%Ta (at. %) aged at T = 973 K for 0.5 h that exceed the theoretical values of the lattice deformation of 8.7 % in tension for crystals oriented along the [001] direction with the γ-α' MT [4]. Maximum value of SME ε_{SME} = 7.8 % observed in the Fe-28%Ni-17%Co-11.5%Al-2.5%Ti (at. %) crystal after aging at T = 973 K for 7 h, which is close to the theoretical values of the lattice deformation of 8.7 %. Found that in complex-alloyed crystals Fe-28%Ni-17%Co-11.5%Al-2.5%TiTa (at. %), the strength properties of the high-temperature phase, the development of stress-induced γ-α' MT, value of the thermal ΔT and stress Δσ hysteresis determined by chemical composition of the γ'-phase particles.

Acknowledgements
This work was financially supported by grant of Russian Science Foundation 14-29-00012.

References

[1] Otsuka K and Ren X 2005 *Progress in Materials Science* **50** 511–678
[2] Otsuka K, Saxena A, Deng J and Ren X 2011 *Philosophical Magazine* **91** 4514–4535
[3] Kokorin V V 1987 Martensite Transformation in Inhomogeneous Solid Solutions *Naukova Dumka* 165
[4] Tanaka Y, Himuro Y, Kainuma R, Sutou Y, Omori T and Ishida K 2010 *Science* **327** 1488–1490
[5] Kireeva I V, Chumlyakov Y I, Kirillov V A, Karaman I and Cesari E 2011 *Technical Physics Letters* **37** 487–490
[6] Chumlyakov Y I, Kireeva I V, Panchenko E Y, Zakharova E G, Kirillov V A, Efimenko S P and Sehitoglu H 2004 *Doklady Physics* **49** 47–50
[7] Kokorin V V and Gun'ko L P 1995 *Metal Physics and the latest technology* **17** 30–35
[8] Chumlyakov Y I, Kireeva I V, Panchenko E Y, Aksenov V B, Kirillov V A, Ovsyannikov A V, Zakharova E G and Sehitogly H 2003 *Russian Physics Journal* **46** 811–823
[9] Chumlyakov Y I, Kireeva I V, Panchenko E Y, Timofeeva E E, Kretinina I V and Kuts O A 2015 Physics of Thermoelastic Martensitic transformation in High-Strength Single Crystals. Shape Memory Alloys: Properties, Technologies, Opportunities *Trans Tech Publications Ltd* 107–173
[10] Omori T, Abe S, Tanaka Y, Lee D Y, Ishida K and Kainuma R 2013 *Scripta Materialia* **69** 812–815
[11] Booth-Morrigon C, Dunand D C and Seidman D N 2011 *Acta Materialia* **59** 7029–7042
[12] Chumlyakov Y I, Kireeva I V, Kretinina I V, Keinikh K S, Kuts O A, Kirillov V A, Karaman I and Maier H 2013 *Russian Physics Journal* **56** 66–74
[13] Chumlyakov Y I, Kireeva I V, Poklonov V V, Pobedennaya Z V and Karaman I 2014 *Technical physics Letters* **40** 747–750
[14] Chumlyakov Y I, Kireeva I V, Korotaev A D, Litvinova E I and Zuev Y L 1996 *Russian Physics Journal* **39** 5–33

Regulation of the burning parameters for energetic materials based on ammonium nitrate and metal nanopowders and nonmetals

V N Popok[1] and A G Korotkikh[2,3]
[1] 659322, Biysk, Russia, Federal research and production center «ALTAI»
[2] 634050, Tomsk, Russia, National Research Tomsk Polytechnic University
[3] 634050, Tomsk, Russia Research Institute of Applied Mathematics and Mechanics, Tomsk State University

E-mail: vnpopok@mail.ru

Abstract. The results of study for the influence of metallic and nonmetallic nanopowders on the combustion parameters of mixed energetic materials based on ammonium nitrate are presented and compared with materials, containing micropowders of metals. The catalytic activity rank of the powders is plotted showing their effect on the combustion characteristics of mixed energetic materials.

1. Introduction

The use of metallic and nonmetallic nanopowders (NPs) as a fuel in mixtures of energetic materials (EMs) with various compositions has a long history [1–3]. Rather extensive material has been collected to date on the thermal–kinetic parameters of NP oxidation and the NP impact on ignition and combustion parameters and the thermal decomposition and combustion of EMs based on ammonium perchlorate (AP), inert hydrocarbonaceous and active combustible binders (CBs), containing primarily nitroesters and energetic supplements, predominantly nitramines [4, 5]. It should be noted that a significant part of compositions used in the studies underlying the EMs does not meet the requirements of processibility and component compatibility of compositions obtained on the basis of metallic and nonmetallic NPs or other solid components of EMs, including oxidizers and energy supplements. The chemical compatibility requirement practically eliminates the possibility of using most components of EMs with NPs of aluminum, as well as transition metal oxides, which are of high interest for the combustion catalysis of EMs. Technological problems associated with the use of NPs are related with the substantial reduction of the critical filling characteristics (filling limit, the degree of filling corresponding to the percolation transition), leading to an increase in the viscosity of EMs of 10 and 100 times more than with EM-based microdispersed powder fillers. In addition to the abovementioned arrangements of the EMs based on AP (relating to rapidly burning materials), the necessary increase in their combustion rate and the reduction of aluminum agglomeration during combustion and of the level of other energy losses is implemented more effectively by methods, which do not require the use of NPs, primarily aluminum, containing usually no more than 90 wt % of the base material. As applied to fast burning EMs based on active combustible binders and nitramines, including ballistic gunpowder, the use of aluminum instead of microfine NP leads to insignificant effects when increasing the combustion rate or even the inhibition of burning. Therefore, it is advisable to focus on the use of metal and nonmetal NP as a primary fuel or additives to EM compositions for fundamental solutions to existing problems of their combustion, including metallized mixtures.

The use of NP in slow burning EMs is of great interest. The high efficiency of using NP in such EMs may be due to the increase in the reaction time of EMs components and products of their thermal decomposition (pyrolysis) with NP, as well as the features of the physicochemical interaction associated with the high catalytic activity and high content of oxide layers on a nanoparticle surface. For example, an oxide layer of aluminum NP predominantly contains a catalytically active γ-modification and an amorphous modification of Al_2O_3 in an amount of 10–20 wt %. Industrial brands of microfine granular aluminum powder (ASD-1, ASD-4, ASD-6, ASD-8) used in EMs contain up to 0.5 wt % of catalytically inert α-Al_2O_3 particles on their surface.

One of the promising applications of NP objects are EMs based on ammonium nitrate (AN), which is widely studied and used in compositions for various purposes, being an ecologically clean and safe chlorine_free oxidizer. The ammonium nitrate, AN-based EMs, and various CBs have low burning rate and temperature values, high values of the lower ignition limit and burning by pressure (2 – 10 MPa), and inad_missibly great losses in energy and mass due to incomplete combustion, as well as an anomalously high agglomeration of metallic fuel and slagging during the combustion of metallized nitrates in EMs. The combustion efficiency of nitrate EMs (diminishing the losses on agglomeration and slagging and increasing the rate of burning) can be increased using nanodispersed powders of metals and nonmetals.

In this work we continue earlier investigations presenting the experimental results on the influence of the type of metallic and nonmetallic powders and their microdisperse equivalents on the burning rate of nitrate energetic materials (EMs).

2. Objects and methods

We investigated regularities of the influence of metallic (Al, Cu, Fe, Ni, Zn, B) and nonmetallic (B, C), aluminum oxide Al_2O_3, and aluminum-magnesium alloy powders with different dispersivities on the combustion rate of nitrate EMs. To obtain a comparative base, we used certified micropowders with a size from units to hundreds of micrometers: aluminum (spherical) ASD-1, ASD-4, ASD-6, and ASD-8; titanium PTEM; tungsten PV-1; iron PZhV-1; copper PMS-1; and nickel PNE-1. These materials were prepared by the electrodeposition of zinc, aluminum-magnesium alloy (PAM) with a weight ratio Al:Mg = 1:1, and aluminum oxide micropowders in α and γ forms. As the nanopowders we used certified metallic powders of aluminum, nickel, and zinc obtained under the electroexplosion of wires and nonmetallic powders of C and B. The choice of α and γ-Al_2O_3 micropowders (MP) for studying the combustion catalysis and thermal decomposition of EMs is motivated by the fact that an oxide film on the MP of Al represents a catalitically "inert" α-Al_2O_3, whereas the oxide film on the NP of Al contains γ-Al_2O_3 used as the catalyst in various chemical processes [2]. In addition, the NPs of Al are used predominantly in most metallized EMs.

The base EMs included SKDM-80 combustible binders (divinyl rubber SKD plasticized by transformer oil), which is very compatible with nearly all metallic and nonmetallic powders [2] in an amount of 12.3 wt %, MP ASD-6 15 wt %, ammonium nitrate and the rest of the substance is a two-component NP mixture with a ratio of coarse to small fraction of 1/1. The effect of the metal fuel type was analyzed upon the full replacement of aluminum powder on the metal powder. The B and C powders were introduced into EMs with the partial replacement of aluminum for connecting the mixtures thereof. In studying the burning processes of the base EMs, the powders were introduced with a concentration of 2 wt % from the mass of the mixture (100%). Experimental data on the burning rate (u) were approximated by the powerlike function of pressure of the form $u(p) = b \cdot p^{\nu}$.

3. Experimental results and discussion

Figure 1 shows the determined burning rates of EMs with different microdispersed metallic fuels depending on pressure (experimental data in the figure are approximated by curves). All EMs are incapable of sustained combustion at pressures below 1-2 MPa (as well as at normal conditions in atmosphere of air). All MPs except for Ti increase the burning rate of the base EMs (at low pressures up to ~5 MPa). For MP of Ti, W, and Cu, parameter b is diminished, whereas zinc and PAM increase the value of b. The exponent v of the power law increases when replacing aluminum (ASD-6) by metallic powders used in this series of experiments. A MP of Ni is similar to a MP of W in regards to its catalytic activity. The PZhV-1 iron powder causes a decrease in the burning rate of the mixture when compared to the base EM and considerable growth in the lower threshold (up to 6-8 MPa) of the pressure of stable ignition and burning the mixture. For the considered EM composition, one can plot the following relation of metallic (alloy) activity corresponding to the increasing burning rate: Fe < ASD-6 ≈ Ti < Ni ≈ W< Cu < Zn < PAM.

Figure 2 samples the data on the impact of metallic MPs and aluminum oxide additives on the burning rate of the base EM. The series of additive activity is fully consistent with the series obtained upon complete replacement of aluminum powder by the metals considered: the most active MP is Zn; the next are the MPs of W and Ti (there is almost no impact of the latter on the rate of base EM burning).

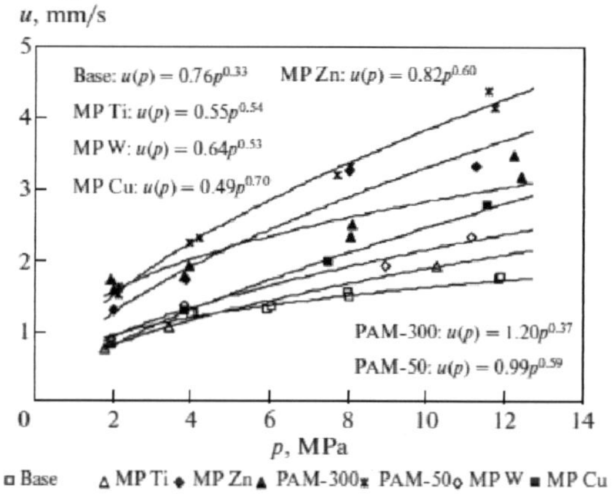

Figure 1. Impact of fine-dispersed metallic fuel on the burning rate of the base EM

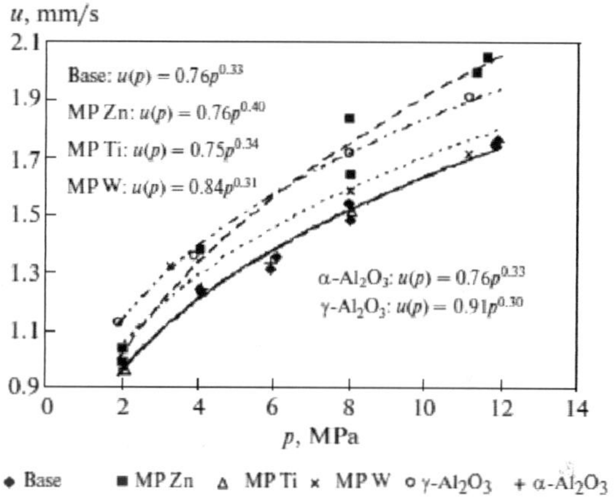

Figure 2. Impact of metallic MP additives (2 wt % above 100 wt % of mixture) and aluminum oxide on the burning rate of the base EM

All other MPs of metals also have an insignificant impact on the rate of base EM burning. This figure also shows the data of EM catalysis by oxides with different crystalline modifications. Additives of W and γ-Al_2O_3 cause the growth of parameter b; other additives of MPs do not affect the value of this parameter, while the values of parameter v are nearly constant. This allows one to make a qualitative conclusion on the effects of W and γ-Al_2O_3 on the processes in condensed phase. The activity of tungsten in the condensed phase (c-phase) is likely caused by its solubility in the melt of AN. A significant influence of γ-Al_2O_3 on the combustion characteristics of nitrate based EMs is qualitatively similar to that observed upon the burning of EMs based on the ammonium perchlorate (AP) as an oxidizer [2].

Figure 3 samples the data showing the impact of metallic and nonmetallic NPs on the burning rate of EMs upon the full or partial replacement of aluminum ASD-6 (EMs with NPs of B and C). As a rule, the intercalation of NP into EMs results in a considerable decrease in the lower pressure of stable ignition and burning up to 0.2-0.5 MPa. However, at a pressure close to atmospheric pressure, only Al and C NPs can maintain the stable combustion of EMs that do not contain other effective combustion catalysts like o-carborane or fast-burning additives. Nanodispersed Cu powder slightly increases the rate of EM burning when compared to a copper micropowder. This is likely caused by the interaction of copper with the melt of ammonium nitrate accompanied by the formation of solutions of basic materials and reaction products (metal, oxide, nitrate, and ammoniates), which is characteristic for the powders of Ni and Zn. The partial replacement of ASD-6 by NP of C (4 wt %) and NP of B (8 wt %) leads to a significant growth in the burning rate of the base EM. Replacing the ASD-6 powder of Al by NP of Al, B, and C increases the burning rate of EMs from three to five times. Additives of Cu nanoparticles result in a decrease in parameter b and a significant increase in parameter ν (by more than two times). Additives of Al NP in EM and the partial replacement of Al MP in the base EM by B and C nanoparticles result in an increase in parameter b by three and more times. Moreover, the NP additives of Al and B lead to an increasing value of ν, while the additive of C practically does not alter this parameter. In addition, the rate of burning of nitrate EMs in the studied interval of pressures corresponds to the AP-based compositions. In general, in the case of nanodispersed powders, the impact of energy efficiency of metals and nonmetals when used as a primary fuel is crucial. One should also take into account strong catalysis of the AN decomposition when using NP of C and γ-Al_2O_3, as well as the solubility of metals and their oxides in the melt of AN. Figure 4 shows data on the impact of metallic NP on the burning of the base EM. The results allow one to plot the following row of metallic NP activity in increasing order of the combustion rate of base EM: NP Ni < (NP Fe ≈ NP B) < (NP Cu ≈ NP Al) < NP C.

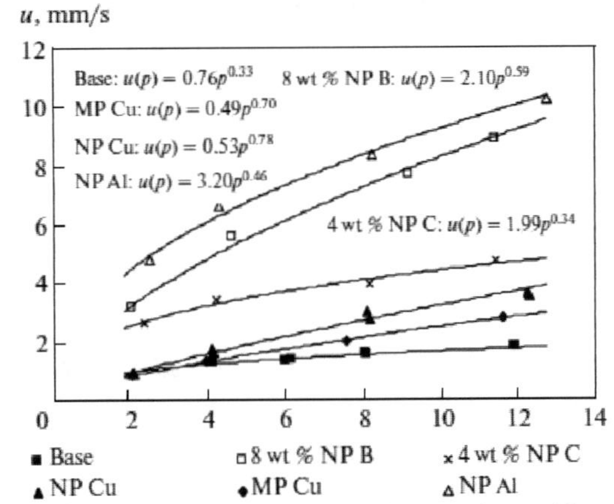

Figure 3. Impact of nanodispersed metallic fuel on the burning rate of the base EM

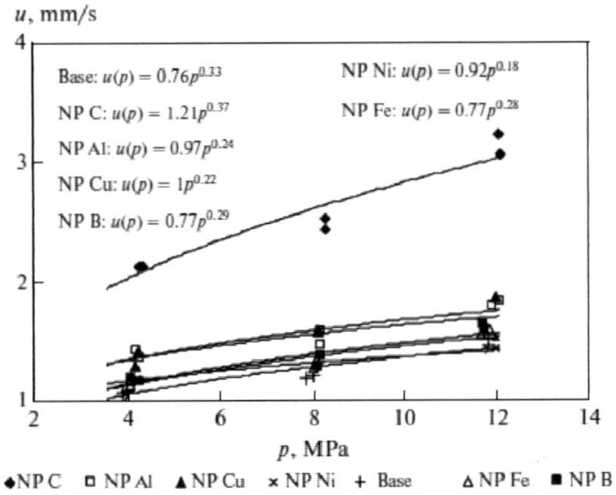

Figure 4. Impact of NP additives (2 wt %) on the burning rate of the base EM

4. Conclusion

The resulting activity row of metallic and nonmetallic NP does not match the series of energetic and catalytic activity upon the combustion and thermal decomposition of EM [6]. With allowance for the area of possible reactions (c-phase and gaseous phase), one can plot the partial activity rows compatible with the rows of catalytic and energetic activity upon thermal decomposition in the mixtures with AN. The aluminum NP additive (2 wt %) has the maximum activity among the metals, increasing the burning rate of EM, but its activity nearly equals the activity of copper NP. Except for the carbon NP, all NP additives diminish the value of parameter ν. Parameter *b* is not altered when using EM with additives of boron and iron NPs; the other NP additives cause an increase in this parameter, indicating the presence of catalysis for reactions in c-phase.

Acknowledgment

The reported study was partially supported by RFBR, research project No. 15-33-50066.

References

[1] Chong F and Li S 2002 *Propellants, Explosives, Pyrotechnics* **27** 34–8.
[2] Beckstead M W, Puduppakkama K, Thakreb P and Yang V 2007 *Progress in Energy and Combustion Science* **33** 497–551.
[3] Arkhipov V A, Bondarchuk S S, Korotkikh A G, Kuznetsov V T, Gromov A A, Volkov S A

and Revyagin L N 2012 *Combustion, Explosion, and Shock Waves* **48** 625–635.
[4] Popok V N, Vdovina N P and Bychin N V 2013 *Nanotechnol. in Russia* **8** 99–107.
[5] Arkhipov V A, Korotkikh A G, Kuznetsov V T, Razdobreev A A and Evseenko I A 2011 *Russian Journal of Physical Chemistry B* **5** 616–624.
[6] Popok V N and Bychin N V 2014 *Nanotechnol. in Russia* **9** 541–548.

Boron carbide morphology changing under purification

I A Rahmatullin[1] and A A Sivkov[2]

[1]Assistant, Energy Institute, Tomsk polytechnic university, Tomsk, Russia
[2]Professor, Energy Institute, Tomsk polytechnic university, Tomsk, Russia

E-mail: riam@tpu.ru

Abstract. Boron carbide synthesized by using coaxial magnetoplasma accelerator with graphite electrodes was purified by two different ways. XRD-investigations showed content changing and respectively powder purification. Moreover TEM-investigations demonstrated morphology changing of product under purification that was discussed in the work.

1. Introduction
Boron carbide attracts a great interest because of low weight, high hardness, high chemical and temperature resistance and neutron absorption capability [1]. According to that it finds a lot of applications in different areas such as wear resistant materials, cutting tools, reinforcing of metal and ceramic composites and neutron absorbers of control rods for nuclear reactors [2]. Thus a lot attempts has been done to find the most effective method of synthesis for producing nanosized boron carbide. The most spread of them are carbothermal reduction [3], zol-gel method [4], mechanochemical synthesis [5], SHS process [6] and gas-phase synthesis [7]. Another attempt to synthesize nanosized boron carbide was done in [8] by using coaxial magnetoplasma accelerator with graphite electrodes [9-12]. Obtained product consists of crystalline boron carbide with a small amount of impurities such as graphite and boron oxide. In the present study different efforts to purify synthesized powder were done and as a result on of them lead to morphology changing without any crystal structure modification.

2. Experimental
Purification of obtained powder was done in two ways. In the first case powder was only washed by hot water to remove boron oxide and in the second case powder firstly was annealed at 600 0C for an 60 minutes to oxidize graphite and then washed also. Slurry mixture of powder in the water was boiled about 20 minutes then precipitated. After that water was pour off and the remaining residue was heated at 100 0C until full drying using Nabertherm P310 furnace. Initial and purified powders were investigated using XRD (Shimadzu XRD7000S, Cu-Ka) and TEM using Philips CM12 electron microscope.

3. Results and discussion
Figure 1 shows XRD data of synthesized and purified powders. XRD-investigations (figure 1a) show that initial powder consist of crystalline boron carbide [PDF4+, # 01-075-0424] as main phase and boron oxide [PDF4+, # 00-006-0297] and graphite [PDF4+, # 01-075-1621] as impurities in a small amount. XRD-pattern (figure 1b) of purified by washing product shows disappearing of boron oxide which forms boric acid dissolved in water by following reaction:

$$B_2O_3 + 3H_2O = 2H_3BO_3 \tag{1}$$

In comparison with initial product intensity of graphite decreases because of boron oxide washing out from boron carbide surface and respectively decreasing of X-Ray radiation absorption by boron oxide.

XRD-pattern of product after annealing shows the same result but graphite intensity increases. The temperature of annealing was chosen on the border of boron carbide and graphite oxidation to decrease boron carbide losses and maximally oxidize carbon. Powder after annealing was covered by boron oxide glass and after washing in a hot water this boron oxide glass was dissolved and free up powder particles from glass cover. Graphite intensity increasing demonstrate that heat treatment temperature was not enough to oxidize the carbon but boron carbide undergoes oxidation and mass lose with decreasing of intensity on XRD-pattern.

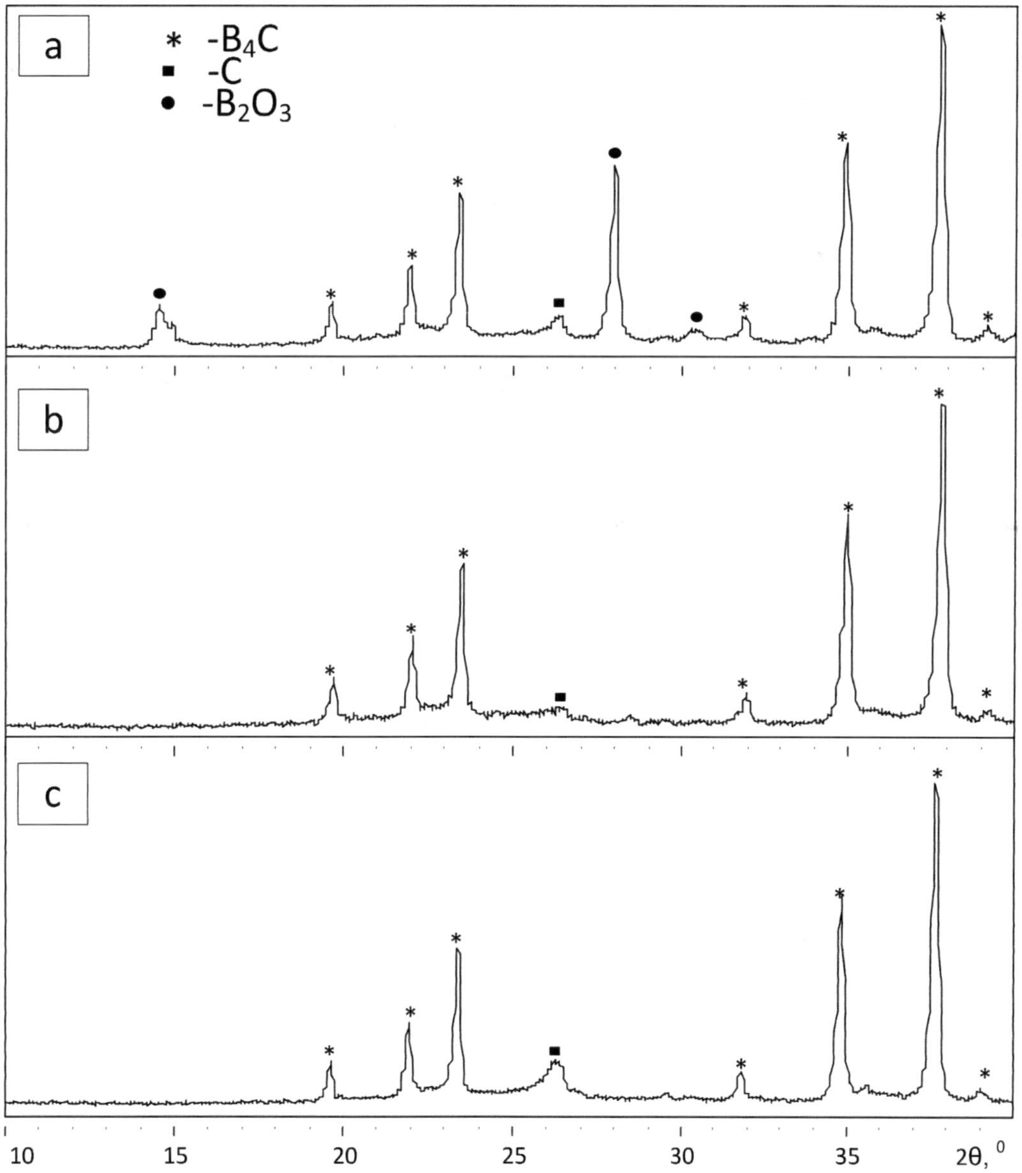

Figure 1. XRD data of initial (a), washed (b) and annealed (c) powder of boron carbide.

IOP Conf. Series: Materials Science and Engineering **93** (2015) 012046 doi:10.1088/1757-899X/93/1/012046

TEM-investigations of powder product shows that particles undergoes morphology changing after heat treatment in comparison with initial powder as it can be seen at figure 2. After annealing particles became smaller and loose classical morphology as on figure 3*a* and 3*b*. Therefore because of annealing influence on particles morphology to determine a role of temperature for morphology additional investigations were done. Powder product was annealed in the range of temperatures from 300 up to 580 ^{0}C to exclude boron carbide oxidation and boron oxide vitrification. As it was shown that product does not support structural transformation it was investigated only by TEM presented at figure 3. Particles treated at 300 ^{0}C does not undergoes any morphology transformation and consist of classical particles but when temperature gets to 400 ^{0}C a low amount of rectangular particles appear in the product. Significant morphology changing occurs at 500 ^{0}C with formation of a large number of crystals with right shape and sizes that do not exceed 100 nm. Also a small amount of large crystals (500 nm) with classical shape are persisting in product.

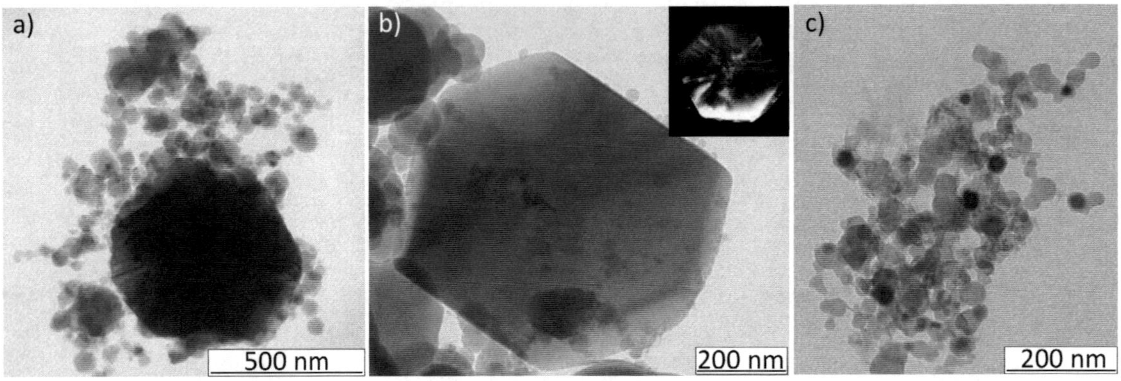

Figure 2. TEM data of initial (a), washed (b) and annealed (c) powder of boron carbide.

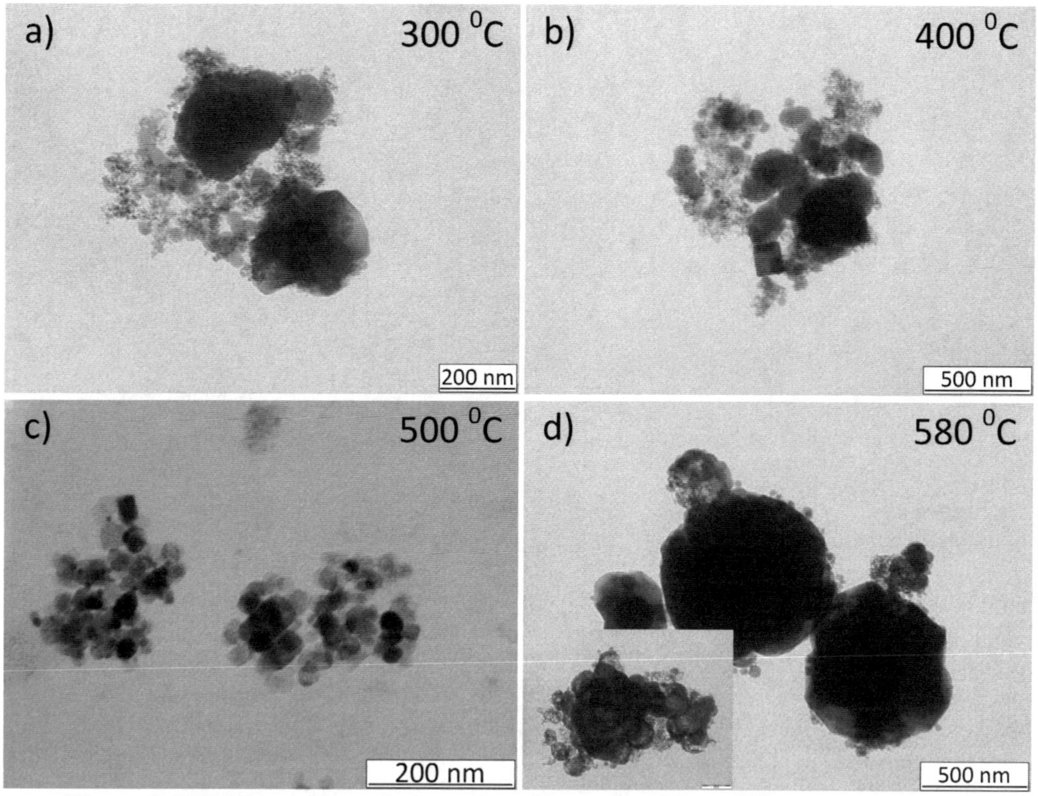

Figure 3. TEM data of annealed powder of boron carbide at different temperatures.

Only large crystals with classical morphology (500 nm) are in product when temperature become 580 ^{0}C. Besides them a new particles are formed at that conditions and have circular shape with thickening on the edges. Also very small fraction is formed with particle sizes less than 40 nm. It is suggested that fine fraction of boron carbide powder oxidize due to the low thermal stability of nanosized particles because of small size.

Conclusion

In accordance with presented data and TEM-capture at figure 4 boron carbide particles might have segmented structure with weak connections in crystal that disintegrate under heat treatment. Moreover synthesis of the same rectangular crystals under lower pressure (P0=0.1 atm.) explained by higher speed of plasma flow because of lower gas resistance. This leads to increasing of crystallization and cooling rates that forms crystals with smaller sizes and different morphology. Increasing of synthesis pressure and decreasing of plasma flow speed respectively leads to increasing of condensation, crystallization and cooling duration. That creates conditions for self-assembling of nanosized rectangular crystals to polyhedral formations with weak connection of them.

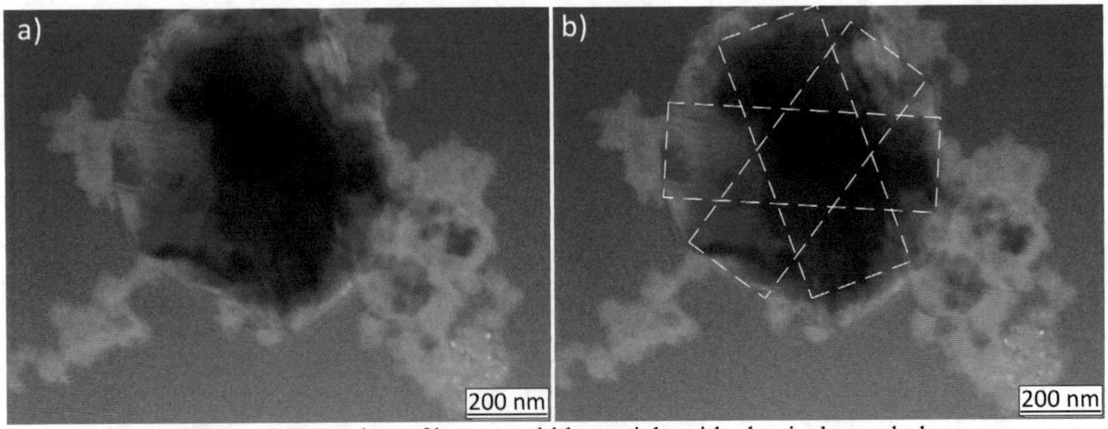

Figure 4. TEM data of boron carbide particle with classical morphology

Acknowledgments
This work was supported by Russian Science Fund project # 15-19-00049.

References

[1] Thvenot F 1990 *J. Eur. Ceram. Soc.* **6** 205
[2] Mondal S and Banthia A K 2005 *Journal of the European Ceramic Society* **25** 287
[3] Kakiage M, Tominaga Y, Yanase I and Kobayashi H 2012 *Powder Technology* **221** 257
[4] Najafi A, Golenstani-Fard F, Rezaie H R and Ehsani N 2011 *J. of Alloys and Compounds* **509** 9164
[5] Sharifi E M, Karimzadeh F and Enayati M 2011 *Advanced Powder Technology* **22** 354
[6] Jiang G, Xu J, Zhuang H and Li W 2009 *Journal of Nuclear Materials* **393** 487
[7] Oyama T and Takeuchi K 1999 *Carbon* **37** 433
[8] Sivkov A, Rakhmatullin I and Makarova A 2014 *Nanotechnologies in Russia* **9** 674
[9] Sivkov A and Pak A, Russian Federation patent № 2431947 Coaxial magnetoplasma accelerator
[10] Pak A, Sivkov A, Shanenkov I, Rahmatullin I and Shatrova K 2015 *Int. J. Refract. Met. Hard Mater.* **48** 51
[11] Shanenkov I I, Pak A Ya, Sivkov A A and Shanenkova Yu L 2014 *MATEC Web of Conferences* **19** 01030
[12] Sivkov A, Saygash A, Kolganova J and Shanenkov I 2014 *IOP Conf. Series Mater. Sci. Eng.* **66** 012048

Stress-induced martensitic transformation in high-strength [236]-oriented $Ni_{51}Ti_{36.5}Hf_{12.5}$ single crystals

N Y Surikov, A S Eftifeeva, E Yu Panchenko and Yu I Chumlyakov

Tomsk State University, Siberian Physical-Technical Institute, Tomsk, 634050, Russia

E-mail: Jet_n@mail.ru, anna_eftifeeva@rambler.ru, panchenko@mail.tsu.ru, chum@phys.tsu.ru

Abstract. The effects of heat treatment on the stress-induced B2-B19' martensitic transformations in the $Ni_{51.0}Ti_{36.5}Hf_{12.5}$ single crystals oriented along [236] direction are studied. It is shown that in the annealed at 1323K for 4 h crystals, the temperature range of superelasticity increase almost twofold from 75K up to 135K as compared to the as-grown single crystal contained disperse particles of H-phase. The [236]-oriented $Ni_{51.0}Ti_{36.5}Hf_{12.5}$ single crystals are characterized with high levels of applied compressive stress up to 1700 MPa in the as-grown state and 1900 MPa in annealed crystals for the completely reversible stress-induced B2-B19′ martensitic transformation with reversible strain up to $|\varepsilon_{SE}|$ =1.4%.

1. Introduction

Alloys of TiNiHf undergoing reversible B2-B19′ martensitic transformation at high temperatures of above 100°C and high level of applied stress of above 1000 MPa [1-4] can find the wide range of practical applications in modern technologies. It is known that precipitation of disperse particles in the Ni-rich TiNi and NiTiHf alloys allows controlling characteristic transformation temperatures, the strength of the B2-phase due to precipitation hardening processes and increase cycle stability of such functional properties as the shape memory effect and superelasticity [1-5]. Up to the present there are no systematic studies of effects of dispersed particles on shape memory effect and superelasticity in high-strength TiNiHf alloys with Ni content greater than 50.5 at.%.

Therefore, the main objective of this work is to investigate dependence of the critical stress level for stress-induced martensitic transformation, temperature range of superelasticity response on microstructure of single crystals of the $Ni_{51.0}Ti_{36.5}Hf_{12.5}$ (at.%) alloy. Carrying out researches on NiTiHf single crystals allows, first, excluding the effect of grain boundaries on the allocation of the dispersed particles and on the development of the martensitic transformation. The second, it allows choosing the crystal orientation along [236] direction, which is characterized by maximum theoretical lattice strain $|\varepsilon_{tr0}|_{[236]}$ = 8.0% at B2-B19′ martensitic transformation in compression [3].

2. Experimental procedure

Single crystals of $Ni_{51.0}Ti_{36.5}Hf_{12.5}$ (at. %) alloy were grown by the Bridgman technique in an inert gas atmosphere. Electro-discharge machining was utilized to cut rectangular compression specimens with nominal dimension of 4 mm x 4 mm x 8 mm. The orientation of the crystal axis was determined in the austenitic state with a DRON-3 X-ray diffractometer using FeKα radiation. The loading axis of the compression specimens was along the [236] crystallographic direction. Mechanical grinding followed by an electrochemical polishing (20% $HClO_4$+ 80% CH_3COOH electrolyte at T = 293 K, U = 18 V) was employed to remove any processing-affected surface layer on specimens. The mechanical tests conducted in the temperature interval 200 K < T < 550 K were performed at testing systems Instron 5969 at a strain rate $\dot{\varepsilon} = 2 \cdot 10^{-3} c^{-1}$. The microstructures were examined with a JEOL 2010 transmission electron microscopes (TEM) operated at a nominal accelerating voltage of 200 kV. Prior to the TEM observation, the 3-mm diameter disk specimens of 0.1μm in thickness were prepared and then electropolished using the twinjet method that utilized electrolyte of composition 20% H_2SO_4+80% Methanol at 10°C and 15–17 V.

3. Results and discussion

Single crystals of $Ni_{51.0}Ti_{36.5}Hf_{12.5}$ alloy are investigated in two structural states: 1) as-grown, i. e. without additional heat treatment; 2) after annealing at T=1323 K for 4 h followed by cooling at a rate of 15-25 $K \cdot min^{-1}$. In as-grown crystals a volume fraction (up to 30 %) of H-phase particles with face-centered orthorhombic lattice (Figure 1 a) is precipitated during the slow cooling in the process of single crystal growth [6, 7]. The TEM bright field image shows the H-phase particles and the characteristic satellites from four crystallographic particles variants of H-phase, including two variants of $\frac{1}{4}[210]_{B2}$ and two variants of $\frac{1}{3}[110]_{B2}$ that are observed on the selected-area diffraction pattern (SAD) (Figure 1 a).

The annealing at T=1323 K results in dissolution of H-phase particles. Figure 1 b shows the TEM bright field image of B2-austenite and residual B19′-martensite after mechanical tests at T ≥ 350 K. This is confirmed by SAD pattern (inset to Figure 1 b), which does not show evidence of any additional spots or streaking that could be due to H-phase particles. Nevertheless, this problem requires additional researches by TEM and X-ray techniques, because we cannot completely exclude the precipitation of nanoscale particles of 3-5 nm during cooling after annealing at 1323 K.

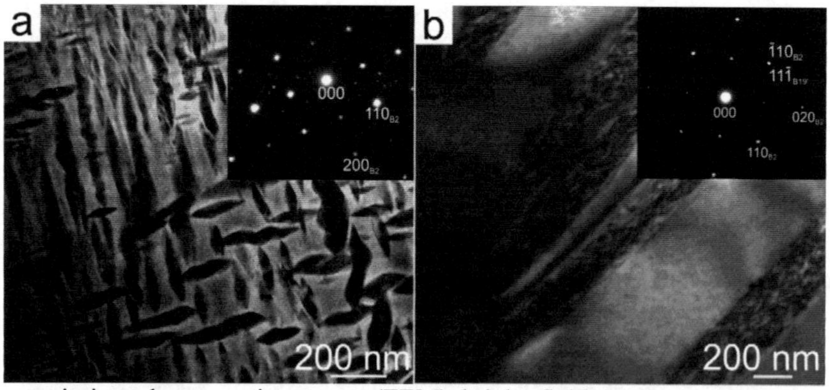

Figure 1. Transmission electron microscopy (TEM) bright field image and corresponding SAD pattern of as-grown $Ni_{51.0}Ti_{36.5}Hf_{12.5}$ single crystal, zone axis is $[001]_{B2}$ (a) and annealed at 1323K for 4 h $Ni_{51.0}Ti_{36.5}Hf_{12.5}$ single crystal, zone axis is $[001]_{B2} \| [011]_{B19'}$ (b).

Figure 2 shows the strain-temperature responses ε(T) recorded in compression tests during cooling/heating of as-grown $Ni_{51.0}Ti_{36.5}Hf_{12.5}$ single crystals oriented along [236] direction. The strain monitored during cooling/heating cycles remains at near zero for the tests with an external compressive stress magnitude |σ| from 0.3 up to 50 MPa. So, during the cooling process a self-accommodated microstructure of B19′-martensite is formed within this stress range. Under larger compressive stresses oriented martensite is grown and its volume fraction increases with external stress level. As a result reversible transformation strain ε_{rev} during cooling/heating is observed, which increases in magnitude from ε_{rev} = 0.5(±0.1) % at |σ |= 75 MPa to ε_{rev} = 0.8(±0.1) % at |σ| = 170 MPa. As expected, the martensitic start temperature (M_s) increases with the applied stress. However, for realization of maximal recoverable strain in TiNiHf crystals the compression stress higher of 170 MPa must be applied [1, 3, 4].

In annealed $Ni_{51.0}Ti_{36.5}Hf_{12.5}$ single crystals during cooling/heating in the temperature range from 77 to 373 K under constant compressive stress from 0.3 to 170 MPa a reversible strain associated with the thermoelastic martensitic transformation is not observed. It can be due to the low martensitic start temperature M_s and high critical stress level for martensite formation in annealed single crystals.

Figures 3 and 4 show the superelastic response of [236]-oriented single crystals in as-grown and annealed states in compression. The temperature dependence of the critical stress level ($|\sigma_{cr}|(T)$) on the test temperature higher M_s for investigated single crystals is presented in Figure 5.

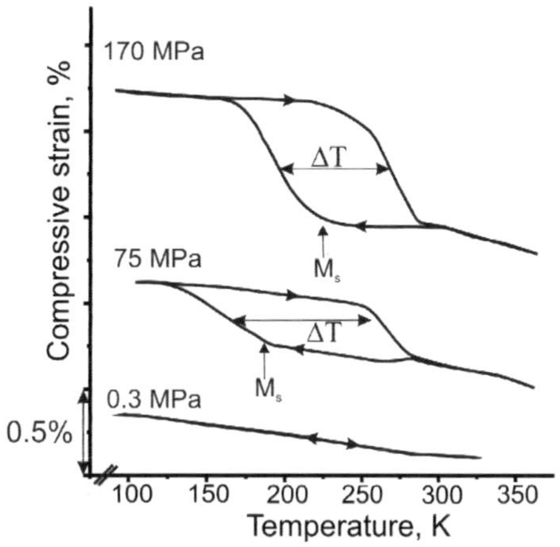

Figure 2. Strain-temperature responses of as-grown $Ni_{51.0}Ti_{36.5}Hf_{12.5}$ single crystals oriented along [236] direction during cooling/heating under constant compressive stress; ΔT is thermal hysteresis.

A linear growth in $|\sigma_{cr}|$ with increase in test temperature is observed both in as-grown and annealed states and can be described by the Clausius-Clapeyron relationship [5]:

$$\frac{d\sigma_{cr}}{dT} = -\frac{\Delta S^{A-M}}{\varepsilon_{tr}^{A-M}} = -\frac{\Delta H^{A-M}}{T_0 \varepsilon_{tr}^{A-M}}, \qquad (1)$$

where $\Delta S^{A-M} < 0$ is the change of entropy associated with the forward transformation, $\Delta H^{A-M} < 0$ is the corresponding change of enthalpy, ε_{tr} is the transformation strain and T_0 is the equilibrium temperature for austenite and martensite.

It is experimentally shown that annealing at 1323 K for 4 h results in reduction of the martensitic start temperature M_s, in increase of the critical stress levels of martensite formation $|\sigma_{cr}|$ as compared to heterophase as-grown crystals (Figures 3-5). So, at T = 250 K the critical stress level of martensite formation for annealed crystals is $|\sigma_{cr}|$ = 918 MPa and it is 2.6-fold higher than $|\sigma_{cr}|$ = 350 MPa for as-grown crystals. In contrast, at higher test temperatures, the critical stress levels for the martensite formation in investigated crystals become close to each other (Figure 5). Such behavior is caused by a difference in the magnitude $\alpha = d|\sigma_{cr}|/dT$.

The annealed crystals are characterized by value twofold lower $\alpha = d|\sigma_{cr}|/dT = 4.7$ MPa·K^{-1} and wider temperature range of stress-induced martensitic transformation as compared to heterophase as-grown crystals with $\alpha = 8.6$ MPa·K^{-1} (Figure 5). As a result, the superelasticity temperature range increases almost twofold as compared to the initial crystals containing H-phase particles. Figure 5 shows superelasticity temperature ranges of ΔT^1_{SE}=75 K and ΔT^2_{SE}=135 K for crystals in as-grown and annealed states, respectively (Figures 3-5).

The analysis of σ(ε)-response shows that stress-induced martensitic transformation and superelasticity are observed at the high levels of applied compression stress from 700 to1900 MPa in $Ni_{51.0}Ti_{36.5}Hf_{12.5}$ single crystals (Figures 3 and 4). Thus, in the studied single crystals the similar inherent features of thermoelastic stress-induced martensitic transformation as in high-strength aged

NiTiHf single crystals and also binary TiNi alloys with Ni content greater than 51.3 at.% in as-grown and aged states are observed [3, 5, 8].

First, in the investigated $Ni_{51.0}Ti_{36.5}Hf_{12.5}$ single crystals the stress-induced martensitic transformation is accompanied by growth of applied compression stress with increase of given strain and $\sigma(\varepsilon)$-curves are characterized by a high strain-hardening coefficient of $\theta=\delta\sigma/\delta\varepsilon=(10\div15)x10^3$ MPa (Figures 3 and 4). A high strain-hardening coefficient θ on $\sigma(\varepsilon)$-curves for stress-induced martensitic transformation is also observed in $Ni_{50.3}Ti_{29.7}Hf_{20}$ single crystals solutionized at 1173 K for 3 h in vacuum and then quenched into water [3].

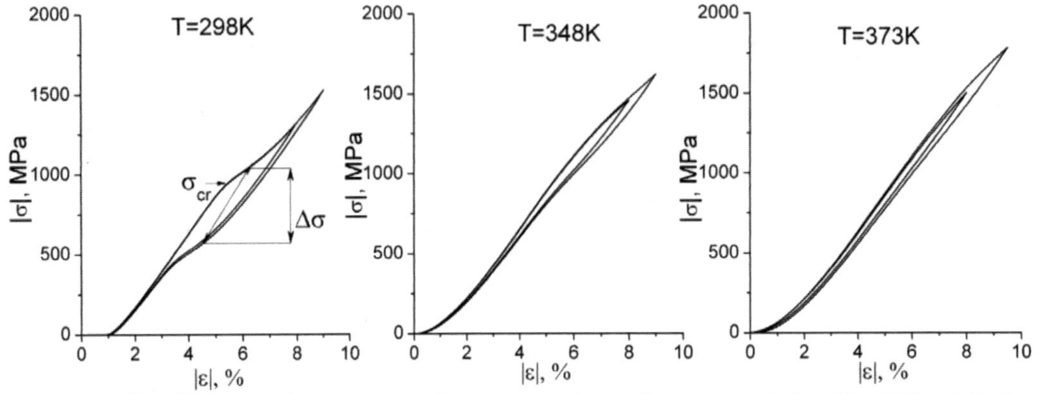

Figure 3. Stress-strain responses in compression of as-grown $Ni_{51.0}Ti_{36.5}Hf_{12.5}$ single crystals oriented along [236] direction at different test temperatures in superelasticity temperature range.

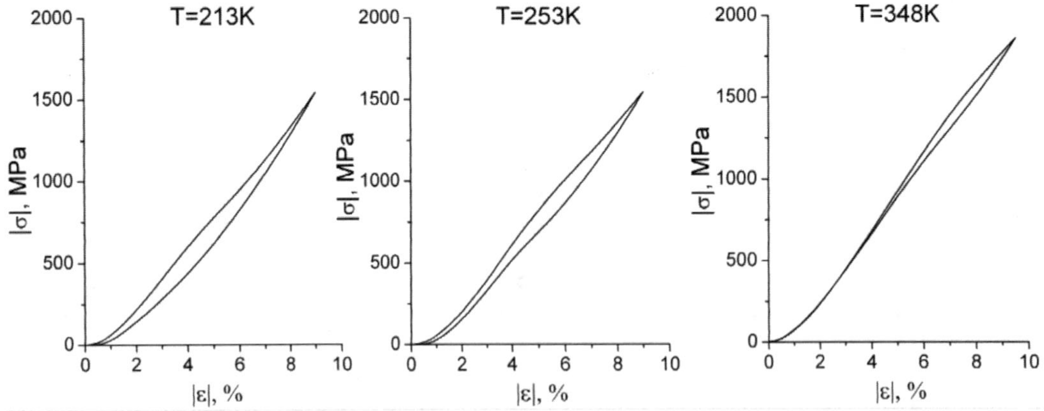

Figure 4. Stress-strain responses in compression of annealed at 1343 K for 4 h $Ni_{51.0}Ti_{36.5}Hf_{12.5}$ single crystals oriented along [236] direction at different test temperatures in superelasticity temperature range.

Second, the investigated $Ni_{51.0}Ti_{36.5}Hf_{12.5}$ crystals are characterized by a small value of reversible strain of $|\varepsilon_{SE}|=1.1\div1.4\%$ as compared to the theoretical value of lattice strain $|\varepsilon_{tr0}|_{[236]} = 8.0\%$ at B2-B19′ martensitic transformation. The strain associated with superelasticity (ε_{SE}) was defined as the maximal recoverable strain level during the stress-induced martensitic transformation. As shown in other investigations [1-4] the experimental values of reversible compression strain do not exceed of 4% in TiNiHf single crystals oriented along directions with maximal value of theoretical transformation strain of 7-8%. The third, Figures 3 and 4 illustrate a strong temperature dependence of the stress hysteresis value. The stress hysteresis $\Delta\sigma$ was determined as the difference of stress levels between the forward and the reverse stress-induced transformation at the middle of the superelasticity curve. Figures 3 and 4 show that value stress hysteresis in heterophase single crystals is almost twice reduced from 490 MPa down to 280 MPa with the increase of test temperature from 298 K up to 348 K. Accordingly, thermal hysteresis ΔT reduced from 100 K to 70 K with the increase of applied stress from 75 MPa to 170 MPa (Figure 2). This behaviour is typical of high-strength single crystals of binary TiNi alloys with disperse particles Ti_3Ni_4 [9].

Physical causes of small experimental values of reversible strain and reduction of stress hysteresis values with growth of the test temperature and level of applied stress require the additional researches. It is assumed that this behaviour can be caused by simultaneous action of several factors.

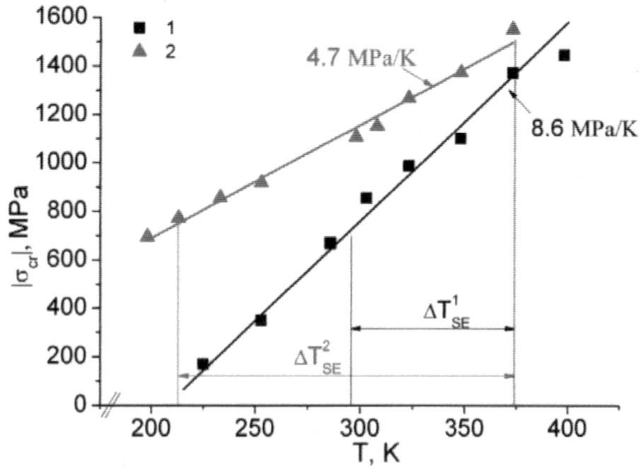

Figure 5. Temperature dependence of the critical stress level σ_{cr} of [236]-oriented $Ni_{51.0}Ti_{36.5}Hf_{12.5}$ single crystals in compression: for as-grown state (curve 1) and annealed state (curve 2).

In heterophase single crystals the decrease of transformation strain associates with the composite effect. When martensitic transformation occurs in the B2-matrix, particles of H-phase do not undergo transformation. Therefore, the volume fracture of material, which undergoes martensitic transformation and, accordingly, a value of reversible transformation strain are reduced.

Based on the thermodynamic description of stress-induced martensitic transformation external mechanical stresses ΔG_{mech}^{A-M} perform work against chemical ΔG_{ch} and non-chemical ΔG_{nonch} component of Gibbs free energy if the test temperature is higher than temperature of chemical equilibrium of phases T_0 [5, 10]:

$$\Delta G_{mech}^{A-M} = \Delta G_{ch}^{A-M} + \Delta G_{nonch}^{A-M} \qquad (2)$$

Here $\Delta G_{nonch} = \Delta G_{rev} + \Delta G_{fr}$ is non-chemical contributions to the total Gibbs free energy. The non-chemical constituent is separated into reversible (elastic strain and surface) energy ΔG_{rev} and irreversible work ΔG_{fr} which is connected with overcoming friction to interfacial motion and dissipation of elastic strain energy.

In as-grown and annealed crystals the stress-induced martensitic transformation occurs with high strain-hardening coefficient θ. Therefore, for increasing the volume fraction of stress-induced martensite we have to increase the driving force of transformation to overcome the growth of the non-chemical contributions (ΔG_{rev} and ΔG_{fr}) to the total Gibbs free energy. Such a mechanism of thermoelastic martensitic transformation is usually observed at multivariant stress-induced martensitic transformation [9] and may be accompanied by the presence of non-transform volumes of austenite in the material that reduces the transformation strain.

Finally, high levels of critical stress of martensite formation from 700 up to 1900 MPa lead to occurrence of the stress-induced martensitic transformation in the conditions of strong elastic strain of austenite and martensite. The elastic compressive strain of B2-austenite along [236] direction can result in monoclinic distortion of cubic lattice and hereby in reducing the distinctions between cubic lattice of B2-phase and monoclinic B19′-martensite. Therefore, a change of austenite and martensite lattice parameters due to their elastic distortion at high critical stress level may reduce the transformation strain and stress hysteresis value with growth of test temperature and level of critical stress at stress-induced martensitic transformations.

4. Summary

It was experimentally established that in the Ni-rich $Ni_{51.0}Ti_{36.5}Hf_{12.5}$ single crystals in as-grown and annealed at T=1323 K for 4 h states the stress-free martensitic transformation at cooling is observed at martensitic start temperature M_s below 225 K as compared to high-temperature TiNiHf alloys with low Ni content of 50.2-50.3 at.% and Hf content of 10-20 at.% [1-4]. Completely reversible stress-induced B2-B19′ martensitic transformation in [236]-oriented $Ni_{51.0}Ti_{36.5}Hf_{12.5}$ single crystals occurs at test temperature up to $T_t \sim 373$ K and high applied compressive stress up to 1700 MPa in as-grown state and 1900 MPa in annealed single crystals with value of reversible strain up to $|\varepsilon_{SE}| = 1.4\%$. The annealing at T=1323 K for 4 h of single crystals results in growth of superelasticity temperature range about twofold as compared to as-grown crystals contained disperse particles of H-phase.

Acknowledgments
Research was financially supported by Russian Science Foundation (grant No. 14-29-00012).

References

[1] Karaca H E, Saghaian S M, Basaran B, Bigelow G S, Noebe R D and Chumlyakov Y I 2011 Compressive response of nickel-rich NiTiHf high-temperature shape memory single crystals along the [111] orientation *Scripta Mater* **65** 577–580
[2] Coughlin D R, Phillips PJ, Bigelow G S, Garg A, Noebe R D and Mills M J 2012 Characterization of the microstructure and mechanical properties of a 50.3Ni–29.7Ti–20Hf shape memory alloy *Scripta Mater.* **67** 112–115
[3] Saghaian S M, Karaca H E, Tobe H, Souri M, Noebe R and Chumlyakov Y I 2015 Effects of aging on the shape memory behavior of Ni-rich $Ni_{50.3}Ti_{29.7}Hf_{20}$ single crystals *Acta Mater.* **87** 128–141
[4] Karaca H E, Saghaian S M, Ded G, Tobe H, Basaran B, Maier H J, Noebe R D and Chumlyakov Y I 2013 Effects of nanoprecipitation on the shape memory and material properties of an Ni-rich NiTiHf high temperature shape memory alloy *Acta Mater.* **61** 7422–7431
[5] Otsuka K and Ren X 2005 Physical metallurgy of Ti–Ni-based shape memory alloys *Progress in Mater. Sci.* **50** 511–678
[6] Santamarta R, Arroyave R, Pons J, Evirgen A, Karaman I, Karaca H E and Noebe R D 2013 TEM study of structural and microstructural characteristics of a precipitate phase in Ni-rich Ni–Ti–Hf and Ni–Ti–Zr shape memory alloys *Acta Mater.* **61** 6191–6206
[7] Yang F, Coughlin D R, Phillips P J, Yang L, Devaraj A, Kovarik L, Noebe R D and Mills M J 2013 Structure analysis of a precipitate phase in an Ni-rich high-temperature NiTiHf shape memory alloy *Acta Mater.* **61** 3335–3346
[8] Sehitoglu H, Jun J, Zhang X, Karaman I, Chumlyakov Y, Maier H J and Gall K 2001 Shape memory and pseudoelastic behavior of 51.5%Ni–Ti single crystals in solutionized and overaged state *Acta mater.* **49** 3609–3620
[9] Chumlyakov Y I, Kireeva I V, Panchenko E Y, Timofeeva E E, Kretinina I V and Kuts O A 2015 Physics of Thermoelastic Martensitic Transformation in High-Strength Single Crystals. *The part in the coll. monograph Shape Memory Alloys: Properties, Technologies, Opportunities. Trans Tech Publications* Eds: N Resnina and V Rubanik 107-174.
[10] Wollants P, Roos J R and Delaey L 1993 Thermally and stress-induced thermoplastic martenstic transformation in the reference frame of equilibrium thermodynamics *Progress in Mater. Sci.* **37** 227–288

IOP Conf. Series: Materials Science and Engineering **93** (2015) 012048 doi:10.1088/1757-899X/93/1/012048

Stress-Induced Thermoelastic Martensitic Transformations and Functional Properties in [011]-oriented NiTiHfPd Single Crystals

A I Tagiltsev[1], E E Timofeeva[1], E Yu Panchenko[1] and Yu I Chumlyakov[1]

[1] Tomsk State University, Siberian Physical-Technical Institute, Tomsk, 634050, Russia

E-mail: antontgl@gmail.com, katie@sibmail.com, panchenko@mail.tsu.ru, chum@phys.tsu.ru, karaca@engr.uky.edu

Abstract. The stress-induced martensitic transformation in the [011]-oriented $Ni_{45.3}Ti_{29.7}Hf_{20}Pd_5$ (at. %) single crystals in as-grown, homogenized and aged states were investigated in compression. It is experimentally shown that heat treatments of single crystals result in increase in martensitic transformation temperatures, two-fold decrease in reversible strain and increase in strain-hardening coefficient. As-grown single crystals demonstrate large temperature range of superelasticity (up to 140 K), large reversible strain (up to 4.3 %) and large work output in comparison with homogenized and aged crystals.

1. Introduction

Alloys with thermoelastic martensitic transformations (MT) are used in wide range of applications. A shape memory effect (SME) and superelasticity (SE) are observed due to the thermoelastic nature of transformation. TiNi is one of mostly usable materials with MT [1]. However, it is known [2] that TiNi alloys demonstrate low transformation and work temperatures (<373 K) and low strength (<700 MPa) that limit its application. Hf and Pd are used as alloying elements to exclude these problems. NiTiHf and NiTiHfPd alloys have high strength properties (critical stress level of stress-induced MT ~1000 MPa) and high damping ability owing to wide mechanical hysteresis (~400 MPa) and promising potential for high damping and high strength applications [3].

It is known that, the control of functionality and mechanical properties of the alloys is possible due to heat treatments. The functional behavior of NiTi alloys can be improved by heat treatments, it is natural to extend such studies to determine the effects of precipitate strengthening on shape memory and functional properties [3]. The aging results in variations of the degree order of high-temperature phase, precipitation of the second-phase particles, which causes changes in MT temperatures, transformation strain, critical stress for stress-induced MT, hardening of the matrix change and so on [2, 4-6]. On these reasons this work is aimed to investigate the effects of heat treatment on stress-induced MT and functional properties in compression in [011]-oriented $Ni_{45.3}Ti_{29.7}Hf_{20}Pd_5$ (at. %) single crystals after different heat treatments.

2. Materials and methods

The single crystals were grown using the Bridgeman method in an inert gas atmosphere. Samples (3 mm x 3 mm x 6 mm) were electro-discharge machined, then mechanically and electrolitically polished in electrolyte 5% $HClO_4$ + 95% CH_3COOH under T=293 K, U=20 B. The sample orientation was verified with DRON-3 X-ray diffractometer. Orientation [011] was chosen due to the highest value of transformation strain that is confirmed by theoretical calculations [7].

The single crystals were investigated in following conditions: as-grown without any additional heat treatments (type-I), homogenized at 1323 K for 4 h (type-II), homogenized at 1323 K for 4 h followed by ageing at 1173 K for 3 h and water quenching (type-III). The homogenization was carried out in quartz tubes in an inert gas atmosphere followed by slow cooling of crystals (cooling rate is about 20-30 $K \cdot min^{-1}$).

The experiments of cooling/heating under constant stresses in range from 0 to 300 MPa (ε(T) response) and the experiments of loading/unloading at isothermal conditions in range from 300 to 473 K (σ(ε) response) were carried out for investigation the stress-induced MT.

Mechanical experiments in compression were conducted on following installations: Instron testing machine 5969; miniature load frame for measuring reversible strain under the constant stress level during cooling/heating. The analysis of single crystal surface was performed by optical microscopy using the universal measuring complex Keyence VHX-2000. Transmission electron microscopy (TEM) was accomplished using electron microscope JEM-2010. Semiquantitative elemental analysis was carried out on scanning electron microscope Supra VP55. Distribution of chemical elements (Ti, Ni, Hf, Pd) was obtained by JEOL 8900 Electron Probe Microanalyzer.

3. Results and discussion

It was observed that as-grown single crystals contain dendrites (figure 1) that is also shown in [7]. Dendrites volume fraction is about 8-9%. The homogenization didn't result in complete dissolution of dendrites (figure 1) but leads to three-fold decrease of dendrites volume fraction. Chemical elements (Ti, Ni, Hf, Pd) distribution was investigated: all elements uniformly allocated both in matrix and in dendrites. Dendrites are richer in Hf (84.6 at.%), poorer in Ti (15.3 at.%) and contain almost no Ni and Pd (0 and 0.1 at. %). The chemical composition of the matrix was defined as Ni = 46.4 at.%, Ti = 30.7 at.%, Hf = 18.3 at.%, Pd = 4.6 at.% that is very close to the nominal composition within 5 % error. It is shown (figure 1(b)), that high-temperature phase of all investigated single crystals has B2 structure. As presented in [8], NiTiHfPd single crystals transform to monoclinic B19' martensite.

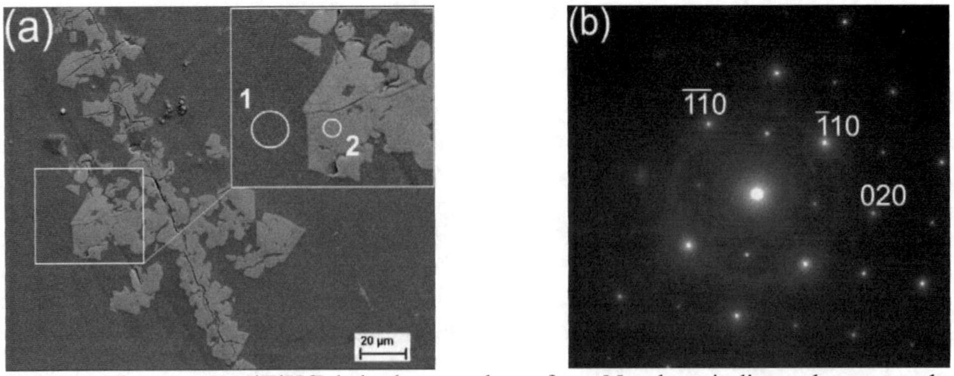

Figure 1. The type-II NiTiHfPd single crystals surface. Numbers indicate the areas where semiquantitative elemental analysis was carried out: 1 – matrix, 2 – dendrite (a); selected-area diffraction pattern of austenite phase exhibiting B2-structure in type-I NiTiHfPd single crystals, zone axis $[001]_{B2}$ (b).

Figure 2 demonstrates ε(T) response for all investigated crystals I-III, received at cooling/heating under the constant stress level. Minimal stress level necessary to destruct the self-accomodation system and to form the oriented martensite at cooling/heating under the constant stress level is

50 MPa. The reversible strain at 50 MPa is about 0.5 % in all investigated crystals. Martensite oriented relative to external stress axis, is going to grow at cooling below T= $M_{s'}$. After $M_{f'}$ the forward transformation is over. At T = $A_{s'}$ the reverse transformation begins. It should be noted, that given strain is reversible at T > $A_{f'}$ and SME is observed. As marked for the ε(T) response in figure 2 transformation strain ε_{rev} and thermal hysteresis ΔT were defined as the horizontal and vertical width of the loops, respectively. The value of reversible strain increases with increase in external stress levels in all studied states I-III (figure 3).

Maximal reversible strain in type-I single crystals is obtained under 250 MPa and equals 4.3 %. This value doesn't change at further increase of applied stress level. It is shown [2, 3, 9, 10], that maximal reversible strain doesn't exceed 4.5 % in different states of NiTiHfPd alloys (homogenization at 1173 K for 72 h, homogenization at 1323 K followed by ageing at 973 K for 72 h, homogenization at 1323 K followed by ageing at 923 K for 3 h).

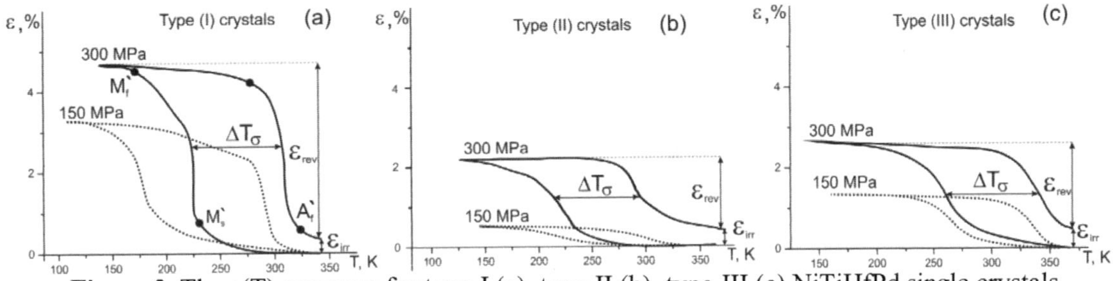

Figure 2. The ε(T) response for type-I (a), type-II (b), type-III (c) NiTiHfPd single crystals.

In type-II and type-III single crystals the reversible strain slowly increases with increase of stresses and saturation such in type-I crystals is not observed. However, the appearance of irreversible strain under 250-300 MPa is indirect evidence of reaching maximal reversible strain values. Reversible strains decrease to 1.7 % and 2 % in type-II and type-III crystals, respectively, in comparison with type-I crystals. The irreversible strain doesn't exceed 0.4% in all investigated crystals.

Work output is an important property of shape memory materials in the design of functional devices such as solid state actuators [3]. In this case, work output can be calculated by multiplying the applied stress by the corresponding reversible strain from the experiments of cooling/heating under constant stresses [3, 11]. Figure 3 shows the work output values as a function of applied stress for all investigated single crystals. Work output increases with applied stress. Type-I single crystals demonstrate the reversible strain two-fold bigger than type-II and type-III, that is why type-I crystals are characterized by higher values of work output (up to 12.8 J·cm^{-3}) compared to types-I and type-III crystals (up to 5-6 J·cm^{-3}). Work output values for NiTiHfPd obtained in this work exceed the same for $Ni_{49.5-x}Ti_{50.5}Pd_x$ (x = 15, 20, 25, 30, and 46 at. pct.) alloys (9 J·cm^{-3} at 300 MPa) [11] and for NiTiPd-20Hf alloys (4-5 J·cm^{-3} at 300 MPa) [3, 8].

Figure 3 also demonstrates the dependence of thermal hysteresis on stress level $\Delta T_\sigma(\sigma)$. Hysteresis is about the same 75-85 K for all studied states I-III at 250-300 MPa when maximal values of reversible strain are achieved. But at lower stress level the hysteresis value depends on the type of crystals. For example, the widest hysteresis 100 K is observed in type-I crystals under 50 MPa. After the heat treatments (type-II and type-III crystals) hysteresis values are 45 K and 70 K under 50 MPa, respectively.

In type-III crystals the hysteresis $\Delta T_\sigma(\sigma)$ almost doesn't depend on stress level and is equal to 75 K(±5 K) vs. type-I and type-II crystals, in which hysteresis values vary nonmonotonically with increase of stress level. Hysteresis values of type-I and type-II crystals increase up to 115 K and 75 K under 150 MPa, respectively.

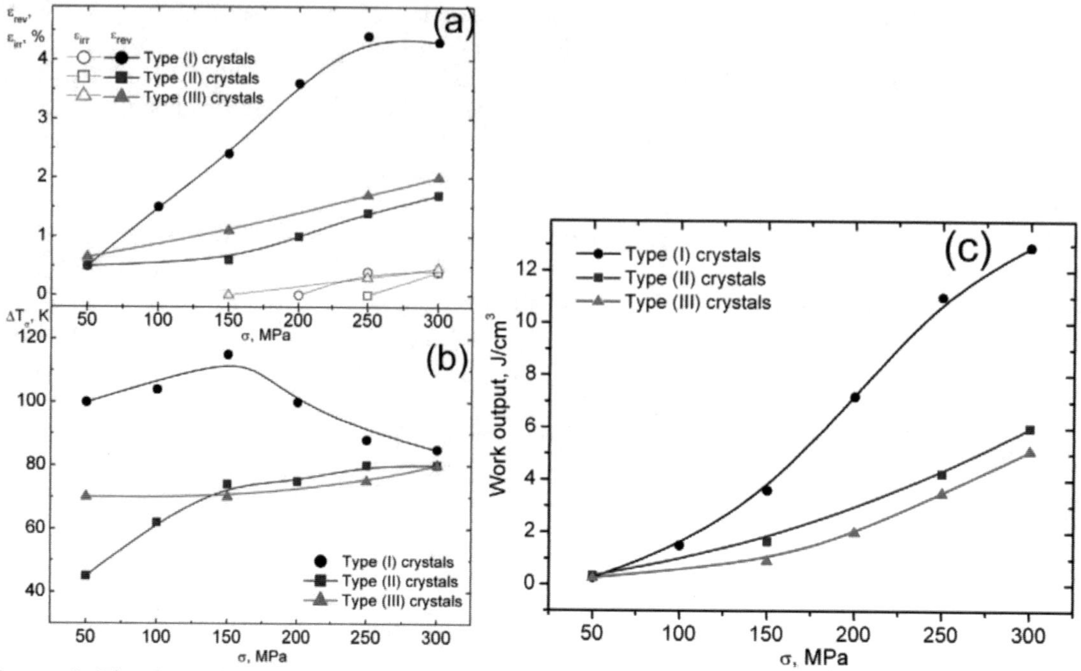

Figure 3. The dependences of reversible strain ε_{rev}, irreversible strain ε_{irr} (a), thermal hysteresis ΔT_σ (b) work output (c) on the applied stresses for type-I – type-III single crystals.

Then at σ>150 MPa, the decreasing of hysteresis is observed in type-I crystals, while hysteresis in type-II crystals doesn't change.

Figure 4 shows the σ(ε) response at loading/unloading in isothermal conditions. Given strain associated with stress-induced MT is reversible at unloading, so SE is observed. It is revealed that reversible strain values, obtained by ε(T) response, are close to values, obtained by σ(ε) response. The form of the σ(ε) curves depends on type of crystals. In type-I crystals the stress plateau with zero strain-hardening coefficient $\theta=d\sigma\cdot(d\varepsilon)^{-1}$ is observed. The θ value is raised up to $7.8\cdot10^4$ MPa with increase in stresses and so the transformation occurs with hardening. In type-III crystals there is no the plateau of stress and $\theta = 8.3\cdot10^4$ MPa.

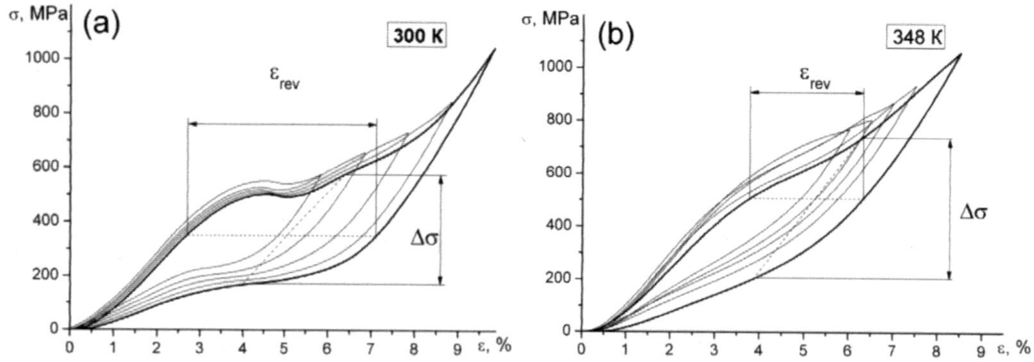

Figure 4. The σ(ε) response for type-I (a) and type-III (b) NiTiHfPd single crystals.

The stress hysteresis value increases with increase in given strain in type-I and type-III crystals. Also the reduction of critical stresses necessary for stress-induced MT is observed for type-I and type-III crystals by 36 MPa and 47 MPa, respectively. These are the evidence of formation of high amount of defects during stress-induced MT. Mechanical hysteresis value at maximal reversible strain is about the same for type-I and type-III crystals and equal to 400 MPa and 460 MPa, respectively.

So, the energy dissipation characterized by stress and thermal hysteresis is similar in type-I and type-III crystals. Indeed the absorbed energy, evaluate as the area inside the hysteresis loop of $\sigma(\varepsilon)$ response, is 17.3 $J\cdot cm^{-3}$ and 16.1 $J\cdot cm^{-3}$ for type-I and type-III crystals, respectively. These values are higher than obtained for aged $Ni_{45.3}Ti_{29.7}Hf_{20}Pd_5$ (at. %) single crystals [10], where the energy absorption capacity didn't exceed 8 $J\cdot cm^{-3}$.

Temperature dependence of critical stresses for stress-induced MT was obtained for all investigated crystals I-III as temperature dependence of critical stress $\sigma_{cr}(T)$ (by $\sigma(\varepsilon)$ response) and dependence of temperature $M_{s'}$ on applied stress level (by $\varepsilon(T)$ response). These dependences coincide and increase in stresses with the growth of temperature can be described by Clapeyron-Clausius equation (1):

$$\frac{d\sigma}{dT} = -\frac{\Delta S}{\varepsilon_{tr}} = -\frac{\Delta H}{T_0 \varepsilon_{tr}} \tag{1}$$

where ΔS – entropy of transformation, ΔH – enthalpy of transformation per unit volume, ε_0 – transformation strain, T_0 – temperature of chemical balance of martensite and austenite phases.

The coefficient $\alpha = d\sigma\cdot(dT)^{-1}$ characterizes the increase of stress level with the growth of temperature and depends on heat treatment: $\alpha_I = 3.4$ $MPa\cdot K^{-1}$, $\alpha_{II} = 5.3$ $MPa\cdot K^{-1}$, $\alpha_{III} = 5.4$ $MPa\cdot K^{-1}$. According to Clapeyron-Clausius equation the dependence of α on heat treatment is simply explained: the bigger value of α correspond to smaller value of transformation strain (figure 3, 5).

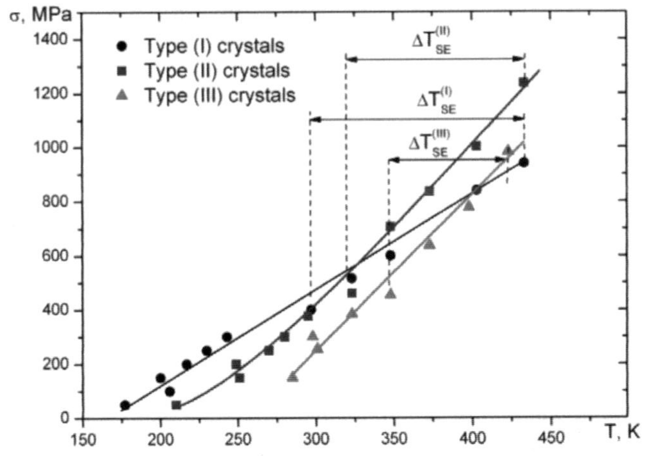

Figure 5. Temperature dependence of critical stress necessary for stress-induced MT in all investigated crystals I-III.

Figure 5 also shows the SE ranges. High-temperature SE (at T>373K) is observed in all investigated crystals and its temperature range depends on heat treatment. The widest SE range $\Delta T_{SE}(I) = T_{SE2} - T_{SE1} = 140$ K is observed in type-I crystals, whereas type-II and type-III crystals are characterized by a narrow SE ranges $\Delta T_{SE}(II) = 115$ K, $\Delta T_{SE}(III) = 75$ K. The temperature of the end of high-temperature SE range T_{SE2} is higher than temperature T_{SE2} for NiTiHfPd polycrystalline alloys [12] and NiTiPd-20Hf alloys [8], therefore, these single crystals can be used as high-temperature alloys with SE.

It is seen from figure 2 and figure 5, that thermal treatments provide the increase of MT temperatures. Increase of MT temperatures, coefficient θ and decrease of reversible strain after heat treatment can be associated with changes of the degree order of high-temperature phase and precipitation of the nano-size second-phase particles [13]. As follows from [3], particles of the second phase appear during aging at 500÷600 K [3]. Slow cooling after the homogenization could be the reason of precipitate of nano-size particle. Particles can provide the modification of chemical composition of alloy, matrix strength, appearance of internal local stresses. Systematic TEM investigations are needed to clarify the reasons of such modification of features of stress-induced MT.

4. Summary

The effect of heat treatment on the features of stress-induced martensitic transformations was investigated in the [011]-oriented $Ni_{45.3}Ti_{29.7}Hf_{20}Pd_5$ (at. %) single crystals. It is experimentally shown, that the heat treatments $Ni_{45.3}Ti_{29.7}Hf_{20}Pd_5$ (at. %) single crystals that consist in homogenization at 1423 K for 4 h (type-II crystals) and followed by ageing at 1173 K for 3 h with water quenching (type-III crystals) result in following changes in comparison with as-grown single crystals: reversible strain reduces 2 times; $\alpha=d\sigma_{cr}\cdot(dT)^{-1}$ coefficient of critical stress growing is increased 1.5 times; the form or SE curves is changed: there isn't stress plateau and strain-hardening coefficient is increased.

Single crystals of $Ni_{45.3}Ti_{29.7}Hf_{20}Pd_5$ (at. %) alloys possess large temperature range of superelasticity (up to 140 K), large reversible strain (up to 4.3 %) and large value of absorbed energy and work output. These crystals can find the application as high-damping materials and solid-state actuator in the aerospace, automotive, and energy exploration industries at high temperatures.

Acknowledgments

This work was supported by Russian Science Foundation grant № 14-29-00012.

5. References

[1] Otsuka K and Wayman C M 1998 *Shape memory materials CUP* 284
[2] Acar E, Karaca H E, Basaran B, Yang F, Mills M J, Noebe R D and Chumlyakov Y I 2013 Role of aging time on the microstructure and shape memory properties of NiTiHfPd single crystals *Mater. Sci. Eng.* A **573** 161–5
[3] Acar E and Karaca H E 2014 Orientation dependence of the shape memory properties in aged $Ni_{45.3}Ti_{29.7}Hf_{20}Pd_5$ single crystals *Intermet.* **54** 60-8
[4] Frick C P, Ortega A M, Tyber J, Maksound A E M, Maier H J, Liu Y and Gall K 2005 *Mater. Sci. Eng.* A **405** 34–49
[5] Meng X L, Cai W, Chen F and Zhao L C 2006 *Scr. Metall.* **54** 1599–604
[6] Zarinejad M, Liu Y and Tong Y 2009 *Intermet.* **17** 914–9
[7] Stebner A P, Bigelow G S, Yang J, Shukla D P, Saghaian S M, Rogers R, Garg A, Karaca H E, Chumlyakov Y, Bhattacharya K and Noebe R D 2014 Transformation strains and temperatures of a nickel–titanium–hafnium high temperature shape memory alloy *Acta Mater.* **76** 40–53
[8] Acar E, Tobe H, Kaya I, Karaca H E and Chumlyakov Y I 2015 Compressive response of $Ni_{45.3}Ti_{34.7}Hf_{15}Pd_5$ and $Ni_{45.3}Ti_{29.7}Hf_{20}Pd_5$ shape-memory alloys *J. Mater. Sci.* **50** 1924-34
[9] Acar E, Karaca H E, Tobe H, Noebe R D and Chumlyakov Y I 2013 Characterization of the shape memory properties of a $Ni_{45.3}Ti_{39.7}Hf_{10}Pd_5$ Alloy *J. Alloys Compd.***578** 297–302
[10] Karaca H E, Acar E, Basaran B, Noebe R D, Bigelow G, Garg A, Yang F, Mills M J and Chumlyakov Y I 2012 Effects of aging on [111] oriented NiTiHfPd single crystals under compression *Scripta Mater* **67** 728–31
[11] Bigelow G S, Padula II S A, Garg A, Gaydosh D and Noebe R D 2010 Characterization of Ternary NiTiPd High-Temperature Shape-Memory Alloys under Load-Biased Thermal Cycling *Metall. Mater. Trans.* A **41A** 3065-79
[12] Karaca H E, Acar E, Ded G S, Basaran B, Tobe H, Noebe R D, Bigelow G and Chumlyakov Y I 2013 Shape memory behavior of high strength NiTiHfPd polycrystalline alloys *Acta Mater.* **61** 5036–49
[13] Hornbogen E, Mertinger V and Wurzel D 2001 Microstructure and tensile properties of two binary NiTi-alloys *Scripta Mater* **44** 171-8

Automatic selection of a subset size at vector fields construction

S V Panin[1,2], P S Luybutin[1,2], V V Titkov[1] and M V Cherepanova[2]

[1]Institute of strength physics and material sciences SB RAS, Tomsk, Russia
[2]National research Tomsk polytechnic university, Tomsk, Russia

E-mail: svp@ispms.tsc.ru

Abstract. An algorithm for selection of the size of a correlation kernel at displacement vector field construction by the method of digital image correlation has been proposed. The algorithm has been tested on simulated and experimental optical images having different texture. The influence of the correlation kernel size and image texture on nose immunity at determining displacements has been studied. It is shown that the proposed algorithm allows to find this size providing the minimum error when determination of displacements and estimation of deformation.

1. Introduction

Optical method for estimation of deformation based on digital image correlation (DIC), consists of two main stages: 1) construction of a displacement vector field, 2) the subsequent calculation of strain components [1]. Most of the research in the area of development of algorithms for constructing displacement vectors are aimed to improve the accuracy and increase noise immunity in determining displacements [2, 3], or increase in speed.

In modern DIC-systems, before loading a speckle pattern is formed on a surface of tested material with a help of paint covering [1] that improves contrast and provides a reliable determination of displacements. In doing so, the shape and size of a speckle can significantly affect the accuracy and noise immunity of displacement measure.

In the method of optical flow determination the subset size (correlation area (SA)) usually corresponds to small fragments of the image which is caused, first of all, the need to reduce computational costs when construction of a complete displacement field, as well as providing a sufficiently high density of displacement vector fields [4, 5, 6] . The presence of noise in the image, deformation of the observed objects, as well as the small subset size causes an appearance of errors in the displacement estimation.

Partially, the problem of their presence can be solved by post-correction of the displacement vector field, such as a spatial filter [7], smoothing [8] and so on. Another way to solve the problem of their influence is to increase the subset size [9, 10]. This size must be sufficient to compare a few portions of images and minimize the noise influences, on the other hand to lead to the minimum smoothing of displacement vector fields.

The subject of this research is the problem of the automatic sizing of the subset in modern DIC-systems containing enough unique and identifiable characteristics (objects in the image) to ensure reliable and accurate determination of displacements.

2. Algorithm of a subset size selection

As indicated above, one of the main stages of the DIC method – is the displacement vector field construction. The algorithm for determining the displacements based on establishing consistency between the areas of two images by calculating the cross-correlation function (CCF) and its extremum seeking. [11] Finding the maximum CCF within the scanning zone is performed line by line by one pixel. The size of the scanning area (*sa*) and the step of vector construction (Figure 1, a) is initially set by the operator. There are following notation in the figure 1: n - the size of the subset side within which the correlation coefficient is calculated; *sa* - the size of the scanning area side; step - step of construction of the vectors; I_e, J_e - the coordinates of the upper left corner of the image area.

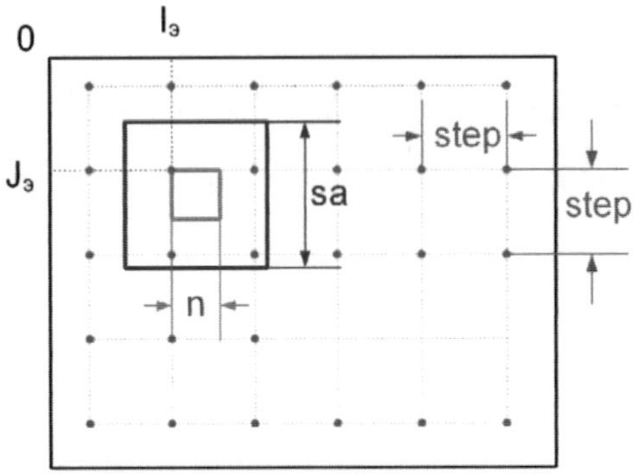

(a)

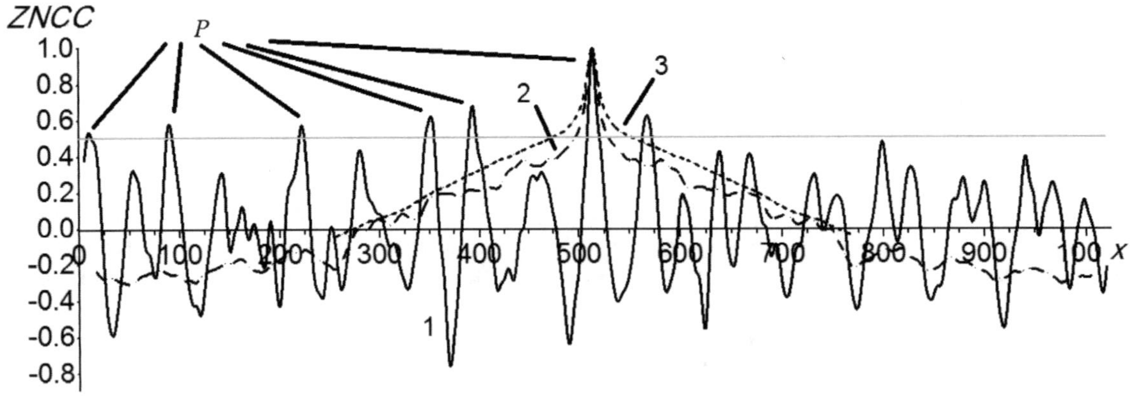

(b)

Figure 1. Illustration of determination displacement: a) principle of the construction of displacement vector fields; b) mapping example of the autocorrelation function for the three values *n*: 1) $n = 8$, 2) $n = 64$, 3) $n = 512$.

The algorithm for determining the subset size for a set of numbers *n* in the range of $0 < n < min\{w/4;\ h/4\}$, where *w* and *h* – width and height of an image, *min* - the operator selecting the minimum values of the arguments (image parameters) is presented bellow in 5 steps:

1. Values of the autocorrelation function are calculated in the horizontal and vertical directions. To calculate the normalized correlation coefficient is used with zero average [4]

$$ZNCC = \frac{\sum(X-\bar{X})(Y-\bar{Y})}{\sqrt{\sum(X-\bar{X})^2 \sum(Y-\bar{Y})^2}}, \qquad (1)$$

IOP Conf. Series: Materials Science and Engineering **93** (2015) 012049 doi:10.1088/1757-899X/93/1/012049

where *X*, *Y* – brightness of comparable areas of the image, $\bar{X} = \frac{1}{n}\sum_{t=1}^{n} X_t$, $\bar{Y} = \frac{1}{n}\sum_{t=1}^{n} Y_t$ – the average values of brightness of the same areas.

2. For calculated values of the autocorrelation function in the horizontal and vertical directions for the parameter *n* varying in the range of 2 < n < 64, the following parameters are calculated:

FS – width of the autocorrelation function (*ZNCC*) at the level of 0.5 pixels (fig. 1, *b*);
N – the number of low-contrast areas of the image.
P – the number of autocorrelation function peaks that exceed the level of 0.5.

3. Steps 1-2 are repeated for any *n*.
4. The minimum value of *n* is selected such for that the following conditions are fulfilled in order to select the subset size:
   - the number of peaks *P* should be equal to 1;
   - the number of areas *N* in the image, for which the condition of low contrast is performed (see above step 2) must be equal to zero.
5. Then, starting since the value *n* selected at the 4th step, one is chosen for which the parameter *FS* is less or equal than that calculated for the next value of *n*. The resulting value is accepted as the required subset size.

3. Methods of testing

In the general statement generation of series model images consists of two stages: 1) formation of the image of a model surface; 2) forming a series of images of the model surface based on the increment of deformation. In total 6 series of images were examined, three of which were the model (simulated).

- Model of multi-layer image [12].
- Model of speckle (paited surface).
- For the third series the spot size was quadrupled.

The following three series were formed of the experimentally obtained images of loaded materials.:

- Aluminum alloy A2024-with sprayed speckle.
- The carbon-carbon composite with a sprayed speckle [13].
- Aluminum alloy A2024 without sprayed speckle [14].

As the result, six series of images were generated, each of which includes 6 frames imitating specimen tension with strain increment of 1% at the final elongation of 5%. Further these series will be named "Series 1", "Series 2" and so on.

The error of deformation estimation was defined by an arithmetic average step-by-step absolute difference of strain fields specified by the model and calculated ε_{xx}:

$$\delta\varepsilon_{xx} = \frac{1}{N}\sum_{i=0}^{N}\left|\varepsilon_{xx\,\text{мод}_i} - \varepsilon_{xx\,\text{расч}_i}\right|.$$

4. Calculation results and discussion

The maximum values of *FS*, *N*, *P* at low values of *n* (Figure 2) explains the small amount of contrast objects (irregularities) in the image which are within the subset area. Therefore:

- number of the area pairs in the images with a correlation coefficient close to unity will be sufficiently large, too (parameter *P*);

- the number of low-contrast areas is also high (parameter *N*);

- a plurality of peaks with an amplitude greater than 0.5, resulting in a higher final value of the parameter *FS* at the autocorrelation function distribution will be illustrated.

The subsequent decline of the parameters *FS*, *N*, *P*, due to increasing *n*, is the consequence of the increasing the number within the subset area of objects (irregularities) of the image (e.g., speckle spots). Accordingly, the number of pairs of image fragments having a correlation coefficient close to 1 (parameter *P*), tends to unity. The number of low-contrast areas will tend to zero (parameter *N*); the number of peaks at the distribution of the autocorrelation functions tends to unity, whereas the peak width should have a minimum value (parameter *FS*). In figure 2 observe various gradient of the characteristic *FS*, *N*, *P* decline: for series 1, 3, 6, there is a gentle decline; for series 2, 4, 5 – sharper. This is due to the degree of image contrast and the number of objects (irregularities) in the image (e.g., speckle size spots, relief on the material surface, and so on.).

For images with sufficient small contrast of the texture (series 1, 6), and also for images having a relatively large area of low contrast (objects) (series 3) the value of n must be greater to reach minimum values of the *FS*, *N*, *P*. For this reason, the character of the parameters with increasing n is smoother. Conversely, for sharper contrast images (series 4 and 5), and images with a small size of speckle spots (series 2) even for relatively small values n these parameters reach their minimum. As a result, a fairly sharp decline of *FS*, *N*, *P* is visible in graphs. In particular, the parameter of the subset size increases from 10 to 48 (table. 1) when the size of speckle spots rises by 4 times on the surface of the colored model (series 2 and 3) which is associated with the low-contrast area increasing (speckle spots). Series 6, as well as series 3 are characterized by a large value of the subset area and the gradual decline in the number of low-contrast areas (Figure 2 *c*, *h*), which is expected, since the surface of the specimen of series 6 is not colored. At the same time for the specimen coated with speckle (series 4), model with colored surface (series 2) and images with a contrast surface (series 1 and 5) there is a sharp decline in the calculated parameters that indicates the higher noise-immunity of the algorithm of subset area determination. Thus, there is a clear relationship between the degree of image contrast, and the nature of the texture of the material surface, and the change in the parameters *FS*, *N*, *P* with increasing value n.

Analysis of the graphs (Figure 2) allows to select the optimal size of the parameter n (table. 1) according to the proposed algorithm. Calculations of dependence of strain measuring error (chapter 3) on the value n were carried out to confirm the optimal choice of SA. On their basis, dimensions of SA were determined with the smallest measure strain error (table. 1). It is clear from the table that these dimensions of SA coincide to that values at the minimum error $\delta\varepsilon_{xx}$ or exceed them.

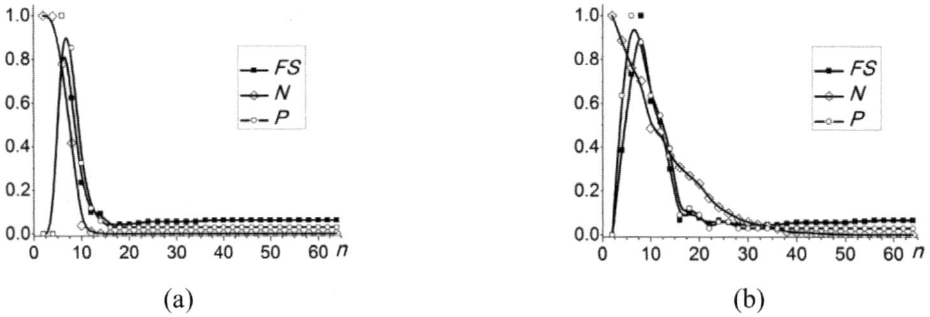

Figure 2. Dependence of parameters *FS*, *N*, *P* of the pad size of correlation n: a) model of multi-layer image; b) aluminum alloy A2024 without sprayed speckle.

Moreover, if SA is greater than the value corresponding to the minimum error $\delta\varepsilon_{xx}$ it is able to get even smaller error for strain determination compared to that with using a smaller SA. This is due to the fact that reducing SA gives rise to a large number of false peaks of the autocorrelation function (Figure 1, b).

From the study it is following that the algorithm makes it possible to conduct successfully image processing as with the model, and with the experimental ones. More sloping graphs of dependence of parameters *FS*, *N*, *P* on SA correspond to the model (series 3) and experimental (series 6) images with little contrast. The rest of the series as model images so and experimental ones with a more contrast texture have cool-changing dependences. Thus, based on the obtained data (Figure 2 and Table. 1) and its analysis it can be concluded that the proposed algorithm can be used effectively in practice.

Table 1. The results of parameter calculation when choosing SA.

Image model	SA				
	According to the algorithm parameters				At the minimum error $\delta\varepsilon_{xx}$
	P	*N*	*FS*	Summ	
1	16	16	16	16	14
2	10	10	10	10	10
3	26	48	48	48	46
4	14	12	16	16	12
5	12	12	12	12	10
6	22	52	52	52	42

5. Conclusions

An algorithm to automatically select the subset size for the problem of constructing vector fields in the evaluation of deformation by the digital image correlation method. The algorithm functions were studied by simulated and experimental data. According to the results of the algorithm for the six series of images with different textures the subset providing a minimum error when estimation of deformation, was determined.

The comparison of revealed dimensions as a result of applying the algorithm to the values of SA when the error of deformation determining reaches to a minimum value has been conducted. The values of SA to which the minimum error corresponds have been found. The comparison of values obtained as a result of the algorithm applying and values which correspond to the minimum error shows a slight deviation of the first to the next higher.

Thus, the developed algorithm is effective to measure deformation of the materials having different relief as for surface images with a sprayed speckle so and for low contrast images (non colored).

Acknowledgements

This study was performed as a part of fundamental research of RAS, RFBR grant No.13-07-00009 "Development of high-speed and error-correcting algorithms for processing and analysis of optical and acoustic signals to the combined method of monitoring the state of loaded materials" as well as the grant of the President of the Russian Federation No. SP-1529.2015.5.

References

[1] Sutton M A, Orteu J-J and Schreier H 2009 *Springer* p 321
[2] Xu L, Jia J and Matsushita Y 2012 *Motion detail preserving optical flow estimation. PAMI* **34(9)** 1744–1757
[3] Sun D, Sudderth E and Black M J 2012 Layered segmentation and optical flow estimation over time. *Computer Vision and Pattern Recognition (CVPR)*, IEEE, pp 1768–1775
[4] Horn B K P and Schunck B G 1981 *Artificial Intelligence* **17** 185–203
[5] Lucas B D and Kanade T 1981 An iterative image registration technique with an application to stereo vision *Proceedings of Image Understanding Workshop* pp 121–130
[6] Anandan P 1989 *International Journal of Computer Vision* **2** 283–310
[7] Panin S V, Titkov V V and Lubutin P S 2013 *Optoelectronics, Instrumentation and Data Processing* **49** 155–163
[8] Panin S V, Titkov V V and Lubutin P S 2014 *Optoelectronics, Instrumentation and Data Processing* **50** 61–67
[9] Hanna K 1991 Direct multi-resolution estimation of ego-motion and structure from motion *In IEEE Workshop on Visual Motion* (Princeton, NJ) pp 156–162

IOP Conf. Series: Materials Science and Engineering **93** (2015) 012049 doi:10.1088/1757-899X/93/1/012049

[10] Stein G P and Shashua A 1997 Model-based brightness constraints: On direct estimation of structure and motion *In IEEE Conference on Computer Vision and Pattern Recognition* (San-Juan) pp 400–406
[11] Panin S V, Syryamkin V I and Lyubutin P S 2005 *Optoelectronics instrumentation and data processing* **41** 42–53
[12] Panin S V, Titkov V V and Lubutin P S 2014 *Optoelectronics, Instrumentation and Data Processing* **50** 139–147
[13] Panin S V, Burkov M V, Byakov A V, Lubutin P S, Hizhnyak S A and Martynyak R M 2012 *Russian Physics Journal* **55** 228–233
[14] Panin S V, Byakov A V, Grenke V V, Shakirov I V and Yussif S A K 2009. *Physical Mesomechanics* **13** 203–211

Microstructure and mechanical properties of CP-Ti subjected to UIT

R R Hairullin[1,2], A I Kozelskaya[2], A V Panin[1,2] and M S Kazachenok[2]
[1] Tomsk, 634050, Russia, National Research Tomsk Polytechnic University.
[2] Tomsk, 634055, Russia, Institute of Strength Physics and Materials Science SB RAS.

E-mail: hairullin@list.ru

Abstract. The ultrasonic impact treatment (UIT) effect on microstructure and mechanical properties of commercial purity titanium (CP-Ti) was studied. The microstrucrure was investigated by X-ray diffraction analysis and optical profilometer New View 6200. The microhardness of CP-Ti specimens was measured by microhardness tester with the Vickers pyramid. It was found that UIT of CP-Ti specimens led to severe plastic deformation of surface layer with a thickness equals to 120 μm. This was accompanied by an increase in compressive residual stresses, a growth of the microhardness, a crystallographic texture change and a rise of density of deformation twins in surface layers of titanium specimens. It was shown that the deformation twins play a determining role in microstructure and mechanical properties changes of the surface layer of titanium specimens.

1. Introduction

It is well-known, that ultrasonic impact treatment (UIT) has been an efficient way to improve the physical and mechanical properties of constructional materials [1-8]. Many articles covered UIT is aimed to study the effect UIT modes on performance properties of steels. For example, it has been shown in [9] that UIT of S45C steel leads to an increase in fatigue life with the growth of a vibration strike number by as much as 33%. Furthermore, Yasuoka et al. [10] have revealed that UIT is appropriate for improving the fatigue strength of SUS304 shafts especially at optimum static load. Moreover, Ye et al. [11] have studied UIT processing of SUS 304. Their results have shown that this process leads to high levels of near- surface compressive residual stresses and dramatic changes in microstructure and hence local and bulk mechanical properties.However, the issue of the UIT effect on the physical and mechanical properties of titanium alloys remains poorly understood. The purpose of this work is to study the effect of ultrasonic impact treatment on microstructure and mechanical properties of commercial purity titanium. It is expected that UIT of CP-Ti allows us to change its texture, values of residual stresses and microhardness.

2. Experimental details

The specimens of commercial purity titanium (0.2 % Al, 0.4 % Zr, 0.3 % Mn, 0.01 % Cr, 0,06 % Si, 0,2 % Fe, 0,02 % Cu and 98.8 % Ti) were annealed for an hour in a vacuum chamber at a temperature of 750 °C and then mechanically polished to a mirror finish. The average grain size of specimens was approximately 100 μm (figure 1). After that, the specimens were subjected by UIT. The UIT of the titanium specimens was carried out by excitation of ultrasonic oscillations in a treating tool. The processing parameters of UIT are presented in table 1.

Figure 1. The surface microstructure of the titanium specimens.

Table 1. The processing parameters of UIT.

Frequency (kHz)	22
Amplitude (μm)	40
Static load (N)	200
Diameter of the deforming tool (mm)	5
Scanning speed (ms^{-1})	0,015

The microstructure of the titanium specimens was studied by optical profilometer New View 6200. X-ray diffraction analysis of specimens was performed by a DRON-7 X-ray diffractometer with Co K_{α}-radiation. The thickness of the modified surface layer was estimated from the measurement of the in-depth microhardness of the titanium specimens after UIT. The measurements were performed by microhardness tester with the Vickers pyramid. The indenter load was equal to 50 g. All the measurements were conducted at room temperature.

3. Results and discussion

According to investigations, deformation twins of a high density are generated in the surface and subsurface regions of CP-Ti specimens subjected to UIT (figure 2). It should be noted, that the twin length is limited by grain size. The deformation twins can intersect with each other. As shown on the figure 2, b UIT of specimens leads to the severe plastic deformation of the material surface. The thickness of the deformation layer is about 120 μm.

Figure 3 shows that the initial microhardness of the titanium specimens is 1300 MPa. However, further UIT leads to an increase in surface microhardness up to 2000 MPa. The thickness of plastic deformation surface layer is about 120 μm that correlates well with the data obtained by microstructural investigations (figure 2).

Figure 4 presents the results of X-ray diffraction analysis. According to obtained data, UIT of CP-Ti specimens causes a substantial decrease in intensity of diffraction peaks and appearance of additional peaks at high angles. This indicates that a crystallographic texture of specimen surface layer is changed. Moreover, a broadening of diffraction peaks is also observed after UIT of CP-Ti specimens (figure 4). It means that a lattice distortions value of specimens is increased. In addition, X-ray diffraction analysis showed that UIT results in a formation of compressive residual stresses (210 MPa) in the surface layer of the titanium specimens.

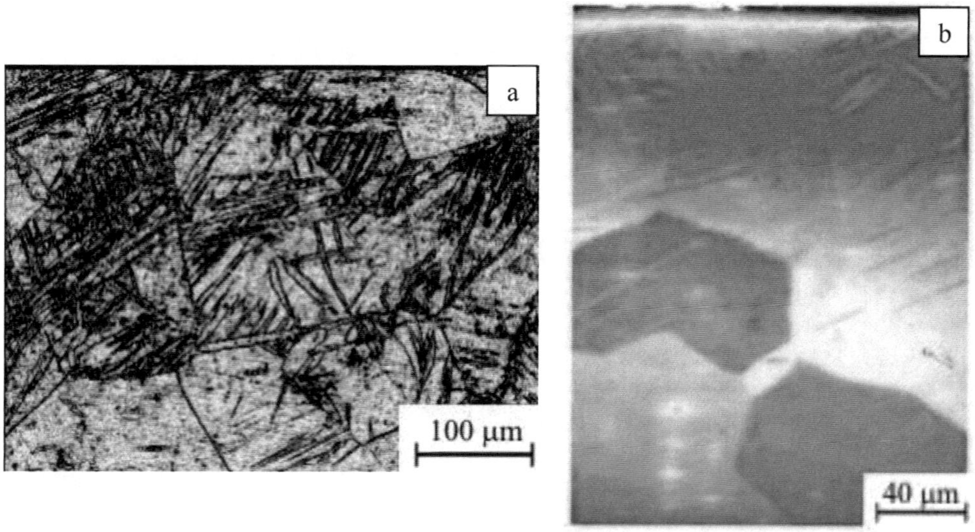

Figure 2. The surface (a) and the cross-sectional (b) microstructure of the titanium specimens subjected to UIT.

The deformation twins that are observed in the surface and subsurface regions of CP-Ti specimens subjected to UIT lead to the reorientation of grains [12,13]. Hence, there is a change of a crystallographic texture in titanium specimens (figure 4). Moreover, the intersection of deformation twins promotes a grain refinement. As a result, the microhardness in the surface layer of titanium specimens is increased.

The formation of compressive residual stresses is associated with severe plastic deformation of the surface layer of titanium specimens under the strikes of the deforming tool vibrating at high frequency.

It is well known, that grain boundaries are an effective barrier for dislocation motion and its propagation deep into the material. As a consequence, the thickness of the hardened layer is determined primarily by the grain size of specimens and limited to one or two grains (figure 2, b).

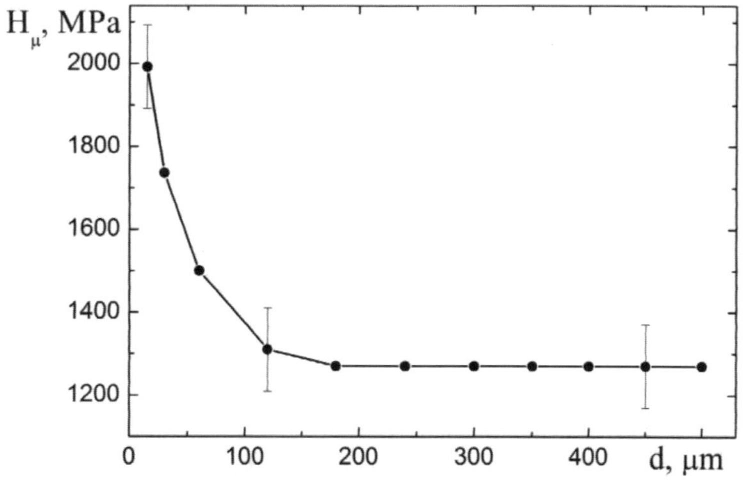

Figure 3. In-depth microhardness H_{μ} of the titanium specimens subjected to UIT.

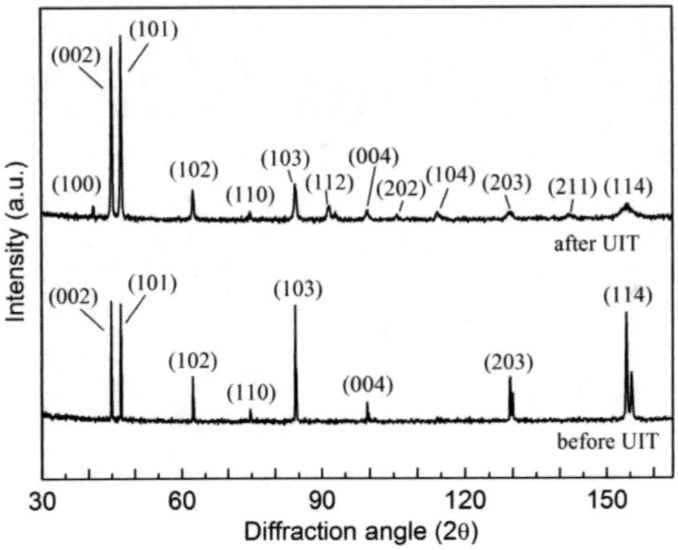

Figure 4. XRD patterns of the titanium specimens before and after UIT.

4. Conclusion

Thus, it was shown that ultrasonic impact treatment of commercial purity titanium specimens led to severe plastic deformation of their surface layer that was accompanied by an increase in compressive residual stresses, a crystallographic texture change and hardening of surface and subsurface regions. The thickness of the hardened layer was about 120 μm and limited by the grain size of titanium specimens. The main reason of the microhardness growth and a change of the crystallographic texture was a formation large numbers of the deformation twins in the surface layer of titanium specimens during ultrasonic impact treatment.

Acknowledgements
The work has been carried out at the financial support of the Russian Science Foundation (grant No. 14-19-00766).

References

[1] Amanov A, Penkov O V, Pyoun Y S and Kim D E 2012 *Tribol. Int.* **54** 106–113
[2] Fan Z, Xu H, Li D, Zhang L and Liao L 2012 *Procedia Eng.* **27** 1718 – 1722
[3] Cherif A, Pyoun Y S and Scholtes B 2010 *J. Mater. Eng. Perform.* **19** 282-286
[4] Liao M, Chen W R and Bellinger N C 2008 *Int. J. Fatigue* **30** 717–726
[5] Wu B, Wang P, Pyoun Y, Zhang J and Murakami R 2012 *Surf. Coat. Tech.* **213** 271-277
[6] Berg-Pollack A, Voellmecke F-J and Sonsino C M 2011 *Int. J. Fatigue.* **33** 513–518
[7] Yang X, Ling X and Zhou J 2014 *Int. J. Fatigue.* **61** 28–38.
[8] Liu Y, Wang D, Deng C, Xia L, Huo L, Wang L and Gong B 2014 *Int. J. Fatigue.* **66** 155–160
[9] Cao X J, Pyoun Y S and Murakami R 2010 *Appl. Surf. Sci.* **256** 6297–6303
[10] Yasuoka M, Wang P, Zhang K, Qiu Z, Kusaka K, Pyoun Y S and Murakami R 2013 *Surf. Coat. Technol.* **218** 93–98
[11] Chang Y, Telang A, Gill A S, Suslov S, Idell Y, Zweiacker K, Wiezorek M K, Zhou Z, Qian D, Mannava S R and Vasudevan V K 2014 *Mater. Sci. Eng. A.* **613** 274–288
[12] Mordyuk B N, Karasevskaya O P, Prokopenko G I and Khripta N I 2012 *Surf. Coat. Technol.* **210** 54–61.
[13] Mordyuk B N, Karasevskaya O P and Prokopenko G I 2013 *Mater. Sci. Eng. A.* **559** 453–461.

The plasma dynamic synthesis of aluminum nitride in system with gaseous and solid precursors

I Shanenkov[1], A Sivkov[1], A Ivashutenko[1] and Yu Shanenkova[1]
[1] Institute of Power Engineering, Tomsk Polytechnic University, Lenin av. 30, Tomsk, Russian Federation

E-mail: Swordi@list.ru

Abstract. Aluminum nitride is widely-used material for semiconductor devices and ceramics production. Despite the large number of known ways to obtain AlN powder, the problem of synthesizing high-purity and nanosized product is still urgent. This paper shows results on plasma dynamic synthesis of aluminum nitride using system based coaxial magneto plasma accelerator. The influence of using gaseous or solid precursors on such characteristics of the final product as phase content and particle size distribution was investigated. According to X-Ray diffractometry AlN phase content is increased in the case of use of solid nitrogen-containing precursor (melamine) in comparison with the use of gaseous nitrogen. The particle sizes distribution histograms are built in accordance with the data of bright-field TEM-images and shown in this paper. The most of particles are less than 100 nm in both experiment but there are some differences, depended on the precursor type, that are also described.

1. Introduction

Aluminum nitride was firstly synthesized in 1877, but only in the late of 20th century this material attracted the attention of scientists due to a number of its unique properties. High thermal conductivity, low dielectric constant, high thermal stability, non-toxic, high-mechanical strength make this material attractive for substrates of semiconductor devices and ceramics production.

But there are several known problems connected with the AlN applications [1]. The sintering process of AlN ceramics is sensitive to the presence of impurities in powder that impacts on the defects appearance. Also the high sintering temperature negatively influences on the obtaining of fully dense AlN. This temperature can be decreased by using powder with the smaller particles in order to increase the particle surface area. Thus, the key issue is how to obtain high-purity and nanosized powder [1-5].

There are many methods to synthesize aluminum nitride both in the form of thin films [6-12] and in the powdered form [12-16]. The synthesis by plasma methods has such advantages as the fast speed of reaction due to the high plasma temperature (~ 10^4 K), the possibility of obtaining ultrafine monocrystalline product due to the high cooling rate (> 10^6 K / s) and the possibility of using precursors without special pre-treatment [1]. The system based on coaxial magneto plasma accelerator can be used to obtain different ultrafine powders by varying energy and construction parameters with high efficiency [17-20].

IOP Conf. Series: Materials Science and Engineering **93** (2015) 012051 doi:10.1088/1757-899X/93/1/012051

This paper shows results on AlN synthesis in the discharge plasma jet generated by coaxial magneto plasma accelerator. The influence of both gaseous and solid precursors on the final product was investigated. Also particle size distribution, particles morphology and phase content were studied.

2. Experimental

Design and function of CMPA-based system were considered earlier [17]. In experiments the CMPA with aluminum central electrode and aluminum electrode-barrel was used. In order to initiate the discharge four Al wires (200 μm thickness) were stretched between electrodes. When the process was started and current, flowing through electrodes, increased, these wires exploded and the arc plasma discharge ignited. This plasma structure was accelerated by forces of conductive and inductive electrodynamics. Aluminum necessary for synthesis was accumulated from the central electrode and inner walls of electrode-barrel due to their electro erosion. Nitrogen was introduced into plasma by two ways: using the gaseous nitrogen atmosphere in the working chamber or using nitrogen-containing precursor put into the plasma formation zone. In the first case the aluminum-containing plasma was interacted with gaseous nitrogen in the second case the aluminum-nitrogen plasma flowed into the chamber space filled with argon atmosphere.

Two experiments were carried out in order to compare the influence of solid and gaseous precursors in this system on such synthesized powder characteristics as phase content, particle morphology, particle size distribution. In the first experiment the working chamber, preliminary evacuated, was filled with nitrogen (N_2) atmosphere at normal conditions. In the second experiment melamine ($C_3N_6H_6$) with the mass of 1 g was chosen as the solid nitrogen-containing precursor due to high nitrogen content and its low binding energy. The chamber was filled with argon atmosphere at normal conditions. Energy parameters of both experiments were as follows: charging voltage – 2.0 kV, charging capacity – 14.4 mF.

Synthesized powders were investigated using such methods as X-Ray diffractometry (XRD) and transmission electron microscopy (TEM). The XRD analysis was carried out using Shimadzu XRD7000S diffractometer with the counter monochromator Shimadzu CM-3121. The database PDF2+ and the software PowderCell 2.4 were used to provide respectively qualitative and quantitative analyses of XRD patterns. The TEM analysis was carried out using Philips CM 30 microscope. Particle size distribution histograms were built in the software ImageJ in accordance with the data of bright-field TEM-images.

3. Results and Discussions

Figure 1 shows XRD patterns of synthesized powders. The product obtained in the experiment with gaseous precursor (figure 1a) strongly differs from the product synthesized in the experiment with solid precursor (figure 1b). Both the main peaks ratio and the phase content noticeable change in dependence on the precursor changing. As for the experiment 1 the powder predominantly consists of aluminum phase with the small presence of hexagonal aluminum nitride (space group No. 186 P63mc). This small content of AlN phase can be explained by low reactivity of pure gaseous nitrogen due to high binding energy.

The powder synthesized in the second experiment also consists of hexagonal aluminum nitride and pure aluminum phases, but in this case there is a presence of Al_4C_3 and melamine in the final product. The quantitative analysis shows the increase of AlN phase content and decrease of pure aluminum. It seems natural due to the fact the energy parameters of experiments were the same but in the case of experiment with solid precursor the binding energy of nitrogen in melamine is less than that of gaseous nitrogen. The aluminum carbide formation can be explained by decomposition of precursor under the plasma influence. Free carbon ions, released after decomposition, react with aluminum, which present in excess, during the cooling down process. The appearance of melamine in the final product is unexpected but not unusual. After opening the system 0.2 g of unreacted melamine was found in the plasma formation zone.

IOP Conf. Series: Materials Science and Engineering **93** (2015) 012051 doi:10.1088/1757-899X/93/1/012051

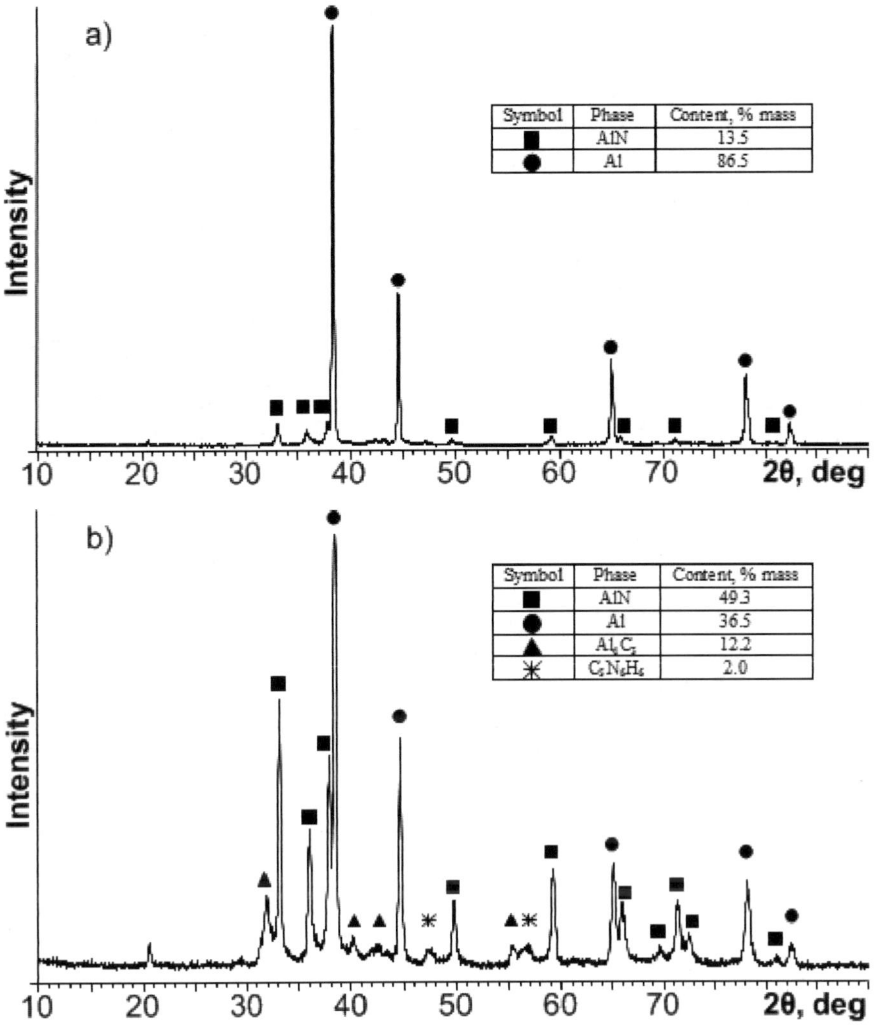

Figure 1. XRD-patterns of synthesized products: a) experiment with gaseous precursor; b) experiment with nitrogen-containing solid precursor.

Apparently due to its known dielectric properties the current flowed not through all volume and some part of melamine didn't convert into the plasma state. A small amount of unreacted precursor from this part could get into the final product that has given a reflection on the diffraction pattern.

The results of XRD analysis are confirmed by data of TEM analysis. Bright-field images of products synthesized in experiments with gaseous and solid precursors (figure 2a and 3a, respectively) are similar in terms of particle morphology. Both products have few clearly-seen types of objects: bright round particles, round particles with dark inclusions, well crystallized particles mostly in the form of hexagons and polygons and ultra-fine particles below 10 nm. The analysis of selected area electron diffractions (SAED), shown in figure 2b and 3b, allows confirming that products predominantly consist of pure aluminum and hexagonal aluminum nitride.

At displacement of the aperture in the area of different bright reflexes dark-field images were obtained. As it is seen in the figures 2c and 2d, obtained from reflexes of AlN in direction of [101] and [110], respectively, the particles with dark inclusions and hexagon particles are shone. The dark inclusions seem to be the regions doped with nitrogen. The round particles and hexagons are attributed to Al and AlN particles, respectively. This result is consistent with previous studies on the growth mechanism of AlN nanoparticles [1].

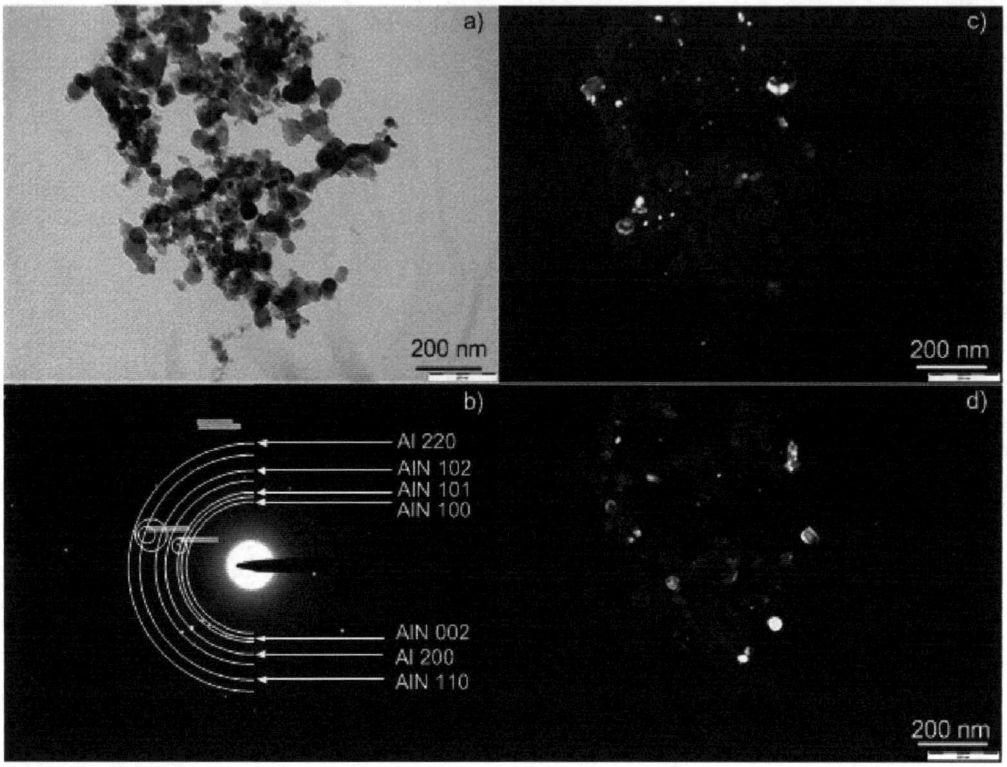

Figure 2. TEM-images of the product synthesized using gaseous precursor: a) bright-field image; SAED; c) dark-field image in the light of AlN reflex in direction [101]; d) dark-field image in the light of AlN reflex in direction [110].

The round Al particles react with the nitrogen in the plasma scope and form into hexagonal-shaped particles, and grain growth occurs by the reaction of nitrogen and Al crust on the surface of AlN particles. It should be noted the product synthesized in experiment one is predominantly consist of round objects attributed to Al that confirm the results of XRD analysis where the Al phase is dominant. In the case of experiment with solid precursor the dominant particles in the product are polygons and hexagons attributed to AlN phase. This result is also in a good agreement with XRD analysis data. Dark-filed images, obtained from reflexes of AlN phase in the direction of [100] and [102] (figure 3c and 3d, respectively) also show the luminescence of hexagons and locally-doped regions.

The visual comparison of bright-field images allows concluding that particles synthesized in experiments with solid precursor have larger sizes. This is confirmed by results of particle size distribution analysis, shown in figure 4. It is seen these histograms are different in terms of particle size maximums. The most of the particles in the product, synthesized in experiment with gaseous precursor, have sizes in the range of 20-80 nm. The particles in the product, obtained in experiment 2, have larger sizes in the range of 40-160 nm without noticeable maximum.

This result can be explained by features of melamine decomposition under the influence of plasma. Hydrogen, which is presented in melamine, is released in the current flow process. Its release increases the heat removal, that impacts on energy parameters of synthesis process. The voltage is increased and the level of current is limited. The limitation of current leads to decrease of plasma temperature and the cooling rate, respectively. The decrease of cooling rate results in increasing of particle sizes.

The decrease of current level due to hydrogen release also explains the presence of unreacted melamine in the plasma formation zone after carrying out the experiment. Thus, the effective conversion of melamine is important task to carry out experiments in this system in terms of increase of aluminum nitride content and decrease of impurity presence. There are several possible ways that can be realized in this system. The first way is to increase the energy parameters in order to increase the current level, but in this case the erosion of aluminum electrodes and its presence in final product will be also increased. The second way is to decrease the solid precursor mass at the same energy parameters. In this case there is a sense of using both solid nitrogen-containing precursor and nitrogen atmosphere with higher pressures. The third way is to use other solid nitrogen-containing precursors or their mix with melamine.

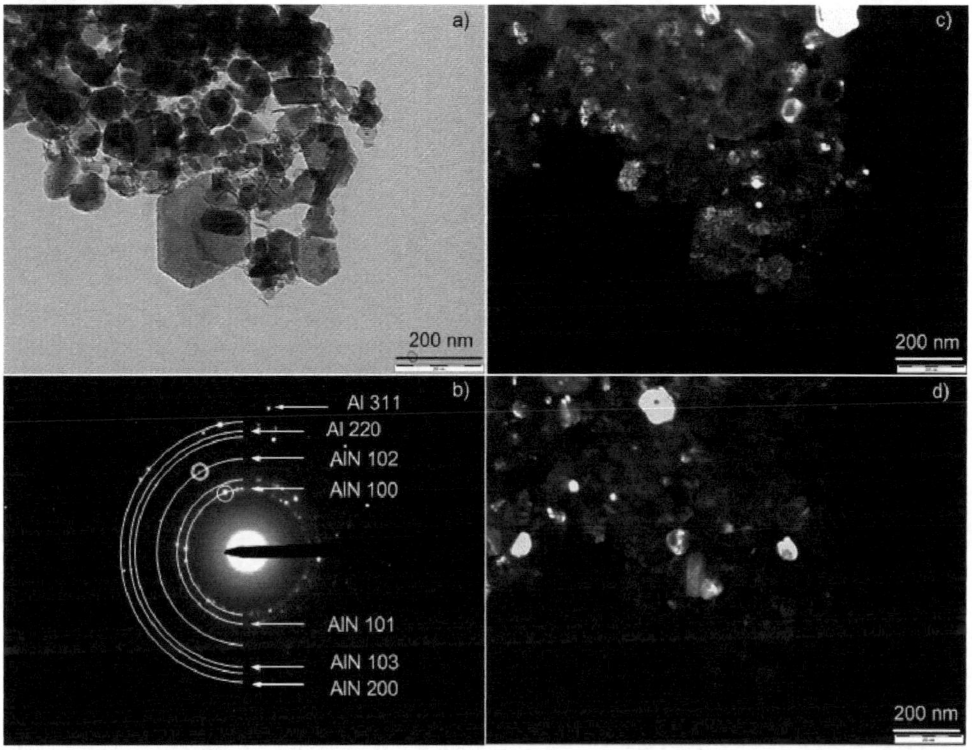

Figure 3. TEM-images of the product synthesized using solid precursor: a) bright-field image; SAED; c) dark-field image in the light of AlN reflex in direction [100]; d) dark-field image in the light of AlN reflex in direction [102].

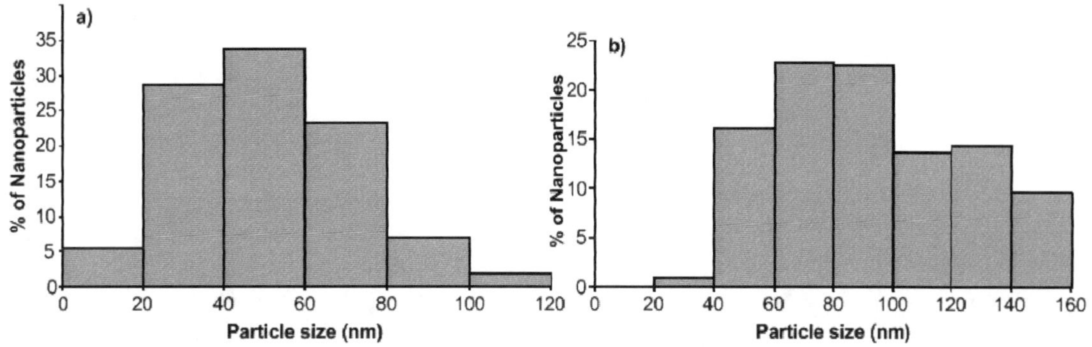

Figure 4. Particle size distribution histograms: a) experiment with gaseous precursor; b) experiment with solid precursor.

Conclusion

The synthesis of aluminum nitride can be realized using plasma dynamic method in the system based on coaxial magneto plasma accelerator. The influence of using solid and gaseous precursors on synthesized powders was investigated. According to XRD and TEM analyses the using of solid nitrogen-containing precursor (melamine) is preferable in terms of aluminum nitride content in the final product. But in this case the problem of impurities presence should be solved. Also the use of melamine influences on the particle size distribution aside increasing due to hydrogen presence. Thus the important task for further investigations in this system is to increase aluminum nitride content and decrease particle sizes.

Acknowledgements

This work was supported by the Russian Science Foundation (grant № 15-19-00049).

References

[1] Kim K, Choi S, Kim J, Cho W, Hwang K and Han K 2014 *Ceram. Int.* **40** 8117
[2] Slack G, Tanzilli R, Pohl R and Vandersande J 1987 *J. Phys. Chem. Solids* **48** 641
[3] Virkar A V, Jackson T B and Cutler R A 1989 *J. Am. Ceram. Soc.* **72** 2031
[4] Otake N, Liu L, Yasuhara T and Kato K 1998 *Jpn. Appl. Phys.* **37** 6128
[5] Oh S M and Park D W 1998 *Thin Solid Films* **316** 189
[6] Bian Y, Liu M, Ke G, Chen Y, Battista J, Chan E and Yang Y 2015 *Surf. Coat. Technol.* **267** 65
[7] Guo Q X, Yoshitugu M, Tanaka T, Nishio M and Ogawa H 2005 *Thin Solid Films* **483** 16
[8] Setter N, Damjanovic D, Eng L, Fox G, Gevorgian S, Hong S, Kingon A, Kohlstedt H, Park N and Stephenson G 2006 *J. Appl. Phys.* **100** 051606
[9] Abdallah B, Al-Khawaja S and Alkhawwam A 2011 *Appl. Surf. Sci.* **258** 419
[10] Bosund M, Sajavaara T, Laitinen M, Huhtio T, Putkonen M, Airaksinen V-M and Lipsanen H 2011 *Appl. Surf. Sci.* **257** 7827
[11] Cibert C, Dutheil P, Champeaux C, Masson O, Trolliard G, Tetard F and Catherinot A, 2010 *Appl. Phys. Lett.* **97** 251906
[12] Schneider M, Bittner A and Schmid U 2015 *Sens. Actuators A Phys.* **224** 177
[13] Lee S H, Yi J H, Kim J H, Ko Y N, Hong Y J and Kang Y C 2011 *Ceram. Int.* **37** 1967
[14] Yamakawa T, Tatami J, Wakihara T, Komeya K and Meg T 2006 *J. Am. Ceram. Soc.* **89** 171
[15] Kanhe N S, Nawale A B, Gawade R L, Puranik V G, Bhoraskar S V, Das A K and Mathe V L 2012 *J. Cryst. Growth* **339** 36
[16] Liu Z-j, Dai L-y, Yang D-z, Wang S, Zhang B-j, Wang W-c and Cheng T-h 2015 *Mater. Res. Bull.* **61** 152
[17] Pak A, Sivkov A, Shanenkov I, Rahmatullin I, Shatrova K 2015 *Int. J. Refract. Met. Hard Mater.* **48** 51
[18] Shanenkov I I, Pak A Ya, Sivkov A A, Shanenkova Yu L 2014 *MATEC Web of Conferences* **19** 01030
[19] Sivkov A, Saygash A, Kolganova J, Shanenkov I 2014 *IOP Conf. Series Mater. Sci. Eng.* **66** 012048
[20] Shanenkov I I, Sivkov A A, Pak A, Kolganova Y L 2014 *Adv. Mater. Res.* **1040** 813

Application of planetary ball mill for manufacturing of shielding composite coatings based on polyester powder paints and carbon fillers

S V Panin[1,2], S Yu Yazykov[1], V I Suslyaev[3] and K V Dorozhkin[3]
[1]National Research Tomsk Polytechnic University, 30, Lenin Avenue, Tomsk, 634050, Russia
[2]Institute of Strength Physics and Materials Sciences of Siberian Branch of Russian Academy of Sciences, 2/4, Akademicheskii avenue, Tomsk, 634055, Russia
[3]National Research Tomsk State University, Lenin Avenue, 36. Tomsk, 634050 Russia

E-mail: svp@ispms.tsc.ru, E-mail: susl@mail.tsu.ru.

Abstract. Blend polyester powder paint and particulate carbon filler in the form of colloidal graphite and the carbon black product were investigated. Powder paint and carbon materials were treated together in a planetary ball mill. The data of the structural analysis and transmission coefficients of electromagnetic waves of terahertz range through the flat shape samples are presented.

1. Introduction

An important condition for the development of radio engineering and electronics is development and study of properties of new composite materials [1]. Due to their combined properties the composite materials based on the metal or carbon-based particles in the polymer matrix are promising for application to problems of electromagnetic compatibility, interference immunity, radio masking and protection of biological objects. Thereby, design materials that absorb or reflect electromagnetic waves (EMW) in the gigahertz and terahertz frequency range it is actual problem [2]. Using of carbon-containing structures for shielding devices by reducing the background level of EMW to safe values was described in the literature [3, 4]. The materials, that actively interact with the EMW (reflecting and absorbing) [5], are used to prevent their harmful effect on the operation of electronic systems, as well as to reduce the radar visibility of aircraft, ships and technological equipment. Using of powder coatings is an effective way for corrosion protection of products. Also, adding the particulate fillers can impart to composites additional functional properties, in particular, the possibility of EMW shielding. The aim of this study is to investigate the ability of electromagnetic waves absorption and reflection of composite coatings based on polyester powder paint and dispersed carbon fillers in the form of graphite and carbon black. [6]

2. Materials and research methods

Flat samples of Mg alloy with a composite coating were used in the studies. Polyester powder paint “Okhteck-1” produced by JSC "Technos-Ohtek", Russia (with mean particle size of 71-50 μm, bulk density of 0.65 g/cm^3) having high corrosion resistance, physical and mechanical characteristics was chosen as the polymer binder. Carbon black "P-267E" (with average particle size of 70 μm, bulk density – 0.23 g/cm^3, specific volume resistivity at the density of 0.5 g/cm^3 – less than 0.002 Ohm/meter) and the colloidal- graphite powder "C-1" (hereinafter CGP) (with average particle size – of 4 μm, bulk density – 0.14 g/cm^3, the specific volume resistivity at a density of 0.5 g/cm^3 – 1.59 ± 0.06 Ohm/meter) (Fig. 1) were used as fillers. The objects of the study were a coating film formed by spraying powder composition onto substrate made of a fluoroplastic with their subsequent separation.

Powder mixture was prepared in a planetary ball mill MF 4/ 0.5 ("Tekhnotsentr" ltd., Rybinsk) at a total frequency of metal bowls rotation 800 rev./min. The weight ratio of the bearing steel balls to the powder made 40:1. Carbon fillers and powder paint were placed in the required quantities in the planetary ball mill where they were subjected to mechanical treatment (MT) during 40 min [7]. The effect of the MT time on the mechanical properties of the powder mixture was evaluated by the data of bulk density measurement. Vibrating sieve with a net size of 160 μm for dispersing large agglomerates of the powder mixture particles to form during the MT processing was used. The fraction of mixture with dispersity less than 50 μm were separated by sieve and then used for electrostatic spraying. Thus, the filler particles in the composition to be sprayed also had the same level of dispersity.

Figure 1. SEM-micrographs of carbon black particles P-267E (*a, b, c*) and colloidal- graphite powder C-1 (d, *e*).

Prepared powder compositions were deposited on the magnesium alloy substrate with a chromium oxide sublayer by electrostatical method using the "Start-50" gun. Further, fluoroplastic (PTFE) plates were used as substrates for deposition coating films and subsequent measurement of reflection and transmission coefficients, as well as permittivity values. During the following sintering at 180 °C for 15 minutes coatings with the thickness of 50 ... 80 μm were formed, and then coatings were delaminated from polytetrafluorethylene (PTFE) substrates.

3. Research methods

The coatings structure was investigated by scanning electron microscope Quanta 200 3D (FEI, USA) in Tomsk State University (Center of shared use of high-tech equipment). Crosscut split of the coatings was made to analyze the morphology. Corrosion tests were carried out in climatic chambers by the UHL 2 regime (GOST 9.401-91, technique No. 13).

Measurements of reflection and transmission coefficients from a flat material layer and spectra of the complex permittivity in the frequency range of 118 ... 258 GHz carried out by the "free space" method in the terahertz spectrometer STD-21 {C1 - C6} (CCU "Center of radio measurements" of TSU) [8, 9]. Terahertz frequency range was selected to study the electromagnetic response in order to determine the possibility of the developed material to be employed as reflecting or absorbing of electromagnetic waves.

Non-contact method were used to measure spectra of the absolute values of the modules and phase components of electrodynamic response: complex reflection coefficient – $R^*(f) = R(f)e^{i\psi(f)}$, and transmission coefficient – $Tr^*(f) = Tr(f)e^{i\varphi(f)}$ from the samples in real time. Operating range was 0.03 – 1.5 THz. Values of the complex permittivity of the material $\varepsilon^* = \varepsilon' - i\varepsilon''$ were calculated using measured modulus and phases of the transmitted and reflected signals. The value ε^* characterizes the formed composite structure, and allows physical mechanisms that determine the electrodynamic properties to be define.

4. Results of the research

The aim of combined treatment of the powder paint and carbon fillers in the planetary ball mill is solving the problem of adding maximum amount of functionalizing filler while ensuring satisfactory mechanical properties of the coating (especially, adhesion). The content of filler particles in the mixture was 12; 15; 17; 20 wt. %; time of the treatment was 40 min [10]. After the MT, change of the bulk density of the powder mixtures was evaluated (Fig. 2). The bulk density has changed from 0.35 to 0.5 g/cm^{3}depending on the filler content, because the bulk density of carbon fillers is smaller than bulk density of the powder paint.

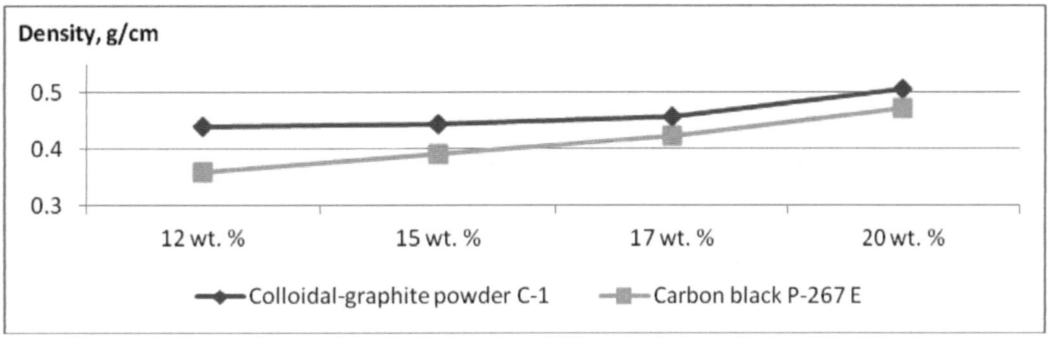

Figure 2. Dependence of the bulk density of "PowderPaint+ CarbonBlack" mixture.

After adding colloidal-graphite powder (CGP) and CB in the powder paint the values of bulk density do not substantially differ. Thus, it is not possible to make far-reaching conclusions by the parameter of bulk density from a point of view of finding the optimal amount of filler to be added into the powder paint. Data of our studies are presented in Fig. 3. The MT is effects on formation of particle fraction with a size more than 100 μm which were absent in the initial components of the mixture. The reason for their occurrence related to agglomerating the paint particles and filler during the MT. An interesting effect was identified after powder mixtures processing. The amount of the fraction suitable for electrostatic spraying is about 65 % for the carbon black being used as the filler and takes about 78% for CGP.

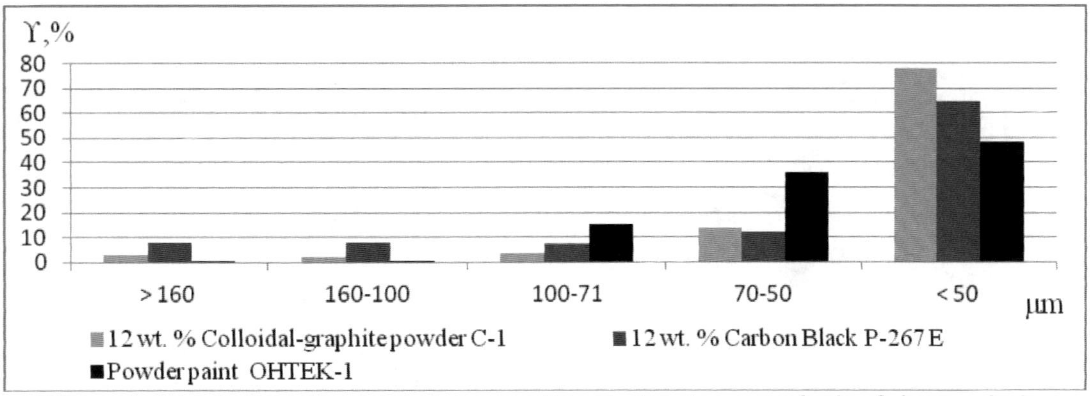

Figure 3. Distribution of powder mixture particles by the size on the basis of sieve analysis.

5. Coating structure and properties

The powder fraction with dispersion less than 50 μm was separated from the powder mixtures processed in the PBM. Then separated fraction was sprayed onto magnesium and fluoroplastic substrates. The degree of functional properties of the coatings was evaluated by measuring the specific volume electric resistance (SVER) (Fig. 5). Analysis of the obtained results shows that SVER of the CGP filled coating is by 15 times less than that of the coating containing carbon black. Most likely, this is due to the fact that the particle size of the CGP is less than 5 μm that is several times smaller than that of carbon black particles (< 70 μm). That is why the content of the particles in the powder fraction used for the spraying (with size less than 50 μm) is greater. This conclusion is also supported by the data presented in Fig. 3. The volume resistivity of CGP is less than volume resistivity of the carbon black powder. That also must affect the SVER of the coatings.

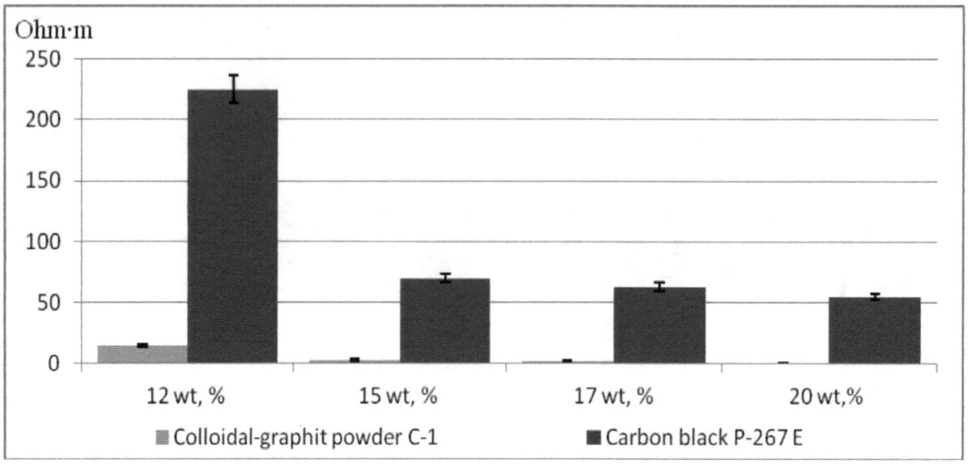

Figure 4. Volume resistance of coatings with various fillers.

Structural investigations carried out for the sprayed coatings with SEM showed that they are substantially non-porous and homogeneous even at the adding 15 wt. % CGP. However, this coating becomes brittle (Fig. 6), so it cannot be used. Adhesion of the coating to the substrate made of magnesium alloy before and after the environmental tests for specimens with filler content of 12 wt. % was performed. The adhesion in relative units is equal to one grade (Fig. 6, a, c).

Climatic tests for the coatings made of the powder mixture with 12 wt. % of the fillers and preliminary treated in the PBM for 40 minutes shown that at coating thickness not less than 70 μm no signs of corrosion are evident. Further increasing of the filler content up to 17 wt. % gave rise to presence of signs of corrosion which is unacceptable from the point of view of operational requirements.

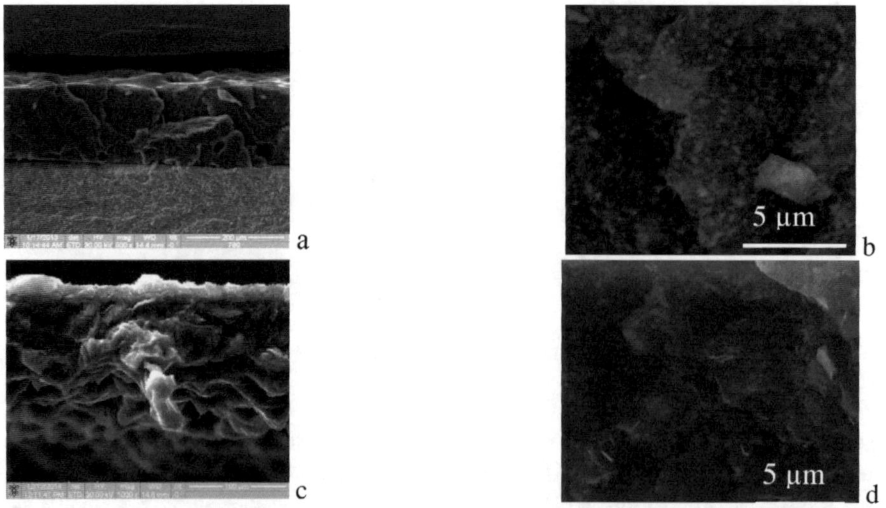

Figure 5. SEM-micrographs of a coating fracture surface at different magnification:
a, b) 12 wt. % carbon black; *c, d*) 15 wt % CGP.

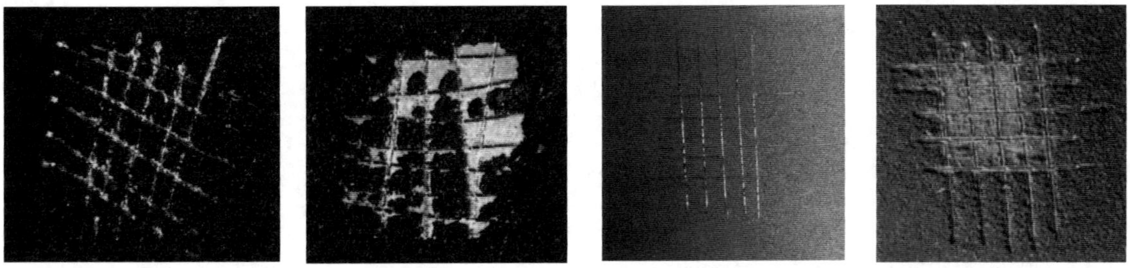

a b c d

Figure 6. Optical micrographs of the coating sprayed onto the substrate of Mg alloy MA-2-1: *a* – 12 wt. % carbon black; *b* – 15 wt. % carbon black; *c* – 12 wt. % CGP; *d* – 15 wt. % CGP.

Spectra of reflection and transmission coefficients and complex permittivity values of three flat shape specimens are in Fig. 7, whose parameters are given in Table 1.

Table 1. Complex permeability values.

Specimen No.	Coating composition	Size of filler particles	Filler content	Coating thickness, mm
1	Powder Paint (PP) OHTEK-1	–	–	0.38
2	PP + BK P-267E	< 71 μm	12 wt. %	0.39
3	PP + BK. P-267E	< 71 μm	12 wt. %	1.33
4	ПК + КГП	< 5 μm	12 wt. %	0.11

The electromagnetic waves is almost completely passed through the paint layer being partially reflected from a smooth surface due to the difference of the electromagnetic characteristics of the sample and the surrounding space (Fig. 7, curve 1). The adding of carbon black powder changes material properties: 45 ... 75 % of the incident power is reflected and partly absorbed by the sample bulk (Fig. 7, curve 2). Increasing the thickness of the layer altered behavior of the spectra of absorbtion and reduce the value of the real component. Particles become more separated by insulating interlayers, than in the thinner layer of the composite. As a consequence, absorptive capacity of the material has increased (Fig. 7, curve 3). This is probably related to the method of coating spraying.

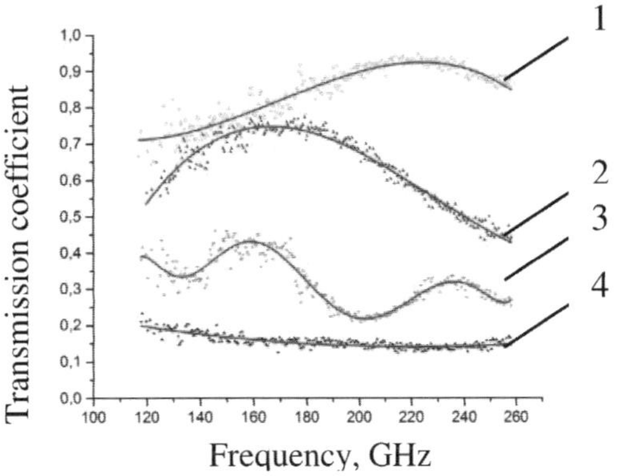

Figure. 7. Electromagnetic response of the “Okhtech” powder paint 1 - specimen № 1, table 1-1; composite «Okhtech + BK P-267E» at different thickness (specimens №2, 3, table 1-2, 3); composite «Okhtech + CGP” (specimen 4 , table 1-4).

Adding of colloidal graphite powder significantly alters properties of the material: 80 % of the incedent power is absorbed by the sample bulk from (Fig. 7). Thus, it is shown that adding of carbon black and CGP into the powder paint changed electrical and physical characteristics of the material. The specimen composed of powder paint with the adding of colloidal graphite with their combined treatment in the PBM can effectively absorb electromagnetic radiation of the terahertz range.

6. Conclusions

1. The results of complex testing simulating various atmospheric and mechanical influences have shown that the developed coating with the content of the carbon black of 12.5 wt. % and the MT time of 40 minutes ensures the formation of the coating film with thickness of 70 μm which exhibits satisfactory functional and protective characteristics.

2. The offered method of forming coatings on parts made of magnesium alloys, combining adding of electrically conductive filler, the treatment of the powder mixture in the planetary ball mill, and electrostatic spraying can be used to remove static electricity from the parts of spacecraft equipment operating in open space.

Acknowledgements

This study was performed in the frame of projects of fundamental research of National Academy of Sciences (2013-2020), and supported in part by grant RFBR number 13-08-98068_r_sibir_a.

References

[1] Villar G 2004 *Maschinenmarkt.* **26**
[2] Kuznetsov V L, Moseenkov S I and Ishchenko A V 2008 *Phys. Stat. Sol.* **2** 2051
[3] Nikolaichuk G A, Ivanov V P and Yakovlev S V 2010 *Elektronika: Nauka, Technologiya, Biznes* **1** 95
[4] Kuleshov G E, Dotsenko O A and Kochetkova O A 2012 *Polzunovskii vestnik.* **2**. 163
[5] Lutsev L V 2008 *Nanotechnika* B **14** 37
[6] Malinovskaya T D, Suslyaev V I, Melentiev S V and Dorozhkin K V 2014 *Izv.vuzov. Fizika* B. **57** 80
[7] Yazykov S Yu, Dammer V Kh, Panin S V and Ovechkin B B 2014 Izvestiya TPU B **325** 105
[8] Suslyaev V I, Kuznetsov V L and Zhuravlev V A 2013 *Russian Physics Journal.* **8** 970
[9] Suslyaev V I, Zhuravlev V A, Dotsenko O A and Babinovich A N 2008 *Russian Physics Journal* **8** 874
[10] Panin S V, Kornienko L A and Poltaranin M A 2014 *Advanced Materials Research* **872** 36

IOP Conf. Series: Materials Science and Engineering **93** (2015) 012053 doi:10.1088/1757-899X/93/1/012053

Statistical control of DOF maintenance condition

T Aleksandrova and E Gromakov
National Research Tomsk Polytechnic University (TPU), 30 Lenin Avenue, Tomsk, Russia, 634050

E-mail: tata09@mail.ru

Abstract. This work aims to develop and rationale statistical and control tools set to ensure correct interpretation of measurements and data received from cluster well in the automated mode. Robust procedure is offered for control chart interpretation that allows maximizing the determination of real faults and minimizing false alarms counts.

1. Introduction

DOF, Digital Oil Field is an automatic control system for oil and gas extraction that ensures the optimization of field maintenance using the information about its state collected from the continuously refined supercomputer geological and hydrodynamic models and equipment condition data [1]. The DOF term is based on the concept of intelligent control that is why digital oil field also referred to as intelligent.

Known realizations of that technology are directed onto automated control of unmanned energy resources extraction with decision-making center of maintenance based on objective data about field.

The aim of this work is to develop and study the composition of the statistical processing and control tools to ensure correct interpretation of measurements and data received in the automated mode from cluster well.

2. Shewhart control charts

Three methods can be used for statistical processing. The first one is based on Neyman-Pearson criterion and represented as Shewhart control chart [2]. It is historically the very first common tool to control the variable parameters of maintenance processes. The second one is based on the repeated application of Wald sequential analysis results [3]. It puts into practice in the form of control charts of cumulative sums. Finally, third approach of violations detection is based on exponential smoothing of statistical results (EWMA charts). Hereinafter the features of the use of the first two methods are considered.

Control chart in DOF as in Production Quality Control is a chart of regular measurements samples of some process indicator ***X***. As ***X*** can be used the output parameter of the energy resources extraction process, such as oil extraction and/or parameters characterizing the state of the equipment (for example, the temperature of the bearing assembly, vibration and etc.). The statistical data collected by telemetry can be combined in the form of samples diagrams on the map ***g(X)***.

Sample on this diagram is the sequence of n independent observations $x_1, x_2, \dots, x_n$ of indicator ***X***, related to some time period (day, month, year).

Suppose that $f(x,\theta)$ describes the distribution of random variable x at a certain value of statistical parameter (average value of sample, spread and others) θ.

Using the control chart the problem of detecting unnatural variability of observed parameter that may be a change of field operating mode. This variability is observed on the basis of changes of selected for chart parameter of distribution function θ of sequential observations.

The features of the energy resources extraction process and state of its technical equipment are the slow changes of variability in operating parameters trends.

We assume that the statistical properties of the time series that characterizes these states and the properties of the reasons that generates its changes remain unchanged or change slowly in the selected intervals of samples formation. Current control is reduced to monitoring of the regular sample observations and detecting of the observed time series θ properties changing from θ_0 to θ_1. The need to change the operational mode of the field may occur at unknown time t_0. We call this event as disorder of the operational process. Useful problems that are solved by the automated system of the DOF are the detection of the disorder at an early stage of its development, the timely control of the operating parameters and the prediction of the operational state of the field.

The generalized statement of the considered range of the practical problems related to the operational state of the field is characterized by random sequence $\{x_1, x_2, \dots, x_n\} = \{x_1^n\}$, that at the time moment t_0 changes its own properties that are uniquely determined by vector of parameters θ, $dim\{\theta\}$= r. Starting from the time t_0 vector of parameters becomes $\theta = \theta_1$. Moment of disorder t_0 is revealed by set criterion for the evaluation of sequence $\{x_1^n\}$, $N \to \infty$, at appearance of the next point x_n.

Each time for each new sample the hypothesis $\theta = \theta_0$, pointing to the absence of the unnatural variability is tested. We call this hypothesis as zero hypothesis and sign it H_0. Competing hypothesis H_1 suggests an unnatural change of the parameter $\theta = \theta_1$. The errors may occur when hypothesis is adopting.

Neyman and Pearson in their work indicated that the checking of the alternative hypotheses may reveal the errors of two kinds which are highly undesirable for making the decision about changing of the operational mode of the field. The error of the first kind is the declination of the hypothesis H_0 (no disorder) while it is true. The error of the second kind is the acceptance of the hypothesisH_0while the competing hypothesis is true H_1 (disorder).

Error of the first kind α risk while the operational parameters are processed may occur when operational state of the field is in statistically controlled state but the monitored parameters are out of control limits at random.

As a result, the actions that are generated by non-existent cause are called as “excessive regulation”.

The error of α risk is equal to the probability of making a wrong decision based on the false alarm. Manager actions based on such kind of alarms lead to unnecessary costs.

Therefore, to statistical processing of operational information about the field from all options of the algorithms that detect the unnatural variability of the indicators that determine the current state and prospects of the field exploitation you need to choose the ones that provide the maximum probability of the correct detection of unnatural variability in case that the probability of the false alarm (α risk) not exceeds the predetermined value.

The error of second kind (β *risk*) occurs when the operational state of the field disorder and the control points do not indicate a lack of control. In this case the corrective action is not performed (lack of control). β *risk* is equal to probability of the hypothesis H_1 acceptance when it is false. Sometimes β *risk* referred to as the risk of real failure or the risk of missing signal.

It is necessary to minimize the probability of these errors through the statistical processing of such important information as the operational state of the field. To solve this problem we can us the criterions of the variability evaluation that are based on the Shewhart control charts.

Let us suppose that ***X*** is a normally distributed random variable with expectation μ and dispersion σ^2. Then the probability that deviation of random variable ***X*** from its expectation μ in absolute value less than the predetermined value $\varepsilon = \varkappa\sigma > 0$ can be calculated by the following formula:

$$P(|X-\mu|<\varkappa\sigma)=P\left(\left|\frac{X-\mu}{\sigma}\right|<\varkappa\right)=P\left(-\varkappa<\frac{X-\mu}{\sigma}<+\varkappa\right)=$$
$$=P\left(\frac{X-\mu}{\sigma}<\varkappa\right)-P\left(\frac{X-\mu}{\sigma}<-\varkappa\right).$$

The desired probability is

$$P(|X-\mu|<\varkappa\sigma)=\Phi(\varkappa)-\Phi(-\varkappa)==2\Phi(\varkappa)-1 \qquad (1)$$

This probability depends only on $\varkappa$. Values of $\Phi(\varkappa)$ for different values of $\varkappa$ can be found with help of special tables [4]. Particularly, using formula (1) we can calculate

for $\varkappa=1$: $P(|X-\mu|<\sigma)=2\Phi(1)-1=2*0{,}84135-1=0{,}6827$;

for $\varkappa=2$: $P(|X-\mu|<2\sigma)=2\Phi(2)-1=2*0{,}97725-1=0{,}9545$;

for $\varkappa=3$: $P(|X-\mu|<\sigma)=2\Phi(3)-1=2*0{,}99865-1=0{,}9973$;

that is 99.73% quality characteristic data will fall in interval $(\mu-3\sigma;\mu+3\sigma)$.

The probability of the results falling out of the interval $(\mu-3\sigma;\mu+3\sigma)$ for statistically stable process is equal to 0.0027 and this value points to the promising application of the Shewhart control charts for evaluation of the operational parameters of the field.

3. ARL for Shewhart control charts

The probability of the operation disturbance recognition is called as sensibility of the control chart. In analysis of the control chart sensibility the average run length (ARL) of the controlled parameters measurement samples.

The mean length of the ARL series of the control chart is a calculated (expected) value of the samples that is necessary to process for the detection of the shift of process mean value. Usually ARL are calculated separately for the zero shift of mean controlled parameter (in-control ARL_0) and non-zero shift (out-of-control ARL_1).

Appropriate metrics of ARL for α error are defined as:

$$ARL_0=\frac{1}{\alpha},$$

and for β error ($\mu\neq\mu_0$) as:

$$ARL_1=\frac{1}{1-\beta}.$$

If there is no real change of process then ARL_0 value must be high.

Since every sample can be associated with time period then the length of ARL_0 may be considered as the prediction depth of the statistically controlled operational state of the field.

In case of statistically uncontrolled exploitation parameter (if the shift is significant) then it is necessary for ARL_1 to have a low value for the fast detection of the process state change.

In other words, when the real state of the process is unsatisfactory the decision rule should give the earliest possible alarm about process change. There can be used different combinations of criteria for determining the absence of control [5].

A significant disadvantage of Shewhart control charts is that they do not allow you to quickly find disorder of the process with a slight shift of the characteristic parameter.

Feature of the methodology of construction of Shewhart control charts is that they give the evaluation of the operational state on the basis of some accumulated array of samples.

4. CUSUM charts

Since Shewhart control charts only take into account the error of the first kind [2] then to improve the reliability of the disorder variability evaluation in addition to control chart we propose to use the Wald criterion [3] that was developed by Page in [6] in the form of sequential probability ratio test for two simple hypotheses:

H_0 (no disorder): $\theta = \theta_1$ and
H_1 (disorder): $\theta = \theta_2$,
where θ is an parameter of the probability distribution function $\omega(x_t/\theta)$ of the observation x_t.

To evaluate the hypotheses on each step of control the probability ratio is calculated as follows

$$\omega(x_t/\theta_2)/\omega(x_t/\theta_1).$$

If

$$B < \frac{\omega(x_t/\theta_2)}{\omega(x_t/\theta_1)} < A,$$

then there is no change of exploitation and we continue the observation of indicator ***X***.

If

$$\frac{\omega(x_t/\theta_2)}{\omega(x_t/\theta_1)} \geq A,$$

then the hypothesis H_0 is declined and H_1 is accepted.

If

$$\frac{\omega(x_t/\theta_2)}{\omega(x_t/\theta_1)} \leq B,$$

then H_0 is accepted.

Constants ***A*** and ***B*** are determined in [8] as

$$A = \frac{1-\beta}{\alpha};$$

$$B = \frac{\beta}{1-\alpha},$$

where α is the probability of the error of first kind; β is the probability of the error of second kind.

In practice, it is easier to calculate the logarithm of the ratio $\omega(x_t/\theta_2)/\omega(x_t/\theta_1)$.

Let the write it as $z_t = \ln\frac{f(x_t/\theta_2)}{f(x_t/\theta_1)}$.

The variable z_t is called as the cumulative sum (*CUSUM*). Through the statistical regulation the values of the cumulative sum z_t accumulates sequentially while adding of samples.

If $\ln B < z_t < \ln A$, then monitoring continues and indicator ***X*** is sequentially evaluated.

If $z_t \geq \ln A$, then the hypothesis H_0 is declined and H_1 is accepted.

If $z_t \leq \ln B$, then H_0 is accepted.

The evaluation of the operational state of the field on the basis of this algorithm is performed each time on the appearance of the sample and unlike the Shewhart control charts there is also performed the current control of the observed operational indicator.

The discussion of the used in construction of *CUSUM* mathematical principles can be found in paper of Woodall [7].

If there is a series of quality variable values $x_1, x_2, \dots, x_n$, then the construction of *CUSUM* would be as follows:

$$S_1 = (x_1 - k),$$
$$S_2 = (x_1 - k) + (x_2 - k) = S_1 + (x_2 - k),$$
$$\vdots$$
$$S_n = \sum_{i=1}^{n}(x_i - k),$$

where x_i is the value of the observed variable;
k is the constant that represents the standard value;
i is the number of the sample.

The calculated and plotted in order of appearance cumulative sums S_n form a *CUSUM* – chart.

The constant k can have any value but often k is equals to the expectation then the k is close to the nominal value of the process variable.

The decision rule then can be formulated as follows.

If the distance between the current value of the cumulative sum and the lowermost one of the preceding points on the graph is greater or equal than h then it is necessary to perform corrective actions because there is a shift of observed parameter mean value.

Thus, the parameters of *CUSUM*-chart are

n is the sample size;

k is the standard value;

h is the interval of the solution.

The shift of the observed parameter mean value may be expressed as $(\mu - k)$. In case of normal distribution of the parameter the shift can be standardized as follows

$$k_{\text{CT}} = \frac{(\mu - k)}{\sigma/\sqrt{n}} = \frac{(\mu - k)\sqrt{n}}{\sigma}.$$

Similarly, the interval of the solution

$$h_{\text{CT}} = \frac{h}{\sigma/\sqrt{n}} = \frac{h\sqrt{n}}{\sigma}.$$

Thus, if the mean value of the process parameter deviation from the nominal value increases then the *CUSUM* increases too. Similarly, if the mean of the process variable decreases then the graph tends downwards.

If we plot the graph of the cumulative sum of deviations from the nominal values for consecutive sample means then even small regular shifts of the mean value of the parameter (minor disorder of operating parameter) would gradually lead to the accumulation of a significant sum of deviations.

In order to establish the control limits in *CUSUM*-charts it was proposed to use the procedure known as the *V*-mask that sequentially moves on the graph after the plotting of the point for the latest sample. We may assume that *V*-mask is an upper and lower control limits for the cumulative sums. However, instead of being parallel to the center line these straight lines are converged at a predetermined angle to the right, thereby forming a shape like a lying letter *V*. Even if one point lies outside of the V then the process is suspected of being out of control.

This type of control charts is particularly well suited for detection of the small regular shifts of the operational parameter that may be overlooked in the application of the classical $\bar{x}$ Shewhart control chart. For example, when due to depreciation of equipment the process slowly “slips” out of control that results the violation of the regulated specifications. The usage of *CUSUM*-charts gives us the monotonically increasing (decreasing) chart of the cumulative sums of deviations from the planned specifications.

CUSUM-chart in comparison with Shewhart control charts reacts on small variability of the operational parameter faster and allows to quickly determine its origin.

As is shown in [7], ARL *CUSUM*-charts are more effective in comparison with ARL Shewhart control charts because they have high values for the processes with zero shift of the controlled parameter mean value and small values for the processes with significant shift of the process parameter from the standard [6].

Sequential state control of the field operational parameter using the both instruments allows us to decrease the number of errors in parameter variability interpretation that leads to increasing of statistical control robustness.

5. Conclusion

The above statistical control tools allow the DOF monitoring of operational state of the field with low level of decisions at least in three dimensions of real time:

1) in operative real-time of daily management of the field;
2) in retrospective archival real-time with depth of established by regulatory documents;
3) in the forecast time that is defined by the strategic objectives of the enterprise.

IOP Conf. Series: Materials Science and Engineering **93** (2015) 012053 doi:10.1088/1757-899X/93/1/012053

References

[1] Lo C 2013 *Making the most of the digital oil field Offshore Technology*

[2] State Standard of Russian Federation P 50779.42 - 99 (ISO 8258 - 91). *Statistical methods for Shewhart control charts. Intro.* 1999-04-15, 1994. Moscow: Standard Publishing House, pp: 99

[3] Wald A 2004 *Sequential analysis* (Mineola, Courier Corporation)

[4] Storm R 2001 *Wahrscheinlichkeitsrechnung, mathematische Statistik und statistische Qualitätskontrolle* (Leipzig, Fachbuchverlag Leipzig)

[5] Wheeler D and Chambers D 1992 *Understanding Statistical Process Control* (Knoxville, SPC Press)

[6] Page E 1954 *Continious Inspection Schemes* (London, Biometrika)

[7] W. H. Woodall, The distribution of the run length of one-sided CUSUM procedures for continuous random variables, *Technometrics* 25 (1983), no. **3**, 295–301

[8] Murdoch J 1979 *Control Charts* (London, Palgrave Macmillan)

IOP Conf. Series: Materials Science and Engineering **93** (2015) 012054 doi:10.1088/1757-899X/93/1/012054

Evaluation of social network user sentiments based on fuzzy sets

E E Luneva[1], P I Banokin[1] and A A Yefremov[1]
[1]Tomsk, 634050, Russia, National Research Tomsk Polytechnic University

E-mail: lee@tpu.ru

Abstract. The article introduces social network user sentiment evaluation with proposed technique based on fuzzy sets. The advantage of proposed technique consists in ability to take into account user's influence as well as the fact that a user could be an author of several messages. Results presented in this paper can be used in mechanical engineering to analyze reviews on products as well as in robotics for developing user communication interface. The paper contains experimental data and shows the steps of sentiment value calculation of resulting messages on a certain topic. Application of proposed technique is demonstrated on experimental data from Twitter social network.

1. Introduction

Today it has become common practice for people to use social networks to express opinions on topics of interest or to write reviews on used products and services. This data allows solving such actual tasks as the identification of potential clients for a particular product or service, forecasting demands and customer expectations for a mechanical engineering product, as well as in robotics for developing user communication interface and public opinion evaluation [1].

Analysis of related scientific papers on the subject of automated assessment of social network user sentiments [2-6] reveals that techniques proposed there do not take into account the relationship between users and limitations associated with the processing characteristics of linguistic hedges frequently occurred in messages. For example, when a user is the author of several messages, according to [2-5] the assessment result depends on a total number of messages, regardless of how many users leave them.

In [7] we represented a method of sentiment analysis that takes into account users' influence within a social network as well as the effect of repetetive messages. The accuracy of the results that the method can provide depends on accuracy of message parsing. The object of this paper is to represent the algorithm of message parsing as well as the application of the method to experimental data from Twitter social network.

2. Theoretical analysis

In order to evaluate social network user's sentiments it's necessary to consider user's relevant messages on certain topic. Considering a sentence as a unit of the message, we assume that each sentence contains positive, negative or neutral sentiments regarding certain subject (product) characteristics K_i from predefined set of characteristics $K=\{K_i\}$, $i=[1...n]$, where n - the number of characteristics. Estimation of a product characteristic implies attributing it with one or several elements of a set of terms $T_i=\{T_{i,j}\}$, $i=[1...n]$, $j=[1...m_i]$, where m_i is the number of terms related to the

i-th characteristic. In order to assess a user's sentiments it is necessary to obtain the sentiment degree for each term $T_{i,j}$, which can be represented as a membership functions of the linguistic variable. In order to generate membership functions of terms automatically, SentiWordNet 3.0 [8] program component was proposed [7].

In order to define membership functions of terms we introduce the universal set U = {*P ("Positive"), O ("Objective"), N ("Negative")*}. Assuming a term could not be positive, negative and objective simultaneously with a membership equals 1, we suggest a normalizing rule $\mu_P + \mu_O + \mu_N = 1$, where μ_P, μ_O, μ_N are values of a membership function for elements *P, O, N* of the fuzzy set "Sentiment degree". Thus, each term can be defined by its membership function on *U*.

When a term is used with linguistic modifier, Zadeh's operations on fuzzy sets (concentration, dilation) should be used [9, 10]. For instance, according to Zadeh, if a term is used with modifier "very" its membership function is obtained by membership function of atomic term raised to the square. However, since $0 <= \mu_k <= 1$, $\mu_k^2 <= \mu_k$, which conflicts with the idea that a term with "very" should have greater maximum membership value than atomic term. Similar reasoning is applied for terms with "somewhat" modifier.

We propose to normalize results of Zadeh's operations according to the following:

$$\mu^*_{\otimes,k} = \frac{\mu_{\otimes,k}}{\sum_{j\in U} \mu_{\otimes,j}},$$

where $\mu_{\otimes,k}$ is a membership value of element k ($k \in U$) to a fuzzy set of a term with a modifier, $\mu^*_{\otimes,k}$ is a normalized membership value; $\otimes$ signifies either concentration or dilation according to Zadeh. Figure 1 depicts membership functions for terms "good", "very good" without normalization and "very good" with normalization.

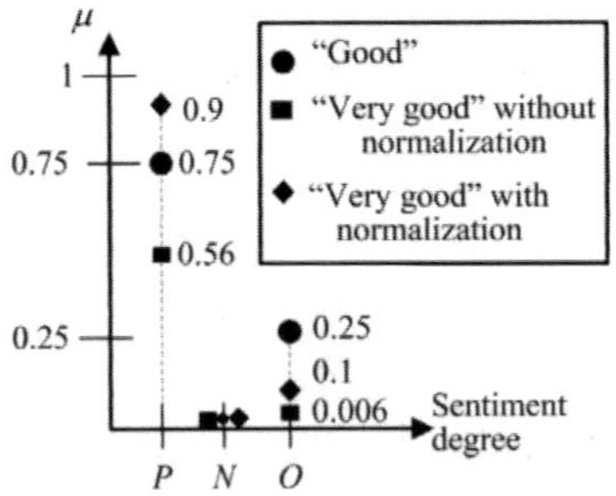

Figure 1. Membership functions of terms "good", "very good" without normalization and "very good" with normalization.

Assuming each message contains a set $L \subseteq \bigcup_{i=1}^{n} T_i$ of terms (both positive and negative), total sentiment score of a message could be calculated as follows:

$$s = \frac{1}{|L|} \cdot \sum_{t_{i,j}\in L} t_{i,j} \cdot (-1)^{\alpha}, \quad (1)$$

where $\alpha = 1$, if a term is positive, $\alpha = -1$, if a term is negative, $t_{i,j}$ are defuzzified (crisp) values of fuzzy terms $T_{i,j} \in L$, and $|L|$ is a cardinality of *L*. The value of *S* falls within the range [-1, 1]. Also, we can use Eq. 1 to evaluate a sentiment degree of a certain characteristic K_i.

Splitting messages into groups according to their sentiment degree requires defining a linguistic variable "Sentiment characteristic" (SC) with term set

Z = {Very Positive (VP), Positive (P), Neutral (NT), Negative (N), Very Negative (VN)}.

Each term is represented by the fuzzy set defined on the range [-1, 1]. Assuming no element of the range [-1, 1] is attributed to different terms from Z with maximal membership value, we introduce a rule concerning membership functions of the terms:

$$\forall x \in [-1 \ldots 1] \ \sum_{z \in Z} \mu_z(x) = 1.$$

With that in mind, the membership functions of fuzzy sets, representing the terms of linguistic variable “Sentiment characteristic”, should be similar to those depicted in figure 2.

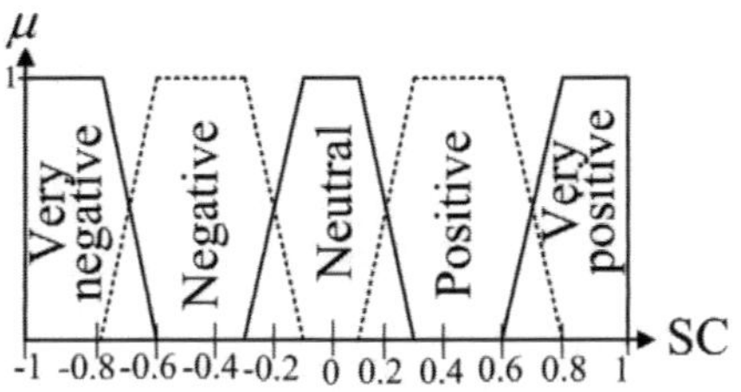

Figure 2. Membership functions of terms of linguistic variable “Sentiment characteristic”.

Total sentiment score of a user’s messages on the subject is evaluated as follows:

$$ss_j = \frac{1}{|S_j|} \cdot \sum_{s_j \in S_j} s_j,$$

where $S_j = \{s_j\}$ is a set of sentiment scores of all messages attributed to j-th user.

Another aspect that should be taken into account is user influence degree. As a measure of this characteristic we propose evaluating closeness centrality metric according to [11]. Considering social network users as experts on a certain subject [12] we could calculate the percentage of user whose assessments contain statements complying with term z of linguistic variable:

$$\beta_z = \frac{\sum_{i=1}^{q} R_i \cdot \mu_z(ss_i)}{\sum_{i=1}^{u} R_i}, \quad (2)$$

where R_i is i-th user influence degree, u – number of users, q – number of users commenting on the subject with attributives corresponded to z.

One of the most difficult tasks in evaluation of social network user sentiments is text processing. Text processing comprises two steps: splitting it into a set of sentences and separate parsing of each sentence. The result of sentence parsing is a word dependencies tree. Word dependencies tree allows finding a set of connected words related to specific feature. Text splitting and sentence processing are implemented in Stanford NLP library [13] which includes a set of classes and training data. Stanford NLP library parses a sentence into a set of pairs. Each pair contains relation name, master and dependent words. Every sentence has a root relation that includes only master word. Usually root relation is a sentence subject. Additionally the library assigns part of speech to each word. Based on parsing result it is possible to create a words tree. Mainly relations between words could be considered separately. For emotion evaluation "advmod" and "amod" relations are significant. "advmod" relation includes adjective and adverbial. "amod" relation contains noun which could be feature name or its synonym and related adjective.

Negation relation should be processed in a different way because negation propagation over sentence parts significantly affects word meanings. During dependency tree building it is possible to assign each node negation value and detect double negations. Primarily double negations occur in complex sentences. Stanford NLP library doesn't trace negation propagation. In our model negation value is assigned during word dependencies tree creation. Negation propagates from master node to all dependant nodes. In the figure 3 an example of dependency tree is represented. Using such tree as input data for sentence emotional analysis it is possible to find characteristic and terms and obtain a sentiment value for the sentence. We propose perfoming sentence emotional analysis in the following way:

- Word tree containing negation values for each node is built.
- The tree is traversed for noun nodes.
- Every noun node is checked if it corresponds to analysed characteristics from K .
- In case of correspondence dependent adjectives and adverbials (terms $T_{i,j}$) are processed.

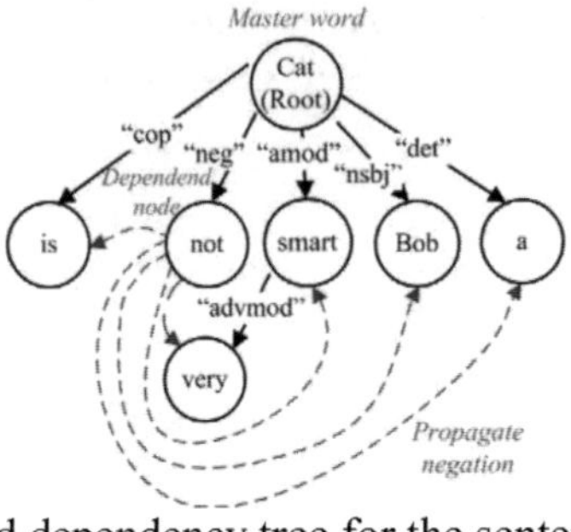

Figure 3. An example of word dependency tree for the sentence "Bob is not a very smart cat" "cop" - copula; "neg" – negation part; "amod" – adjective; "nsbj" – noun subject; "det" – determiner; "advmod" – adverb.

3. Experimental analysis

This section is devoted to experimental data analysis. Experiments based on the theoretical reasoning from the previous section were conducted. A software application was developed for this purposes. The application downloads a set of tweets related to specific topic and calculates SC. figure 4 shows component architecture of the software.

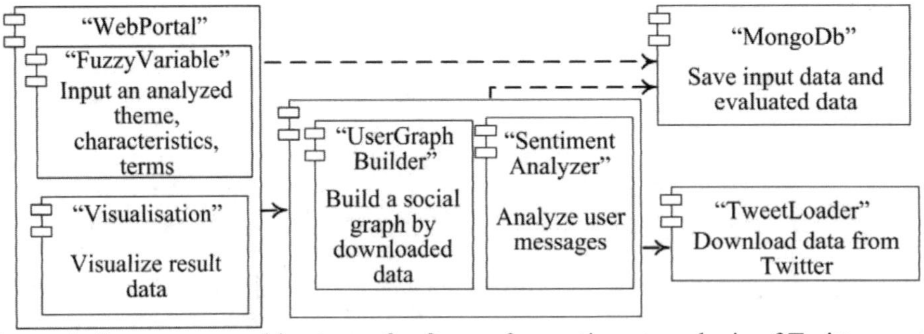

Figure 4. Component architecture of software for sentiment analysis of Twitter users.

Additionally, data about users' relations was analysed and social graphs were build. A social graph from a set of created graphs is shown on figure 5. If value of link weight is missing, it equials 1. Sentiment score of a message s could be calculated by a mock object or by automatic message content processing depending on component configuration (figire 4, SentimentAnalyzer component). The value generated by the mock object varies from -1 to 1. If a message is reposted, the original value of sentiment score is used. Table 1 presents input data containing messages from users 1-10.

Closeness centrality and *ss* metric were calculated for every user because of the fact that one user could be an author of several messages. Table 1 shows the data based on which the social graph was built (figure 5). Values of *ss* metric are used for evaluation of membership degree μ to fuzzy terms of linguistic variable SC. Table 1 uses the following denotation: $m_{i,j}$ – *j*-message of *i*-user; $rm_{i,j}$ – message retweet $m_{i,j}$.

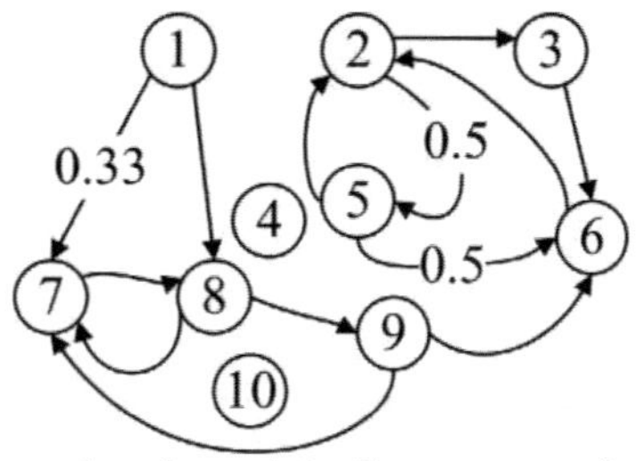

Figure 5. An example of graph built on a set of twitter messages.

Table 1. Calculated parameters for *u* users

u	Abbreviation of a message	s_j	ss_i	R_i	SC / μ
1	$m_{1,1}$	0.76	0.76	0.22	VP / 0.86 P / 0.14
2	$m_{2,1}$	0.53	0.53	0.13	P / 1
3	$rm_{2,1}$ $m_{3,1}$	0.67 0.53	0.6	0.13	P / 1
4	$m_{4,1}$	-0.34	-0.05	0.09	NT / 1
5	$m_{5,1}$ $rm_{3,1}$	0.24 0,53	0,39	0.13	NT / 0.16 P / 0.84
6	$m_{6,1}$ $m_{6,2}$	0.7 0,34	0.52	0.13	P / 1
	$rm_{8,2}$	0.78			
7	$m_{7,1}$	0.80	0.81	0.17	VP / 1
	$m_{7,2}$	0.84			
8	$m_{8,1}$ $m_{8,2}$	0.43 0.78	0.61	0.19	P / 0.98 VP /0.02
9	$m_{9,1}$	0.15	0.15	0.20	NT / 0.91 P/0.09
10	$m_{10,1}$	-0.47	-0.47	0.09	N / 1

Using data from Table 1, it is possible to calculate how users' sentiments are distributed among different SC. Figure 6 shows visual representation for data from Table 1. Table 1 uses the following denotation: $m_{i,j}$ – *j*-message of *i*-user; $rm_{i,j}$ – message retweet $m_{i,j}$.

Results depicted in figure 6 allows for the conclusion that users' sentiments on the analyzed topic vary significantly: from negative value to a highly positive. Nevertheless, it is possible to infer that more than a half of users have positive attitude on the analyzed issue.

Further research should focus on developing faster and more accurate algorithms [14] for semantic processing of sentences and user influence evaluation provided that data is aggregated from various social networks. Data from various social networks could be obtained via Web Observatory infrastructure while sentiment analysis tool could be listed in Web Observatory [15] analytics and visualizations tools catalog. Challenging tasks are social relationships detection, unified social graph development and users' profile matching across different networks. "Analyzer" component interface should be modified to expose its functionality via standardized and documented REST interface allowing developers to create their own visualization tools (figure 4).

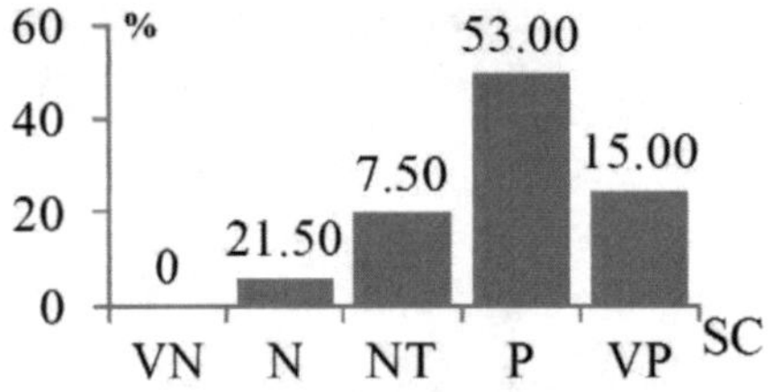

Figure 6. Users' sentiments distribution on different SC.

4. Conclusion

The algorithm of message parsing for social network user sentiments evaluation was described in the paper. Sentiment characteristic evaluation was performed on data obtained from Twitter. A series of experiments were conducted using a software application developed as a part of the research. Experimental results confirmed applicability of presented technique for analysis of users' emotions. Evaluation of social network user sentiments were obtained taking into account user influence within a social network as well as the effect of repetetive messages. Proposed technique could be applicable in the areas of potential customer targeting, demand prediction, evaluation of opinion ratings on a product or a service.

References

[1] Information on http://www.statisticbrain.com/
[2] Dalal1 Mita K and Zaveri Mukesh A 2014 *Applied Computational Intelligence and So. Computing Volume* **735942** 1–9.
[3] Bollen J, Mao H and Zeng X 2011 *Journal of Computational Science* **2** 1-8.
[4] Pimpalkar A, Wandhe T, Swati R M and Minal K 2014 *Smiley's International Journal of Computing and Technology* **1** 39-44.
[5] Haque M and Rahman T 2014 *International Journal of Computer Science, Engineering and Information Technology (IJCSEIT)* **4** 33-48.
[6] Information on http://sentiment.christopherpotts.net/lexicons.html
[7] Luneva E E, Banokin P I, Yefremov A A and Tiropanis T 2015 Method of Evaluation of Social Network User Sentiments Based on Fuzzy Logic *Preprint / HTRA 2015*
[8] Denecke K 2008 *Proc. International Conference on Data Engineering* **4498370** 507-512.
[9] Zadeh L A 1975 *Information Sciences* **8** 199-249.
[10] Van Leekwijck W and Kerre E E 1999 *Fuzzy Sets and Systems* **108** 159-178.
[11] Freeman L C 1979 *Soc. Networks* **1** 215-236.
[12] Rowe G and Wright G 2001 Expert Opinions in Forecasting. Role of the Delphi Technique, (Boston: Kluwer Academic Publishers).
[13] Information on http://nlp.stanford.edu/software/dependencies_manual.pdf
[14] Song Q and Wang X F 2012 *Journal of the University of Electronic Science and Technology of China* **41** 176-184.
[15] Tiropanis T, Hall W, Shadbolt N and De Roure, D, Contractor N 2013, Hendler J, The web science observatory, *IEEE Intelligent Systems* **28** 100-104

IOP Conf. Series: Materials Science and Engineering **93** (2015) 012055 doi:10.1088/1757-899X/93/1/012055

The interval-parametric synthesis of a linear controller of speed control system of a descent submersible vehicle

S A Gayvoronskiy[1], **T Ezangina**[1], **I Khozhaev**[1]
[1]Department of Automatics and Computer Systems, National Research Tomsk Polytechnic University, Lenin Avenue 30, Tomsk, Russia

E-mail: eza-tanya@yandex.ru

Abstract. The algorithm for the definition of the interval settings of the linear regulator ensuring its robust stability and admissible oscillation was developed. The algorithm is based and constructed on the sufficient conditions binding the interval coefficient of the characteristic polynomial of the system and its regulator parameters. The application of this algorithm for the definition of the interval coefficient of the transfer function was also considered in the given paper. Performance of the algorithm was tested by construction of localization regions of the roots of the interval polynomial upon the determined intervals of the controller parameters.

1. Introduction

One of the main objectives of the robust control theory is the analysis and synthesis of the linear control systems with unknown parameters which are located in certain ranges. The solution of the problem relating to the analysis of the stability of the specified system is possible on the basis of the necessary and sufficient criterion applying the Kharitonov theorem [1]. The overview of the methods concerning problem solution of control system synthesis with the interval - indefinite parameters is presented in papers [2-5]. Despite the available results, the investigation in the area of the synthesis of the robust control system with interval - indefinite parameters is still crucial. Thereby, the problem concerning the interval - parametric synthesis of the robust regulators where the location of the system poles is in the left semiplane at any variations of the coefficient of the characteristic polynomial of the system is of great importance.

The aim of the given work is to develop the technology intended to determine the interval parameters of the robust regulator which ensure the robust stability of the system at any interval coefficients of the control object.

2. Algebraic conditions for stability and oscillation of the interval polynomial

Let us assume that the automatic control system contains an object $W_{co}(s)=\dfrac{B(s)}{A(s)}=\dfrac{[b_c]s^c+[b_{c-1}]s^{c-1}+...[b_0]}{[a_j]s^j+[a_{j-1}]s^{j-1}+...[a_0]}$, where $\underline{a_j}\le a_j\le\overline{a_j}$, $\underline{b_c}\le b_c\le\overline{b_c}$ and linear regulator is $W_c\left(s,\vec{k}\right)=K\left(s,\vec{k}\right)/s$, where $\vec{k}$ is the vector of adjustable parameters. Let us write down the characteristic polynomial of the control system

$$P\left(s,\bar{k}\right)=B(s)K\left(s,\bar{k}\right)+sA(s)=([b_c]s^c+[b_{c-1}]s^{c-1}+...[b_0])*([k_m]s^m+[k_{m-1}]s^{m-1}+...+[k_0])+s([a_j]s^j$$
$$+[a_{j-1}]s^{j-1}+...[a_0])=\sum_{i=0}^{n}[p_i(\bar{k})s^i] \quad (1)$$

Let us introduce the coefficient parameters of stability λ_i [6] formed by the quadruples of the adjacent polynomial coefficients (1): $[\lambda_i]=\dfrac{[p_{i-1}][p_{i+2}]}{[p_i][p_{i+1}]}$, $i=\overline{1,n-2}$. The interval automatic control system is stable if the following conditions are satisfied

$$\overline{\lambda_i}=\frac{\overline{p_{i-1}}\,\overline{p_{i+2}}}{\underline{p_i}\,\underline{p_{i+1}}}\le\lambda^*,\ i=\overline{1,n-2},\ \lambda^*\approx 0.465 \quad (2)$$

Stability domains constructed upon these conditions will be located inside the exact stability region found by any necessary and sufficient condition. Restriction of this field is insignificant and occurs through discarding points, which refer to systems with a minor stability factor.

The inclination of the system to the oscillations is characterized by the index of oscillation δ_z [6-7] inequalities for the determination of unknown parameters of the regulator. The oscillatory character $[\delta_z]$ for the interval system is determined on the basis of the coefficients of the interval characteristic polynomial [1] $[\delta_z]=\dfrac{[p(\vec{k})_z^2]}{[p(\vec{k})_{z-1}][p(\vec{k})_{z+1}]}$, $z=\overline{1,n-1}$. Sufficient condition for a given degree of oscillation is formulated on the basis of oscillation indexes δ_z [6-7]. To locate the roots of the interval characteristic polynomial (1) in a given angular sector $\pm\varphi$, we must select controller settings, which would enable the following conditions

$$\delta_z=\frac{\underline{p(\vec{k})_z^2}}{\overline{p(\vec{k})_{z-1}}\,\overline{p(\vec{k})_{z+1}}}\ge\delta_d, z=\overline{1,n-1}, \quad (3)$$

where δ_d - is an admissible oscillation index.

Robust stability and robust oscillation of the interval characteristic polynomial can be estimated upon the conditions (2) and (3).

3. The algorithm of the definition of the regulator interval parameters

Suppose that at least two leading coefficients of the characteristic polynomial (1) are known. Let us create the inequality systems to determine the unknown parameters of the regulator using the conditions (2), (3).

$$\begin{cases}\overline{\lambda_i}=\dfrac{\overline{p(\vec{k})_{i-1}}\,\overline{p(\vec{k})_{i+2}}}{\underline{p(\vec{k})_i}\,\underline{p(\vec{k})_{i+1}}}\le\lambda^*,\ i=n-2,n-3,...1\ \lambda^*\approx 0{,}465;\\ \underline{\delta_z}=\dfrac{\underline{p(\vec{k})_z^2}}{\overline{p(\vec{k})_{z-1}}\,\overline{p(\vec{k})_{z+1}}}\ge\delta_d, z=n-1,n-2,...1.\end{cases} \quad (4)$$

We will obtain the equations for the limits of the regulator coefficients $\bar{k}$ on the basis of the system (4).

$$\overline{k_m}=\frac{\underline{a_{z-1}^2}-\delta_d\overline{a_{z-2}}\,\overline{a_z}}{\delta_d\overline{b_c}\,\overline{a_z}},\ z=g,g-1,...0, g=n-l; \quad (5)$$

$$\overline{k_m} = \frac{\overline{a_{i-2}}\,\overline{a_{i+1}} - \lambda(\underline{a_{i-1}}\,\underline{a_i})}{\overline{b_c}\,\overline{a_i}}, \ i = g-1, g-2, \ldots 0, g = n-l; \tag{6}$$

$$\underline{k_m} = \frac{\overline{a_{i+2}}\,\underline{a_i} - \lambda^{*}\delta_d\,\underline{a_{i+1}}\,\overline{a_{i+1}}}{\overline{a_{i+2}}\,\overline{b_c}}, \ i = g, g-1, \ldots 0, g = n-l; \tag{7}$$

$$\underline{k_m} = \frac{-\underline{a_i}}{\underline{b_c}}, \ i = g, g-1, \ldots 0, g = n-k; \tag{8}$$

$$\overline{k_{m-1}} = \frac{(\underline{a_i} + \underline{b_c}\,\underline{k_m})^2 + \delta\overline{a_{i+1}}(\overline{a_{i-1}} + \overline{b_{c-1}}\,\overline{k_m})}{\delta\overline{b_c}\,\overline{a_{i+2}}}, \ i = g-1, \ldots 0; \tag{9}$$

$$\overline{k_{m-1}} = \frac{\overline{a_{i+2}}(\overline{a_{i-1}} + \overline{b_{c-1}}\,\overline{k_m}) + \lambda\underline{a_{i+1}}(\underline{a_i} + \underline{b_{c-1}}\,\underline{k_m})}{\overline{b_c}\,\overline{a_{i+2}}}, \ i = g-1, ..0; \tag{10}$$

$$\underline{k_{m-1}} = \frac{-(\underline{b_{c-1}}\,\underline{k_m} + \underline{a_{i-1}}) \pm \sqrt{\delta b_{c-1}\overline{k_{m-1}}(\overline{a_i} + \underline{b_{c-1}}\,\overline{k_m})}}{\underline{b_c}}, i = g-1, \ldots 0; \tag{11}$$

$$\underline{k_{m-1}} = \frac{(-\overline{b_{c-1}}\,\overline{k_{m-1}}\,\overline{a_{i+1}} - \lambda((\underline{a_i} + \underline{b_c}\,\underline{k_m})(\underline{a_{i-1}} + \underline{b_{c-1}}\,\underline{k_m}))}{\underline{b_c}\lambda(\underline{a_i} + \underline{b_c}\,\underline{k_m})}, i = g-1, \ldots 0. \tag{12}$$

The algorithm intended to determine the limits of the regulator parameters of the control system is constructed on the basis of the equations (5) - (12). These limits ensure the robust stability of the system. The given algorithm includes the following stages:

1. The specification of the coefficients of the control object and the admissible index of oscillation δ_d.
2. The polynomial formation (1).
3. The check of the fulfillment of the problem specification (2), (3) for the known coefficients of the polynomial (1).
4. The calculation of the upper limits of the parameter k_m on the basis of equations (5), (6) and the selection of the maximum value.
5. The calculations of the bottom limit of the parameter k_m on the basis of the equations (7), (8) and the selection of the maximum value.
6. Similar calculation of the limits of other parameters on the basis of the equations (9) – (12).
7. The check of the fulfillment of the problem specification (2)-(3) for the evaluated values of the limits of the regulator parameters.

4. Application of the algorithm in the interval-parametric synthesis of the robust controller

Let us consider the operability of the reduced algorithm by the example of interval-parametric synthesis of speed control system of a descent submersible (DS) vehicle. The description of the given system is presented in work [8]. The block diagram of the system constructed on the basis of mathematical description [8] is illustrated in Figure 1.

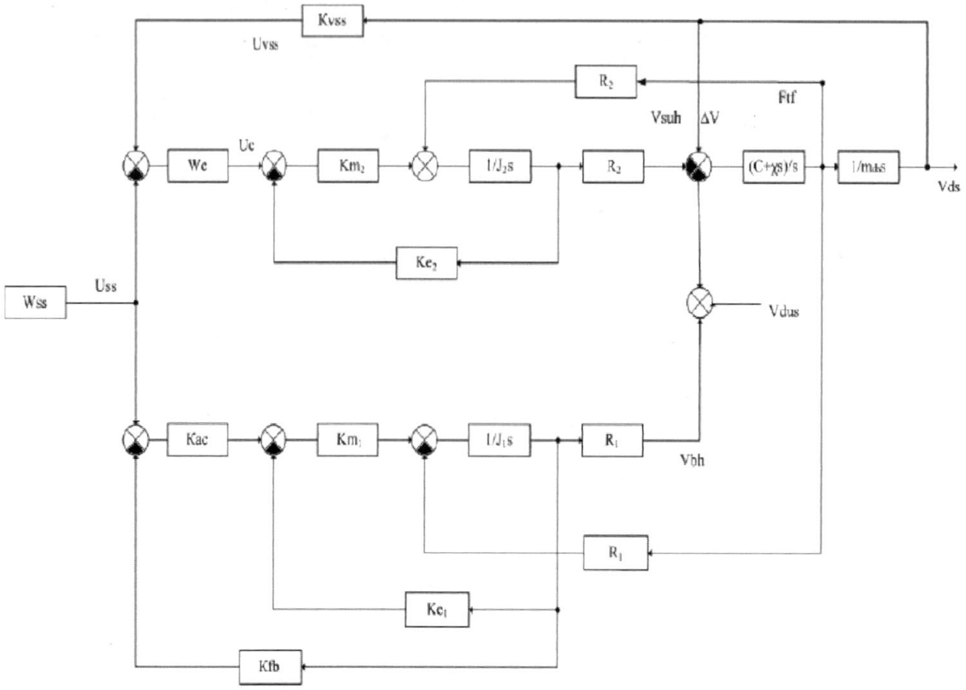

Figure 1. Structure diagram of the speed control system of a DS.

The system has the following constant parameters [8]: *Kvss =0,2[Vs/m]* – transfer constant of the vertical speed sensor of DS; $Km_1=Km_2=0,3[Nm/A]$ – transfer constant of the engines; $Ke_1=Ke_2=1[Vs/rad]$ – transfer constants of counter-emf of the engines of boat hoist (BH) and shock-absorbing hoist SAH; $J_1=100\ [kgm^2]$, $J_2=0,5[kgm^2]$ – are relatively the moments of inertia of BH and SAH; $R_1=0,2[m]$ *and* $R_2=0,1[m]$ – are relatively the drum radii of BH and SAH; $K_{fb}=0,21[Vs/rad]$ – feedback coefficient on speed of BH. The length of the rope and the DS mass are considered as the interval parameters of the system. Speed loop of BH has a typical structure including speed feedback and proportional speed controller $K_{ac}=20000$. The aim of this loop is to provide the required dynamic of BH in the descent and ascent mode of DS by means of the selection of the parameters settings *Kac* and *Kfb* (Figure 1).

The length of the cable from the interval *[2; 20] m* changes with the change of the depth of the descent of submersible vehicle and hence its parameters are equal to æ= *(6·104/l)*, *C=(6·106/l)*. Weight m_{ds} of the descent submersible vehicle can also undergo some changes by the lifting of various objects from the bottom and is within the range *[300; 350] kg*.

The specified factors lead to the variation of dynamic properties of speed control systems of the descent submersible vehicle that has negative impact on the damping of oscillation of the descent submersible vehicle and can cause the system performance loss. In connection with above mentioned factors it is necessary to impart robust properties to the system which ensure the permissible operation quality at any possible variations of unstable parameters.

As a result of block diagram transformation shown in Figure 1 the transfer function of open-loop system is as follows:

$$W_{ol}(s)=\frac{(b_0+b_1s+b_2s^2)(k_1+k_2s)}{s(a_4s^4+a_3s^3+a_2s^2+a_1s^1+a_0)},$$

where

$[a_4]=[600000;875000],[a_3]=[28921800;\ 35447625]\ [a_2]=[2394400080;\ 2738425575]$

$[a_1] = [25429505400;\ 29034005400.0000]$, $a_0 = 2160540000$, $b_0 = 43210800$, $b_1 = 4032108$, $b_2 = 36000$.

We need to determine the setting intervals of PI-controller: k_1, k_2 which provide that the automatic control system is robustly stable and conditions for oscillation at given δ_d=1,4.

We write the interval characteristic polynomial of this system, as follows

$$P(s,\vec{k}) = [p_5]s^5 + [p_4]s^4 + [p_3(k_2)]s^3 + [p_2(k_1.k_2)]s^2 + p_1(k_1.k_2)s + p_0(k_1), \tag{13}$$

where $[p_5] = [a_4]$; $[p_4] = [a_3]$; $[p_3] = [a_2] + [k_2]b_2$; $p_2 = [a_1] + ([k_2]b_1 + [k_1]b_2)$; $p_1 = [a_0] + ([k_2]b_0 + [k_1]b_1)$; $p_0 = [k_1]b_0$.

To solve this task, we elaborate the system in the form (4):

$$\begin{cases}
\overline{\lambda_1} = \dfrac{b_0 \overline{k_0}(\overline{a_2} + b_2 \overline{k_1})}{(\underline{a_0} + b_0 \underline{k_1} + b_1 \underline{k_0})(\underline{a_1} + b_1 \underline{k_1} + b_2 \underline{k_0})} \le 0,465; \\
\overline{\lambda_2} = \dfrac{(\overline{a_0} + b_0 \overline{k_1} + b_1 \overline{k_0})\overline{a_3}}{(\underline{a_1} + b_1 \underline{k_1} + b_2 \underline{k_0})(\underline{a_2} + b_2 \underline{k_1})} \le 0,465; \\
\overline{\lambda_3} = \dfrac{(\overline{a_1} + b_1 \overline{k_1} + b_2 \overline{k_0})\overline{a_4}}{(\underline{a_2} + b_2 \underline{k_1})\underline{a_3}} \le 0,465; \\
\underline{\delta_1} = \dfrac{(\underline{a_0} + b_0 \underline{k_1} + b_1 \underline{k_0})^2}{b_0 \overline{k_0}(\overline{a_1} + b_1 \overline{k_1} + b_2 \overline{k_0})} \ge 1,4; \\
\underline{\delta_2} = \dfrac{(\underline{a_1} + b_1 \underline{k_1} + b_2 \underline{k_0})^2}{(\overline{a_0} + b_0 \overline{k_1} + b_1 \overline{k_0})\,(\overline{a_2} + b_2 \overline{k_1})} \ge 1,4; \\
\underline{\delta_3} = \dfrac{(\underline{a_2} + b_2 \underline{k_1})^2}{(\overline{a_1} + b_1 \overline{k_1} + b_2 \overline{k_0})\overline{a_3}} \ge 1,4; \\
\underline{\delta_4} = \dfrac{(\underline{a_3})^2}{(\overline{a_2} + b_2 \overline{k_1})\overline{a_4}} \ge 1,4;
\end{cases}$$

Solving this system and using an algorithm to determine the unknown parameters of the regulator $[k_0] = [1.385 \times 10^6; 1.111 \times 10^8]$ $[k_1] = [2.74 \times 10^5; 1.824 \times 10^7]$

This result is checked by constructing the localization regions of the roots the interval polynomial (13) (figure 2).

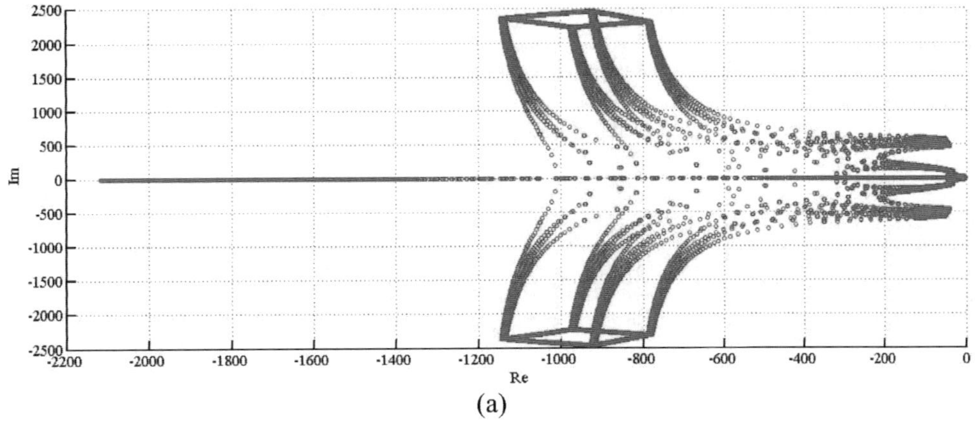

(a)

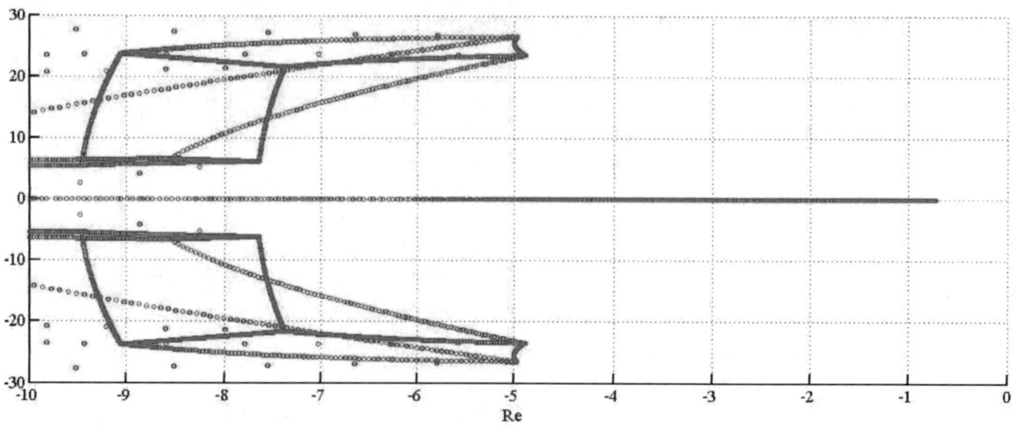

(b)

Figure 2. (a) Regions of root localization; (b) regions of root localization on a larger scale.

Thus, the system with the interval objects under consideration will be robust – stable at any parameters of PI-regulator from the computed and specified ranges.

5. Conclusion

The algorithm intended to determine the limit of parameters of the regulator ensuring the robust stability of the system with the interval control objects based on the robust modification of the coefficient method was developed in the given paper. The efficiency of this algorithm was checked on the interval-parametric synthesis of the robust PI- regulator and was confirmed by the graphs of the areas of the location of roots of the characteristic polynomial.

Acknowledgments

The reported study was partially supported by RFBR, research project No 15-01-99447

References

[1] Polyak B.T., Scherbakov P.S. 2002 *Robust stability and control* (Moscow: Science) p 303

[2] Yu, J., Yan, Z., Wang, J. and Li, Q 2014 Robust Stabilization of Ship Course Via Convex Optimization *Asian Journal of Control* **16** 871-877.

[3] Yogesh V. Hote, J. R. P. Gupta, and D. Roy Choudhury 2010 Kharitonov's Theorem and Routh Criterion for Stability Margin of Interval Systems *International Journal of Control, Automation, and Systems* **8** 647-654.

[4] Yong Wang , Shu Liang 2013 Two-DOF Lifted LMI Conditions for Robust D-Stability of Polynomial Matrix Polytopes *International Journal of Control, Automation, and Systems* **11**,636-642.

[5] Keel L.H. 1999 Robust stability and performance with fixed-order controllers *Automatica* **35** 1717-1724.

[6] Gayvoronskiy S.A., Ezangina T.A. 2013 Bundled Software for the Desing of Interval Dynamic Systems *Applied Mechanics and Materials* **446 – 447** 1217-1221.

[7] Gayvoronskiy S.A., Ezangina T.A. Robust control of complex dynamic units with interval parameters *2nd Int. Conf. Systems and Computer Science (ICSCS) (Villeneuve d'Ascq, France, 26-27 Aug. 2013)* pp 201-204.

[8] Ezangina Tatiana Al., Gayvoronskiy Sergey An. 2014 The control of the descent submersible in the confused sea environment *Applied Mechanics and Materials* **536-537** 1200-1206.

IOP Conf. Series: Materials Science and Engineering **93** (2015) 012056 doi:10.1088/1757-899X/93/1/012056

Method of preliminary localization of the iris in biometric access control systems

N Minacova[1] and I Petrov[1]
Altai State University,
61 Lenin av., Barnaul, 656049, Russia

E-mail: PetrovIV90@gmail.com

Abstract. This paper presents a method of preliminary localization of the iris, based on the stable brightness features of the iris in images of the eye. In tests on images of eyes from publicly available databases method showed good accuracy and speed compared to existing methods preliminary localization.

1. Introduction

At the present time, biometric authentication systems have found a wide acceptance as a result of technological advances and increase in computing capacity. These systems allow to identify the users by certain biometric parameters. A biometric parameter may be a fingerprint, a hand shape, an iris pattern and so on. Basic requirements for a biometric parameter include versatility, stability within a longer period of time, and access for scanning. Parameters of texture determined by mathematical and physical methods used for inhomogeneous structures [1,2].

The iris identification is not a trivial task and includes several phases:

- taking eye pictures with a special scanner;
- iris localization in the picture;
- normalization of the localized iris;
- extraction of feature vectors;
- comparison of feature vectors.

One of the most difficult and labor-intensive phases of the recognition is the iris localization in the picture. The reason is that the position, the angle of rotation and slope of the iris are not stable and can vary greatly. Besides, the localization process gets complicated by factors such as pupil dilation, closing a part of the iris by lids, eyelashes, and fine white lines. At this moment, there are many iris localization methods and algorithms. One of the best known and proven methods is using the Daugman's integro-differential operator to approximate the inner and outer iris boundaries. There are other methods used for iris localization as well, for example, the Hough transformation method for searching the circumference, the method of active circuits and so on.

For most of these methods, it is necessary to establish the initial and terminal conditions, such as the initial searching point (it may coincide with the approximate pupil center), the minimum and

[1] To whom any correspondence should be addressed.

maximum radii of the inner and outer iris boundaries, and the search area. Without these parameters the search is not efficient.

The work objective is to develop a new method of the preliminary localization by pinpointing the location of the iris in an effort to increase efficiency.

Tasks:

1. To analyze current methods of the preliminary iris recognition.
2. To develop a method to increase the iris recognition.
3. To apply the method on the iris test pictures from databases publicly available.

In several cases, the method of brightness' projection [3] is used to define the initial search point. The essence of the method: the pupil of most people has the form closely resembling a circle, which allows searching for it as a dark circle on the image.

The clusterization of images is performed [4]. On the image of an eye, the areas of eyebrows, skin, sclera, and iris are differentiated. This allows narrowing the search area and increasing its accuracy.

2. Proposed approach

In this publication, a new method of the preliminary search for the iris is offered. The essence of the method: the iris on the image has a series of consistent characteristics that can be used for a preliminary search. It is well known, that the area surrounding the iris has a higher brightness in most cases. At the same time, the pupil inside the iris is darker. Such brightness variations between an eye's elements allow selecting a steady pattern, on which to base the search for an iris.

An eye image considered a two-dimensional function of the brightness $I(x, y)$. Let us think of a two-dimensional function as a set of one-dimensional functions $I_y(x)$. In other words, a two-dimensional image is divided into a sequence of horizontal stripes.

The signal has to undergo a low-pass filtering to eliminate noise and undesirable details. For this purpose the Gaussian filter can be used.

The upper boundary of the brightness range must be cut off as well to eliminate the influence of flares in the search.

However, not the brightness function $I_y(x)$, but its first derived function $I_y'(x)$, is the most informative. This function allows identifying the boundaries of transition between the sclera and the iris, and between the iris and the pupil. (picture 1).

On the derived function $I_y'(x)$ the sequence of peaks is searched for. The peaks have to meet the criteria as follows:

1. The derivate has a negative value on the boundary of transition from the light sclera to the darker iris and from the iris to the darker pupil, the opposite is true as well:
$I_y'(x_1), I_y'(x_2) < 0, I_y'(x_3), I_y'(x_4) > 0$
2. The boundaries of transition from the sclera to the iris, from the iris to the left and right pupil must have values close to the module's:
$\left\| I_y'(x_1) \right| - \left| I_y'(x_4) \right\| < \varepsilon, \left\| I_y'(x_2) \right| - \left| I_y'(x_3) \right\| < \varepsilon$,
where ε is some value that determines an acceptable deviation level. This value is due to heterogeneity of illumination (under ideal conditions ε is very close to zero).
3. The relation between the diameters of the iris and the pupil is established as:
$k_{min} d_{ir} < d_p < k_{max} d_{ir}$,
where k_{min}, $k_{\max}$ are the coefficients that define the maximum possible pupil dilation and the

minimum possible contraction of the pupil with respect to the diameter of an iris. They depend on anatomic features of the eye [5].

Besides, proximate to $I_y^{'}(x)$, for $I_{y-l}^{'}(x)$ and $I_{y+l}^{'}(x)$, where l is a certain constant, the criteria quoted above should be met as well, while the values of x_1 and x_2 with $I_{y-l}^{'}(x)$ and $I_{y+l}^{'}(x)$ should be greater than with $I_y^{'}(x)$. In the same way, the values of x_3 and x_4 should be smaller.

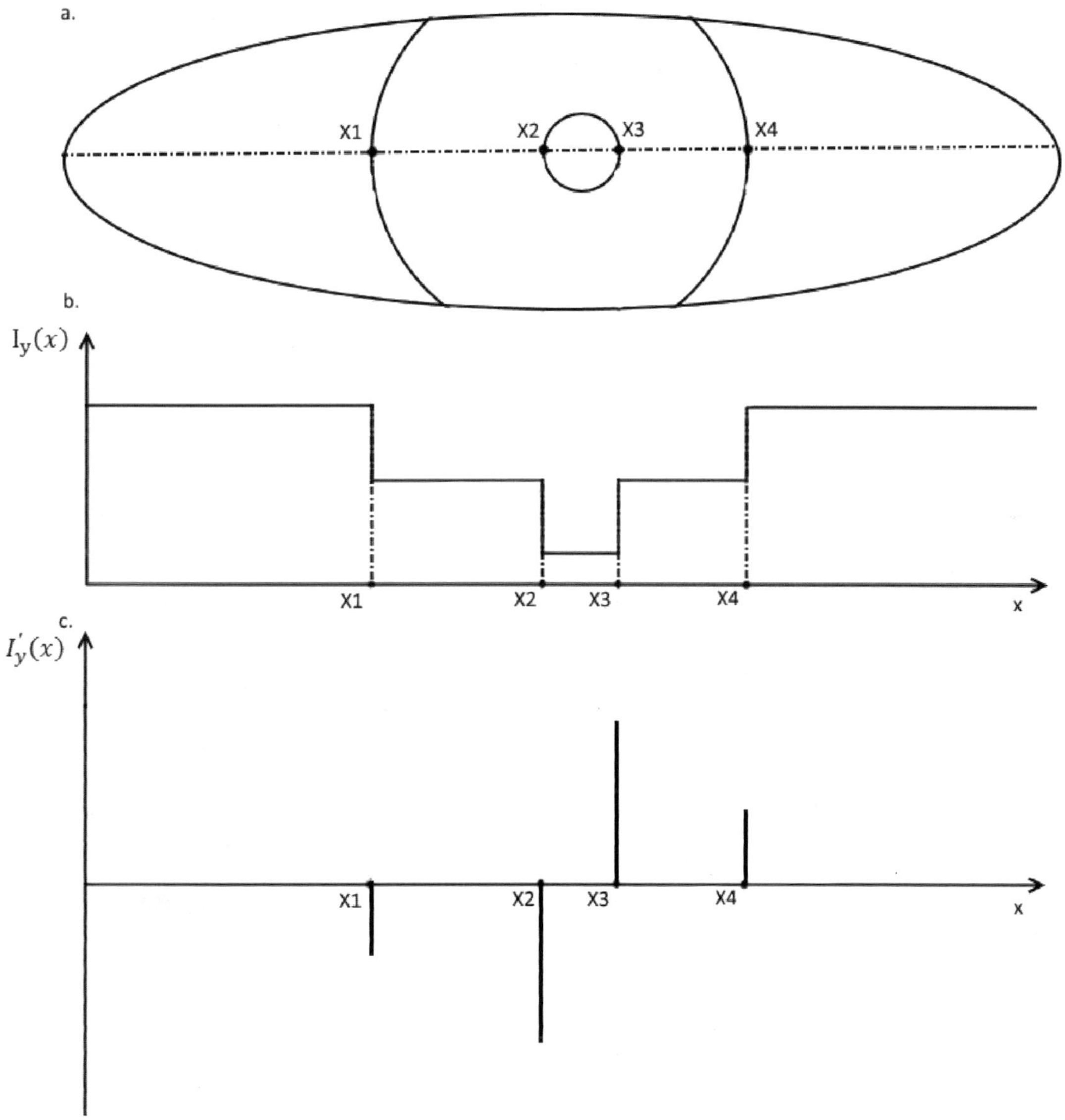

Figure 1. a) The artistic rendition of the iris with the line $I_y(x)$ that goes through the pupil's center. b) Diagram of values of the $I_y(x)$ function. c) Diagram of values of the derived function $I_y(x)$.

After the determination of the pattern and the points x_1, x_2, x_3, x_4, the assumed values of the iris boundaries are calculated: the coordinates of the center (x_p, y_p) and the radius r_p of the pupil, the coordinates of the center (x_i, y_i) and the radius r_i of the outer boundary of the iris.

The result of the method described above should produce a list of prospects $(x_p, y_p, r_p, x_i, y_i, r_i)$. After that, the boundaries are specified for each candidate by the Daugman's integro-differential operator, whereupon candidates with the maximum value of the Daugman's operators are picked up.

The essence of the method, as applied to the real eye image, is shown on picture 2. The areas of the sclera, the iris, and the pupil can be extracted on the brightness function diagrams. On the derivate diagram, the peaks by the coordinates that coincide with the inner and outer boundaries of the iris can be seen. The obtained values allow identifying the areas of search for the iris, as well as the approximate coordinates of the pupil and the iris.

3. Results and discussion

The database system [6] was used for the purpose of the test. The Daugman's integro-differential operator was applied as the specifying method for localization.

For checking purposes, the tests were performed using the databases UTIRIS [7] and CASIA Iris Image Database (version 1.0) [8]. For each image, the automatic iris localization was performed. The results of the automatic and manual localizations were compared. The outcome of experiment is shown in Table 1.

Table 1. The results of testing of the method.

Name of the database	Number of shots for the test	Accuracy, %
UTIRIS	100	97
CASIA Iris Image Database (version 1.0)	100	90

When testing the method on images from the CASIA Iris Image Database (version 1.0) the accuracy was lower. This is due to the fact, that in some cases the boundaries between the sclera and the iris are scarcely distinguishable. For this reason they are considered false extrema during the analysis.

During the experiment, the speed of localization of the iris of the methods of brightness projection and of the offered method was compared. The Daugman's integro differential operator was applied as the specifying method for localization. The results are shown in Table 2.

Also during the check of localization speed, iris was compared to the preliminary use of the method of brightness projection and the offered method

Table 2. The results of testing the speed of the method.

Name of the database	Average processing rate, ms	
	Brightness' projection	Offered method
UTIRIS	65	48
CASIA Iris Image Database (version 1.0)	52	25

As Table 2 shows, the localization speed of the offered method is higher. This is because the boundary conditions for the search are set more precisely.

4. Conclusions

The offered method of the preliminary search allows narrowing the iris recognition area. This method decreases the time of the current recognition methods.

IOP Conf. Series: Materials Science and Engineering **93** (2015) 012056 doi:10.1088/1757-899X/93/1/012056

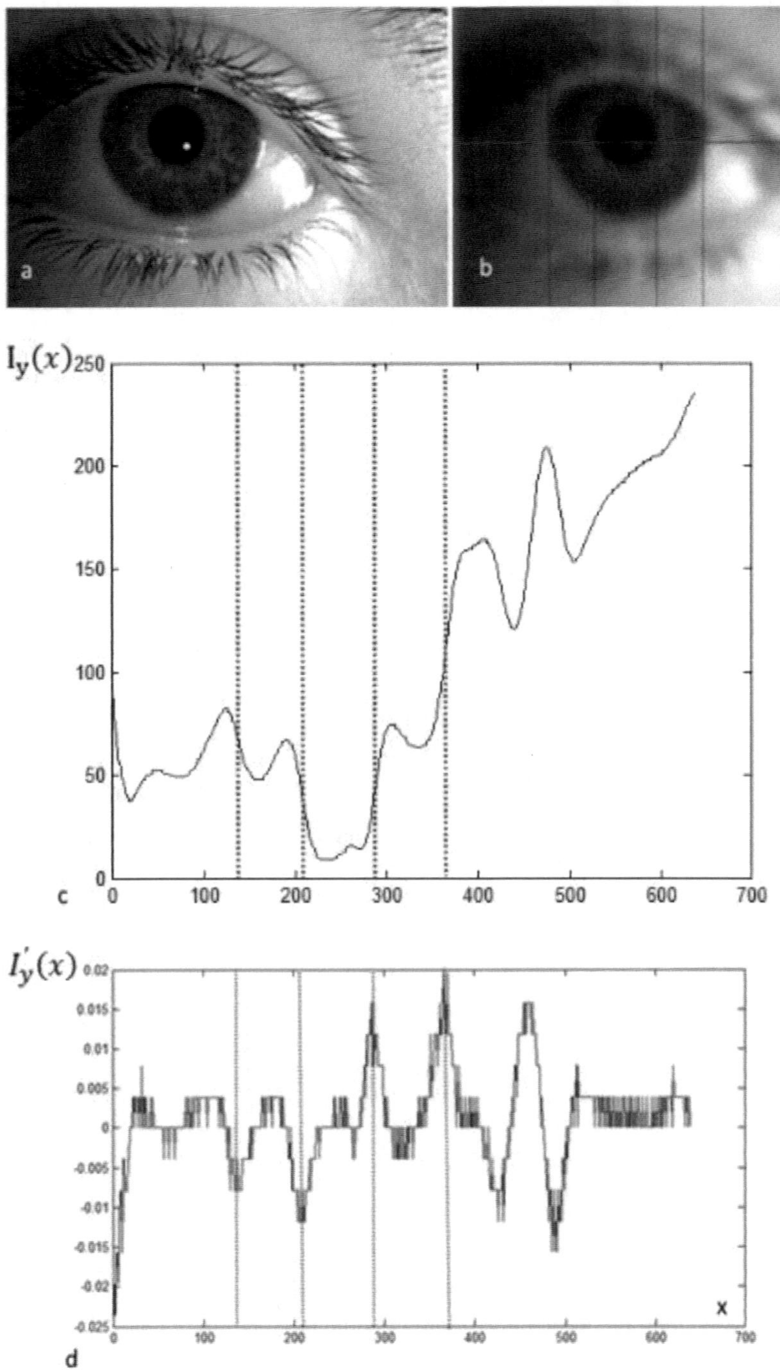

Figure 2. How the method works. a) The initial image of the eye. b) The smoothed image of the eye with marked boundaries. The red line marks a single line taken for the analysis. c) The brightness diagram from the coordinate x taken for the band of the image. d) The diagram of the brightness function is derived by the coordinate x taken for the band of the image.

The test results have shown, that the application of the offered method, compared to the method of brightness' projection, allows speeding up the work involved in iris recognition.

IOP Conf. Series: Materials Science and Engineering **93** (2015) 012056 doi:10.1088/1757-899X/93/1/012056

References

[1] Gonzalez R C, Woods R E 2002 Digital image processing *Prentice Hall,* New Jersey 793

[2] Minakova N N Karpov S A Ushakov V Ya 2000 Texture Method of Studying the Resistance Properties of Disperse-Filled Elastomers *Russian Physics Journal* **43 (10)** 821-8

[3] Sanz J L C Hinkle E B Jain A K 1988 Radon and projection transform-based computer vision *Springer-Verlag,* Berlin 123

[4] Tieniu Tan Zhaofeng He Zhenan Sun 2010 Efficient and robust segmentation of noisy iris images for non-cooperative iris recognition *Image and Vision Computing* **28 (2)** 223-30

[5] Birich T A Marchenko L N Chekina A Yu 2007 Ophthalmology *Vysheyshaya Shkola,* Minsk (in Russian) 576

[6] Minakova N N Petrov I V 2014 Information system analysis barely visible texture of the iris in the visible range *Proceedings of TUSUR* (in Russian) **32 (2)** 105-8

[7] Mahdi S Hosseini Babak N Araabi and H Soltanian-Zadeh 2010 Pigment Melanin: Pattern for Iris Recognition *IEEE Transactions on Instrumentation and Measurement* **59 (4)** 792-804

[8] http://biometrics.idealtest.org/dbDetailForUser.do?id=4, CASIA. Iris images database

IOP Conf. Series: Materials Science and Engineering **93** (2015) 012057 doi:10.1088/1757-899X/93/1/012057

Algorithms and software for modules of automatic neurocomputer control of smart electromechanical system such as hexapods

V A Romanchuk
Department of Informatics and Computer Technology, Ryazan State University named for S.Yesenin, Ryazan, Russia

E-mail: v.romanchuk@rsu.edu.ru

Abstract. The article discusses issues in developing algorithms and software for specialized computing devices based on neuroprocessors, to be used in automatic control of electric-mechanical system modules (in this case study, a hexapod) in a mode that is close to real-time. The practical implementation employed an NM6406 neuroprocessor based on an MC 51.03 tool module and an MB 77.07 microcomputer, developed by the Module Research Centre.

1. Introduction

Controlling dynamic objects involves two modes: operation and standby, due to the rigid necessity of high-speed processing of great amounts of data. In fact, no computing control system today can ensure such processing of data at a set tempo, when such input is fed from multiple sensors; therefore, computing operations have to be parallelized into additionally connected specialized computational devices (SCDs). Such task sharing supports the unsteady condition, without slowing down the basic computing process. Two primary tasks require real-time problem-solving: the computation of control action and integration of a differential equation system. Transforming the differential system into a system of algebraic equations (vector-matrix) is convenient, as far as it enables its implementation in digital computing, namely in SCDs: their processor modules will simultaneously perform accumulative multiplication actions with multiple operands.

To develop and implement the corresponding mathematical content, algorithms and software, we propose to activate a selected set of hardware tools: neuroprocessor devices based on special-class processors (Generation 6), widely used in data processing. The choice of neuroprocessors considered such parameters as the speed of parallel processing, availability of accumulative addition, and high energy efficiency [1, 2]. In our case, we chose the NeuroMatrix 640x processor family, manufactured by the Module Research and Development Centre (Moscow, Russia) [3].

The input data parameters for development of algorithms and software are as follows:

- definition of digital signals from sensors;
- description of algorithms for generation of management signals;
- a mathematical model implemented in a SIMULINK module model of the MATLAB software package.

The source data of algorithms developed include the following:

- definition of sensor signals received after conversion of analog signals from the hexapod into digital form (via an analog-digital converter).

The output data of algorithms developed include the following:
- control signals, converted into analog (via a digital-analog converter);
auxiliary and maintenance signals.

2. Mathematical formalization of SCD operational tasks

The flowchart of an SCD is shown in figure 1. The device's functional purpose is to calculate optimal parameter values for a shared controller and to calculate quotients (k1, k2, k3) of PID controllers of the hexapod's movement engine, to ensure seizure-free and smooth motion of the hexapod's platform.

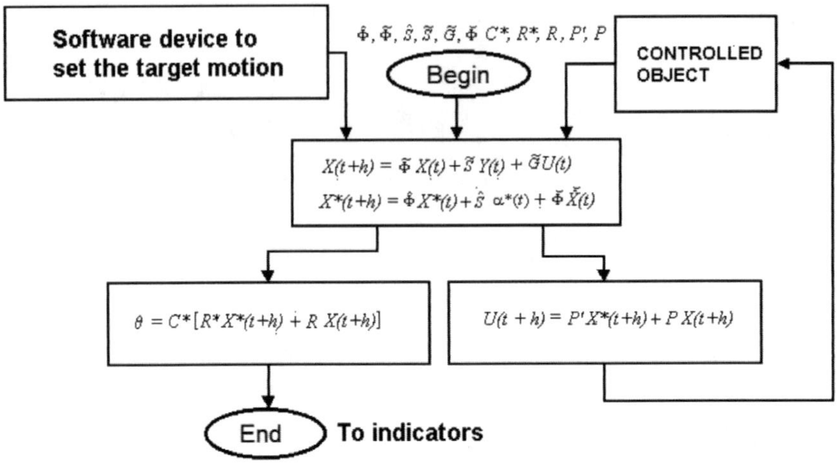

Figure 1. SCD flowchart algorithm.

The chart uses the following symbols:

$\hat{\Phi}$, $\tilde{\Phi}$, $\hat{S}$, $\tilde{S}$, $\tilde{G}$, $\breve{\Phi}$ — operational matrixes received from the parameter matrixes of the hexapod and from those of the correction filters;

P', P — parameter matrixes of correction associations (6x100);

$\alpha^*(t)$ — program vector function (18x1);

$Y(t)$ – vector function of observation (signals received from the measuring system) (30x1);

$X(t)$, $X^*(t)$ – vector function of the system's current state and of its programmed state at Step i of its control (100x1);

$X(t+h)$, $X^*(t+h)$ – the same at Step $(i+1)$ (100x1);

$U(t)$ and $U(t+h)$ – vector of control influences at Steps i and $(i+1)$ of control (30x1);

h –step of integration (~0.001) ;

θ – deviation value of the hexapod's focal axis from the object's line of sight (3x1);

C^*, R^*, R — permanent matrixes sized (3x3, 3x100, 3x100).

We can say that the basis of a neuroprocessor unit is an emulation of a formal neuron, which is mathematically a total of operations of addition and multiplication, i.e., "accumulative multiplication".

From the hexapod control algorithm, it becomes clear that the main procedure in the SCD is the multiplicationя of the matrixes by the vector:

$$\Phi \times X = \begin{bmatrix} \Phi_{11} & \Phi_{12} & \cdots & \Phi_{1n} \\ \Phi_{21} & \Phi_{22} & \cdots & \Phi_{2n} \\ - & - & - & - \\ \Phi_{m1} & \Phi_{m2} & \cdots & \Phi_{mn} \end{bmatrix} \times \begin{bmatrix} x_1 \\ x_2 \\ \vdots \\ x_n \end{bmatrix} = \begin{bmatrix} \Phi_{11}x_1 + & \Phi_{12}x_2 + & \cdots & +\Phi_{1n}x_n \\ \Phi_{21}x_1 + & \Phi_{22}x_2 + & \cdots & +\Phi_{2n}x_n \\ - & - & - & - \\ \Phi_{m1}x_1 + & \Phi_{m2}x_2 + & \cdots & +\Phi_{mn}x_n \end{bmatrix}$$

One of the vector result components is the sum of the item products of matrix lines by the corresponding items of the vector-column. This sequence of operations should be repeated as many times there are lines in a matrix. To calculate the product of matrixes by the vector, it is necessary to perform multiplication of operations $T_{\text{увм}} = m \times n$, where m is the number of line in the matrixes; n is the number of columns, and $N_{\text{смв в}}$ is the number of addition operations, defined by the formula: $N_{\text{смв}} = m \times (n-1)$ addition operations, or approximately 3,000 pairs of "multiplication – addition" operations.

3. Development of algorithm support of the SCD

As the conducted experimental research employed an NM6406 neuroprocessor based on an *MC 51.03* toolkit and an *MB 77.07* microcomputer, we have to say that parallelizing is made effective due to hardware support of the vector-matrix multiplication operation in the NeuroMatrix NM6406 neuroprocessor. All arithmetic computing directly related to computation of control influences is performed on a vector coprocessor. As the vector node allows handling data with variable width, it is convenient to allocate 32 bits to the integral part and 32 bits to the real part.

For development of algorithms and of program code, subprograms of the functional items (of the mathematical model of a hexapod-controlling SCD) were implemented [1]. So, the chief operation of the SCD consists in multipliucation of matrixes which can be split into a couple of elementary operations «multiplication – addition» (accumulative multiplication). For this operation, a vector coprocessor is used, i.e., a shared matrix node, to perform operations of accumulative multiplication with arithmetic and logical operations, masking, and enabling vectors and matrixes.

The accumulative multiplication is performed in the operational matrix of the neuroprocessor [4]:

$$Z_i = Y_i + \sum (X_i W_{ij}); i = \overline{1,m}; j = \overline{1,n},$$

where X_i is a data item sent to the vector coprocessor input;

Y_j is the subsum accumulated at the previous stage of weighted addition, or the remainder from the previous operation;

W_{ij} – the weight quotient, stored in a dedicated cell of the operating matrix of the processor;

m – the number of columns in the operative matrix of the processor;

n – the number of lines in the operative matrix of the processor.

Data input at *X* should be multiplied by the number of the matrix cells, they are further added column-wise, which means that in case of overflow there occurs a loss of sign bits. Data from the *Y* input are added item-wise with the multiplication product, resulting from operations with the data of the *X* input.

To perform weighted addition, we must first export W_{ij} weighted quotients into the matrix.

Partitioning of matrixes into lines is defined in the *sb2* operational registry. It provides preliminary partitions of 64-bit input data words that are sent to the *Y* input. Operational matrix partitioning into column is set in *nb2* operational registry. The same registry defines partitioning of 64-bit data at the *Y* input and is also preliminarily entered as a word in the registry. The same registry defines the computing bit rate that will move to the AFIFO buffer registry in the end. Thus, input data (operands) and output values are packed into 64-bit words during partitioning. All operations in the matrix are performed as parallel within one stroke.

The accumulator of two or more operands is implemented via a vector unit of the processor and operations of weighted addition (Figure 1) with void multiplication, i.e., a "single" shadow and operational matrix (where each 64x32 bit is filled with single bits). The effectiveness of this operation depends on the bit rate of input data, as the computing speed is increased via matrix partitioning into rows and columns, because work is performed simultaneously on several operands. One important property of a vector processor is its ability to work with operands of various length.

Activation functions are implemented with the help of the processor's vector unit and operations of weighted addition (Figure 1). Subprograms of two activation function types are implemented: the threshold function and the overflow function. In activation units computation is performed on packed data words. Activation units allow using activation functions on all items of a packed word simultaneously. The main role in controlling the activation functions is that of the *f1cr* and *f2cr* registries. The activation units are located between the masking device and the operational matrix or the vector calculator. The activation function may be performed on input *X* or *Y* data, or on both inputs.

All elemental basic operators are implemented in macros form in the neuroassembler language.

4. Development of software for SCD operation

For work convenience with various source data a task model was implemented, to automatically control electric and mechanical systems in the developed *Visual Programming* subsystem of the *NP Studio* software platform (figure 2).

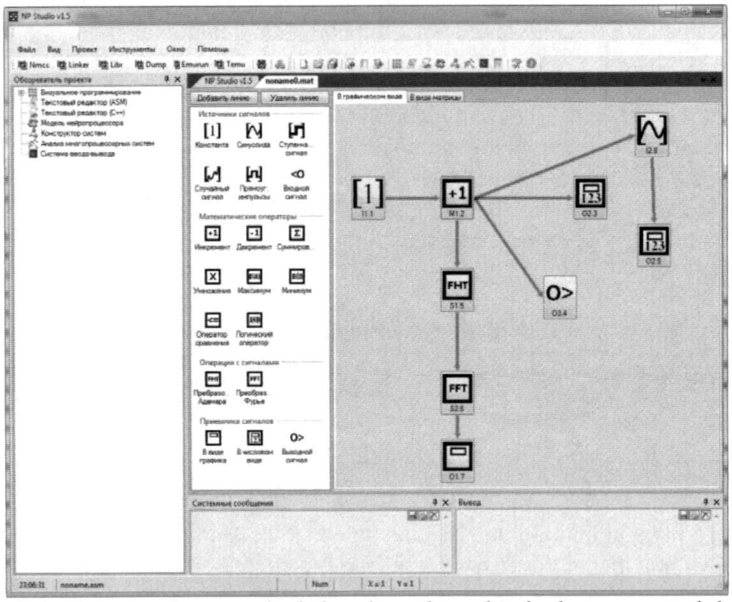

Figure 2. Task model of automatic control of electric and mechanical system modules in he developed *Visual Programming* subsystem of the *NP Studio* software platform.

The chief item in the subsystem is the notion of *a functional unit*, connectable to other functional units to perform certain functions.

The workspace of the software application is divided into functional parts:

1. The area of functional units, instance of which may be transferred to the modeling workspace with the drag-and-drop function. The same area contains control items for visualization and deletion of connections between items. When added to the workspace, each item gets a unique ID, which is a concatenation of the item category, the item number within the category and the index number of the item, e.g., "O2.9" defines an item from category Output Signals, Number, with "9" as its unique ID.

2. The workspace (which, in its turn, can be represented as visual items and connections between them, or as a matrix of connections between items (Figure 3).

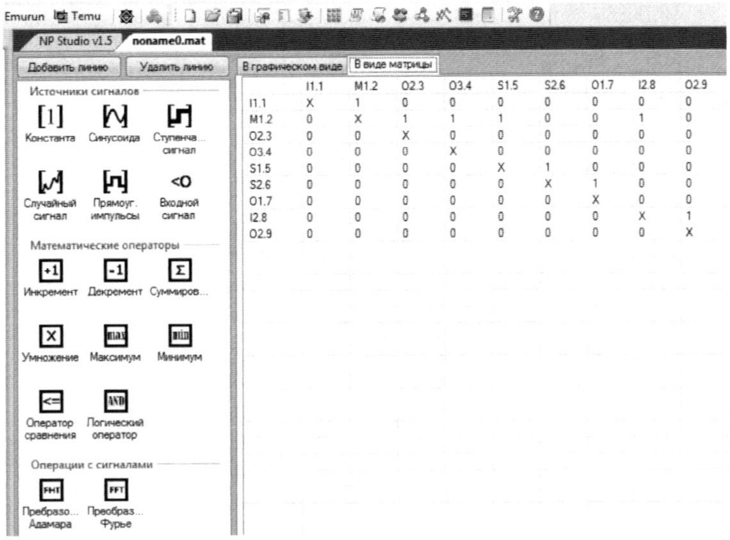

Figure 3. Task model of automatic control of electric and mechanical system modules in the developed *Visual Programming* subsystem as matrix connections.

5. Conclusion

The article proposes algorithms and software for a specialized computational device based on neuroprocessors, to be used in automatic control of electric-mechanical systems (in this case study, a hexapod) in a mode that is close to real-time, employing a novel computing device, an *NM6406* neuroprocessor, developed by the *Module* Research Centre. The research proves effectiveness of the neuroprocessor in performing tasks of automatic control of electromechanical system modules, through parallel processing and consequent speed gain, use of the *accumulative addition* operation, and low power consumption.

References

[1] Ruchkin V, Romanchuk V and Sulitsa R 2013 Clustering, restorability and designing of embedded computer system based on neuroprocessors *Proceedings of the 2nd mediterranean conference on embedded computing (*Montenegro: Budva*)* 58-62

[2] Kolinummo P, PASI Pulkkinen P, Hämäläinen T and Saarinen J 2000 *Neural Processing Lett.* **12** 171-82

[3] Izeboudjen N, Larbes C and Farah A 2014 *Artif Intell Rev* **41** 491–534

[4] Pande S, Morgan F, Cawley S, Bruintjes T, Smit G, McGinley D, Carrillo S, Harkin J and McDaid L 2013 *Neural Process Lett.* **38** 131–153

Results of research on development of an intellectual information system of bankruptcy risk assessment of the enterprise

E Telipenko[1], T Chernysheva[2], A Zakharova[3] and A Dumchev[4]
[1] Cand.Tech.Sci., associate professor of information systems NR TPU, Tomsk, Russia
[2] Cand.Tech.Sci., associate professor of information systems NR TPU, Tomsk, Russia
[3] Cand.Tech.Sci., associate professor of information systems NR TPU, Tomsk, Russia
[4] Master's Department of Information Systems NR TPU, Tomsk, Russia

E-mail: kochetkovaev@mail.ru

Abstract. The article represents research results about the knowledge base development for the intellectual information system for the bankruptcy risk assessment of the enterprise. It is described the process analysis of the knowledge base development; the main process stages, some problems and their solutions are given. The article introduces the connectionist model for the bankruptcy risk assessment based on the analysis of industrial enterprise financial accounting. The basis for this connectionist model is a three-layer perceptron with the back propagation of error algorithm. The knowledge base for the intellectual information system consists of processed information and the processing operation method represented as the connectionist model. The article represents the structure of the intellectual information system, the knowledge base, and the information processing algorithm for neural network training. The paper shows mean values of 10 indexes for industrial enterprises; with the help of them it is possible to carry out a financial analysis of industrial enterprises and identify correctly the current situation for well-timed managerial decisions. Results are given about neural network testing on the data of both bankrupt and financially strong enterprises, which were not included into training and test sets.

1. Introduction

Currently bankruptcy risk assessment is of great importance. The urgency of the issue resulted in development of numerous methods and methodologies for bankruptcy risk assessment with some degree of possibility. However there are a lot of problems here. These are methods that could assess unsuccessful outcome which are almost absent. The majority of those methods failed for a number of reasons. Consequently, an urgent problem is development of systems that can overcome challenges of existing models and take into account our country specifics.

2. A process analysis of knowledge base development of an intellectual information system for bankruptcy risk assessment of the enterprise

Connectionist simulation is made in several steps.

At the first step an index range for further study is determined. For bancruptcy risk assessment the authors propose to use 10 indexes that characterise financial and economic activity of the enterprise: x1 is working capital ratio; x2 is acid test ratio; x3 is absolute liquidity ratio; x4 is liquid capital ratio in assets; x5 is concentration ratio of capital loans; x6 is long term fund raising ratio; x7 is borrowed and own funds ratio; x8 is turn-round of float; x9 is turn-round of owned capital; x10 is product profitability.

Correctness of this index system is proved by the research "Decision support system for bankruptcy risk management of the enterprise".

At the second step the authors sampled companies with accurate financial accounting on the basis of which they did calculations of the given indexes. Several problems are solved here:

1) Search for time quasi-homogeneous data.

Since 2011 till 2013 there was relative macroeconomic stability: inflation was about 6-7%, fluctuation of the dollar with respect to the ruble was about 10 %. Therefore it is possible to say that 2011 – 2013 time period gives quasi-homogeneous financial statistics for financial analysis.

2) Search for quasi-homogeneous data at inter-branch level.

Difference in index values and their norms from branch to branch caused reduction of studied enterprises to one basic Russian Classification of Economic activities (OKVED). During this research the authors decided to process financial accounting of the enterprises with basic OKVED 29 – Machinery and equipment production apart from OKVED 29.6 – Arms and weapons production.

This OKVED was selected due to two main reasons: 1. The proposed 10 indexes for the research obtained during previous studies were selected for industrial enterprises; 2. Industrial sector is an object of state focused attention being a key sector for the Gross Domestic Product formation and influenced by crisis phenomena of different levels.

3) Search for one-type data.

To get one-type data of enterprises the authors bought quarterly subscribing to the information system "SKRIN" where they exported data from of financial accounting of 155 enterprises since 2011 till 2013 [1].

At the third step for calculating values of marked indexes the authors instantiated two program modules in Delphi XE3: the first module realized parsing of xls files with financial accounting data; the second module realized data pre-processing, i.e. normalizing. For holding calculated index values the authors established database on My SQL Server 5.3.

At the fourth step a type of neural network was chosen, it was a three layer perceptron with back propagation algorithm as a training one. The reason for this choice was relative simplicity of realization together with scalability and a wide range of tasks that could be completed with the help of perceptrons.

At the fifth step the authors built and trained neural network by means of the program Deductor Studio, which had a good tool set for data analysis and building artificial neural networks.

At the sixth step the authors did validity check of obtained neural model for bankruptcy risk assessment of the enterprise on the basis of testing with the help of data set trained in the program Deductor Studio. At the end of testing they obtained results that sufficiently reflected the state of things at analyzed enterprises.

3. Structure of intellectual information system (IIS) and knowledge base

The interaction mechanism of the basic IIS subsystems is as follows:

1) A knowledge expert or engineer enters financial accounting data into the system by means of a special form, which allows inserting data by hand or loading the file in xls format.

2) Data pre-processing, i.e. normalizing, is realized. Heterogeneous quantities can be as input so output. It is evident that connectionist simulation results should not depend on units of measurement of the quantities. The network should interpret their values consistently, all the input and output quantities should be normalized to one gauge. To increase rate and quality of training it is necessary to realize optional pre-processing that adjusts classification of values before training. Normalizing to one gauge is provided by norming of every variable on the scatter range of its values.

IOP Conf. Series: Materials Science and Engineering **93** (2015) 012058 doi:10.1088/1757-899X/93/1/012058

At that it is necessary to be oriented not to extreme values but to typical ones, i.e. data statistic characteristics as medium and dispersion. In this case the majority of data has one gauge, i.e. typical values of all variables are comparable.

3) After pre-processing the data is recorded on the database MySQL.

4) The connectionist model is built (trained) on the basis of available data. Then this model is used for bankruptcy risk assessment of the enterprise. The processed information and information processing method in the form of the connectionist model form the knowledge base of the intellectual information system. So in general it is possible to say about two-level knowledge representation in IIS: the first level is factual information, data; the second level is neural network that identifies the way of manipulation of factual information.

5) Decision output is done by means of the output subsystem, which cooperates with knowledge base during interpretation of methods and database data.

6) Problem solving and work with a model (specimens) and data is done in a special module – working space. Query description (soled problem, data and methods from database) are presented in working space.

7) The subsystem illustrates the way of a solved problem (or why is has not solved the problem) and used knowledge. It makes easier system testing for an expert and increases confidence of a user in obtained results.

8) The term "user interface" covers all aspects of cooperation between the user and IIS. In this system an interface on the base of cooperation formation is used. The user enters data or an order into special forms (boxes). Form names (or report/table manes) are promptings for enter. Computer can give some output as a result and ask the user if it is necessary to continue an interactive process [2].

4. Theoretical bases and mathematical models of knowledge base

For building artificial neural network the authors selected 125 enterprises with basic OKVED 29.

One quarter of these enterprises is bankrupts, i.e. these enterprises were adjudged bankrupts by the Supreme Commercial Court of the Russian Federation or against which they instituted bankruptcy proceedings.

The authors calculated input data values, i.e. 10 indexes, on the basis of financial accounting annual data for 2011-2013. So they obtained 347 training specimens (irregular units of observation were deleted from the population).

There is only one output variable, it is bankruptcy. The variable id dyadic: 0 means that the enterprise is "healthy", 1 means that it is bankrupt.

Network configuration and its training is done with the help of the program Deductor Studio.

Choosing an optimal quantity of neural processing elements on the core layer is made by means of the theorem of Arnold – Kolmogorov – Khekht-Nilsen.

Therefore, taking into account made calculations it is possible to conclude that a quantity of neural processing elements on the core layer should be in the interval [3;39]. Calculated a mean value the authors have: N optimal = 22 neural processing elements.

During Building and training the network the authors used the following parameters:

- training algorithm – back propagation of error;
- significance initialization – automatical;
- quantity of input factors – 10;
- quantity of output factors – 1;
- quantity of neural processing elements on the core layer – 22;
- quantity of training specimens – 347;
- training set – 95% (330 bars);
- testing set – 5% (17 bars);
- activation function type – sigmoid;
- training rate – 1;
- gauge of acceptable mistake (the stopping down condition) – 0,05;

- mean mistake for training set – 0,01 (1 %)
- mean mistake for testing set – 0,047 (4,7 %)
- training period – 5 мин. 14 сек.
- quantity of training stages – 22047;
- taken into account from training set – 98,55%;
- taking into account from testing set – 94,44% [3].

5. Algorithm of information processing during bankruptcy risk assessment

Algorithm of information processing during bankruptcy risk assessment with the help of the connectionist model is presented by means of functional models and includes the following stages.

At the first stage one realizes data origination for network training. Then they make parsing of xls files (financial accounting – forms No 1, 2), which contain data for calculations, calculate 10 financial and economic indexes, which are the most crucial from the point of view of bankruptcy risk assessment of the enterprise.

At the second stage data pre-processing – data normalization – is realized. Neural network works with time-series values from 0 till 1. Input index values greatly exceed the limit. To use neural network it is necessary to normalize input time-series, save prepared data in MySQL database.

At the third stage training of artificial neural network is realized, i.e. model building. This stage includes three main steps: data input with further data diffusion to output directions, finding and back propagation of the error and significant adjustment; and this when network characteristics are already collated by means of an experiment.

At the fourth stage they assess bankruptcy risk of the enterprise. It is supposed that for this operation they need just data input and its data diffusion to output directions [4,5].

6. Mean values of 10 indexes for industrial enterprises

Obtained values will help in financial analysis of industrial enterprises and correct definition of the situation for timely management decision-making.

Mean index values for financially stable enterprises:

1) working capital ratio – 2,2;
2) acid test ratio – 1,1;
3) absolute liquidity ratio – 0,38;
4) liquid capital ratio in assets – 0,44;
5) concentration ratio of capital loans – 0,41;
6) long term fund raising ratio – 0,05;
7) borrowed and own funds ratio – 0,86;
8) turn-round of float – 7,38;
9) turn-round of owned capital – 2,16;
10) product profitability – 10,55%.

7. Testing of connectionist model

The authors tested obtained connectionist model with the data of bankrupt enterprises that were not taken into training and testing sets. For example, for OJSC "Izhevsk machine-building plant", Izhevsk city, which was declared a bankrupt in first quarter 2012, they obtained the following results: the first quarter 2010 – risk level is 0,07 (absence of bankruptcy); the first quarter 2011 – 0,99 (the plant is bankrupt); the first quarter 2012 – 1,0 (the plant is bankrupt). The authors also tested the model with the data of financially stable enterprise (it was not taken into training and testing sets) CJSC "Sibkabel", Tomsk city. The results were as follows: the first quarter 2010-2012 – 0 (absence of bankruptcy). It is necessary to say that basic OKVED of CJSC "Sibkabel" is 31.1 what shows that the obtained model can be applied to enterprises of other branches.

A possible solution to this problem is usage of connectionist technologies, which are a high-efficiency tool.

Conclusion

At the result of the research the authors obtained connectionist model for bankruptcy risk assessment of the enterprise on the basis of new 10 indexes of enterprise activity, which are the crucial for bankruptcy risk assessment.

Calculations are made for industrial enterprises with basic OKVED – 29. It eliminates the problem of inter-branch disagreements when one assesses index values for identifying bankruptcy risk of the enterprise.

Acknowledgments

This work was supported by Russian foundation for basic research as a part of the project “Knowledge base development for an intellectual information system of bankruptcy risk assessment of the enterprise”, the project № 14-01-31208.

References

[1] Telipenko E and Iavorskii M 2014 *Economy and Entrepreneurship* **7** 509-514.
[2] Telipenko E 2014 *Modern technologies of decision making support in economy: Conference proceedings Yurga, 28-29 April, 2014. Tomsk: TPU* pp 197-199.
[3] Telipenko E Zakharova A and Iavorskii M 2014 *In the world of scientific discoveries* **4** 128-135.
[4] Laurene V 1993 *Pearson* p 461.
[5] Chernysheva T. Yu. 2013 *Applied Mechanics and Materials* **379** 220-223.

Automated control system for a mashing process

E Teterin[1,a] **and V Rudnickiy**[1,b]
[1]Tomsk Polytechnic University, Lenina av. 30, Tomsk, 634050, Russia

E-mail: [a]badscrool@gmail.com, [b]rudnickiw@tpu.ru

Abstract. The goal of this paper is to describe a system for a mashing process, which is the first part of brewing beer. The mashing is a procedure where the fermentable (and some non-fermentable) sugars are extracted from malts. The program part based on LabVIEW, which is used to control NI CompactRIO. The main target of the project is to reach a predefined levels of the temperatures and maintain it during the pauses. When the necessary break time is ended the system is ready to go to the new value. The precise control of the temperatures during the breaks is one of the critical factors that define the texture and alcohol content of the beer. The system has two tanks with resistors PT100 in both of them, heat exchanger (coil), heater and pump. The first tank has heating element in order to rise the temperature in the other one. This project has practical solution with all explanations and graphs which are proven working ability of this control system.

1. Introduction

Heating is a common process that is used in almost all areas of life. For instance, heating procedure of mash may be used in beer, kvass and other alcohol factories [1]. A lot of serious manufacturers include automation and control of process into their production in order to get competitive advantage. Furthermore, it provides safety, environmentally friendly conditions and reliably for all operations. Consequently, there is an essential to automate heating processes. To supply this need, nowadays exists various types of intellectual appliance and software for them. The project includes the ways and set of devices to solve the problem in the automation. In order to design and construct effective control system, it is necessary to find out what and how all components are going to control system functions. So, the system has sensors, controller (CompactRIO), and final control elements (controlled device) [2].

As efficient controller in this project was chosen CompactRIO which was made by National Instruments company by way of one of the best variant to find solution. It is worked with NI LabVIEW which has reconfigurable I/O (RIO) architecture and includes a lot of different functions that were made primarily for technical measuring and program applications. This controller introduces an embedded real-time manufactured device which is representing the possibility to fast program, change and actuate CompactRIO with easy-to-use graphical instruments. This represents security and productivity for any applications and controller may be remotely managed with Ethernet. However, CompactRIO is not so flexible because it must be programmed only with LabVIEW. On the other hand, LabVIEW is strong graphical language that is introducing the flexible and the fast method to create a program. Software which was mentioned above is perfect for different kind of measurements. It is able to help programmer to focus on the difficult tasks and create invisible for user graphical complexity [3].

2. The system functions

The system has the pump to manage the temperature in the tank which has not the heat exchanger and to deliver mash to the second one container when the temperature does not reach the predefined level. The voltage for the pump is 12 VDC and can be controlled with PWM. The second tank has heat exchanger. Wort is pumped through the heat exchanger. For some reason, it seems that such system is irrational, but due this fact that mash is isolated from the water it makes the project is logical because often some technological process requires solution like this. The system has two resistance thermometers (PT100) to control the temperature which were installed in both containers. They are sensors to measure a level of temperature by correlating the temperature with the resistance of an element of sensor. The element has a length coiled wire wrapped around a platinum core. The heating resistor (230V AC, ca. 2 kW) in the heating water container is controlled with a relay. Simplified scheme of the control system is pictured in the figure 1. It is not precise prototype, but there is shown basic principles and it has all essential moments in order to understand automated mashing process control.

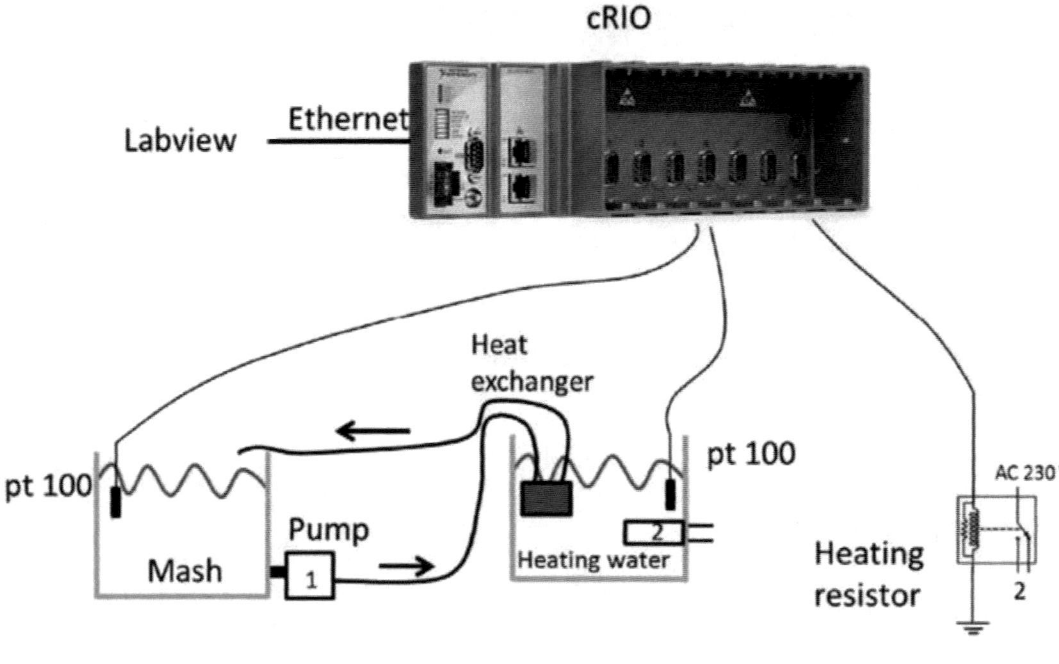

Figure 1. The basic features of control heating system.

The control system has the following functionalities:

1. The heating resistance resistor is controlled with LabVIEW and through the fixed time the program plot the graphs of the temperatures from tanks. The application may open and read any required temperatures and break times profile from a text file (txt, doc and so on);
2. Pump is programmed using VHDL (Very high speed integrated circuits language) and the operation is based on PWM (pulse-width modulation);
3. The system has opportunity for emergency stop;
4. The temperature of water must not be higher than the limit in the temperature profile;
5. The system is automated. Means that it has possibility for a little participation of human;
6. The application is able to print out the temperature through some intervals into a log file.

3. LabVIEW programming

IOP Conf. Series: Materials Science and Engineering **93** (2015) 012059 doi:10.1088/1757-899X/93/1/012059

The majority piece of system functionality is based on LabVIEW programming. As the first step to solve the problem, there was created correlation between Celsius degree and electrical resistance. By using equation of Callendar- Van Dusen the program can converse the resistance to temperature, as in equation (1).

$$R_t = R_0(1 + At + Bt^2) \tag{1}$$

where R_t is the resistance at temperature t, $A = 3.9 \cdot 10^{-3} °C^{-1}$ and $B = -5.8 \cdot 10^{-7} °C^{-2}$.

Realization the general part of the system requires the method of state machine, which has 7 states. This method is a technique which is using in programming in a lot of different languages. It is essential and flexible instrument because a coding becomes more efficient, simple and understandable [4]. The application consists from the next states:

1. «Read Array». The stage was made to load temperature profile and set starting value. Moreover, this state begins to indicate all parameters. In order to save time and resources, as soon as program was started, the state has been already turned on pump and heater (figure 2).

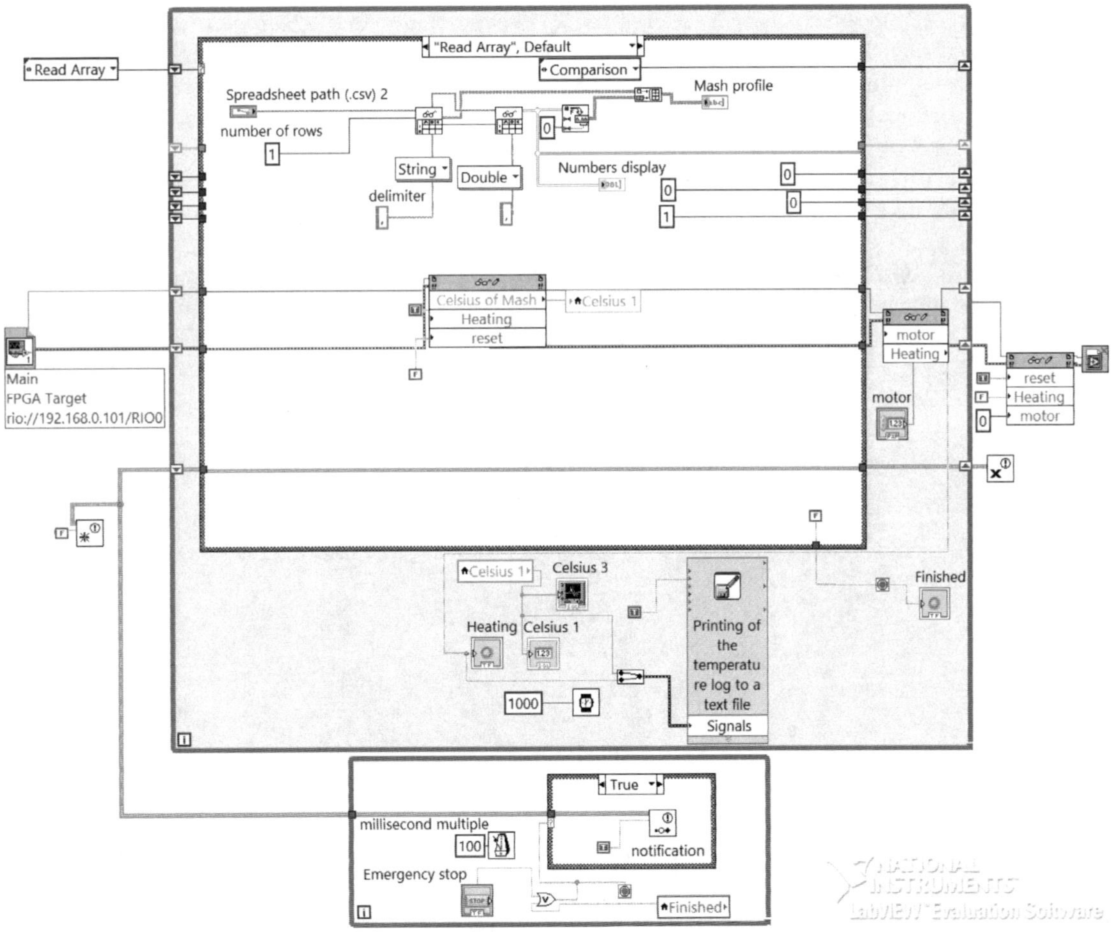

Figure 2. Realization of state machine in LabVIEW.

2. «Comparison». On this state ideal temperature which was getting from the profile compare with actual temperature from container. After that the program is going to start indicate pause time and ideal temperature value to manage the control system better.

3. «Start heating and pumping». The main idea of this stage that if the ideal temperature is higher than real temperature (condition from the previous state is true) than the program is going to leave the pump and heater is turned on and go to the previous state. Otherwise the program will go to the next stage.
4. «Stop heating». As was said above, the system will go to the next state in case if the condition from the «Comparison» state is false. It means application will turn off the heater, however the pump will stay turned on in order to supply the temperature level.
5. «Start timer». The state is responsible for the break time. This is the most important stage because all technological processes depend on how accurately the temperature is maintained. This state has complex structure because inside of it the program includes while loop in order to support the temperature during the pause time. The most complicated moment is connected with stopping loop during the pause time because there is two parallel loops which should work simultaneously. Furthermore, here exists an indicator which is able to show how much time left. Because of necessity of controlling pump and indicators inside of while loop, a problem with indicators and pump can be solved with «local variable».
6. «Increase the index». This case is made to check actual value of temperature and go to the next value from the temperature profile if it is necessary. The program will comparing the number of rows which were multiplied by 2 with sum of the previous index which increased by 1.
7. «Finished». This is the last stage of the program if the application stops without emergency case. On this state will turn off pump, heater and clear all indexes. However, in case of emergency stop the program after all operations will return on the state that is why this stage is so important.

LabVIEW introduces special block which is called «Write to a measurement file» to realize printing out data into a text file. Moreover, using an operator «Stop control», which is connected with Boolean operator «Or», there is opening opportunity to create the possibility for emergency stop. Due to this combination the program can stop either in common way or in emergency stop if it is necessary. To stop both parallel loops there was used block «Notification». It provides the real time control for the procedure in all operations included the pause times [5]. An interface of the applications is show in the figure 3.

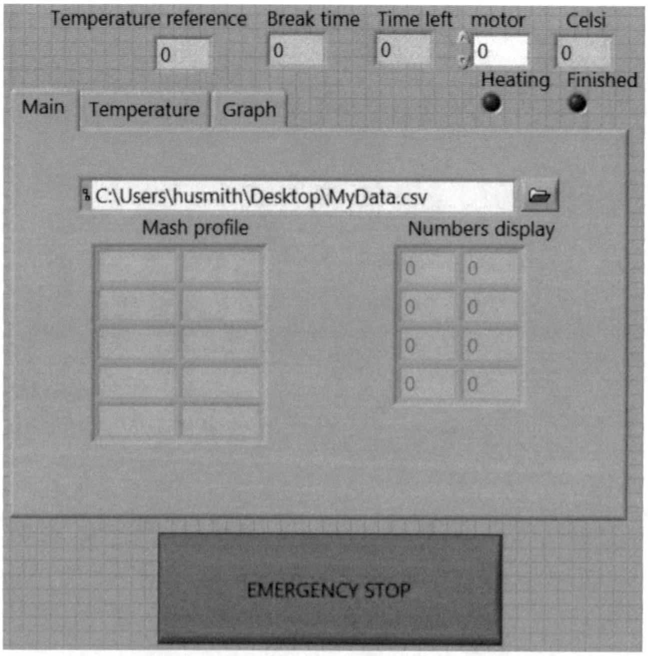

Figure 3. Interface of the program.

4. Operation of the pump

The pump is based on PWM technology. It is programmed with VHDL and after that this code was integrated into LabVIEW by using special functions «IP-block» and «Read/Write Control» in the FPGA (part of coding in LabVIEW). The general idea of VHDL code is only to control PWM value of the pump's motor. It has three inputs; «mclk», «reset» and «motor». «Mclk» means clock signal, and «reset» is reset signal. «Motor» is vector (7 down to 0) which is put manually. It decides the maximum constant of counter as 255. If «reset» signal is activated, output value pump directly goes to 0. If it is not, there are two stages; «update» and «pwm». In «update» stage, it gets the duty cycle from the input value «motor» and move to «pwm» stage. In «pwm» stage, the counter increases by 1 for every rising edge of clock and if it reaches the value of motor, the output value «pump» turn on to 1 unless the value of pump is 0. When counter reaches 255 (maximum constant), initialize count value and go to «update» stage again. The figure 4 represents the simulation of the pump in VHDL.

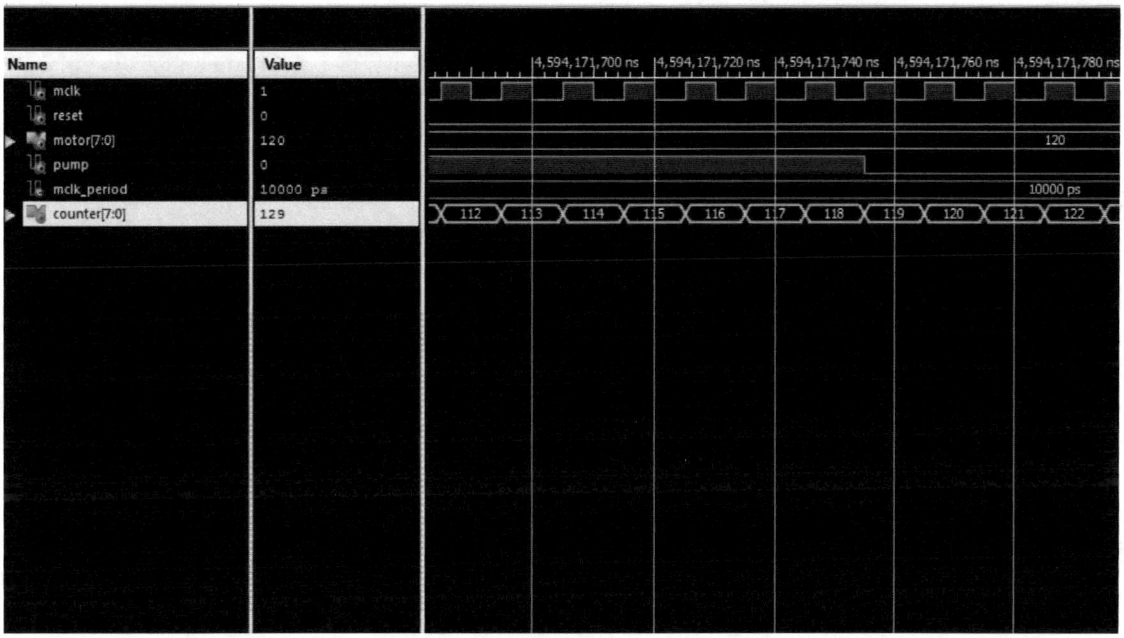

Figure 4. Operation of the pump.

5. Conclusion

The project is represented the working solution of control the mashing process. Practical tests of the system in the laboratory was proven the program is working in the right way. LabVIEW and CompactRIO are one of the best way to solve the problems like this. All results, which was reached and explained in this paper, can be useful in a lot of branches of production. Implementation of such tasks required knowledge in a number of technologies including: electrical systems, instrumentation, real time data acquisition, process modeling, control structure development, and controller tuning.

6. References

[1] Malyshenko A M and Semyonov N M 1999 Using of indexes of fundamental characteristics of controllable objects for raising resistance to failure of automatic control system *The 3th Russia-Korean Int. Symp. on Science and Technology*. **1** 5

[2] Bitter R, Mohiuddin T and Nawrocki M 2007 *LabVIEW: Advanced Programming Techniques* (Boca Raton: CRC Press) p 163

[3] Sharma K 2011 *Overview of Industrial Process Automation* (Waltham: Elsevier) p 210

[4] Marlin T 1995 *Process Control: Designing Processes and Control Systems for Dynamic Performance* (New York: McGraw-Hill) p 453
[4] Malyshenko A M 1997 Automatic control systems with redundant dimension of a vector of control: theoretical and applied aspects *The 1st Korea-Russia Symposium on Science and Technology.* **1** 49
[5] Kachin O S 2007 *Diagnosing the mechanical state of the collector-brush unit of the high-speed electric machine at resource tests* (Tomsk University Publishing) pp 112-116

Development of strapdown inertial navigation system with MEMS sensors, barometric altimeter and ultrasonic range meter

I S Kholopov
Associate Professor,
Ryazan State Radio Engineering University, Ryazan, Russian Federation

E-mail: kholopov.i.s@rsreu.ru

Abstract. The results of strapdown inertial navigation system (SINS) tests with 9 degrees of freedom MEMS sensor MPU-9150 (triaxial accelerometer, gyroscope and magnetometer), pressure sensor LPS331 and ultrasonic range meter HC-SR04, implemented on the FPGA Altera Cyclone-II evaluation board DE1 is considered. SINS measures the spatial coordinates and altitude relative to the starting point, the orientation angles and distances to obstacles along the way. It is shown that the relative error of the spatial coordinates estimation does not exceed 1.1% in interval of some minutes.

1. Angular coordinates estimation

The classical parameters that allow to define the angular position of a rigid body uniquely are three Euler - Krylov angles: yaw ψ, pitch θ and roll φ. As the result of rotation in three-dimensional space is not invariant to the sequence of turns, so we consider the case when the resulting rotation is specified by a sequence of rotations $\psi \to \theta \to \varphi$ [1].

It is suitable to use quaternions – 4-component vectors $\mathbf{q} = [q_0, q_1, q_2, q_3]^{\mathrm{T}}$ – for solving the problem of integrating of measured gyro angular velocities:

$$\mathbf{q} = q_0 + \mathrm{i}q_1 + \mathrm{j}q_2 + \mathrm{k}q_3,\ \|\mathbf{q}\| = 1,$$

$$\mathrm{ij} = \mathrm{k},\ \mathrm{ji} = -\mathrm{k},\ \mathrm{jk} = \mathrm{i},\ \mathrm{kj} = \mathrm{i},\ \mathrm{ki} = \mathrm{j},\ \mathrm{ik} = -\mathrm{j},\ \mathrm{i}^2 = \mathrm{j}^2 = \mathrm{k}^2 = -1, \tag{1}$$

$$\begin{aligned}
q_0 &= \cos(\psi/2)\cos(\theta/2)\cos(\varphi/2) - \sin(\psi/2)\sin(\theta/2)\sin(\varphi/2),\\
q_1 &= \cos(\psi/2)\cos(\theta/2)\sin(\varphi/2) + \sin(\psi/2)\sin(\theta/2)\cos(\varphi/2),\\
q_2 &= \sin(\psi/2)\cos(\theta/2)\cos(\varphi/2) + \cos(\psi/2)\sin(\theta/2)\sin(\varphi/2),\\
q_3 &= \cos(\psi/2)\sin(\theta/2)\cos(\varphi/2) - \sin(\psi/2)\cos(\theta/2)\sin(\varphi/2).
\end{aligned} \tag{2}$$

For a right-handed coordinate system vectors used in MPU-9150 (Y-axis is directed forward, X-axis is directed right and Z-axis is directed up) the quaternion rate of change according to [2] is

$$\dot{\mathbf{q}} = \frac{1}{2}\begin{bmatrix} -q_1 & -q_2 & -q_3 \\ q_0 & -q_3 & q_2 \\ q_3 & q_0 & -q_1 \\ -q_2 & q_1 & q_0 \end{bmatrix}\begin{bmatrix} w_y \\ w_x \\ w_z \end{bmatrix}, \tag{3}$$

where $[w_x, w_y, w_z]^T$ is the gyroscope data.
Integration of (3) gives

$$\mathbf{q}_{i+1} = \mathbf{q}_i \circ \mathbf{q}_{rot}, \tag{4}$$

where the symbol "$\circ$" denotes the quaternions multiplication according to the rules (1), $\mathbf{q}_{rot} = [\gamma + i\Delta_y + j\Delta_z + k\Delta_x]$ is quaternion rotation during the time interval T_i between points of time i and $(i + 1)$, $\|\mathbf{q}_{rot}\| = 1$, $\Delta_k = 0{,}5w_kT_i$, $\gamma = [1-(\Delta_x^2+\Delta_y^2+\Delta_z^2)/8]^{1/2}$, $k = \{x, y, z\}$ or [2]

$$\mathbf{q}_{i+1} = \frac{\mathbf{q}_i + \mathbf{q}_{rot}}{\|\mathbf{q}_i + \mathbf{q}_{rot}\|}. \tag{5}$$

Rodrigues - Hamilton parameters and Euler - Krylov angles are associated by relations

$$\psi = \mathrm{atan2}\left(q_0q_2 - q_1q_3, q_0^2 + q_1^2 - 0{,}5\right),$$

$$\theta = \mathrm{atan2}\left(q_0q_1 - q_2q_3, q_0^2 + q_2^2 - 0{,}5\right), \tag{6}$$

$$\varphi = \arcsin\left(2q_1q_2 + 2q_0q_3\right).$$

The accelerometer and the magnetometer are used to determine the initial values of Euler - Krylov angles:

$$\varphi_a = -\mathrm{atan2}\left(a_x, a_z\right), \quad \theta_a = \mathrm{atan2}\left(a_y, a_z\cos\varphi_a - a_x\sin\varphi_a\right), \tag{7}$$

$$\psi_m = \mathrm{atan2}\left(H_x, H_y\right), \tag{8}$$

where $H_x = m_x\cos\varphi_a + m_z\sin\varphi_a$, $H_y = m_x\sin\varphi_a\sin\theta_a + m_y\cos\theta_a - m_z\cos\varphi_a\sin\theta_a$ and $[a_x, a_y, a_z]^T$ and $[m_x, m_y, m_z]^T$ is the accelerometer and magnetometer data respectively.

As the expression (8) allows to determine the deviation from the Earth magnetic pole, not from geographic, so it must be also taken account of magnetic declination by evaluating the course in (8) [3].

2. Estimation of travelled distance, altitude and distance to the obstacles

In the presence of the speed sensor according to the method of dead reckoning the course from the known current coordinates $[X_i, Y_i, Z_i]^T$ at a discrete point of time i and the ground speed vector V_i projections $[V_{xi}, V_{yi}, V_{zi}]^T$ the coordinates along the axis X, Y and Z at next discrete point of time $(i + 1)$ can be estimated at the following:

$$[X_{i+1}, Y_{i+1}, Z_{i+1}]^T = [X_i, Y_i, Z_i]^T + T_i[V_{xi}, V_{yi}, V_{zi}]^T,$$

where $V_{zi} = V_i\sin\theta_i$, $V_{yi} = V_i\cos\psi_i\cos\theta_i$, $V_{xi} = V_i\sin\psi_i\cos\theta_i$.

Except the inertial sensors in the considered SINS were also used the ultrasonic range meter to determine the distance to obstacles and barometric pressure sensor for measuring the altitude.

To reduce the measurement error the sound velocity c was constantly adjusted depending on the temperature change:

$$R(t_d, t) = c(t)t_d/2 = (331 + 0.6t)t_d/2, \tag{9}$$

where t_d is the delay time of the ultrasonic signal.

The atmospheric pressure sensor was used to determine the altitude H of the object (in meters) relative to the start point [4]:

$$H = 44307.7\left[1 - \frac{P(H)}{P_a}\right]^{0.190284}, \quad (10)$$

where $P(H)$ is the pressure on the current height in mbar and P_a = 1013 mbar is the atmospheric pressure.

3. Fusion of navigation information from various sensors

For SINS on MEMS sensors are generally used either the explicit complementary filter [5, 6] or AHRS algorithm [7]. In [8] it is shown that the use of AHRS by the high level of accelerometer noise (for example, by driving on uneven surfaces) gives a more noisy estimate of the roll and pitch than the use of the algorithms [5] or [6], that's why the explicit complementary filter from [8] was used for the developed SINS with proportional and integral gains $k_p = 0.28$ and $k_i = 0.015$ respectively.

4. Experiment results

During the research of the developed SINS (its block diagram indicating the types of sensors and communication protocols is shown at figure 1), a number of experiments was made: the passage of wheeled platform attached to it SINS through a closed loop with a avoiding obstacle from the left and the right side and with a single stop at start point (figure 2), movement of the platform in a vertical plane, distancing from the obstacles and approach to it. The accelerometer, the gyroscope and the magnetometer were previously calibrated according to the procedures [9-11]. Because the area where the experiment was carried out attended by a large number of massive metal objects the magnetometer data for course correction is not used. Assembled under the scheme of incremental encoder, the speed sensor produced a pulse signal for every 7 mm of travelled distance. The output data rate of the inertial sensors and the speed sensor was 100 Hz, and the output data rate of the ultrasonic range meter and the pressure sensor was 25 Hz.

For MEMS gyro temperature drift compensation used simple linear approximation [12]:

$$b(t) = b(t_0) + \beta_t (t - t_0),$$

where $b(t)$ is gyro bias, t_0 is start temperature and β_t is a linear term coefficient determined during gyro calibration.

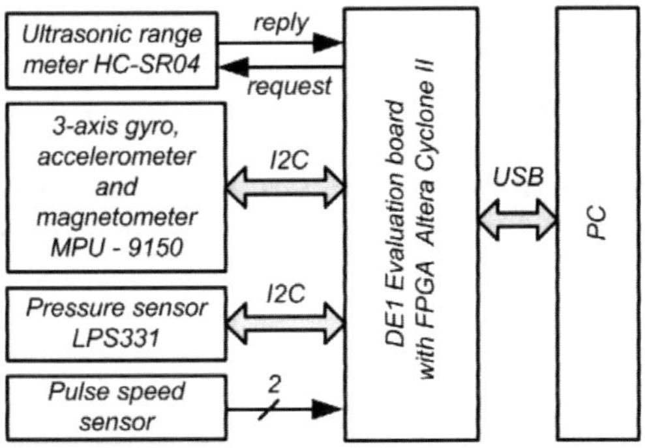

Figure 1. Block diagram of the investigated SINS.

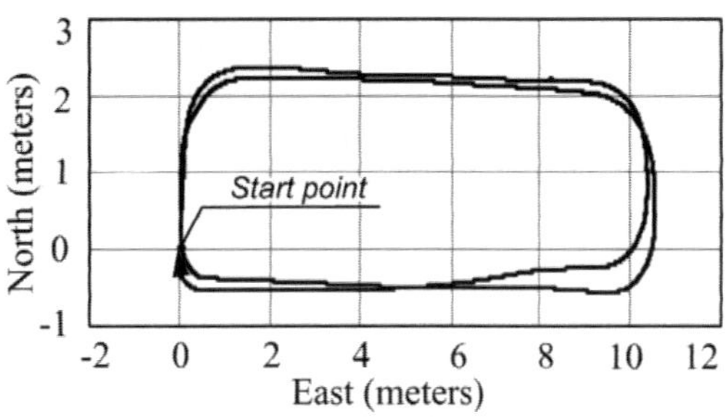

Figure 2. Track built by SINS.

Estimates of yaw, pitch and roll as well as distance to the obstacles are shown in figures 3-5. Simple threshold filter that the values in the rate gyro signal which lie under a certain threshold are filtered out and set to zero [13] is not used.

IOP Conf. Series: Materials Science and Engineering **93** (2015) 012060 doi:10.1088/1757-899X/93/1/012060

High level of pitch and roll fluctuations (figure 4) is due to bad quality (roughness) of asphalt surface on the area where experiment was carried out.

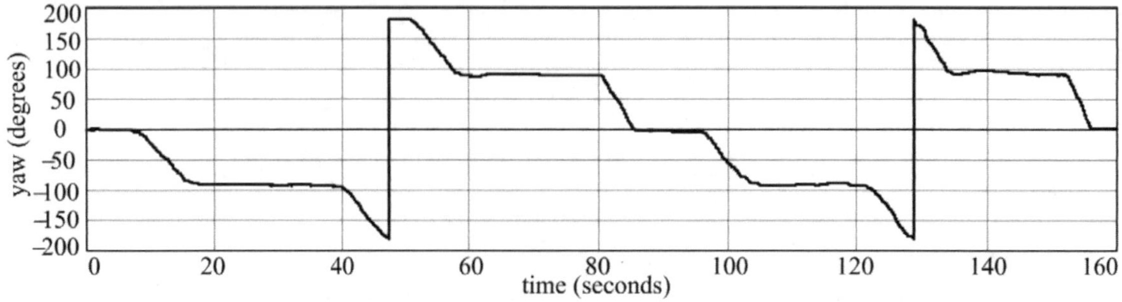

Figure 3. Yaw estimation.

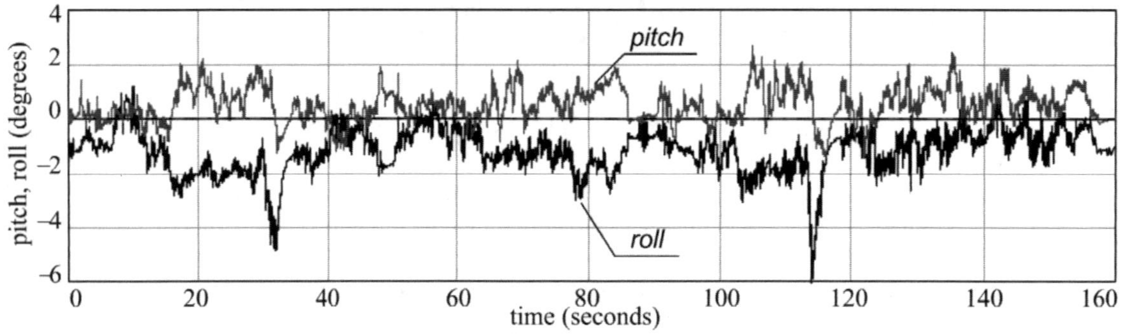

Figure 4. Pitch and roll estimation.

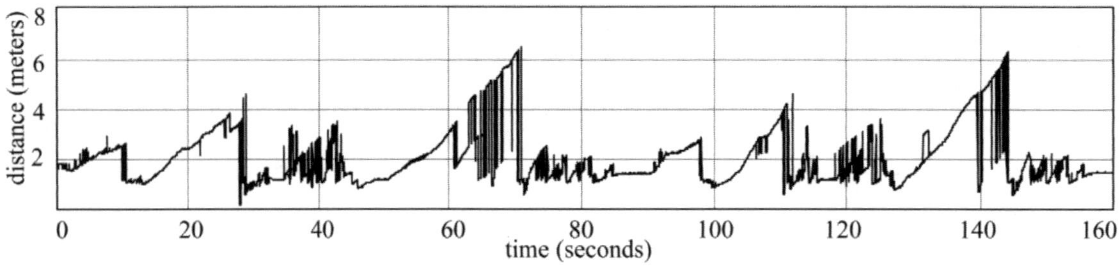

Figure 5. Distance to obstacle estimation.

Approximately similar results were obtained for the twice more velocity of the platform.

By the relative error on the end of moving obtained result the worse than result in [6] (where yaw correction by magnetometer signals is used), but comparable with the results obtained in [14] (where additional correction by optical system is used) and [15] after 50 meters from start point.

5. Conclusion

The results of several experiments have shown:

1) in the range of several minutes and by vibration less than 1 m/s^2 the root mean square error (RMSE) of angular coordinates is less than 0.9° and the relative spatial coordinates error is less than 1.1%;

2) at a distance up to 2.5 m (9) provides an absolute measurement error less than 10 mm and the RMSE less than 3.2 mm;

3) RMSE measurement of barometric altitude (10) by temperature correction according to [4] is less than 0.8 m.

Acknowledgments
This work has been accomplished with the assistance of the Ministry of Education and Science of Russian Federation within the agreement on granting a subsidy No. 14.574.21.0012 (RFMEFI57414X0012).

References

[1] Chelnokov Yu N 2006 *Quaternion and Biquaternion Models and Methods of Mechanics of Solids and Their Applications* (Moscow: Fizmatlit)
[2] Kuipers J B 1998 *Quaternions and Rotation Sequences* (New Jersey: Princeton University Press)
[3] Magnetic Declination [Electronic resource]. – URL: http://magnetic-declination.com/ (request date: 05.03.2015).
[4] Hardware and Software Guidelines for Use of the LPS331AP [Electronic resource]. – URL: http://www.st.com / st-web-ui / static / active / en / resource / technical / document / application_note / DM00064026.pdf (request date: 11.04.2015).
[5] Mahony R and Hamel T 2006 *Proc. Int. Conf. Robotics Automation (ICRA)* (Orlando) pp 2170-75
[6] Mahony R, Hamel T and Pflimlin J-M 2008 *IEEE Trans. on Automatic Control* pp 1203-18
[7] Madgwick S O H, Harrison A J L and Vaidyanathan R 2011 *IEEE Int. Conf. on Rehabilitation Robotics Rehab Week* (Zurich) pp 1-7
[8] Fakhri A, Zhou Z and Hu J 2014 *2nd Int. Conf. on Research in Science, Engineering and Technology (ICRSET'2014)* (Dubai) pp 86-91.
[9] Kian S T, Awad M, Dehghani A and Zahedi S 2011 *Proc. of the World Congress on Engineering* vol. III (London).
[10] Wang L and Wang F 2011 *J. of Intelligent Engineering and Systems* **4** 32
[11] Feng W, Liu S and Yang S 2013 *J. of Information & Computational Science* **10** 1551
[12] Prikhodko I P, Trusov A A and Shkel A M 2013 *Sensors and Actuators A: Phys.* **201** 517
[13] Hong S K and Park S 2008 *Sensors* **8** 7287
[14] Guarnieri A, Pirotti F and Vettore A 2013 *Sensors* **13** 1510
[15] Liu S, Maher Atia M, Gao Y and Noureldin A 2015 *Micromachines* **6** 196

IOP Conf. Series: Materials Science and Engineering **93** (2015) 012061 doi:10.1088/1757-899X/93/1/012061

A sensorless initial rotor position's estimation for permanent magnet synchronous machines

I Krasnov [1], S Langraf [1], I Odnolopylov[1]and V Koltun [1]
Russia, Tomsk, Lenin ave, 30, National Research Tomsk Polytechnic University

E-mail: kiu@tpu.ru

Abstract. Permanent magnet synchronous motors for the effective start require information about the initial position of a rotor. In this regard, most systems use position sensors, which substantially increase entirely a cost of an electrical drive [1-3]. The aim of this article is to develop a new method, allowing determining the absolute angular position of the permanent magnet synchronous motors' rotor [4,5]. With a certain voltage pulses applied to the motor, its stator is magnetized by currents leakage in the windings. This allows using a special algorithm to calculate the absolute position of the rotor without using any motor parameters [6]. Simulation results prove the simplicity and efficiency of this method for determining an initial position of the permanent magnet synchronous motors' rotor. Thus, this method can be widely used in the electrical industry.

1. Introduction

Recently the drive with a vector control, based on a synchronous motor with permanent magnets, is increasingly used in various industries because of their advantages. There is an Urgent problem of sensorless Permanent Magnet Synchronous Machines (PMSM) control, as it allows reducing the cost of the structure and the size of the drive [7].

Many foreign scientists have been working on this issue. The most significant contribution to the study of methods for determining the initial position of the rotor is made by such scholars as Japanese T. Noguchi from Nagaoka University of Technology, Ying Yan and Jian Guo Zhu of the University of Technology, Sydney, Australia. In addition, at the company ZAO "EleSy" (Tomsk, Russia) the possibility of sensorless PMSM control applying is currently studied. Known algorithms allow operating a motor without using mechanical sensors for high and medium motor speeds, but the question remains of determining the initial position of the rotor at start-up. The main problem on this issue is to find an algorithm which would be at the same time accurate enough and simple to implement in a frequency converter.

Thus, the aim of this work is to develop an algorithm determining the initial position of the sensorless PMSM. The object of the study is a sensorless controlled driver within PMSM.

For performing PMSM modeling, which takes into account the saliency of the rotor and the stator core saturation, should be developed. The algorithm using for determining an initial position of the rotor without using of mechanical sensors should be carried out before motor starts. Thus, this algorithm has to be fast and simple to be implemented.

From the practical point of view, presented algorithm can be applied both by newly design and by modification of existing motor drives within a frequency converter.

2. Model of a Permanent Magnet Synchronous Machine

Schematic design and construction of a synchronous motor with permanent magnets are shown in Figure 1 [8].

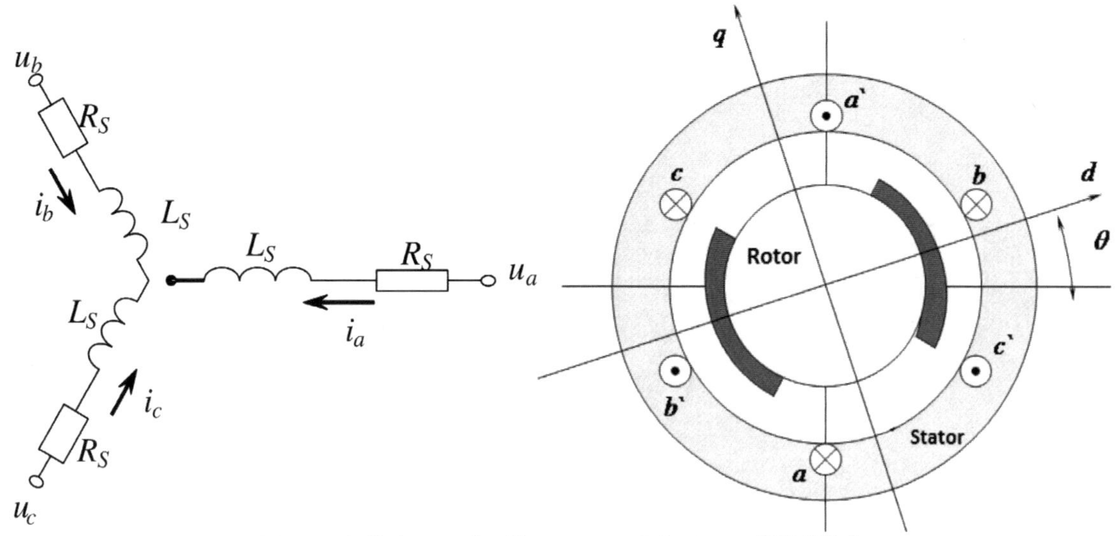

Figure 1. Schematic diagram and design of PMSM.

PMSM model, created by the blocks MATLAB Simulink, is presented in the Figure 2.

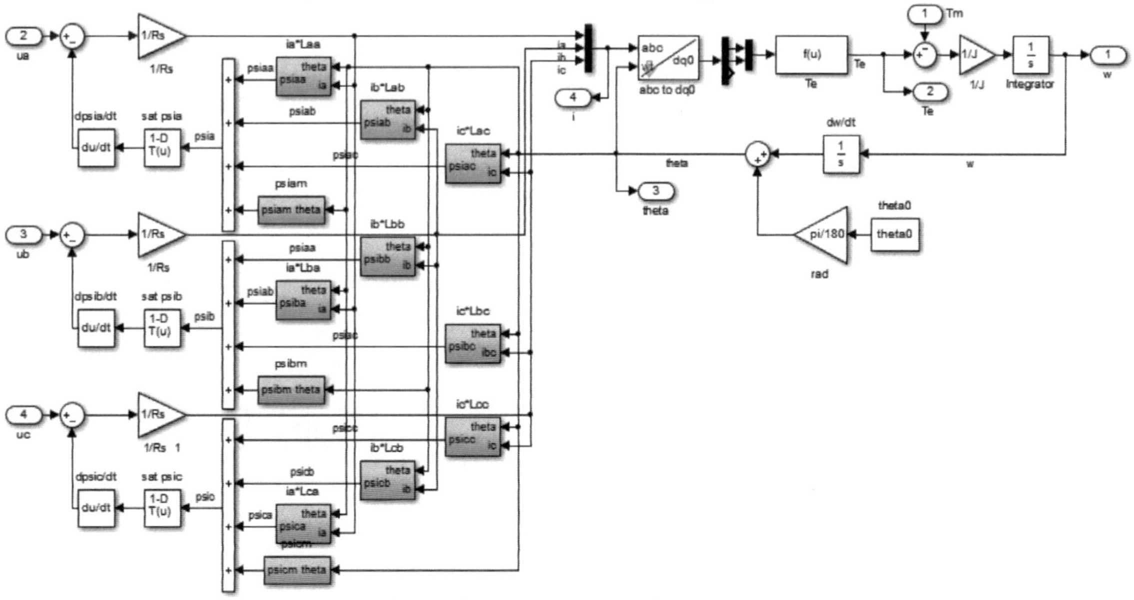

Figure 2. The contents of the PMSM.

The Next step is checking an adequacy of developed model (in comparison with the built-in Simulink model. To make it practical the scheme shown in Figure 3, was constructed in MATLAB Simulink).

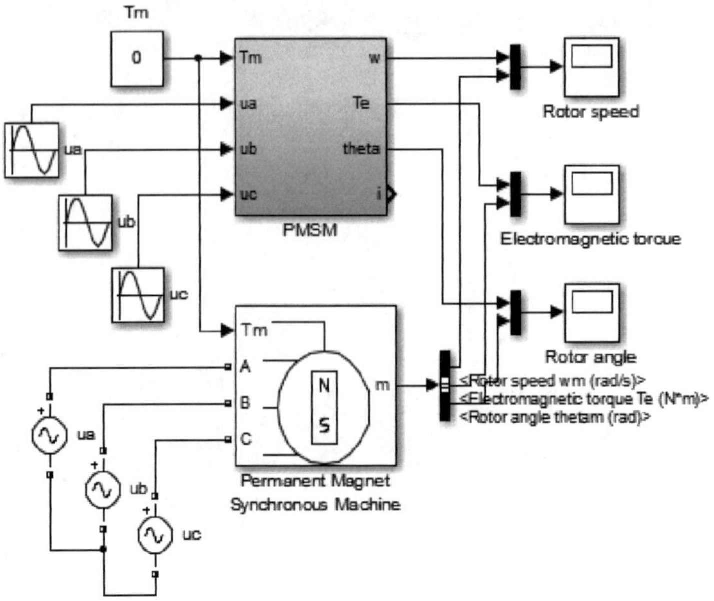

Figure 3. A comparison of the model with a built-in MATLAB Simulink.

The resulting graphics of velocity are shown in Figure 4.

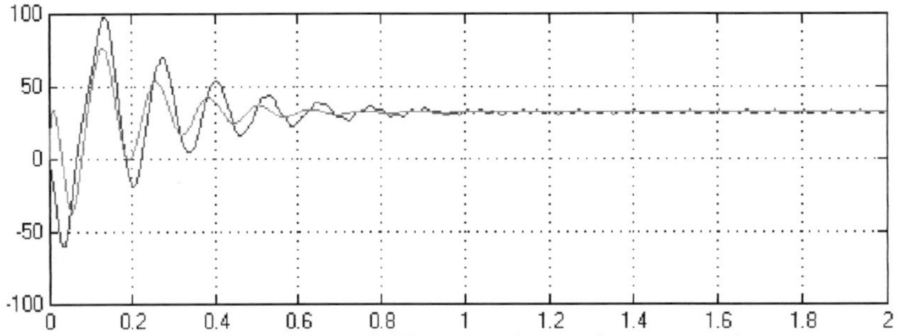

Figure 4. The waveform of rotor speed.

According to the obtained graphics it is possible to conclude that the model of a synchronous motor with permanent magnets works properly, and it can be used for further research algorithms to determine the initial position of the rotor.

3. Determining methods a rotor's initial position

Each phase of the motor is fed alternately firstly by positive and then negative voltage pulses [6]. At the same time maximum and minimum values of pulse currents depending on the angle of rotation by a certain law rotor can be measured. Thus, we have the input data for calculation: I_{Ap}, I_{Bp}, I_{Cp} – maximum values of positive pulses, I_{An}, I_{Bn}, I_{Cn} – maximum negative pulses.

First of all, we define the average current value:

- for positive pulses

$$I_{0p} = \frac{I_{Ap} + I_{Bp} + I_{Cp}}{3}, \tag{1}$$

- for negative pulses

$$I_{0n} = \frac{I_{An} + I_{Bn} + I_{Cn}}{3}. \quad (2)$$

Then, the deviation from the mean value for each phase is determined:

$$\Delta I_{Ap} = I_{Ap} - I_{0p} \quad (3)$$

$$\Delta I_{Bp} = I_{Bp} - I_{0p} \quad (4)$$

$$\Delta I_{Cp} = I_{Cp} - I_{0p} \quad (5)$$

$$\Delta I_{An} = I_{An} - I_{0n} \quad (6)$$

$$\Delta I_{Bn} = I_{Bn} - I_{0p} \quad (7)$$

$$\Delta I_{Cn} = I_{Cn} - I_{0p} \quad (8)$$

After that the sector where the north pole of the rotor is placed can be determined.

**Table. Cases for (θ`) defining** The Sector is defined by the angle shift $\theta$` and determined by the maximum value of current deviation from the mean value.

p.p	Maximum value (A)	Sector θ` (rad)
1.	ΔI_{Ap}	0
2.	ΔI_{Bp}	$\frac{2\cdot\pi}{3}$
3.	ΔI_{Cp}	$\frac{4\cdot\pi}{3}$
4.	ΔI_{An}	π
5.	ΔI_{Bn}	$\frac{5\cdot\pi}{3}$
6.	ΔI_{Cn}	$\frac{\pi}{3}$

Then, the angle within a sector is determined by two remaining values with a help of the following equation:

$$\theta = \frac{\cos\left(\frac{2\cdot\pi}{3}\right)}{\sin\left(\frac{2\cdot\pi}{3}\right)} \cdot \frac{\Delta I_1 - \Delta I_2}{\Delta I_1 + \Delta I_2} \text{ radian.,} \quad (9)$$

where $\Delta I_1, \Delta I_2$ the currents in other two phases.

The angle of rotor's rotation is defined by the sum of the angle that characterizes the sector, where the north pole of the rotor is situated, and the angle calculated by the formula [9]:

$$\theta_r = \theta\text{`} + \theta \text{ [rad]}, \quad (10)$$

Model, constructed in MATLAB Simulink to analysis the developed algorithm, is shown in Figure 5.

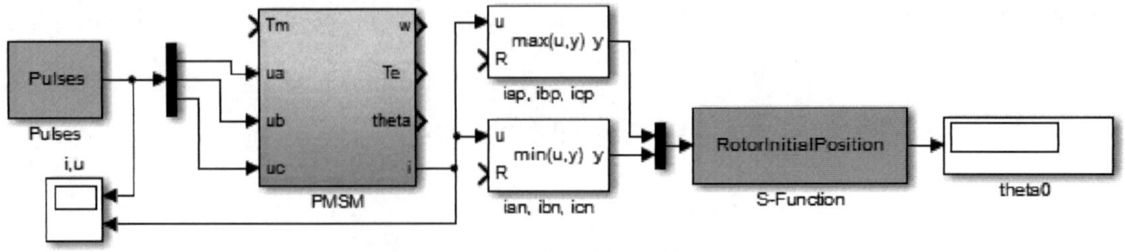

Figure 5. Model in MATLAB Simulink.

The inputs of the block PMSM are supplied by voltage pulses with the form and duration, as shown in Figure 6.

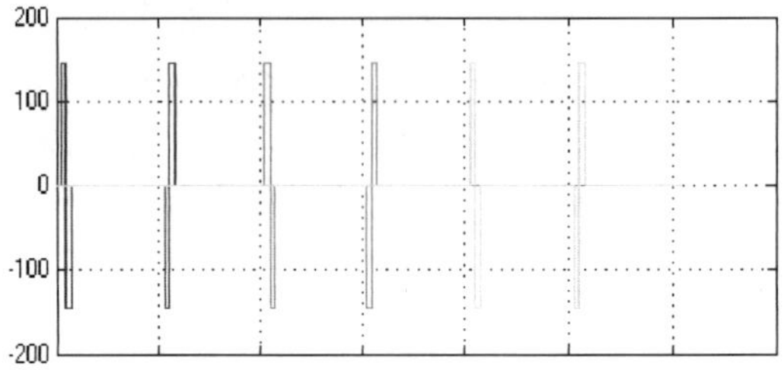

Figure 6. The voltage pulses at the input of the motor.

Oscillograms of the currents depend on the position of the rotor. The angles of rotor's rotation 45 and 180 degrees [10] are shown in Figure 7.

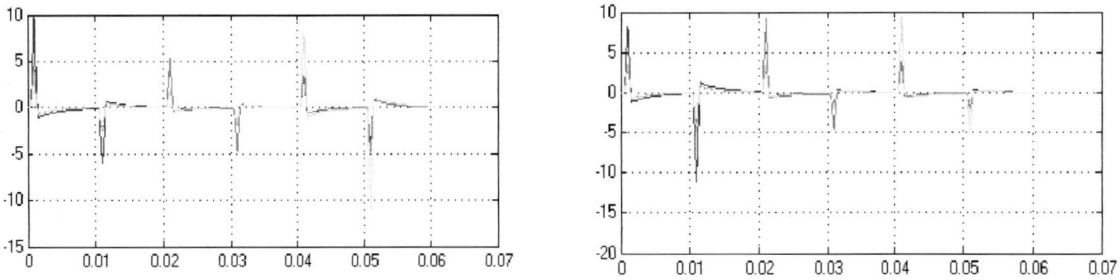

(a) (b)

Figure 7. The oscillograms of current for rotation angles: a) 45 degrees; b) 180 degrees.

After all minimum and maximum values of the currents, measured in the windings of the motor, have been transmitted to the input of "Rotor Initial Position", the calculation of rotor position starts.

For different values of angle of rotor's rotation the experiments were carried out. According to that we constructed the real and calculated position of the rotor. Moreover an angle's error determining histogram was received and data are reflected in Figure 7 [11].

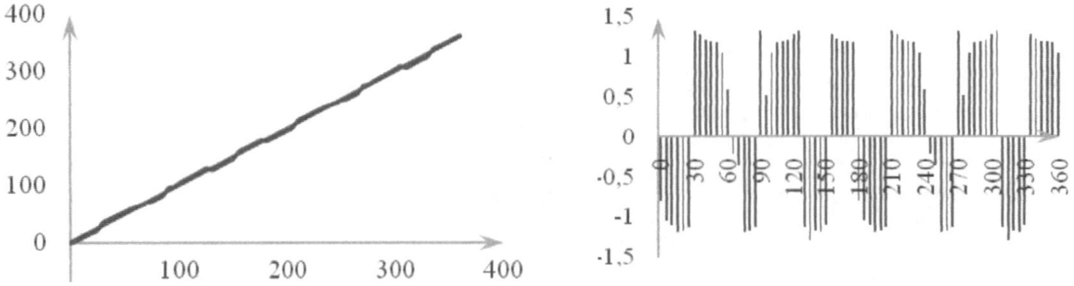

(a) (b)

Figure 8. The actual and calculated position of the rotor, degrees (a) and the determining error (b).

4. Conclusion

This work presents a technique to calculate the absolute position of the initial rotor angle of synchronous machines with permanent magnets. It is possible to measure by applying voltage pulses that partially drive the stator iron into saturation of the rotor's absolute position for a given pole pair. The scheme is computationally simple and does not need the information of any motor parameters.
Total error of measurement is less than 5%. So, this is a valid value, and the proposed algorithm that can be used in industry [12].

Implementing this technique to sensorless control algorithms while a motor turns around will allow a totally sensorless control scheme operating the motor reliably from zero speed up to full speed. Research to implement sensorless technology on a process of application is currently being processed.

References

[1] Vogelsberger M, Grubic S, Habetler T and Wolbank T 2010 *IEEE Trans. Ind. Electronics* **57** 365-374

[2] Raca D, Garcia P, Reigosa D, Briz B and Lorenz R 2010 *IEEE Trans. Ind. Applications* **46** 167-178

[3] Hammel W and Kennel R 2010 *Symposium on Sensor less Control for Electrical Drives* p 56-63

[4] Paulus D, Landsmann P and Kennel R 2011 *Symposium on Sensorless Control for Electrical Drives* 41-46

[5] HuaY, Sumner M, Asher G, Gao Q and Saleh K 2011 *IET Electric Power Applications* **5(4)** 359-370

[6] Wang Yi, Guo N, Zhu J, Duan N, Wang S, Guo Y, Xu W and Li W 2010 *IEEE Transactions On Magnetics* **46(6)** 2016-2019

[7] Bolognani S, Calligaro S, Petrella R and Sterpellone M 2011 *Symposium on Sensor less Control for Electrical Drives* 64-73

[8] Leidhold R 2011 *IEEE Trans. Ind. Electronics* **58(12)** 5371-537

[9] Schmidt P, Gasperi M, Ray G and Wijenayake A 1997 *Proc. IEEE-IAS Annu. Meeting* 463-549

[10] Haque M, Zhong L and Rahman M 2002 *Proc. IEEE Power Electron. Spec. Conf.* **1** 276-281

[11] Parasiliti F, Petrella R and Tursini M 2000 *In: IAS 2000 Annual Meeting and World Conference on Industrial applications of Electrical Energy* p 8-12

[12] Jang J, Sul S, Ha J, Ide K and Sawamura M 2003 IEEE Transactions on Industry Applications 1031-1039

Information system for maintaining a database of geobotanical descriptions while studying a landscape

K Shchukova[1]
Russia, Tomsk, 634050, National Research Tomsk Polytechnic University

E-mail: embrre@yandex.ru

Abstract. Nowadays, there is a need in developing a common information space for providing a centralized data storage of geobotanical descriptions while studying a landscape and custom software for collecting, storing and processing semi-structured data. In the paper, the results of information system development for maintaining a geobotanical database are discussed. The database was developed via the bottom-up design approach and ER-model (Entity-Relationship). The analysis results of geobotanical data obtained from field studies and the database structure of geobotanical descriptions while studying a landscape are considered. At the moment, the information system for maintaining the geobotanical database is applied by the Institute of Steppe, Ural Branch of Russian Academy of Sciences.

1. Introduction

A xylium and phytocoenosis inventory process is one of the most important scientific tasks because the results of studies provide to analyze biological diversity and perform ecological vegetation monitoring of the areas being explored.

The growing popularity of computer-based activities has encouraged not IT-specialists in Vegetation Science to use a wide range of various software tools for solving their scientific fundamental and applied problems. In the field studies of sample areas, a large amount of geobotanical data is constantly being accumulated. In order to provide a central data storage, it is necessary to ensure a common information space. The custom software has to have a user-friendly interface and sufficient functional capabilities for data processing and agile reporting.

Our analysis of similar information systems focused on geobotanical and ecological data processing has shown that the best-known products are TurboVEG, Juice and IBIS. The systems are intended for maintaining a phytocoenotic database. They also allow to significantly automate analysis and classification processes of different plant communities.

However, the software products are not fully satisfied by the user requirements. Unfortunately, there is no opportunity to modify these systems because they are proprietary or based on obsolete technologies.

Thus, the main aim of the study is to create a geobotanical database (DB) while studying a landscape and an information system (IS) for its using intended for collecting, storing and processing tremendous volumes of geobotanical data. The functional capabilities of the IS were discussed in cooperation with the Institute of Steppe, Ural Branch of Russian Academy of Sciences (UB RAS).

[1] To whom any correspondence should be addressed.

2. Geobotanical data analysis and database development

Nowadays, most specialists in Geobotany apply a rich variety of theoretic and practical methods to perform geobotanical descriptions studies.

In the field investigations while studying a landscape, they fill out special forms of geobotanical descriptions, create documents with the results of the chemical soil analysis in Microsoft Word and take pictures of a landscape being studied.

The use of common forms provides to standardize and unify a traditional procedure of geobotanical description handling. The geobotanical data obtained from the field studies are primarily used for analyzing and comparing phytocoenotic data.

The information about ecotope features and geographical coordinates of sample areas is gained while studying a landscape. In order to determine exact geographical coordinates and an altitude of areas being studied, GPS devices are primarily used.

While studying a phytocoenosis, all vegetation species are classified according to the stratification of the plants. A vegetation layer is a structural part of a vertical phytocoenosis building.

The woodland structure includes the following layers: Canopy/Tree Layer (A), Understorey/Shrub Layer (B), Field Layer (C) and Ground Layer (D).

In order to describe various vegetation species, the following quantitative characteristics are used depending on a particular layer: abundance evaluation according to the scale of O. Drude, a plant cover, landscape aspects and a common plant cover.

The characteristics obtained from field studies are recorded in special forms of geobotanical descriptions for each layer and sublayer. A name of vegetation species is identified according to dominate plant species of each layer.

On the basis of described forest and herbaceous vegetation characteristics, all the plants are classified into the following categories: a life form, an ecological group, a coenotic group, latitudinal and longitudinal groups, a particular value. In addition to the vegetation classification, a statistical analysis of species, communities and families is carried out for various geobotanical descriptions.

In the process of database design, the following information sources provided by the Institute of Steppe, UB RAS, were used: geobotanical descriptive forms created in MS Word, documents of chemical soil analysis created in MS Word and Excel, summary tables of phytocoenotic data generated in MS Excel, landscape images of studied areas, the results of biomorphological and phytocoenotic vegetation analysis described in documents of MS Word and Excel, GPS tracks including a point name, geographical coordinates, an altitude and etc.

In general, these forms contain the following characteristics of forest and herbaceous communities: a number of a geobotanical description, a value of a sample area, descriptions of vegetation layers, names of plants, geographical coordinates of a studied area, a description of relief features, a description of soil properties, the results of chemical soil analysis, human and animal impacts, a total plant cover; vegetation species, plant heights, abundance evaluation according to the scale of O. Drude, phonological stages, aspects, location features and etc.

Then, a taxonomic, biomorphological and phytocoenotic analysis is performed by using the obtained geobotanical data [1].

At the stage of conceptual database scheme design, semantics, typical structures of phytocoenotic descriptions, and relationships between geobotanical objects were identified. In the process of geobotanical data analysis, information inaccuracy and incompleteness have been founded out in all kinds of documents provided by the Institute of Steppe, UB RAS. Semi-structured data representation has led to the conflict with data unification and formalization requirements of IT.

The infological analysis has revealed that the most common type of relationships between objects of geobotanical descriptions is the many-to-many relationship. For instance, various vegetation species may live in the same area, and vice versa. A single abundance evaluation according to the scale of O. Drude may be characterized by several descriptions of woody species, and vice versa.

IOP Conf. Series: Materials Science and Engineering **93** (2015) 012062 doi:10.1088/1757-899X/93/1/012062

Geobotanical objects are also related by the one-to-many relationship. A plan community may include a lot of plant families, but each family consists of various species. Vegetation layers may involve a large number of sublayers that may be related to a lot of vegetation descriptions.

While designing the physical data model, the database of geobotanical descriptions has been successfully developed. This database structure allows to expand a scope of work without any information system modification. A database framework was developed via the bottom-up design approach and entity-relationship (ER) diagram. The developed database consists of 22 tables in total including 10 lookup tables.

In order to distinguish different types of tables, the terms 'lookup table' and 'domain object' were introduced. A lookup table is a table storing reference data. These tables are used for storing data frequently used by users. For example, lookup tables of vegetation layers and phytocenosis types allow to eliminate a manual data input. Due to the use of lookup tables, the time for data search can be significantly reduced.

The developed database includes the following lookup tables: phytocenosis types, vegetation layers and sublayers, plant families, communities and etc.

A 'domain object' is a table containing useful data of the field being studied with a more sophisticated framework in contrast with a lookup table.

The database also involves the following domain objects: a geobotanical description for a specific sample area, a description of a vegetation layer and sublayer, and etc.

In order to store chemical soil analysis files created in MS Word and Excel and obtained images, special tables in the database were generated.

In all database tables, surrogate keys were introduced to ensure uniformity across the database framework. The alternative keys have been generated for table attributes which are candidate keys [2].

3. Development features and functional capabilities of the information system

In order to describe a functional mechanism of IS in considerable details, the IDEF0 Functional Modeling method was applied for effective visualizing of IS functional capabilities [3].

The top level of the IDEF0 model is shown in Figures 1.

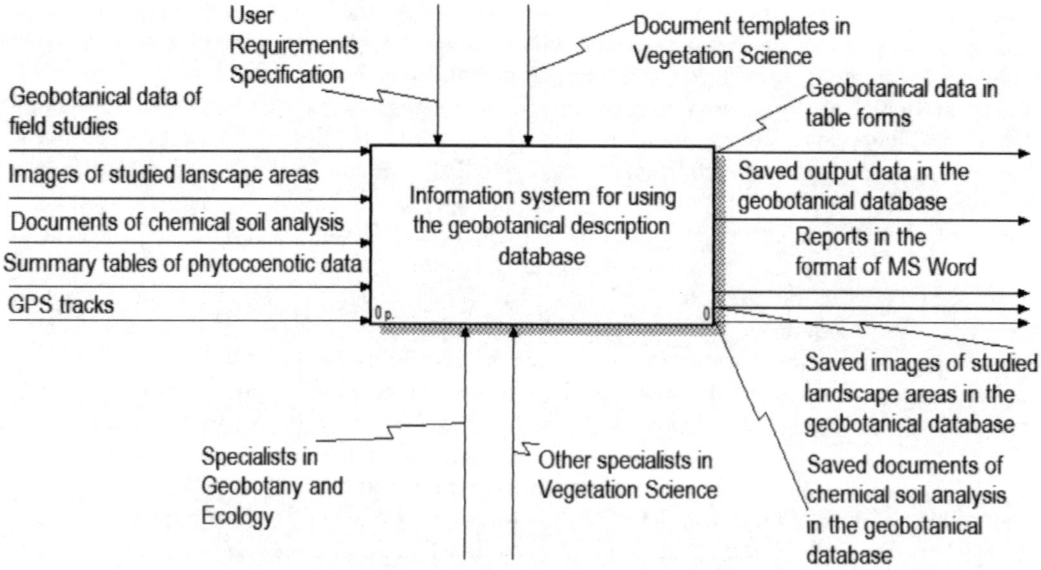

Figure 1. The top level of the IDEF0 model.

The system consists of four subsystems (Figure 2):

1. Data handling subsystem. It is intended for transforming input data into a database format.

2. Data storing subsystem. This subsystem provides a centralized data storage in the database of geobotanical descriptions.
3. Reporting subsystem. The subsystem allows to generate a report in a necessary format.
4. Data visualization subsystem. It is used for visualizing various geobotanical data, for example, in table forms.

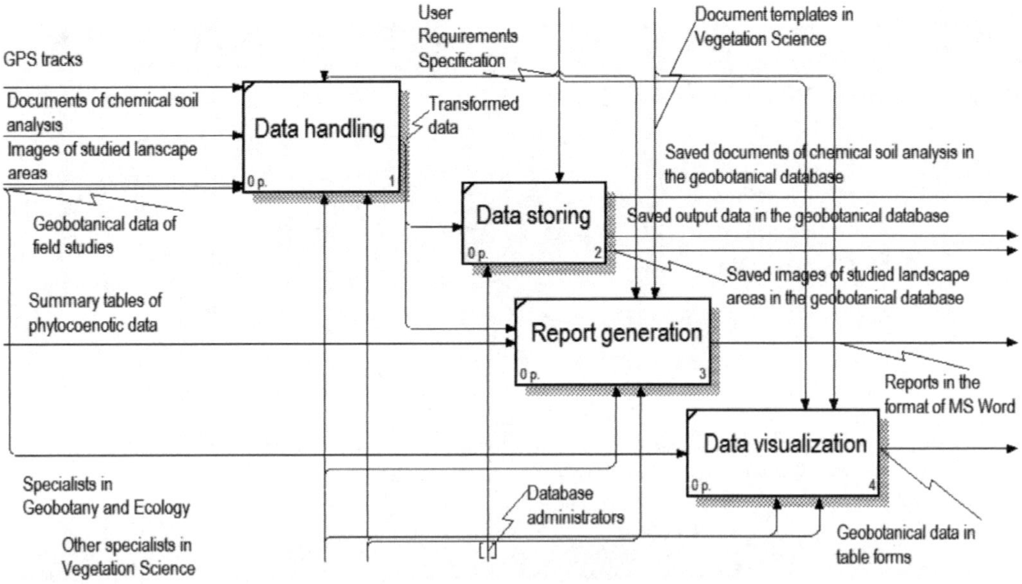

Figure 2. The decomposition of the IDEF0 model.

The information system for geobotanical database maintenance has been implemented in Microsoft Visual Studio 2013 by the C # language. MS SQL Server 2008 R2 Express was used as a database management system (DBMS). The use of ADO.NET helps to accelerate the process of data access functionality encapsulation and eliminate a lot of mistakes caused by a human factor.

In order to connect with MS Word, the Microsoft.Office.Interop.Word and Microsoft.CSharp libraries were added in a project. The interface of the information system includes various drop-down lists and lookup tables providing a quick data access. All input user's data is controlled by special software tools developed for checking input data and spelling. The IS has a user-friendly interface and offers efficient tools for processing, manipulating and extracting geobotanical data. The system operation does not require specific knowledge and skills in IT. Hence, it may be applied by any specialists in Geobotany and Vegetation Science. The IS is intended for intensifying a routine work with tremendous volumes of the customer data.

4. Conclusion

Due to the use of enabling technologies, the information system for geobotanical database maintenance with sufficient functional capabilities has been successfully developed. At the moment, this system is used by the Institute of Steppe, UB RAS. In the future, the DB framework and IS functional capabilities are being planned to expand in order to carry out a biomorphological and phytocoenotic vegetation analysis and set up a connection with GPS devices and Google Earth.

References

[1] Korchagin A A 2012 *The Field Geobotany. Vegetation Building* vol 5 (Moscow: Book On Demand) pp 200-320
[2] Date C J 2006 *An introduction to Database Systems* (Book On Demand) p 1328
[3] Jesse R and Cohn R 2015 *IDEF0* (Book On Demand) p 105

IOP Conf. Series: Materials Science and Engineering **93** (2015) 012063 doi:10.1088/1757-899X/93/1/012063

The Adaptive Control of Accuracy at Centerless Grinding of Rolling Bearings

O A Yalovoy[1], O V Zakharov[1] and A V Kochetkov[2]
[1] Yuri Gagarin State Technical University, Saratov, Russia,
[2] Perm National Research Polytechnic University, Perm, Russia

E-mail: zov17@mail.ru

Abstract. The method of adaptive control of accuracy at centerless grinding is developed. This method is based on statistical modeling of Monte Carlo and considers basic data of measurement of roundness of details. Results of application of the method for control of the accuracy of processing of rings of rolling bearings are given.

1. Introduction

Continuous increase of requirements to the accuracy of geometrical characteristics of details is observed in modern mechanical engineering. Therefore adaptive control of processing accuracy became actual. At centerless grinding the greatest difficulty is caused by receiving a precise form of a detail [1]. For precision details, for example rolling bearings, it is difficult to provide values of roundness within 0.0003-0.002 mm. It is known [2, 3] that a form deviation is usually a result of violation of a trajectory of the relative movement of a detail and a tool. For rolling bearings' details except traditional error fraction (elastic movements, thermal deformations, vibrations) the significant contribution is made by a hereditary drawback of geometry.

Centerless grinders work in an automatic cycle and represent poorly governed systems. The organization of control for output parameters by means of feedback is difficult. Such machines on motionless support as the models SIW-5 and SWaAGL-50 (Germany) have limited opportunities for geometrical adjustment change and cutting modes. On the other hand static control for the size of such machines is more difficult because of nonrigid kinematic communications and power short circuit of contact between a detail, a tool and basing machine elements. The arising errors of basing depend as on parameters of adjustment of machines as on an initial error of details' geometry.

2. The method of adaptive control of accuracy at centerless grinding

The main objective of processing accuracy control on grinders is to provide the maximum correction of roundness. The solution of this task is carried out on the basis of optimum adjustment of the machine within the parameters setting the relative provision of a detail 1, a grinding wheel 2 and basing elements – faceplates 3 with a magnetic clip and two motionless support 4, 5 (figure 1). The main complexity is that deviations of a detail's form have stochastic character and there is no real possibility for the best adjustment.

Statistical simulation (the Monte Carlo method) may be used here [4]. In setting up a machine tool with minimum basing error, this approach involves simulation of the stochastic input data (shape deviations of the blanks) and multiple implementation of the analytical basing model. Simulation yields probability characteristics whose numerical values are equal to the results of solving the determinate problem. Statistical analysis of a series of basing-error values permits the determination of the influence exerted on the machining accuracy of a batch of blanks by the setup parameters of the machine tool.

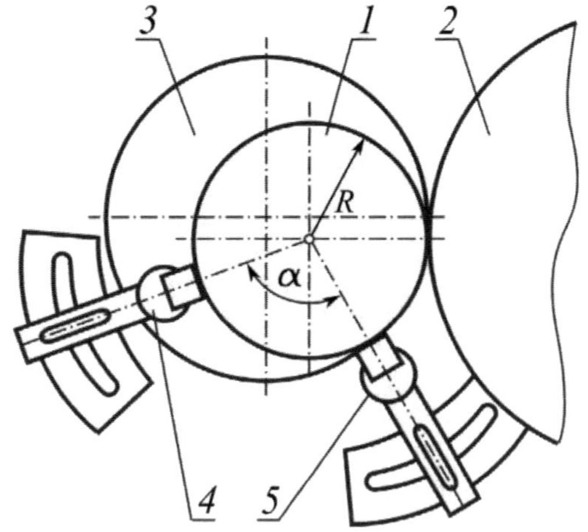

Figure 1. Calculating model at centerless grinding.

The corresponding algorithm is shown in figure 2. The initial data include the blank parameters *BP* (radius r_0 of the mean profile circumference; number p of harmonics; distribution parameters and boundaries of the ranges of the amplitudes a_n and initial phases φ_n of the harmonics), the setup parameters S of the machine tool (the angle α between the motionless bearings), and the number m of blanks in the batch. The variation of α in optimization is bounded on the basis of the machine tool characteristics.

In the first stage, the geometric shape deviations in the batches of blanks are simulated. First, sequences of uniformly distributed numbers random z_i are generated for each blank j, as a function of the number of profile harmonics ($i = \overline{1,m}$). The values of z_i are converted to the required distribution law for the amplitude a_{ni} and initial phase φ_{ni} of each harmonic. As a result of summation, the profile r_j of the blank is formed [5, 6].

The cross section of the blank with periodic shape deviations is described by a trigonometric polynomial of the form

$$r = R + \sum_{n=2}^{p} a_n \cos(n\varphi - \varphi_n), \tag{1}$$

where R is a Least Squares Circle radius; a_n, φ_n is the n-harmonic's amplitude and epoch angle; p is a maximum number of harmonics considered; φ is a polar angle.

The generation of random numbers with a specified distribution is based on the inversion method [4]. In this method, a sequence of random numbers zi uniformly distributed in the interval [0, 1] is formed and then transformed:

$$x_i = F^{-1}(z_i), \tag{2}$$

where $F^{-1}(z_i)$ is the inverse of the distribution function of the random quantity x_i.

Experiments show that the amplitudes a_n of the harmonics are best described by a β distribution, while the initial phases φ_n are best described by an equal-probability law.

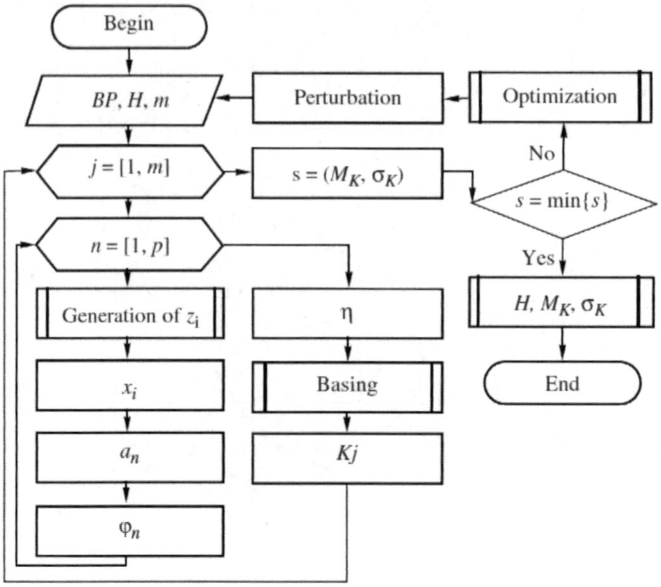

Figure 2. Simulating algorithm for machine-tool setup (Monte Carlo method).

The probability-density function of the β distribution takes the form

$$f(x_i) = \frac{\Gamma(\eta+\mu)}{\Gamma(\eta)\Gamma(\mu)} x_i^{\eta-1} (1-x_i)^{\mu-1}, \qquad (3)$$

where Γ is a gamma function; η and μ are the parameters of the β distribution.

The distribution function of the initial phases of the harmonics takes the form

$$F(x_i) = (x_i - a)/(b-a), \qquad (4)$$

where b and c are the boundaries of variation of φ_n.

Experiments show that there are strong correlations between the amplitudes of some harmonics; these must be taken into account in simulation. Therefore, for random errors x_1 and x_2 with different distribution functions $F_1(x_1)$ and $F_2(x_2)$, mathematical expectations m_{X1} and m_{X2}, and mean square deviations σ_{X1} and σ_{X2}, it is expedient to switch to the random quantities z_1 and z_2 uniformly distributed in the interval [0, 1], by means of the transformations

$$z_1 = F_1\left(\frac{x_1 - m_{X1}}{\sigma_{X1}}\right); \; z_2 = F_2\left(\frac{x_2 - m_{X2}}{\sigma_{X2}}\right). \qquad (5)$$

The linear correlation coefficient is [7]

$$k = \frac{12}{n-1}\sum_{i=1}^{n}\left(z_{1i} - \frac{1}{2}\right)\left(z_{2i} - \frac{1}{2}\right). \qquad (6)$$

This simplifies the generation of correlated random quantities with different distributions and eliminates the dependence of k on the form of these distributions.

In the second stage, we calculate the basing error for each blank from the analytical model [7]. Then we find the basing characteristic K_j for each blank j in the batch.

The third stage is statistical analysis of the K values calculated for all the blanks in the batch. As a result, we obtain the mathematical expectation M_K and the mean square deviation σ_K. Then, we optimize the setup parameters H of the machine tool so as to minimize M_K and σ_K.

Analysis of the results obtained by the proposed simulation algorithm show (figure 2) that K is best described by a normal or log-normal distribution. (The log-normal distribution appears when there is a correlation between groups of harmonic amplitudes.) The probability-density function is uniquely determined by M_K and σ_K, which are selected as the parameters to be optimized. Numerical experiments indicate that M_K and σ_K are a minimum at the point corresponding to a specific setup angle α.

3. Results of experimental research

The proposed method is verified at Saratovsky Bearing Plant (EPK Saratov) in grinding bearing rings 5-830900AE1.02 on SWaAGL-50 machine tools. The tolerance for noncircularity of the annular channel is 0.0012 mm; the surface roughness Ra = 0.32 μm. In grinding the wheel and the blank move in the same direction, at speeds of 35 m/s and 35 m/min, respectively.

The data of monitoring of roundness of a ring of the bearing received during 2010 are given in figure 3. Correction coefficients (the roundness relation before and after fair grinding) for average value of K_X and a standard deviation of K_σ are presented. The vertically shaped line shows processing time border at standard and adaptive adjustment of the machine.

The optimum adjustment of the grinder has stabilized the average value and the standard deviation of roundness in party irrespective to the initial error of details' form. Dimensionless coefficients of correction of K_X and K_σ increased accordingly from 0.6 – 1.4 to 1.1 – 2.75. Along with that a stable decrease of a standard deviation of roundness on an absolute value is noted. These results are consistent with the data in Refs [8. 9].

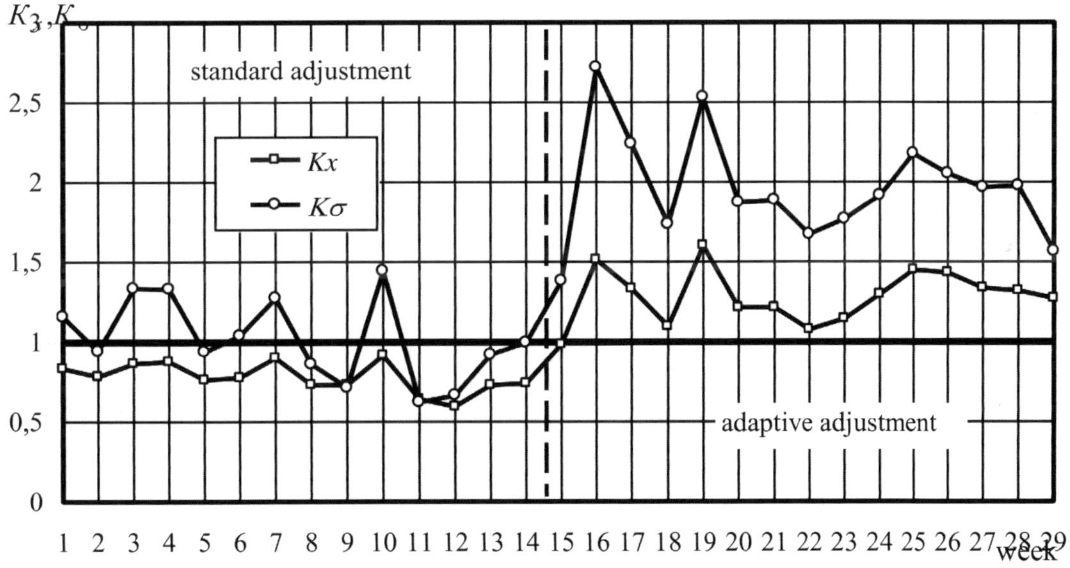

Figure 3. Monitoring of roundness of a ring of the bearing.

4. Summary

Application of adaptive control of accuracy on the basis of statistical modeling opens new opportunities for adjustment of centerless grinders. Further development of grinders is connected with creation on their basis mechanotronic systems for ensuring optimum processing of details of rolling bearings.

Acknowledgments
The research work has been done due to the support of the grant of the President of Russian Federation (MD-1377.2014.8).

IOP Conf. Series: Materials Science and Engineering **93** (2015) 012063 doi:10.1088/1757-899X/93/1/012063

References

[1] Bazrov B M 1984 *Computer Calculation of Machine Precision* (Moscow: Mashinostroenie)
[2] Fil'kin V P and Koltunov I B 1971 *Progressive Methods of Centerless Grinding* (Moscow: Mashinostroenie)
[3] Zakharov O V 2006 *Minimizing the Shaping Errors in Centerless Abrasive Machining* (Saratov: SGTU)
[4] Sovetov B Ya and Yakovlev S A 1998 *Modeling Systems* (Moscow: Vysshaya Shkola)
[5] Zyuzin A A, Kaz'min B N and Yurov M D 2014 *Russian Engineering Research* **34** 212-215
[6] Yanushkin V N, Kolyada Yu B and Krushnyak N T 2014 *Measurement Techniques* **57** 643-646
[7] Labutin S A and Pugin M V 2000 *Measurement Techniques* **43** 918-922
[8] Muzakkir S M, Hirani H and Thakre G D 2013 *Tribology Transactions* **56** 1060-1068
[9] Vasin A N, Iznairov B M and Bochkarev A P 2014 *Russian Engineering Research* **34** 112-114

Evaluation of reliability of a perforated pipe when working in the stream of liquid metal coolant.

D B Belousova [1], **D S Samokhin** [1]

[1] Obninsk Institute for Nuclear Power Engineering of the National Research Nuclear University MEPhI, Obninsk, Russia

E-mail: das9026@yandex.ru

Abstract
The article describes the approach to the assessment of indicators of reliability of products, having in its composition of perforated pipes. The idea of writing this article was born during the practical work on the reliability of the mass transfer performance evaluation system, which is part of the system to maintain the concentration of oxygen in a systems perspective SVBR-100 . The article provides an overview of the currently existing, approaches to estimate the reliability of the products containing a perforated structure, as well as theoretical and practical part, to the use of the proposed approach.

1. Introduction

The use of heavy liquid metal coolants in power plants due to several problems. One of them is ensuring corrosion resistance in contact with these fluids structural materials and preventing cases of removing toxins from the circuit or its individual parts[1-4].

Feature of nuclear power plants (NPP) is a large range of different items and often their uniqueness. This determines the requirements for models in parametric estimation of reliability of nuclear power plants and research the nature of all calculations carried out. Studies that are conducted in order to examine the stochastic properties of the performance element of nuclear power plants, allow you to select the model to evaluate the reliability of the parameter.

To remove solids from the coolant often at the points of the technological scheme of the nuclear power unit, where necessary and possible, establish a grid-hole filters. The accumulation of solid particles on the filters last clog and become inoperative, which can lead to the accident of the nuclear power unit. Consequently, the required periodic inspections and replacement of filters that is associated with the need to shut down the reactor and conduct repair works of large scale.

The requirements of reliability and economic efficiency of nuclear power plants lead to the need of optimal inspection and replacement of filters. For this optimization perforated pipe requires the study of the dynamics of clogging.

2.Analytical formulas for determining reliability assessment

2.1 A model of the dynamics of clogging of the mesh and holes in the pipe

Let to some moment of time t on the filter got n particles and flux particles are distributed evenly throughout the filter. Depending on the ratio of the sizes of particles and holes of the filter when particles on the filter can occur, the following two events, leading to clogging of the filter:

1) a particle exceeds the size of the hole of the filter, and the particles on the filter leads to clogging of the openings;

2) the particle does not exceed the size of the hole of the filter, but its getting to the border between the openings of the filter leads to the fact that particles are captured on the filter and block the portion of the passage section of the holes (safety margin).

Suppose that an event A is that when hit by one particle on the filter hole of the filter will be clogged, and the distribution of particle sizes is such that with probability P_1 the sizes of the particles exceed the dimensions of the hole and with probability $P_2=1-P_1$ is not exceeded.

Then the probability of event A is $P(A) = P(A/1)P_1 + P(A/2)P_2$, where $P(A/1)$ is the conditional probability that the particle will clog the hole with the larger hole sizes; $P(A/2)$ is the same probability, but assuming that the sizes of the particles smaller than the holes. The probability $P(A/1)$ and $P(A/2)$ are determined by the filter design [5].

Let N filter holes. The probability that any of the holes are not clogged any of the n particles, there are

$$\Omega=\left(1-\frac{P(A)}{N}\right)^n$$

According to the classical definition of probability Ω is the share of the remaining free holes of the total number N. When the equality of the areas of the holes, the probability of Ω is a fraction of the free flow cross-sections of the filter.

If we denote

$$\chi= -\ln\left(1-\frac{P(A)}{N}\right) \quad , \quad \chi>0 \tag{1}$$

then

$$\Omega=\exp(-\chi n) \tag{2}$$

Thus, the share of the remaining missed flow area of the perforated pipe decreases exponentially with increasing number of n particles trapped on the filter for some time t.

As can be seen, and (1) and (2), the share of the remaining missed flow section Ω is essentially dependent on the probability $P(A/1)$ and $P(A/2)$. These probabilities are determined by the design of the perforated pipe.

The pipe has openings which perform the filtering function and is called hole filter. At the end of the pipe there is a grid – mesh filter.

2.2 Mesh filter.

Let the filter mesh with the threads of the net that make up the squares with sides $2a$ in figure 1. If we neglect the thickness of the yarn, it is clear that for particles with a largest linear size exceeding $\sqrt{8}a$,

$$P(A/1)=1$$

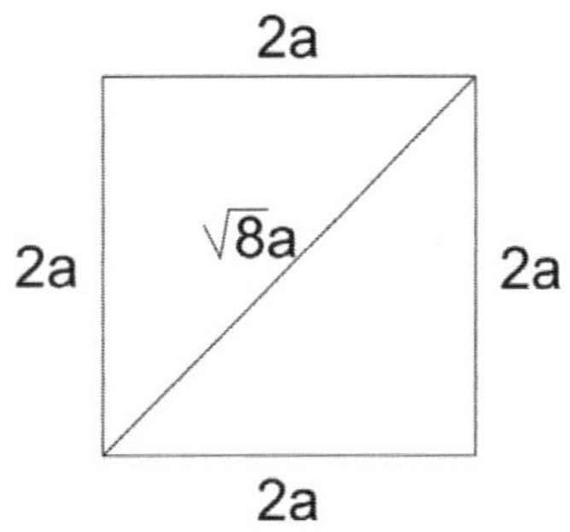

Figure 1. One of the holes of a strainer

It is possible to show that the particles with the largest linear size, not the superior $\sqrt{8a}$ could be kept in the filter probability

$$P(\mathrm{A}/2) = 1 - \left(1 - \frac{S}{\pi \cdot a}\right)^2$$

where S is the parameter of the projection of the particles on the plane of the filter.

2.3 Perforation

The perforations may be round, square or oval, offset, oblong or straight rows of holes. It can also be decorative, in the form of a specific pattern of holes of different shape and size. The shape of the holes are manufactured in accordance with specific technical needs of the customer.

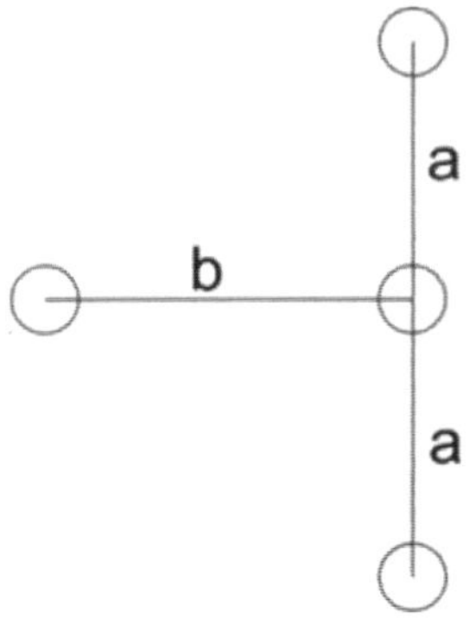

Fugire 2. The holes of the perforated pipe and the distance between them

In the perforated pipe has holes, they perform the filter function and distribution of the particles. Let the hole radius r arranged on a plane of a regular rectangular lattice with the distance between the centers of the holes is equal to a and b for each of the lattice directions. Let the filter fall of spherical particles of diameter d. Using the concept of geometric probability, it is easy to show that in this case

$$P(\mathrm{A}/1) = \frac{\pi(2r+d)^2}{4ab}\left[1 + \frac{\pi(\alpha+\beta)}{180} - \sin\alpha - \sin\beta\right]$$

$$P(\mathrm{A}/1) = \frac{2\pi rd}{ab} + \frac{\pi(2r+d)^2}{4ab}\left[\frac{\pi(\alpha+\beta)}{180} - \sin\alpha - \sin\beta\right] \tag{3}$$

where α,β is the central angles formed by the intersection of the circles of radius $r+d/2$, the centers of which are located respectively at a distance a and b from each other (figure 2), and

$$\sin\frac{\alpha}{2}=\begin{cases}\sqrt{1-\dfrac{a^2}{(2r+d)^2}} & \text{when } 2r+d>a\\ 0 & \text{when } 2r+d\le a\end{cases}$$

$$\sin\frac{b}{2}=\begin{cases}\sqrt{1-\dfrac{b^2}{(2r+d)^2}} & \text{when } 2r+d>b\\ 0 & \text{when } 2r+d\le b\end{cases}$$

From (3) it is clear that not always P(A/1)=1 [5].

2.4 Stochastic model of the process of clogging of the perforated pipe.

Assume that the flow rate through the pipe is on average constant and the particles are evenly distributed throughout the volume of the coolant. Then the growth in time of the number of particles entering the perforated pipe, occurs on average at constant speed $\upsilon > 0$, which imposed some random component $v(t)$ [5]. This component is determined by random fluctuations in the number of particles per unit volume of coolant and random fluctuations of the flow. Thus, the change in time of the number of particles deposited on the perforated pipe, can be simplified to describe the stochastic differential equation of the form

$$\frac{dn}{dt}=\upsilon+\nu(t) \tag{4}$$

Further suggest that $v(t)$ is a stationary Gaussian random process, the type of white noise, and for which $M\cdot\nu(t)=0$ and $\operatorname{cov}(\nu_t,\nu_{t+\tau})=\sigma^2\cdot\delta(t)$

Equation (4) describes a Markov random process $n(t)$ with a constant coefficient of demolition and υ diffusion σ^2. For transitional plane $w(n,t/n_0)$ is a Markov process $n(t)$ is the Kolmogorov equation has the form [5]:

$$\frac{\partial w\left(n,\dfrac{t}{n_0}\right)}{\partial t}+\upsilon\cdot\frac{\partial w\left(n,\dfrac{t}{n_0}\right)}{\partial n}-\sigma^2\cdot\frac{\partial^2 w\left(n,\dfrac{t}{n_0}\right)}{\partial n^2}=0 \tag{5}$$

Because in a physical sense $n\ge 0$, the boundary conditions for equation (5) as follows:

$$w(n,0/n_0)=\sigma\cdot(n-n_0) \tag{6}$$

$$\left.w(n,t/n_0)-\frac{\sigma^2}{\upsilon}\cdot\frac{\partial w(n,t/n_0)}{\partial n}\right|_{\substack{n=0\\ n=\infty}}=0 \tag{7}$$

The solution of the boundary value problem (5) – (7) has the following form:

$$\begin{aligned}w(n,t/n_0)=&\frac{1}{2\sigma\cdot\sqrt{\pi\cdot t}}\cdot\left[e^{-\frac{(n-n_0-\upsilon\cdot t)^2}{4\cdot\sigma^2\cdot t}}+e^{-\frac{\vartheta\cdot n_0}{\sigma}-\frac{(n-n_0-\upsilon\cdot t)^2}{4\cdot\sigma^2\cdot t}}\right]\\ &-\frac{\upsilon}{2\cdot\sigma^2}\cdot e^{-\frac{\upsilon\cdot n}{\sigma^2}}\cdot\left[1-\Phi\left(\frac{\upsilon\cdot\sqrt{t}}{2\cdot\sigma}+\frac{n+n_0}{2\cdot\sigma\cdot\sqrt{t}}\right)\right]\end{aligned} \tag{8}$$

where $\Phi(x)=\dfrac{2}{\sqrt{\pi}}\displaystyle\int_0^x e^{-z^2}dz$ is the error integral.

One-to-one relationship (2) between Ω and *n* at the same time allows you to record that Ω is also a Markov process described by the stochastic differential equation

$$\frac{\partial \Omega}{\partial t} = \alpha \cdot \Omega + \Omega \cdot v_1(t)$$

where $\alpha = \chi \cdot \upsilon$; $v_1(t)$ is a Gaussian stationary process (white noise) . For a Markov process $\Omega(t)$ from (2) we can obtain the expression for the transition density $w(\Omega, t/\Omega_0)$, which has an entry similar to (8) , if we introduce the variable $\omega=\ln\Omega$:

$$w(\omega,\, t/\omega_0) = \frac{1}{2\sigma \cdot \sqrt{\pi \cdot t}} \cdot \left[e^{-\frac{(\omega-\omega_0+\alpha\cdot t)^2}{4\cdot s^2\cdot t}} + e^{-\frac{\alpha\cdot\omega_0}{s^2} - \frac{(\omega-\omega_0-\alpha\cdot t)^2}{4\cdot s^2 \cdot t}} \right] - \frac{\alpha}{2\cdot s^2} \cdot e^{-\frac{\alpha\cdot\omega}{s^2}} \cdot \left[1 - \Phi\left(\frac{\alpha\cdot\sqrt{t}}{2\cdot s} + \frac{\omega+\omega_0}{2\cdot s\cdot\sqrt{t}} \right) \right] \quad (9)$$

where $s = \chi \cdot \sigma$.

The expression (9) allows to investigate the dynamics of clogging of the perforated pipe, which in the model is completely determined by the values of α and s .These values can be estimated from the results of the inspection pipe.

However, the complexity of the expression (9) for the transition density $w(\omega, t/\omega_0)$ –complicates its use for the estimation of the model parameters α and s . In these circumstances you can do this by using an approximate expression for $w(\omega, t/\omega_0)$, which is taken as the fundamental solution of the Kolmogorov equation (5) for the process $\omega(t)$, i.e.

$$f(\omega,\, t/\omega_0) = \frac{1}{\sqrt{2\pi t}\cdot s} \exp\left\{ -\frac{(\omega-\omega_0-\alpha\cdot t)^2}{2s^2 t} \right\} \quad (10)$$

To determine the coefficients of α and demolition diffusion s using the maximum likelihood method

$$\left. \begin{aligned} &\left. \frac{\partial \ln \prod_i f(\omega, t/0)}{\partial \alpha} \right|_{\substack{\omega=\omega_i \\ t=t_i}} = 0; \\ &\left. \frac{\partial \ln \prod_i f(\omega, t/0)}{\partial s} \right|_{\substack{\omega=\omega_i \\ t=t_i}} = 0; \end{aligned} \right\} \quad (11)$$

here $f(\omega, t/0)$ is determined by the expression (10) ; ω_i – operational data pipe inspection at time t_i . It is assumed that after inspection and cleaning, the pipe is clean.

The solution of the systems (11) estimates of α and s are determined from the expressions

$$\hat{\alpha} = \sum_{i=1}^{M_0} \omega_i \Big/ \sum_{i=1}^{M_0} t_i \quad (12)$$

$$\hat{s}^2 = \frac{1}{2M_0} \left[\sum_{i=1}^{M_0} \frac{\omega_i^2}{t_i} - \left(\sum_{i=1}^{M_0} \omega_i \right)^2 \Big/ \sum_{i=1}^{M_0} t_i \right] \quad (13)$$

where M_0 is the number of inspections of the filter.

Next, using the static data on the results of inspections of the pipe, estimate the coefficients of drift and diffusion and solve the problem of the definition of reliability

IOP Conf. Series: Materials Science and Engineering **93** (2015) 012064 doi:10.1088/1757-899X/93/1/012064

3. Evaluation of reliability of the perforated pipe.

As an example, reliability assessment, take any existing pipe, the service life of which is 9 years. As a rule, during the period of operation is checked for defects (ω), then there is cleaning or replacement. Assume that operating according to the regulations, the inspection of the perforated pipe 1 times in 1.5 years. For 9 years, during which time carried out 6 inspections , there have been 3 defect. We use the formulas (12) and (13) to determine the reliability assessment. Substituting our data, we see that:

$\alpha = 0.33$ [1/year] (the failure rate of a perforated pipe)

$S = 0.29$ (probability of failure)

Then find the uptime, using the formula:

$$S(t) = e^{-\alpha t}$$

substituting our values, we obtain that

$$0.288 = e^{-0.33t}$$

$t = 3.75$ [year] (uptime)

4. Conclusion

Developed computational model of a perforated pipe, which allows to analyze the reliability characteristics of a perforated pipe when pumping through it of the liquid coolant. According to the result obtained by the example, it can be concluded that the perforated pipe is not high quality, with experience in the operation. Completed work confirms the possibility of estimating the reliability characteristics of a perforated pipe with the aim of selecting the optimal design and parameters of its work.

References

[1] Zukov A, Kuzina J, Blozerov V 2011 .Reactors with heavy liquid metal coolants and some thermohydraulic *Izvestiya wysshikh uchebnykh zawedeniy. Yadernaya energetika (Proseedings of Universities. Nuclear Power)* **3** pp 100 - 112

[2] Legkikh A, Martynov P, Askhadyllin R 2013 . Calculation of mass exchangers for providing specific oxygen in heavy liquid metal coolants *Izvestiya wysshikh uchebnykh zawedeniy. Yadernaya energetika (Proseedings of Universities. Nuclear Power)* **1** pp 80 - 91

[3] Askhadyllin R, Martynov P, Rachkov V, Legkikh A 2014 Calculation and experimental research in support of mass exchangers for providing specific oxygen regime in in heavy liquid-metal coolants(Pb, Pb-Bi) *Izvestiya wysshikh uchebnykh zawedeniy. Yadernaya energetika (Proseedings of Universities. Nuclear Power)* **1** pp 160 - 169

[4] Martynov P, Asadullin R, Simakov A, Lung Stated A , 26.08.2013; publ. 10.04.2015 Mass transfer apparatus *patent 2547104 of the Russian Federation, INC B01F1/00*

[5] Strekowski V, Salnikov N 1990 *Probabilistic forecasting performance of nuclear power unit components* (Moscow :Energoatomizdat) pp 416

IOP Conf. Series: Materials Science and Engineering **93** (2015) 012065 doi:10.1088/1757-899X/93/1/012065

Analysis of Wigner energy release process in graphite stack of shut-down uranium-graphite reactor

E V Bespala[1], A O Pavliuk[1], S G Kotlyarevskiy[2]
[1]Tomsk Polytechnic University, Tomsk, Russia
[2]JSC « Pilot and Demonstration Center for Uranium-Graphite Nuclear Reactor Decommissioning», Seversk, Tomsk region, Russia

E-mail: bespala_evgeny@mail.ru

Abstract. Data, which finding during thermal differential analysis of sampled irradiated graphite are presented. Results of computational modeling of Winger energy release process from irradiated graphite staking are demonstrated. It's shown, that spontaneous combustion of graphite possible only in adiabatic case.

1. Introduction

The problem of irradiated graphite shutdown reactors is essential in justifying the safety of operations in preparation to decommissioning and decommissioning of nuclear graphite-moderated reactors [1]. One of the most important steps to solve this problem is a detailed analysis of the issues associated with the effect of the accumulation and release of stored energy (Wigner) in irradiated graphite [2,3]. The parameters that characterize the accumulation and release of stored energy in graphite, determine the conditions and the possible consequences of its release (thermal effects) in the graphite stack [4].

2. Computation of heat effects

During researches [5] took samples of graphite by volume of graphite stack (figure 1) of shut-down the reactor with water cooling, which is operated in a flow mode. At the same time covering the entire range of damaging neutron fluence and temperature of irradiation. These factors determine the parameters of accumulation and release of stored energy. Coring of graphite sleeves was made using hollow cutters after removing them from the graphite stack. Tackling cross-cutting core samples from the walls of graphite blocks was carried out directly from the graphite stack using a special device UGB-3. The direction of coring is horizontal. Selected core length of 57 mm and a diameter of 8 mm was cut into four pieces evenly along the length. Next, from the obtained core pieces were manufactured the same form samples as for the sleeves.

The most critical in terms of values of the possible thermal effects due to self-sustained release of stored energy are the most "cold" (irradiation temperature 50–80°C) graphite sleeves from the top of the graphite stack (figure 2, curve 2). For them the following conditions (1) of the possibility of self-sustaining release of stored energy for adiabatic conditions is realized:

$$\frac{dH(T)}{dT} > C(T) \tag{1}$$

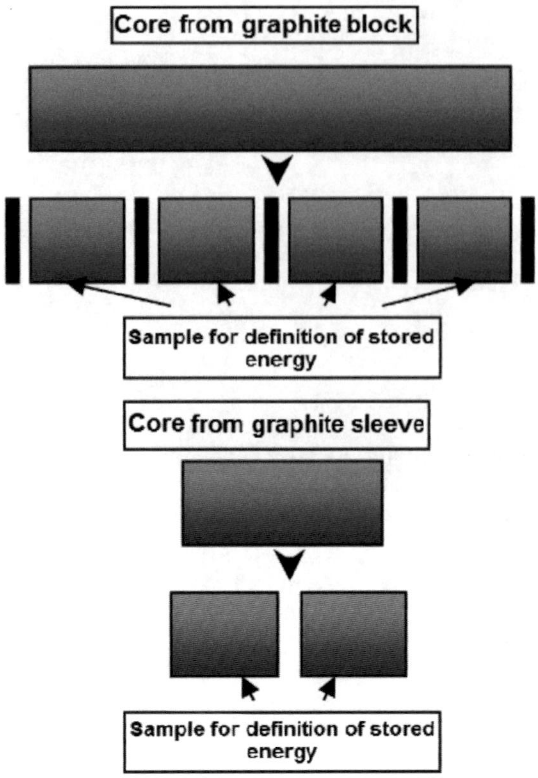

Figure 1. Scheme of manufacture of core samples that were extracted from the blocks and sleeves of shutdown reactor graphite stack.

The graphite block in the entire volume of the graphite stack contents stored energy is substantially lower than in the sleeves. Therefore, self-sustaining release of energy is impossible (figure 3, curves 2 and 3). This is because the block was irradiated at a higher temperature. It can only be forced annealing.

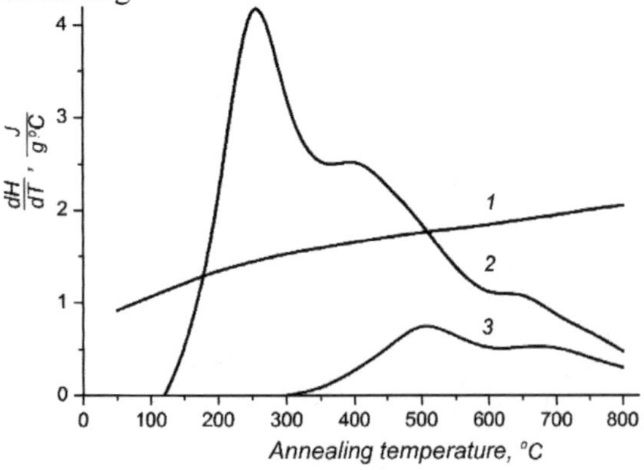

Figure 2. The spectrum of energy Wigner release in sleeve: 1 – the specific heat capacity of graphite; 2 – the intensity of energy Wigner release for sleeve from the top of the graphite stack; 3 – the intensity of energy Wigner release for sleeve from central part of the graphite stack.

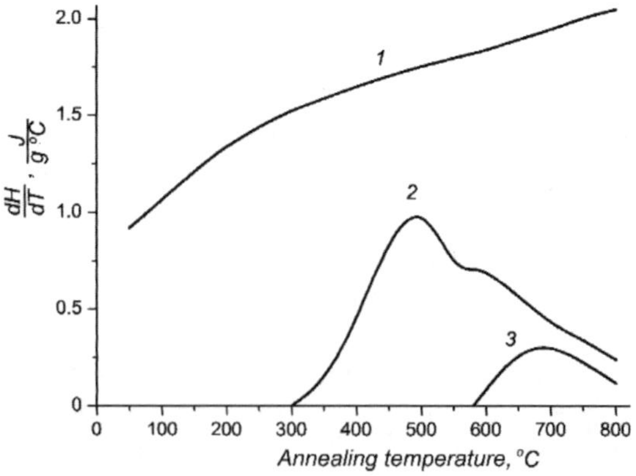

Figure 3. The spectrum of energy Wigner release in graphite block: 1 – the specific heat capacity of graphite; 2 – the intensity of energy Wigner release for sleeve from the top of the graphite stack; 3 – the intensity of energy Wigner release for sleeve from central part of the graphite stack.

It can only be forced annealing. Typical distribution curves of the integral value of the stored energy adjustment sleeves for the graphite stack and blocks shown in figure 4 and 5.

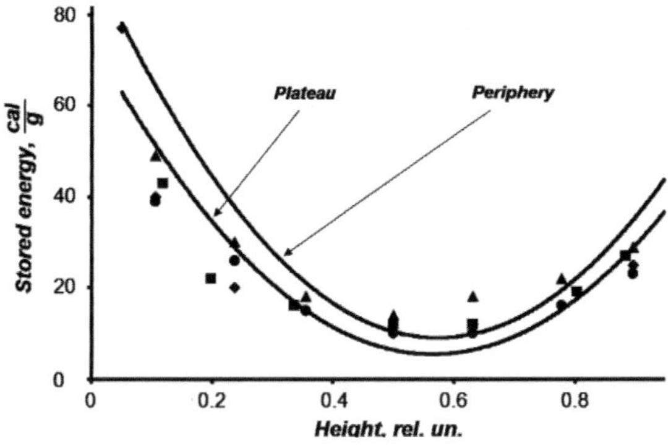

Figure 4. Distribution of the integral value of the stored energy adjustment graphite stack shutdown reactor for block.

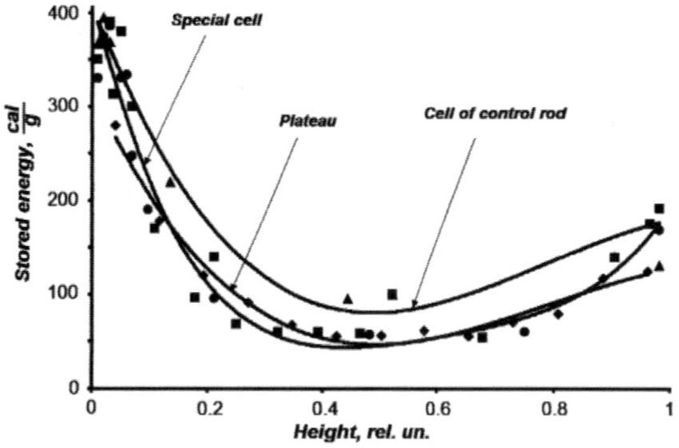

Figure 5. Distribution of the integral value of the stored energy adjustment graphite stack shutdown reactor for sleeve.

Compare the results of the experimental determination of the parameters of energy Wigner (figure 2 and 3) and its integral value (figure 4 and 5). It can be concluded that the most critical area in terms of a possible warming of graphite due to release the stored energy is the upper part of the graphite stack.
Analysis of the magnitude and dynamics of thermal effects for the most critical areas of the graphite stack is made by solving the non-stationary heat equation by methods described in [6]. The parameters characterizing the release of stored energy in the simulated processes are defined on the basis of its release spectra obtained experimentally (figure 2 and figure 3).
Calculations have shown that as a result of energy Wigner growth temperature of graphite to a value higher than the temperature at which it is possible burning of graphite (T_b~700°C), is observed only in the case adiabatic heating most «cold» graphite sleeves (figure 6). This requires the graphite heat up due to external sources of heat to a temperature $T_{initial}$>150°C.

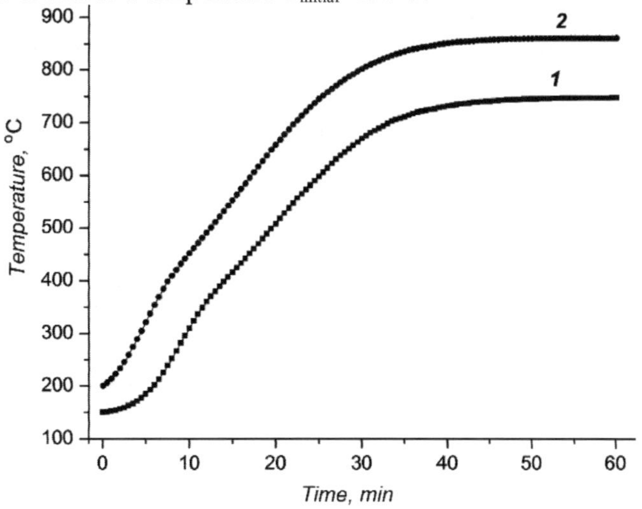

Figure 6. The dynamics of growth temperature of the "cold" graphite sleeve during release of the Wigner energy under adiabatic conditions: 1 – $T_{initial}$=150°C; 2 – $T_{initial}$=200°C.

Obviously, in practice, to arrange a complete absence of heat exchange of graphite sleeve with environment is practically impossible. In particular, the sleeves, located in the graphite stack have thermal contact with the graphite block. The content of the stored energy in the blocks is significantly lower than in the sleeves. Therefore, self-sustaining release of stored energy is impossible. Heat dissipation is also due to convective heat exchange with air on the surface of the sleeve hole. Therefore, in real conditions due to the thermal effects of energy Wigner will be lower.

When calculating the thermal effects in the graphite stack was used by a number of conservative assumptions. The problem was formulated for the equivalent radially symmetric cell (figure 7), of the infinite layers of graphite stack with a grid spacing of 200 mm.

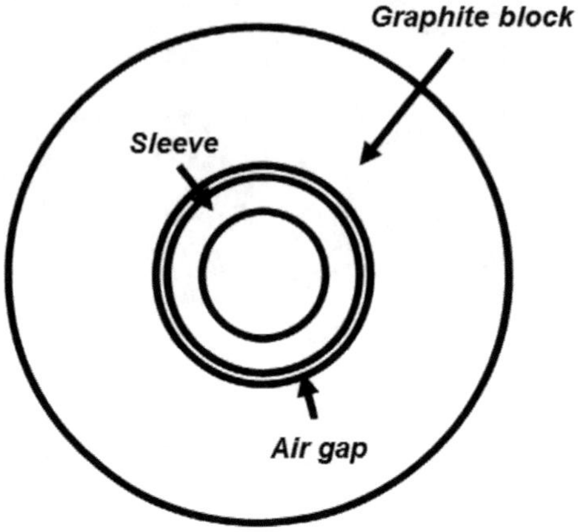

Figure 7. Scheme of equivalent radially symmetric cell graphite stack.

Stack has been preheated to a temperature $T_{initial}$ by external heat sources. This temperature is higher than the temperature of the beginning of a self-sustaining release of stored energy in the sleeves (~150°C). The value and parameters of energy Wigner release for sleeves were selected the most critical (the respective sleeves of the top of the stack). The heat sink through holes of cells graphite stack was set to zero. The air gap between the sleeve and the block has been set as conservatively assumed to be 0.5 mm (0.1–0.4 mm really). The result of the calculation of the dynamics of the temperature field for the equivalent cell graphite stack is shown in figure 8.

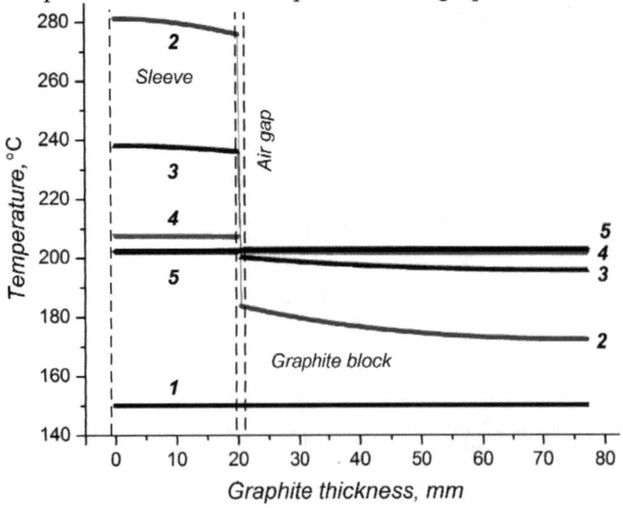

Figure 8. The dynamics of the temperature field in the cell of graphite stack during Wigner energy release: 1 – t = 0 min; 2 – t = 20 min; 3 – t = 40 min; 4 – t = 60 min; 5 – t = 80 min

The results show (figure 8), the maximum heat effect due to release of Wigner energy in the sleeves observed after about 20 minutes after the start of warm-up and amounts not exceeding ~130°C (respectively, the temperature of sleeve is no higher than 280°C). Heat effect the steady after 80 minutes for the entire system is about 50°C, and thus the temperature of all graphite (sleeves and blocks) is set equal to 200°C. The stored energy of the graphite block does not contribute to heat generation. Because steady-state temperature of the graphite block is about 200°C, and the temperature began to release Wigner energy in the graphite blocks with the lowest temperature of the radiation is $T_{initial}$ 250–280°C. Moreover, even in case of achievement block temperature $T_{initial}$ above, heat effect will be negligible. Because self-sustaining release of stored energy in graphite blocks is impossible. It will wear quickly of damped character.

3. Conclusion

Thus, release of stored energy for the most critical areas of the graphite stack of shut-down reactor even in several conservative assumptions will not result in heating of the graphite to a temperature exceeding T_b~700°C. Thus, excess temperature graphite to possible ignition of a substantial amount - at least 420°C at the moment of maximum heat effect due to release of Wigner energy.

References

[1] Izmestiev A, Pavliuk A, Kotlyarevskiy S 2015 *Advanced Materials Research* **1084** pp 613–619
[2] Bell J C, Bridge H, Cotrell A H et al 1962 *Phil. Trans. Roy. Soc. London* **254/1043** p 35
[3] Neubert T J, Lees R B 1957 *Nucl. Sci. and Eng.* **2/3** p 748
[4] Bridge H, Kelly B T, Gray B S 1962 *Nuclear Engineering* **7/70** p 91
[5] Tsyganov A, Kotslyarevskiy S, Pavliuk A et al 2008 *Izvestiya of Tomsk Polytechnic University* **4/2** pp 32–38
[6] Dostov A 2005 *Thermal physics of high temperatures* **43** pp 267–273

IOP Conf. Series: Materials Science and Engineering **93** (2015) 012066 doi:10.1088/1757-899X/93/1/012066

City snow's physicochemical property affects snow disposal

V O Dovbysh[1], A V Sharukha[1], P V Evtin[1], and S V Vershinina[1]

[1]Tyumen State Oil and Gas University, Tyumen, Russia

E-mail: Vadim932@mail.ru

Abstract
At the present day the industrial cities run into severe problem: fallen snow in a city it's a concentrator of pollutants and their quantity is constantly increasing by technology development. Pollution of snow increases because of emission of gases to the atmosphere by cars and factories. Large accumulation of polluted snow engenders many vexed ecological problems. That's why we need a new, non-polluting, scientifically based method of snow disposal. This paper investigates polluted snow's physicochemical property effects on snow melting. A distinctive feature of the ion accelerators with self-magnetically insulated diode is that there.

1. Introduction

As is well-known snow it's a type of atmospheric widespread precipitation consisting of small ice crystals and falling out in certain meteorological conditions. Singularity of snow cover it's ice layers which effects on snow characteristics [1].

Snow cover has a clearly defined layering because of different meteorological conditions during the deposition of snow. These characteristics are the main reason of snow inhomogeneity, but snow is generally thought to be homogeneous if we want to find out basic characteristics.

Table 1. A number of key parameters represented

Characteristic	Unit	Notation
humidity	% (by weight)	W
density	kg/m³	ρ
temperature	°C	T
heatcapacity	J/K	C
blend composition :		
- salt content	% (by weight)	S
- solids content	% (by weight)	G

Snow pack humidity is an indicator of liquid phase volume, and has an importance at determining the disposal optimum. Humidity generally depend on snow temperature, therefore also depend on atmospheric temperature; and for new-fallen snow humidity can vary between 1% and 25%. The calculating humidity, depending on temperature, formula: $W = (k_1 - k_2) \cdot T$

W-snow humidity; T-snow temperature (in modulus); k_1 - coefficient, depending on basic snow characteristics, 1/oC, k_1 = (12; 15); k_2 - coefficient, depending on external factors, k_2 = (1,5; 2).

This formula is correct for temperatures vary between the 0 deg. C and the 8 deg. C. Ground snowpack falls and is shoveled away at this temperature. So we will use this form to calculate snowpack humidity. Humidity can be finding by centrifugal process with a calorimetric method or by the measurement of different chemicals dissolution rate. But all of these measurements are time-sapping or difficult to unity [2].

Because of these reasons the international snow classification divides snow into dry, damp, wet, soggy, and slush.

The calculating density formula: $p = \frac{m_c}{V_в}$

m_c – mass of snow; $V_в$ – water volume after thawing.

Density largely affects the disposal process, for this reason this parameter value must be true. Snow density be depending on different conditions can vary between 1 t/m3 end 0, 9 t/m^3. Naturally snow density is about 0,2t/m^3, new-fallen snow density - 0,1 t/m^3, snow slush at a snow dump is about 0,6 – 0,8t/m^3. Statistics about snow density, which take away from Tyumen streets, mean density is 0,3 t/m^3. Snowpack temperature is a base parameter to determinate how snowpack characteristics vary with its parameters. Usual the snow temperature is about the atmospheric temperature.

2. Another section of your paper

Snow component composition greatly varies with heat required to thawing. There are a lot of glaciological and petro physical studies about snowpack depending on pollution. But these studies don't carry inference about salts and mud depending on the snow disposal.

Researches show a high level of city snow pollution. A mean of common pollution agents are greatly changed for different city sectors (Table 2).

Snow samples include a lot of suspended solids, slow bio oxidizable organic compounds and hardness salts. Chlorides content exceed MPC 7 - 15-fold; sulphates content – 5-fold. Toxic metal ion concentration (ferrum, manganese, lithium, zinc, cuprum, molybdenum, cobalt, cadmium) exceed MPC from 1 to 46 times. Oil-products and phenols content exceed MPC from 25 to 150 times and from 1 to 4 times respectively [3,4].

Table 2. A mean of common pollution agents

Agents	Dimension	Rate
Suspendedsolids	mg/l	121 – 732
Ammonium	mg/l	0,33-9,6
Chlorides	mg/l	16,11-338,2
Sodium	mg/l	13,76-442,4
Potassium	mg/l	22,3-110,9
Ferrum	mg/l	0,68-1,45
Manganese	mg/l	0,11-0,67
Zink	mg/l	0,013-0,09
Plumb	mg/l	15,23-37,23
Oil-products	mg/l	1,4-50,2
Detergent	mg/l	0,23-1,012

A date analysis has shown that main volume of pollution agents consist of suspended solids; oil-products and chlorides: the mean suspended solids concentration equal to 426,5 mg/l occurs with 27%

probability; the mean oil-products concentration equal to 25,6 mg/l occurs with 33% probability; the mean chlorides concentration equal to 177,2 mg/l occurs with 28% probability.

For a theoretical estimate of interphasic heat exchange processes going in the snow melting chamber, a mathematical model basic on actual heat exchange conditions was created. In this model we deal with a problem of heat exchange between liquid and a separate fragment of solid body in the form of a cube for clarity. Also we evaluate the effect of local heat exchange processes sum on interfusion heat exchange integral characteristic [5,6].

We shall deal with the problem of heat exchange between liquid and a separate fragment of solid body. Suppose the solid cube with side $\sqrt[3]{V}$ (v – volume of solid) is immersed in liquid. The liquid temperature (T_f) is higher than solid fusion temperature, and melting taking place at temperatures between T_{1s} and T_{2s}, $T_{2s}>T_{1s}$. In this case molten zone, closed by surfaces which shift deep into the solid body during the time, is arisen. In the molten zone fusion heat release; and location of molten borders depends on balance between the heat for fluxes on borders and the heat essential to melting.

In the snow melting chamber, containing mix of heated liquid and ice fragments, a low permanent current is maintained. Therefore a rate of heat exchange processes in the perpendicular direction of flux is stronger than in right direction of flux. So, for the system: liquid - solid body, we can suppose that temperature (T) function of time (t) and one space coordinate – x, in the perpendicular direction of fluxT = T(x,t).

Consequently our problem of heat exchange between liquid and a separate fragment of solid body in the form of a cube changes to a problem of monadic heat exchange between liquid and an unlimited ice plate, and the melting process is going symmetric by two opposite faces of ice plate. Let's line x =0 - is a location of one plate face at the time point t=0; X1 (t), X2 (t) - a molten borders formula at the time point t, at melting temperatures T_{1s} and T_{2s} [7].

If we use indexes 1, 2 and 3 for three molten zones we'll get a set of equations for required functions.

$$\frac{\partial T_1}{\partial t} = a_1^2 \frac{\partial^2 T_1}{\partial x^2} \; ((t \geq 0, x \leq X_1(t) \qquad (3)$$

$$\frac{\partial T_2}{\partial t} = a_2^2 \frac{\partial^2 T_2}{\partial x^2} \; ((t \geq 0, X_1(t) \leq x \leq X_2(t)) \qquad (4)$$

$$\frac{\partial T_3}{\partial t} = a_3^2 \frac{\partial^2 T_3}{\partial x^2} \; ((t \geq 0, X_2(t) \leq x)) \qquad (5)$$

$a_j = \frac{K_j}{C_j \rho_j}$-a temperature conductivity coefficient for city snow component (j),

K_j - a thermal conduction coefficient for city snow component (j),

C_j - a specific heat for city snow component (j),

ρ_j - density for city snow component $j = (\overline{1,3})$.

Formula (6) was written subject to the following restrictions.

Suppose that heat input, depending on the second zone temperature, at a steady pace, than heat input by melting will be calculated approximately, using following meaning of specific heat:

$C_2 = C_1 + \frac{\lambda}{T_{2s} - T_{1s}}$,

C_1- liquid specific heat, λ – specific heat of melting.

As follows from experiment results, city snow in the snow melting chamber amount to snow-iced pack with so huge quantity of multicomponent mixture, that we can't neglect their effect on the melting process. In the first case, neglecting an interaction heat effect, we will suppose additive and admission of mix thermo physical properties:

$$K_j = \sum_i K_{ji} l_i, \; C_j = \sum_i K_{ji} l_i, \; \rho_j = \sum_i K_{ji} l_i \qquad (7)$$

l_i – mass concentration of i- pollution (i= 1….n) in city snow.

Solution of thermal conduction (8)-(10):

$$T_1 = A_1 + B_1 \, \Phi(\frac{x}{2a_1\sqrt{t}}) \qquad (8)$$

$$T_2 = A_2 + B_2\, \Phi(\frac{x}{2a_2\sqrt{t}}) \tag{9}$$
$$T_3 = A_3 + B_{31}\, \Phi(\frac{x}{2a_3\sqrt{t}}) \tag{10}$$

A_j, B_j $j = (\overline{1,3})$-constants, $\Phi(Z) = \frac{2}{\sqrt{\pi}} \int_0^Z e^{-\xi^2} d\xi$- error function. Constant sat equations (8)-(10) deduce from first and boundary conditions. Let's measure outcomes of ice melting in the snow melting chamber. If ice fragments cube with volume and side, than a border square, and mass of melted ice per second will be: $\Delta m_i = 2\frac{\alpha x_1}{at}(v_i) \cdot \rho$. Let us denote number of ice fragment by n_i, size between (v_i) and $(v_i + \Delta v)$.If we suppose that volume of ice fragments in mix follow the normal distribution law with characteristics $\mu(v) = \bar{v}$ и $S(v) = \sigma$, than the random variate density will be:

$$f(v) = \frac{1}{\sigma\sqrt{2\pi}} e^{-\frac{(v_1 - v)^2}{2\sigma^2}} \tag{11}$$

Number of ice fragments with size between (v_i) and ($v_i + \Delta v$) will be:

$$n_i \approx f(v_i)\Delta v_i = \frac{1}{\sigma\sqrt{2\pi}} e^{-\frac{(v_1 - v)^2}{2\sigma^2}} \tag{12}$$

Than mass of melted snow per second will be:

$$M \approx \sum_1^N \Delta m_i = 2\frac{\alpha x_1}{at} \rho \sum_1^N n_i v_i^{\frac{2}{3}} = 2\rho \frac{\alpha x_1}{at} f(v_i) v_i^{\frac{2}{3}} \Delta v \tag{13}$$

Sum in the left-hand member of an equation (13) is integral sum for continuous function $f(v_i)v_i^{\frac{2}{3}}$.

So, for a snow mass estimation melting per second, taking into account the three sigma rule, we use next equation:

$$M \approx 2\rho \frac{\alpha x_1}{at} \frac{1}{\sigma\sqrt{2\pi}} \int_{\bar{v}-3\sigma}^{\bar{v}+3\sigma} e^{-\frac{(v_1 - v)^2}{2\sigma^2}} av \tag{14}$$

For analysis we use the program «Maple-7», realizing a chord method for system of nonlinear transcendental equations. The following are results for every phase of program scheme. System of equations was changed to dimensionless form by using typical test value at a temperature of 0 degrees for all dimensional quantities.

Graph 1 displays trend of the thermal conductivity coefficient loading (a) of snow-iced pack, taking from city area, in model cube volume, using test data of winter 2013/2014 at Tyumen, depending on total density (ρ) and total thermal conduction coefficient (K) at a constant heat capacity.

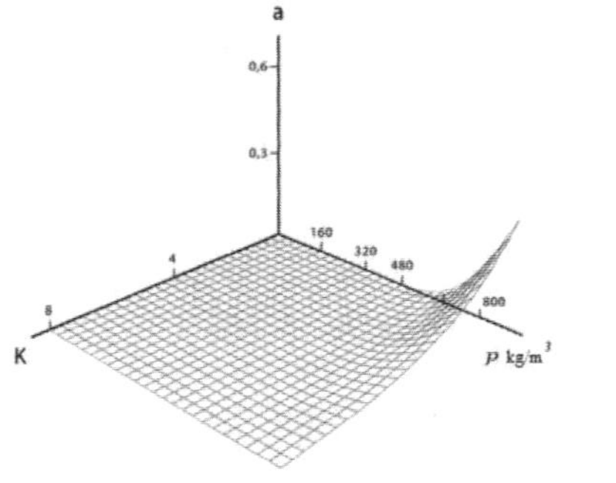

Figure 1. The thermal conductivity coefficient loading (a) of snow-iced pack depending on total density (ρ) and total thermal conduction coefficient of pollute snow (K)

As far as density is valuation of pollution, at graph №1 we can see that exists a critical pollution level changing snow thermo physical properties. From now on it should be figure on any snow melting chambers. Test data show total snow-iced pack pollution, taking from city area, exceeds critical level – 2-fold. But fast growth of thermal conductivity; mean proportional decreasing of thermal capacity. And, as we can see at the graph, snow pack density have a little effect on the growth of thermal conductivity. So, for melting same volume of snow we can reduce sewage drain flow or use lower temperature drain. As follows from analytical results at graph №2, melting temperature - density relationship with pollution rising is a no monotonic function with max, which is evidence of even one liquidus line at phasing diagram of multiphase system – city snow, telling about a new phase origin - chemical mixture, with melting temperature beyond the natural snow melting temperature.

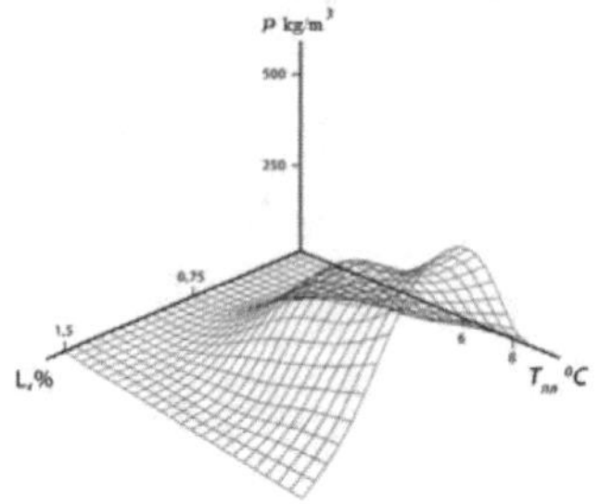

Figure 2. Dependency graph $T_{пл}(L)$, phase 3-solid body

It's important that increase of melting temperature requires increase of sewage drain flow, as follows from analytical results at graph №2. Turn attention to analysis results at graph №3. As we can see, relationship between solid body melting temperature and thermal capacity with total pollution, (uniform dependence for natural snow), in this case have two local maximums and onelocal minimum. It shows two facts. First, if city snow has pollutants, applying by deicing agent, a layer structure with two non overlapping range of melting will come up.

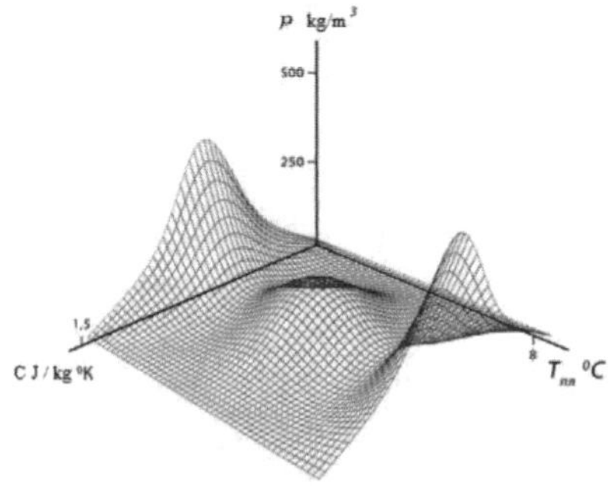

Figure 3. Dependency graph $T_m(C)$, phase 3-solid body

Second, if such structure comes up in snow melting chamber, one part of snow-iced pack will melt completely at set temperature point of sewage drain, but another more heatproof part can create a cork in interceptor. In that case if sewage drain, like coolant, has characteristics designed for more heatproof part, we have to control flow in real time.

3. Conclusion

As follows from analytical results at graph № 1-3, snow pollution leads to nonlinear and non monotonic changing of the melting temperature by density, and density is an additive function of pollution. In certain pollution value clear-cut zones of peak snow-iced melting temperature are seen at all graphs. Obviously, for so pollute snowpack melting need greater heat flow than for natural snow.Let's view a snow disposal. Basic method of snow disposal is removal by special automotive equipment. This method is very expensive (several billions rubles for one wintertime), and pollutes urban ecosystem.

References

[1] Balovnev VI Belyaev, MA Machinery maintenance and repair of city and highways. Textbook for high schools. - Ed. 2nd. - Omsk "Omsk printing house", 2005. - P. 10-11
[2] Khramenkov SV, Pakhomov AN Bogomolov MV Danilovich DA Romachkin OV Pupyrev EI Koretsky VE Systems of removal of snow from the use of urban sewage. Water supply and sanitary engineering, 2008, №10. 1.5 pp
[3] Peaceful AN, Skvortsov LS Pupyrev EI Koretsky VE Comm-functional environment. Encyclopedic Reference. M .:. Prima-Press-M., 2007. 50.5 pp
[4] ColbeckS. C 1973. Theory of metamorphism of wet snow. Res. Rep. 313, US Army Cold Reg. Res. Eng. Lab., Hanover, N. H.
[5] SP 32.13330.2012. Rulebook. Sewer. External networks and facilities
[6] Kawashima K., Endo T., Takeuchi Y. A portable calorimeter for measuring liquid-water content of wet snow/Editor: D.M. McClung//Papers from the International Symposium on Snow and Avalanches held at Chamonix Mont-Blanc, France, 26-30 May 1997, Annals of Glaciology Volume 26. P. 103-106
[7] Heat exchangers: a tutorial/BE Baygaliev, AV Clicks, AB Yakovlev, PY Gortyshov.-Kazan: Publishing House of Kazan. state. tehn. University Press, 2012.-180 p.ISBN 987-5-7579-1773-3

Comparison of the calculated and experimental data of the extracted electron beam profile

I A Miloichikova, A Povolná, S G Stuchebrov, G A Naumenko
Tomsk Polytechnic University, Tomsk, Russia

E-mail: miloichikova@gmail.com

Abstract. The current commercial use of electron accelerators grows in research, industry, medical diagnosis and treatment. Due to this fact, the creation of a model describing the electron beam profile and shape is an actual task. The model of the TPU microtron extracted electron beam created in the program "Computer Laboratory (PCLab)" is described and compared with experimental results in this article. The value of the internal electron beam divergence determination is illustrated. The experimental data of the electron beam profiles at the selected distances from the output window are analysed and compared with the simulation data. The simulation data of the electron beam profiles are shown.

1. Introduction

Nowadays, the X-ray sources and the particle accelerators have a large number of practical applications, particularly medical applications, but they are also used for a variety of biological and industrial applications [1-4]. Particle accelerators have an amazingly broad range of beam parameters for different applications such as sterilization of food products, pulsed X-ray radiography, external beam radiotherapy and plasma heating for fusion reactors [1-5]. In all these applications it is necessary to have an accurate representation of the electron beam profile and shape and be able to predict the beam parameters in accordance with the particular use. Therefore the creation of a model describing the electron beam profile and shape is a contemporary task.

The majority of programs use for the simulation of electron beams the Monte Carlo method or the macro Monte Carlo method [6-8]. In this paper the simulation is carried out by applying the Monte Carlo method.

In the present research the theoretical analysis of the TPU microtron extracted electron beam was carried out. The model of the accelerator electron beam profile and shape was developed in the framework of the program "Computer Laboratory (PCLab)" and was compared with the experimental results. The following characteristics of the electron beams are analysed in this research: the internal beam divergence and the electron beam profile and shape.

2. Materials and methods

2.1. Emitting source

In this experiment, the TPU microtron extracted electron beam, with the following characteristics was used as the emitting source: beam size at output ≈ 2.0 mm^2; electron energy – 6.1 MeV.

2.2. Simulation program

For the model creation of the TPU microtron extracted electron beam, the program "Computer laboratory (PCLab)" version 9.6 was used. Simulation is carried out using the Monte Carlo method. The software package of "Computer laboratory (PCLab)" allows the calculation of the propagation process of photons, protons, electrons and positrons in matter with specified characteristics [9].

2.3. Experiment geometry

For the first measuring, the scan frame was placed at the distance equal to 10 cm from the output window and the horizontal and vertical profiles of the microtron extracted electron beam were measured. The scan frame was shifted by additional 5 cm away up to a distance 35 cm from the output window in each subsequent measurement. The Faraday cup was located at a distance of 60 cm from the output window.

In the simulation, the normal plane disc (diameter – 2.0 mm) monoenergetic electron source with energy of 6.1 MeV corresponding to the actual TPU microtron beam was used. The source was placed in front of the beryllium output window (diameter – 40 mm; thickness – 50 μm). The beam shape analysis was carried out in the air.

The figures 1, 2 show the experimental setup and the calculation geometry, correspondingly.

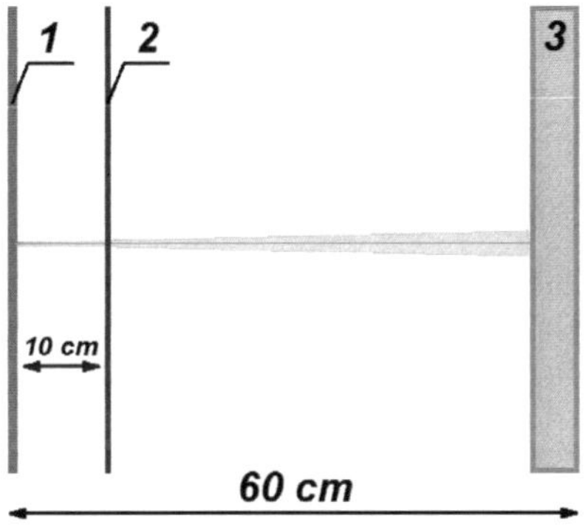

Figure 1. Scheme of the experiment: 1 – beryllium output window, 2 – scan frame, 3 – Faraday cup.

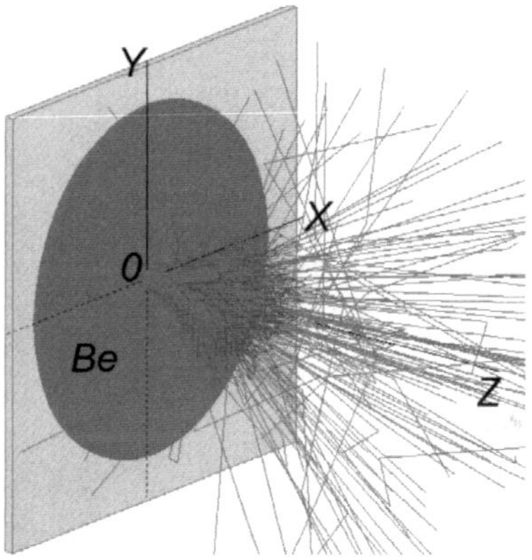

Figure 2. The calculation geometry: blue lines – electron path; red lines – photon path.

3. Results and discussions

The horizontal and vertical profiles of the microtron extracted electron beam were measured at specified distance from the microtron output window. Data from all measurements were analysed and fitted with the Gaussian function. The horizontal and vertical profiles were averaged and the half widths at half maximum were determined for those profiles.

The figure 3 shows an example of experimental data approximation with the Gaussian function for the horizontal profile of the microtron electron beam. The beam profile is normalized to the maximum value. The frame distance from the microtron output window is equal to 15 cm.

The angle of internal divergence of the microtron electron beam α was obtained by linear approximation of the half width at half maximum values – $\alpha = 2{,}5°$.

The figure 4 shows the determination of the internal divergence angle of the microtron electron beam α by linear approximation, where R is the frame distance from the microtron output window and d is the half width at half maximum value.

Knowing the internal divergence of the beam, the beam model was developed, followed by comparing beam profiles obtained experimentally and beam profiles obtained by simulation.

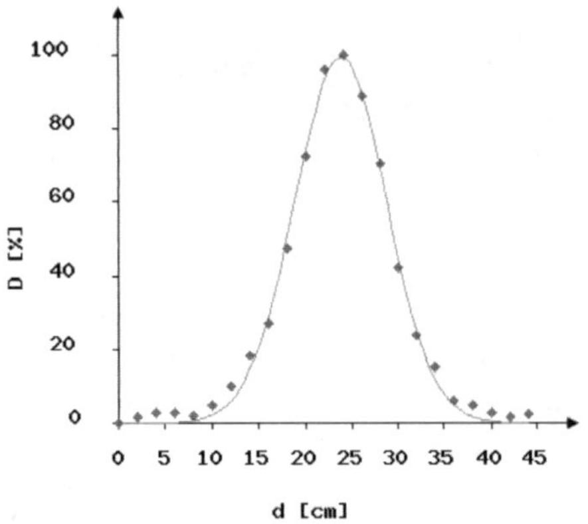

Figure 3. Horizontal profile of the microtron extracted electron beam: ♦ – experimental results; — – approximations result.

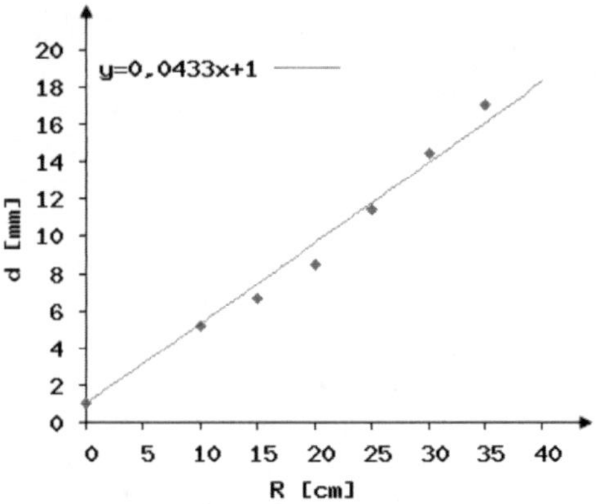

Figure 4. Determining the angle of internal divergence of the microtron electron beam.

In the figure 5 the comparison of electron beam profiles obtained experimentally and electron beam profiles obtained by simulation are shown for three different scan frame distances from microtron output window.

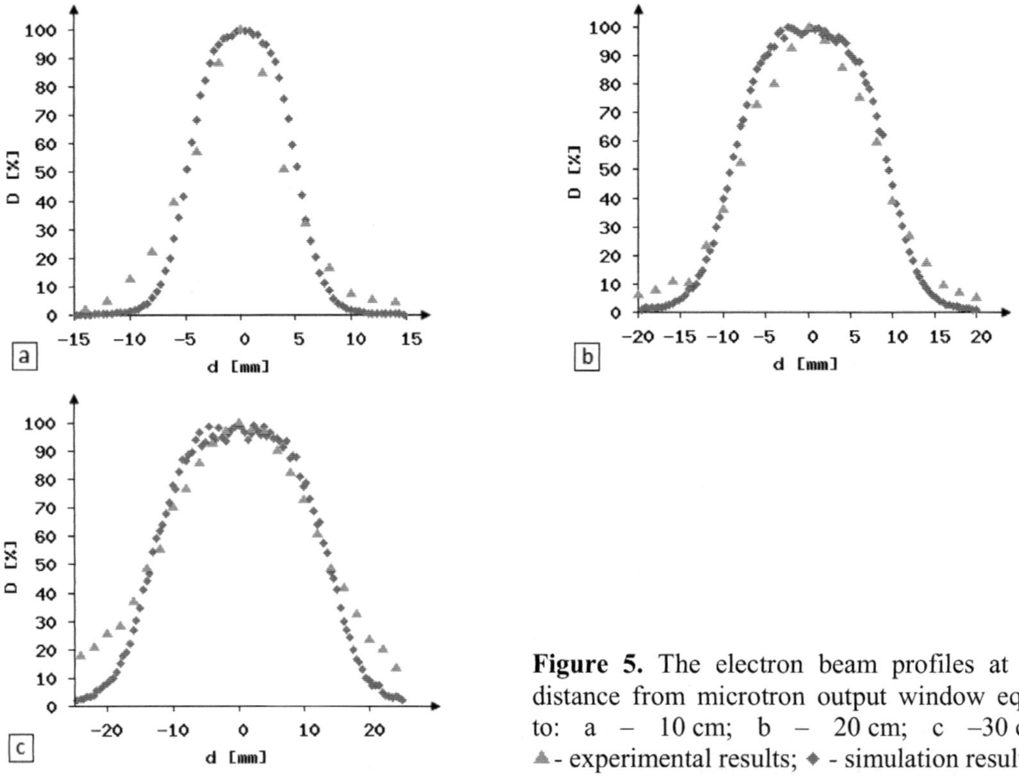

Figure 5. The electron beam profiles at the distance from microtron output window equal to: a – 10 cm; b – 20 cm; c –30 cm; ▲- experimental results; ♦ - simulation results.

The figure 5 illustrates that results from the model are in a good agreement with the experiment.

The simulation data of the TPU microtron extracted electron beam profile at the 0 cm, 5 cm and 15 cm distance from the output window are shown in the figure 6. The dose results were averaged and normalized to the maximum simulation dose in the layer.

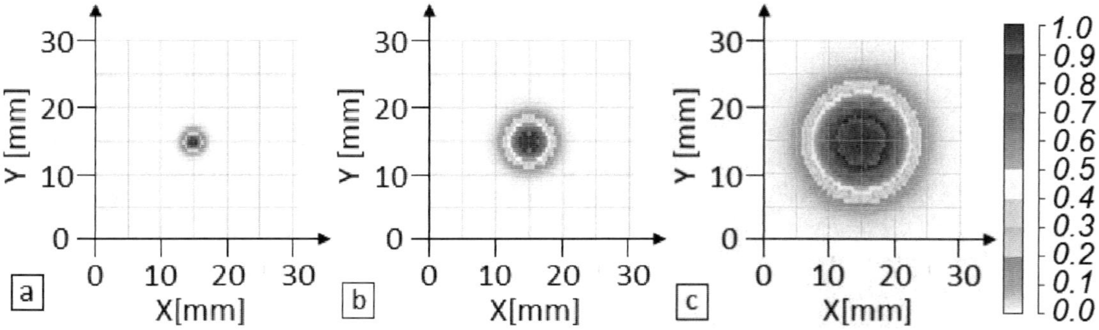

Figure 6. The TPU microtron extracted electron beam profile: a – at the 0 cm distance from the output window; b – at the 5 cm distance from the output window; c – at the 15 cm distance from the output window.

4. Summary

The theoretical model of the TPU microtron extracted electron beam was calculated in the simulation program "Computer laboratory (PCLab)" and compared with experimental data in this paper. The obtained results show the eligibility of this program for the real electron beams analysis and therefore the program can be used later, for example in the development of the accelerator's collimating system.

The theoretical model allows to estimate the dose distribution and the electron beam size at the selected distance from the output window, making it faster and easier to determine these characteristics than the actual measurements. The depth dose distribution allows computing the radiation burden values in the electron beam propagation direction.

Acknowledgements

This work was partly supported by Competitiveness Enhancement Program of Tomsk Polytechnic University.

References

[1] Cherepennikov Yu and Gogolev A 2012 Method to reduce radiation exposure in the medical X-ray diagnostics *7th International Forum on Strategic Technology (IFOST - 2012)* **2** pp 244-248

[2] Humphries S 2012 Principles of charged particle acceleration. (US: Courier Corporation) 592 p

[3] Gogolev A, Stuchebrov S, Wagner A, Cherepennikov Yu and Potylitsyn 2012 Acoustic "pumping effect" for quartz monochromators *J. Phys.: Conf. Ser.* **357** 012031

[4] Information on http://www-pub.iaea.org/MTCD/publications/PDF/P1433_CD/datasets/papers/ap_ia-12.pdf

[5] Cherepennikov Yu and Gogolev A 2014 Device for X-ray spectral absorption analysis with use of acoustic monochromator *J. Phys.: Conf. Ser.* **517** 012037

[6] Ro Ch-U, Osán J, Szalóki I, DeHoog J, Worobiec A and Van Grieken R 2003 A Monte Carlo Program for Quantitative Electron-Induced X-ray Analysis of Individual Particles *Anal. Chem.* **75(4)** pp 851-859

[7] Neuenschwander H and Born E J, Szalóki I, DeHoog J, Worobiec A and Van Grieken R 1992 A macro Monte Carlo method for electron beam dose calculations *Phys. Med. Biol.* **37(1)** 107

[8] Righi S et al. 2013Dosimetric characteristics of electron beams produced by two mobile accelerators, Novac7 and Liac, for intraoperative radiation therapy through Monte Carlo simulation *Journal of Applied Clinical Medical* **14** pp 6-18

[9] Bespalov V I 2015 Computer laboratory (Version 9.5) (TPU: Tomsk) 115 p [in Russian]

IOP Conf. Series: Materials Science and Engineering **93** (2015) 012068 doi:10.1088/1757-899X/93/1/012068

Feasibility of resonant diffraction radiation from inclined gratings for a nondestructive beam diagnostics.

A P Potylitsyn[1], A R Mkrtchyan [2,3], V R Kocharyan [4,5], A I Novokshonov [6]

[1] Professor, Tomsk polytechnic university, Tomsk, Russia
[2] Professor, Tomsk polytechnic university, Tomsk, Russia
[3] Professor, Institute of Applied Problems of Physics of NAS RA, Yerevan, Armenia
[4] Assistant, Tomsk polytechnic university, Tomsk, Russia
[5] PhD, Institute of Applied Problems of Physics of NAS RA, Yerevan, Armenia
[6] PhD student, Tomsk polytechnic university, Tomsk, Russia

E-mail: novokshonov@tpu.ru

Abstract. Characteristics of the resonant diffraction radiation (RDR) generated by ultrarelativistic particles passing near a tilted grating have been considered. Angular distribution of RDR can be used for diagnostics of a beam divergence. Such a technique is non-invasive practically.

1. Introduction

One of the known technique for low-emittance electron beam diagnostics is using of optical transition radiation (OTR) which is generated by passing of an electron beam through the thin foil. Obviously that the process of multiply scattering of electrons passing through such a foil makes the emittance worse. Moreover electron beams of modern accelerators with micron transverse sizes interacting with the target can lead to changing of target optical characteristic and that has affected negatively on accuracy of measurements.

Recently years methods of nondestructive diagnostics based on radiation of charged particles in external field (synchrotron radiation, scattering of laser radiation) or on processes of electron Coulomb field scattering during its flight near optical inhomogeneity (diffraction radiation, Smit-Parcell radiation) were developed extensively [1,2]. In this case energy losses of an electron due to radiation only are negligibly small that allows to consider such methods as nondestructive ones.

In article [3] authors measured characteristics of optical diffraction radiation (ODR) during the flight of an electron through the slit in the inclined target and they showed possibility of ODR using for a beam transverse size measurements. It is well known that the spectrum of diffraction radiation is continuous. But if the beam is passed in the vicinity conductive target with periodical deformed surface the radiation spectrum contains set of spectral lines with wavelengths depending on observation angle [4].

2. Calculations and simulations

In this article we propose to use inclined grating with the period from submicron to millimeters (depends on investigated wavelength range) instead flat slit target. In this case when the resonance condition is fulfilled (see below) there will appear narrow spectral lines in spectrum were line wavelengths will be determined not only by observation angle but the inclined angle of grating too [4].

This fact we suggest to use for the initial beam angular divergence measurements. This is the reason that this radiation can be called the resonant diffraction radiation (RDR).

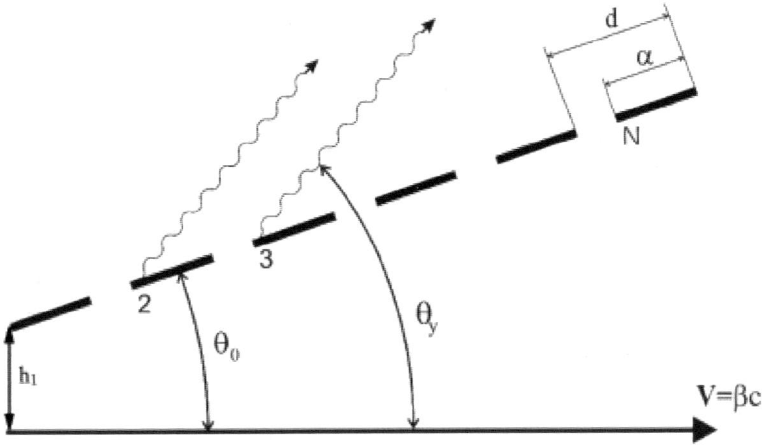

Figure. 1. Scheme of resonance diffraction radiation generating.

In Fig. 1 the geometry of RDR generating in the simplest grating is showed. This grating consists of ideal conducting strips with vacuum gaps between. Intensity of radiation for this grating is calculated by the formula [4]:

$$\frac{dW_{RDR}}{\hbar d\omega d\Omega} = \frac{dW_{DR}}{\hbar d\omega d\Omega} F_{\text{strip}} F_N, \tag{1}$$

here ω is a frequency, $dW_{DR}/\hbar d\omega d\Omega$ is a spectral-angle intensity of diffraction radiation from flat half-infinite target. The factor F_{strip} describes diffraction radiation from a single strip:

$$F_{strip} = 4 \cdot (\sinh^2 \alpha + \sin^2 \varphi), \tag{2}$$

$$\alpha = \frac{a\pi \sin \theta_0}{\lambda \gamma} \sqrt{1+\gamma^2\theta_x^2}, \tag{3}$$

$$\varphi = \frac{a\pi\left[\cos(\theta_y - \theta_0) - \cos\theta_0 / \beta\right]}{\lambda}. \tag{4}$$

Here β = v/c - speed of an electron in units of speed of light, γ – Lorentz-factor, λ – wavelength, θ_0, θ_y are determined in Fig. 1. The projection angle θ_x is counted from the plane of figure. For grating consisting of *N* periods interference factor F_N is writing as following:

$$F_N = \exp[-(N-1) \cdot \alpha_0] \cdot \left(\frac{\sin^2(N\varphi_0 / 2) + \sinh^2(N\alpha_0 / 2)}{\sin^2(\varphi_0 / 2) + \sinh^2(\alpha_0 / 2)} \right), \tag{5}$$

$$\varphi_0 = \frac{2\pi d\left[\cos\left(\theta_y - \theta_0\right) - \cos\theta_0/\beta\right]}{\lambda}, \tag{6}$$

$$\alpha_0 = \left(\frac{2\pi d \sin\theta_0}{\gamma\lambda}\right)\sqrt{1+\gamma^2\theta_x^2}\,. \tag{7}$$

It should be noted that a structure of the formula (1) is similar to the formula describing intensity of resonant transition radiation from a stack of foils with taking into account radiation absorption [5].
For a parallel flight ($\theta_0 = 0$) formula (5) becomes the known expression: $F_N = 2\pi N\delta(\varphi_0 - 2k\pi)$,

$$F_N = \frac{\sin^2(N\phi_0/2)}{\sin^2(\phi_0/2)} \to \qquad F_N = 2\pi N\delta(\varphi_0 - 2k\pi), \tag{8}$$

with $N \to \infty$.
As was shown in the work [4] in the case of "long" grating $N \gg 1$ the generalized dispersion relation connecting wavelength, tilt angle θ_0 and observation angle θ_y can be written in the following manner:

$$\lambda_k = \frac{d\left[\cos(\theta_y - \theta_0) - \cos\theta_0/\beta\right]}{k}, \tag{9}$$

where k is diffraction order.
Spectral-angular distribution of DR intensity from an inclined perfectly conducting target for charge flying at distance h_1 (see Fig.1) describing by the formula:

$$\frac{d^2W_{DR}}{\hbar d\omega d\Omega} = \frac{\alpha}{4\pi^2}\exp\left[-(\omega/\omega_c)\sqrt{1+\gamma^2\theta_x^2}\right]\left\{y^{-2}(1+\cos\theta_0)\times\left[1-\cos(\theta_y-\theta_0)\right]+\right.$$
$$\left.+2\theta_x^2\left[1-\cos\theta_0\times\cos(\theta_y-\theta_0)\right]\right\}\left[(\gamma^{-2}+\theta_x^2)\sin(\theta_y/2)\times\sin^2(\theta_0-\theta_y/2)\right]^{-1}, \tag{10}$$

Here $\alpha = 1/137$ is the fine structure constant, $\omega_c = \gamma c/2h_1$ is the characteristic frequency of DR.

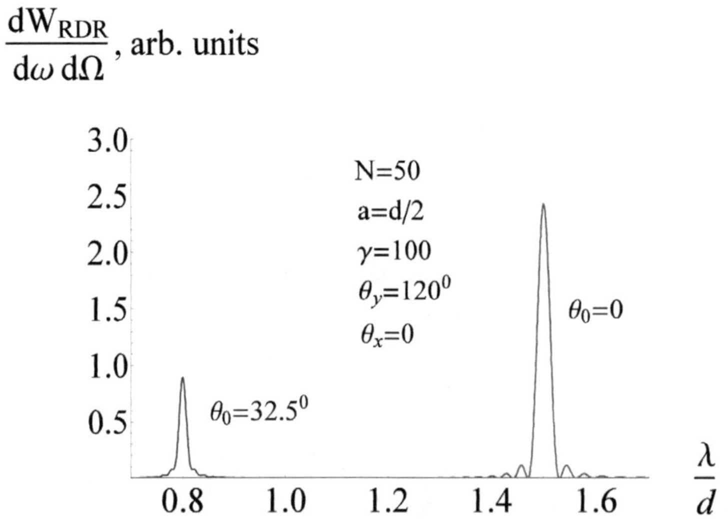

Figure. 2. Spectrum of RDR for two grating angles $\theta_0 = 0$ – right one and $\theta_0 = 32.5^\circ$ with the fixed observation angle $\theta_y = 120^\circ$.

IOP Conf. Series: Materials Science and Engineering **93** (2015) 012068 doi:10.1088/1757-899X/93/1/012068

Fig. 2 shows the spectrum of RDR calculated for parameters: $\omega / \omega_c = 1$; $\gamma = 100$; $N = 50$; $\alpha = d/2$; $\theta_y = 120^\circ$ for parallel flight of electrons ($\theta_0 = 0$, Smit-Parcell radiation) and $\theta_0 = 32.5^\circ$ (RDR). In both cases quasi monochromatic radiation is generated for the grating with *N* periods it the condition $\gamma\lambda > Nd\cos\theta_0$ is fulfilled:

$$\frac{\Delta\lambda}{\lambda} \sim \frac{1}{N}.$$

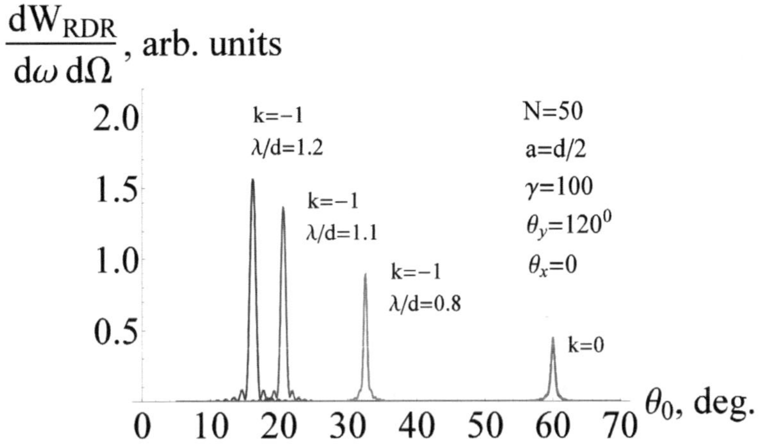

Figure 3. RDR intensity dependence on tilt angle θ_0 for observation angle $\theta_y = 120^\circ$ and different wavelengths.

On Fig. 3 one can see so-called orientation dependencies – dependencies of RDR intensity with $\lambda = \text{const}$, $\theta_y = \text{const}$ with changing of grating angle θ_0. Three left peaks correspond to diffraction order $k = -1$ for different wavelengths whereas maximum for $\theta_0 = 60^\circ$ corresponds to zero order of diffraction (k=0) (mirror reflection $\theta_y = 2\theta_0$). Angular distribution of RDR for $\lambda / d = 0.8$ and $\theta_0 = 32.5^\circ$ is shown on Fig. 4. Angular width of maximum $\Delta\theta_y = 0.9^\circ$ for $N = 50$ can be determined by number of grating periods (condition $\gamma\lambda > Nd\cos\theta_0$ is satisfied).

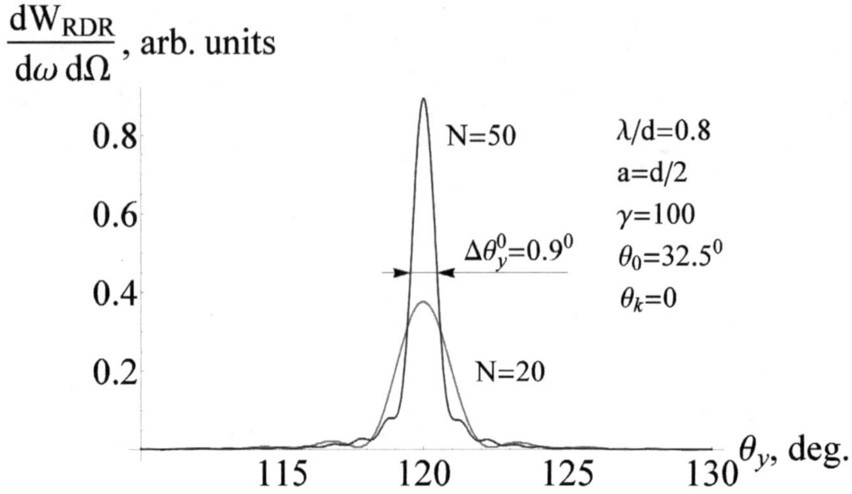

Figure. 4. Angular distribution of RDR for wavelength $\lambda = 0.8d$ from a grating with number of periods $N = 50$ and $N = 20$.

All results shown above were obtained for an electron beam. Obviously that such a divergence neglecting by angular divergence leads to angular distribution broadening. Simulation results for a beam divergence approximated by the Gauss distribution with parameter $\sigma = 0.9^0$ (16 mrad) is shown in Fig. 5 in comparison with ideal case. Both distributions are normalized to unity. Therefore using of RDR generated during the flight of ultrarelativistic electrons near inclined grating with a reasonable number of periods (N = 50 in this case) allows to measure a divergence of initial beam measuring RDR angular distribution with fixed wavelength. In real conditions of experiment it is possible to use filter with a finite bandwidth. The broadening of RDR angular distribution due to this factor can be simulated without problems.

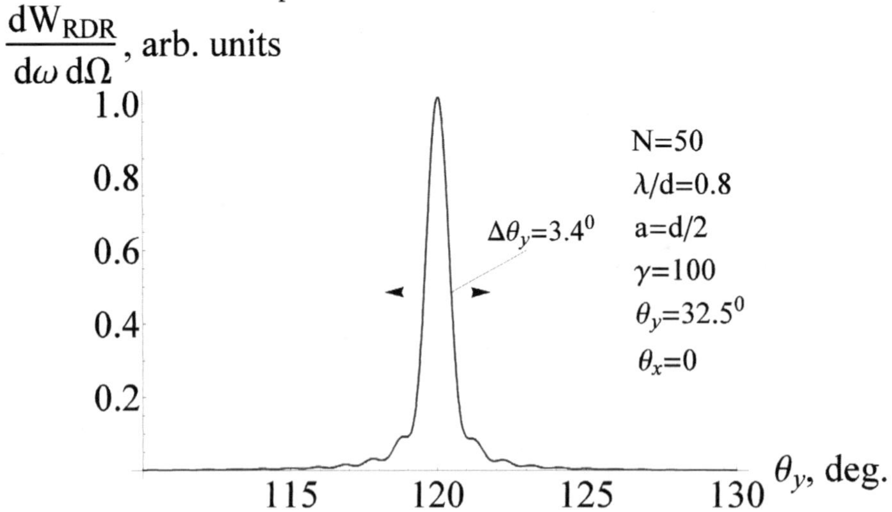

Figure. 5. Angular distribution of RDR for wavelength $\lambda = 0.8d$ from grating with a number of periods $N = 50$ for ideal electron beam (continuous curve) and for divergent beam ($\sigma = 0.9^\circ$ - points)

IOP Conf. Series: Materials Science and Engineering **93** (2015) 012068 doi:10.1088/1757-899X/93/1/012068

3. Conclusion

In conclusion were note that optical transition radiation from optical grating [6] where number of periods is was investigated in experiment for electron beam energy E0 successfully.
This work was partly supported by the Russian Ministry of Education and Science within the program "Nauka" Grant # 3.709.2014/k

This work was particularly supported by the grant of the Ministry of Education and Science of the Russian Federation within program "Nauka".

References

[1] Smith S.J., Purcell E.M. Phys. Rev. **92** (1953) 1069.
[2] Potylitsyn A.P., Ryazanov M.I., Strikhanov M.N., Tishchenko A.A. Diffraction radiation from relativistic particles // Berlin: Springer, 2010, 278 p.
[3] Karataev P., Araki S., Hamatsu R. et al. Phys. Rev. Lett. **93** (2004) 244802.
[4] Potylitsyn A.P., Karataev P.V., Naumenko G.A. Phys. Rev. E. **61** (2000) 7039.
[5] Garibyan G.M., Shi Ian. X-ray transition radiation. Erevan, Academy of Science USSR, 1983.
[6] Henry P., Haeberle O., Rullhusen P. Phys. Rev. E. **60** (1999) 6214.

Development of the device for study of the skin characteristics

[1]N Turgunova ,[1]A Aleinik and [2]J Hlubik
1Tomsk Polytechnic University, Tomsk, Russia
[2]Czech Technical University in Prague, Prague, Czech Republic

E-mail: ousa@tpu.ru

Abstract. The skin is the largest organ of the human body that performs many specialized functions. The main kind of the skin study is histological analysis, so the development of non-invasive methods for investigating the characteristics of the skin is very important. For a more detailed study of the skin the combined infrared spectroscopy and bioimpedance method has been proposed so that to examine the structure of the skin in more detail.

1. Introduction

The skin is the largest organ of the human body and performs many specialized functions. It consists of three main layers - the epidermis, dermis and hypodermis. The skin acts as a barrier between the body and the environment and provides immune protection against the dangerous pathogens. Skin cancer is one of the most common types of cancer [1-3]. White-skinned people are more susceptible to this type of the disease, due to the low content of melanin in the skin, which performs the protective function [4]. The main type of the skin study is a histological analysis, so the development of non-invasive methods for investigating the characteristics of the skin seems to be relevant.

Infrared spectroscopy and bioimpedance analysis are non-invasive techniques which are widely used in various applications, ranging from the determination of the quality of wood samples, metal corrosion detection or to study tissue samples for medical diagnostics. In particular Halter et al. used bioimpedance method for detection of prostate cancer tissue [5]. The sensitivity in these cases ranged 81-95% and a specificity of 49-64%. However, in these studies micro-invasive electrodes were used, designed specifically to overcome the top layer of the skin.

Ewaga applied infrared spectroscopy for study urea and water content in the human skin, while Shuler et al. used the method for determining the oxygen content in the leg after acute injury [6, 7]. Authors have reached 70-98% accuracy in the differentiation of benign and malignant neoplasms. However, they excluded the case of melanoma from their study.

For a more detailed study of the skin it has been proposed to combine the two methods, and to develop the appropriate type of electrode. Such combination will provide more information on the state of the skin, than by using one of the methods. It is therefore necessary to develop a non-invasive device for fast and reliable studies of the skin characteristics.

2. Materials and methods

The main part of the device is bioimpedance spectrometer which measures the impedance of tissues in the frequency range 100 Hz - 100 kHz. As a main element, the bioimpedance spectrometer chip AD5933 is selected, which represents an impedance converter. For a more accurate and stable measurements, the Howland current source is selected. The main requirement to the current source is high output impedance as compared to the load resistance. Additional requirements are stability, load capacity and low sensitivity to the various components of the circuit. The study of the load characteristic of the current source is shown in Figure 1. As can be seen, the current source maintains a stable current value when the load changes from 100 ohms to 1 Mohm.

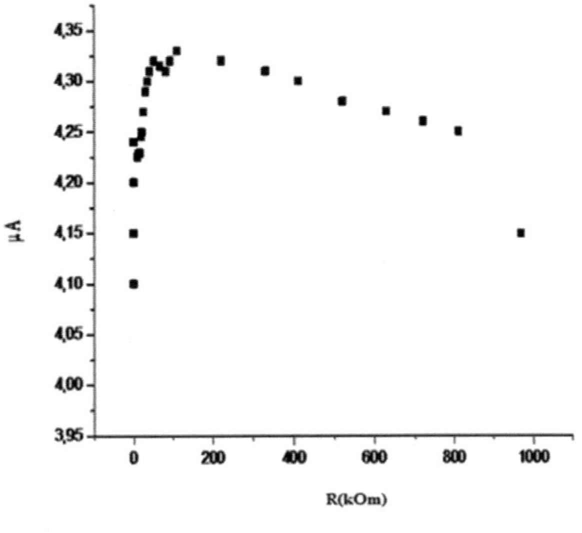

Figure 1. Load characteristics of the current source.

According to [8] the cell may be represented by an equivalent circuit, consisting of two resistors and one capacitor. The equivalent circuit has a characteristic frequency at which the phase reaches its peak.

$$\begin{cases} f = \dfrac{1}{2\pi c}\sqrt{\dfrac{R_1 + R_2}{R_1 R_2^2}} \\ \theta_{peak} = -arctn\left(\dfrac{1}{2}\sqrt{\dfrac{R_2^2}{R_1(R_1 + R_2)}}\right) \end{cases} \tag{1}$$

The calculated values are - the phase reaches its peak at θ_{peak} = -32,6 ° and frequency f = 100 kHz. Further, the equivalent circuit dependence on the frequency was measured in 30 frequency points in the frequency range 1 kHz-100 kHz. The result at each point is averaged over 50 measurements, which reduces errors. The measurement results are shown in Figure 2.

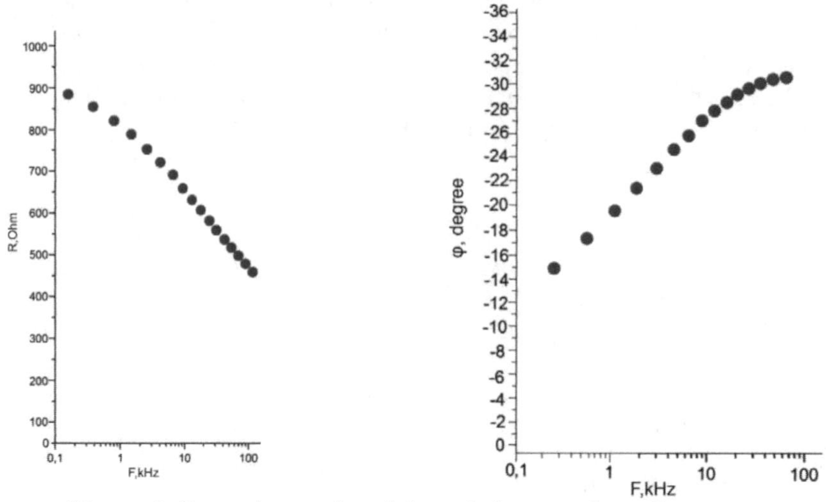

Figure 2. Dependence of module and phase angle on frequency

As seen from the graph, the maximum phase value = -32.6 ° is reached at 100 kHz. The result matches the calculated data. It follows that the device can measure the resistance of the tissue in the frequency range 1 kHz-100 kHz.

Oxygen content in the skin is the important factor, determining the correct cell metabolism. For controlling the oxygen content in the skin the method using two light sources with a wavelengths of 660 and 1300 nm is selected.

The device consists of the receiving part, optical converter and the system for collecting and processing data. The main control element is a microcontroller Atmega16. The signal is processed by a 10-bit ADC. Each LED is controlled via a multiplexer. The main objective is to select the AC signal with the exception of its DC component. This task is performed by using an operational amplifier. As smoothing filter the moving window is used.

The sensor is a cylinder. In the middle of the sensor are two light emitting diodes with a wavelength of 660 and 1300 nm, and a photodiode. Along the edges two electrodes for bioimpedance analysis situated.

Before measurement the skin was wiped with saline for 1 minute. First measurement was done with normal blood flow. Then, rubber band was imposed on the forearm for 1 minute. This was followed by a second measurement. Further measurements were taken every 30 seconds, until the full restoration of blood flow.

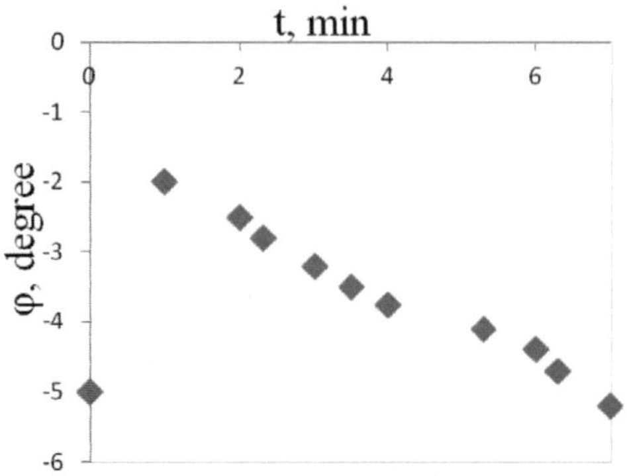

Figure 3. The dependence of the phase angle on the blood flow

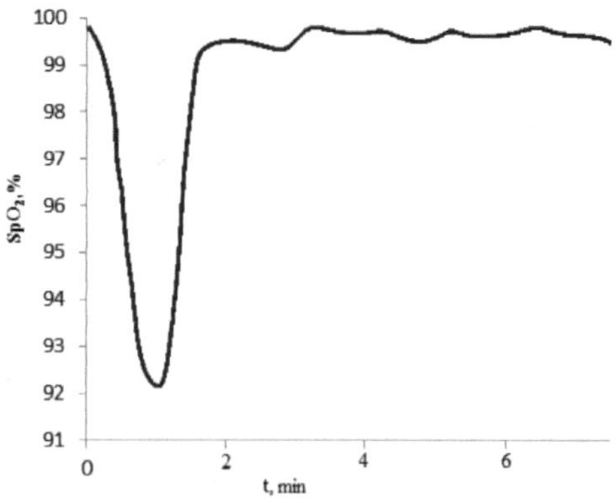

Figure 4. Oxygen saturation curve

3. Results and discussions

During hypoxia there is a lack of oxygen in the cells, which are not able to provide enough energy to the ionic channels and pumps. Since water molecules easily penetrate through the membrane, the interior of the cell contains a larger amount of water which leads to an increase in size, occupying more space in the extracellular space. Consequently, the current in the extracellular space becomes limited and the resistance at low frequency increases, which reduces the phase angle (Fig.4.).

From Figure 3. it is clear that approximately in the first minute there is a reduction of oxygen saturation in cells.

4. Summary

Combining the two methods made it possible to improve the reliability of measurements using non-invasive methods, as confirmed by the measurements. In the future, it is expected to carry out studies for the diagnosis of malignant tumors of the skin.

Acknowledgements

This work was partly supported by Competitiveness Enhancement Program of Tomsk Polytechnic University.

References

1. So P-L Skin Cancer (New York: Infobase Publishing)
2. Helfand M, Mahon S M, Eden K B, Frame P S, Orleans C T 2001 *Am J Prev Med.* **20** 47-58
3. Markovic S N, Erickson L A, Rao R D, Weenig R H, Pockaj B A, Bardia A, et al.2007 *Mayo Clin Proc.* **82** 490-513
4. Garbe C, Eigentler T K. 2006 *Melanoma Res.* **17** 117-27.
5. Halter R J, Schned A, Heaney J, Hartov A, Schutz S, Paulsen K D 2008 *J Urol.* **179** 1580-1586
6. Egawa M 2009 *Skin Res Technol* **15** 195- 198
7. Shuler M S, Reisman W M, Whitesides T E, Kinsey T L, Hammerberg E M, Davila M G, et al. Near-Infrared Spectroscopy in Lower Extremity Trauma. 2009 *Journal of Bone and Joint Surgery-American Volume*. **91A** 1360.
8. Neves C E and Souza M N 2000 *Physiol. Meas.* **21** 395-408.

The use of fly ash the thermal power plants in the construction

R S Fediuk and A M Yushin

Far Eastern Federal University, Vladivostok, Russia

E-mail: roman44@yandex.ru

Abstract. The problems of ecological and radiation safety of the construction of man-made waste like fly ash thermal power plants were researched. The chemical composition of TPPs ashes of Primorsky Territory was studied, defined their specific effective activity of natural radionuclides. The most modern research methods were used - differential thermal analysis, thermogravimetry, X-ray analysis. It was revealed that the ash of the Primorskaya TPP and Partizanskaya TPP has exceed the permissible parameters of radioactivity, so not suitable for use in construction. Ashes of Vladivostok TPP-2 and Artem TPP of Primorsky Region on parameters radioactivity and chemical composition have suitable for use as a filler in the concrete.

1. Introduction

The use of technogenic waste in addition to saving material and natural resources, protects the environment. Therefore, during the design phase of the construction of the object should be possible to use a man-made waste. At the same time, once made a focus on the environmental aspect, you need carefully to select building materials.

In this paper we will focus on the study of the properties of fly ash thermal power plants as an additive in cement composite binder, and special attention will be paid to the assessment of its radioactivity.

2. The use of fly ash of thermal power plants

Waste heat power plants mainly divided into two categories: ash and fly ash, different way to remove [1-3]. According to [4], fly ash, it is a more effective additive in cement compositions than ash. In this case, the amount alite phase is increases and aluminate is decreases compared with a control sample, respectively, increases the strength of the cement stone [5-10].

The objects of study are fly ash of the largest thermal power plant Primorye Territory: Vladivostok TPP, TPP Artem, Primorye SRPS and Partizansk power station. An important factor is the ability to separate the selection of dry ash that is currently implemented in these thermal power plants [11-15].

The composition and structure and ash depends on a set of concurrent factors: the type and morphological features of combusted fuel fineness in the course of its preparation, the ash content of the fuel, the chemical composition of the mineral part of the fuel, the temperature in the combustion zone, the residence time of the particles in this zone, and others (figure 1).

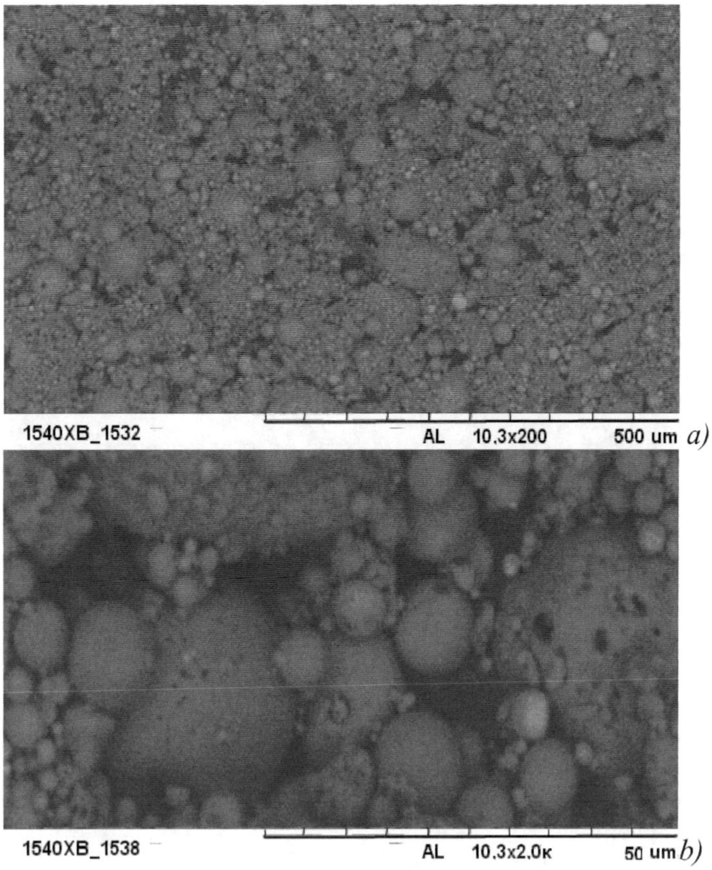

Figure 1. The micrographs of fly ash Vladivostok TPP-2. Magnification 200 and 2000 times

The laboratory research have set the basic characteristics of the fly ash, determine the possibility of its use for the production of polymer-mineral construction materials.

The main criteria for the selection of fly ash for study the possibility of its use as a component of the composite binder:

- Shape of the particles;
- Particle size distribution;
- Specific surface area;
- Packing density;
- Chemical composition.

Samples of fly ash produced from a large silo capacity, where it could be shipped for delivery to consumers or, depending on the filling of the silo through the mixing device in the pressure system to move to the ash dump. After the selection of the private trials were combined, mixed thoroughly and quartering reduced to laboratory sample.

Laboratory samples of fly ash were sieved through 008; as a result of the screening on the sieve remained a significant part of the unburned coal particles, magnetite beads, and microspheres; ash, pass through a sieve, it was selected for the study and preparation of concrete.

Chemical analysis of the sol (Table. 1) is made in accordance with GOST 10538-87. This standard applies to brown and black coal, anthracite, oil shale, peat and coke and establishes methods for determining the fuel in the ash of silicon dioxide (SiO_2), iron oxide (Fe_2O_3), aluminum oxide (Al_2O_3), magnesium oxide (MgO), calcium oxide (CaO), potassium oxide (K_2O), sodium oxide (Na_2O), phosphorous oxide (P_2O_5), titanium dioxide (TiO_2), sulfur trioxide (SO_3), mixed manganese oxide (Mn_3O_4).

Table 1

The chemical composition of ashes TPPs the Primorsky Territory

		TPPs			
		Primorskaya	Vladivostok	Artem	Partizansk
The predominant type of coal		Luchegorsk and Bikin	Primorsky lignite (Pavlovsky section)	Black coal	Neryungri black coal
Content of elements on an oxide basis, %	SiO_2	55,3	63,0	48.1	47,4
	TiO_2	0,5	0,5	0,0	0,9
	Al_2O_3	12,6	21,4	29,3	22,3
	Fe_2O_3	10,7	7,5	6,5	19,6
	CaO	12,5	3,4	9,7	4,8
	MgO	3,5	2,1	1,8	2,8
	K_2O	1,0	1,3	1,2	0,1
	Na_2O	0,4	0,3	0,2	0,4
	SO_3	3,4	0,6	2,3	1,62
	CaO_{free}	1,0	0,4	<0,1	no
	LOI	2,3	1,4	0,6	<5

The data on the chemical composition of the ash indicate that the contents of the individual oxides in the fly ash derived from the combustion of various kinds of pulverized coal has significant deviation. It determines the difference in the properties of ash and possible areas of use in the manufacture of building materials.

Ash depending on the total effective specific activity of natural radionuclides A_{eff} less 370 Bq / kg is applied to the production of materials, products and structures, residential and public buildings.

The results of determination of specific effective activity of fly ash on spectrometric complex "USC Gamma Plus" are shown in Table. 2.

Table 2

Determination of specific effective activity of fly ash TPPs the Primorsky Territory

Parameter	The result of the measurement (A), Bq / kg			
	Primorskaya TPP	Vladivostok TPP	Artem TPP	Partizansk TPP
Activity ^{40}K	496.9±101	392±89	342±68	516.9±101
Activity ^{232}Th	153.6±20.3	31.5±19.7	29.5±15.7	193.2±22.3
Activity ^{226}Ra	163.1±9.36	37.63±6.32	27.23±5.93	113.1±6.37
$A_{eff} = A_{Ra} + 1,31A_{Th} + 0,085A_K$	>398	80±30	93±20	>410

Ashes of Vladivostok TPP and TPP Artem belong to the first class of materials (less than 370 Bq / kg) in accordance with GOST 30108-94 "Building materials and products. Determination of the specific effective activity of natural radionuclides". This material can be used for all kinds of construction work.

Ashes of Primorskaya TPP and Partizansk TPP areexceed the permissible parameters of radioactivity, so not suitable for use in construction.

Thus, by GOST 25592-91 "Mixes slag thermal power plants for concrete" the ashes of Vladivostok TPP and TPP Artem are selected for further research. The high ash content in Al_2O_3 (to 29.3%) and SiO_2 (63%) can cause crystallization like mullite compounds. The loss on ignition of less than 1.5% say a small amount of residual fuel that minimizes manufacturing processes shrinkage during firing.

Thermal studies were performed on raw by thermogravimetric analyzer Shimadzu DTG-60H at a rate of temperature rise of 20°C / min, in the range of 20-1100°C. The results of thermal analysis are presented graphically in figure 2.

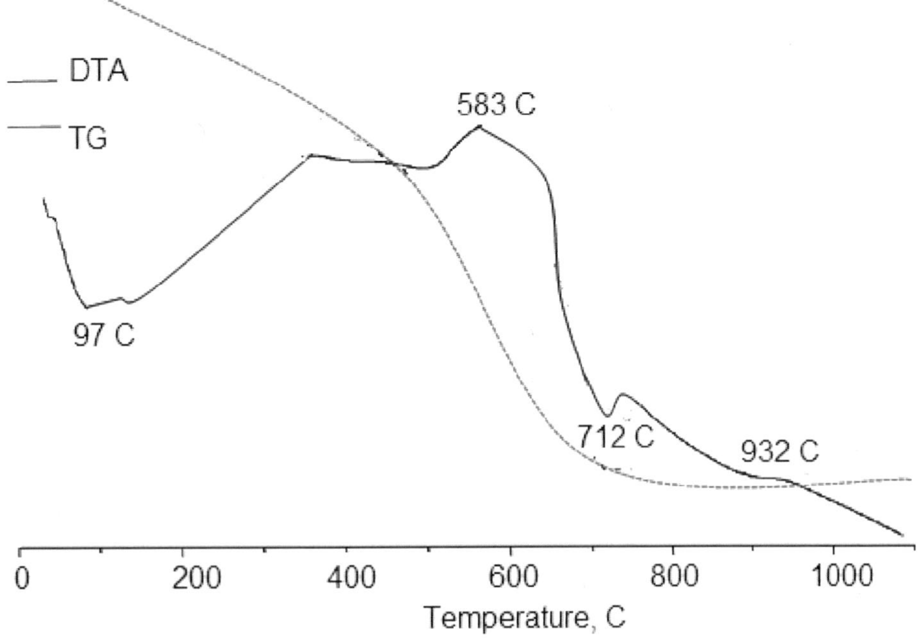

Figure 2. The results of DTA and TG of Vladivostok TPP fly ash

During heat treatment the ash in the range of 40-200 ° C, is the loss of water adsorbed surface area of the particles. Carbonate decomposition is observed at a temperature of 712°C. Burning residual fuel - 500-700°C. The nature and intensity of loss of mass during this indicates an amount of unburnt residues are particles of coal and coke and semi-coke residue. Relatively small exotherm with a maximum of 932°C reflects the crystallization of compounds in mullitopodobnyh aluminosilicate phase.

XRD analysis showed that the ash except quartz is determined clearly expressed mullite phase (figure 3).

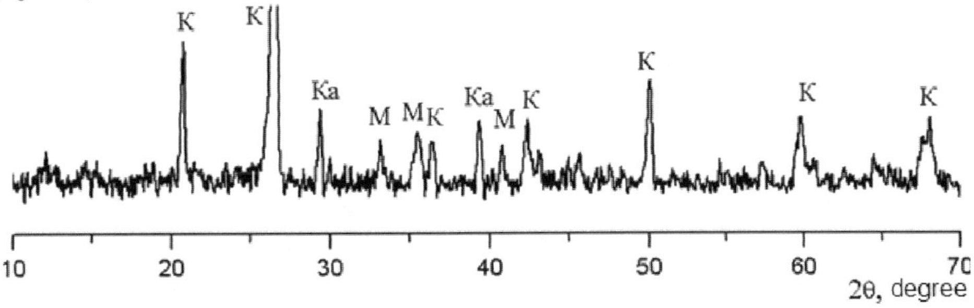

Figure 3. XRD results of Vladivostok TPP fly ash. K - quartz, Ca - calcite, M - mullite

3. Conclusion

The test material is relatively homogeneous, multicomponent. According to the composition closest to aluminosilicates, due to the high content of silicon and aluminum oxides to 80-90%, of which about 2/3 of silicon oxide. At the same time, fly ash has almost no unburned particles, in which tend to concentrate harmful components.

Ash consists of crystalline and amorphous phases. The crystalline phase contains quartz, feldspar, mullite, et al., The amorphous phase is represented primarily in the form of glass.

Thus, the fly ash consists of minerals that are widely used separately as filler dispersed in composite binders. Therefore, we can assume that the ash CHP Primorsky Region on chemical composition suitable for use as a filler in the cement composition.

According to the classification of GOST 24640-91 selected are low-calcium fly ash (acidic) and serve as additives, the material composition of the components (active mineral additives having pozzolanic properties).

According to the classification committee RILEM, you can add that both selected fly ash (Vladivostok TPP and TPP Artem) pozzolanic have normal activity.

Hydration products binders have high adhesion beans silica component, which has a significant amount of structural defects, which also contributes to the growth of a preliminary mechanical activation in the preparation of the composite binder, which allows us to strengthen and improve the processes of hydration of clinker minerals, and leads to the formation of more tumors by increasing the active nucleation sites on the particle surface.

References

[1] Ichiko Misumi, Jun-ichiro Kitta, Hiroyuki Fujimoto, Satoshi Gonda, Yasushi Azuma, Keisuke Maeda, Tomizo Kurosawa, Yoshiyasu Ito, Kazuhiko Omote, Yoshinori Nakayama and Hiroki Kawada 2012 *Meas. Sci. Technol.* **23** 015002

[2] M Tabuchi, H Tameoka, T Kawase and Y Takeda 2011 1757-899X **24** 012007

[3] H Nakao, Y Yamasaki, J Okamoto, T Sudayama, Y Takahashi, K Kobayashi, R Kumai and Y Murakami 2014 1742-6596 **502** 012015

[4] R V Ovchinnikov *diss. ... cand. tech. sciences* 2014

[5] J del Cerro, J Manchado, F J Romero and M C Gallardo 2012 *Meas. Sci. Technol.* **23** 035003

[6] Camila Barreneche, Aran Solé, Laia Miró, Ingrid Martorell, A Inés Fernández and Luisa F Cabeza 2012 *Meas. Sci. Technol.* **23** 085606

[7] Md Jashim Uddin, N H Md Yusoff, O Anwar Bég and Ahamd Izani Ismail 2013 *Phys. Scr.* **87** 025401

[8] Hideyuki Nakano, Akira Kouchi, Shogo Tachibana and Akira Tsuchiyama 2003 0004-637X **592** p 1252

[9] J M Cuevas, R Rubio, J M Laza, J L Vilas, M Rodriguez and L M León 2012 *Smart Mater. Struct.* **21** 035004

[10] Zhi Chen, Dai Mochizuki, Masato M Maitani and Yuji Wada 2013 *Nanotechnology* **24** 265602

[11] Chadwick A.V., Düvel A., Heitjans P., Pickup D.M., Ramos S., Sayle D.C. and Sayle T.X.T. 2015 *IOP Conference Series: Materials Science and Engineering* **80**

[12] Polat R., Demirbofa R. and Khushefati W.H. 2015 *Construction and Building Materials* **81** pp 268-275.

[13] Raheek I Ibrahim Wong Z.H. and Mohammad A.W. 2015 *IOP Conference Series: Materials Science and Engineering* **80**

[14] Baeza F.J., Galao O., Zornoza E. and Garcés P. 2013 *Materials* **6 (3)** pp 841-855.

[15] Al-Rahman L.A. and Raja R.I. 2012 *Attenuation of Noise By Using Absorption Materials and Barriers: A Review* **2 (7)** pp 1207-1217.

IOP Conf. Series: Materials Science and Engineering **93** (2015) 012071 doi:10.1088/1757-899X/93/1/012071

JOULE HEATING EFFECTS ON QUARTZ PARTICLE MELTING IN HIGH-TEMPERATURE SILICATE MELT

V Vlasov[1,2], G Volokitin[2], N Skripnikova[2], O Volokitin[2], V Shekhovtsov[2]

[1]National Research Tomsk Polytechnic University, Applied Physiscs Engineering Dept.,Tomsk, 634050, Russia
[2]Tomsk State University of Architecture and Building, Applied Mechanics and Materials Science Dept., Tomsk, 634003, Russia

E-mail: volokitin_oleg@mail.ru

Abstract. This work is mostly focused on the melting process model simulation of quartz particles having the radius within the range of 10^{-6}–10^{-3} m. The melting process is simulated accounting for the heat generation at an electric current passage through a quartz particle.

1. Introduction

As a rule, natural sand is presented by silica grains (SiO_2). α-Quartz is the most widespread form of silicon dioxide [1]. Among silica-based materials, vitreous silica has a special place. It possesses a number of valuable physicochemical properties, such as heat resistance, refractoriness, chemical and radiation resistance, and transparency within a wide wavelength range. Methods for the production of silica sand-based products differ from those accepted for the common glass technology. This is because the super high viscosity of silica melt even at temperatures exceeding 2000°C and its higher volatility [2-3]. The low-temperature plasma (LTP) used for silicate melting possesses a high energy concentration and 3000–5000°C temperature that will allow achieving the required viscosity of silicate melt and providing its uniform heating [4-7].

This work is mostly focused on the melting process model simulation of quartz particles having the radius within the range of 10^{-6}–10^{-3} m. The melting process is simulated accounting for the heat generation at an electric current passage through a quartz particle.

2. Experimental

Quartz particles are suggested to be melt in the plasma apparatus designed for the production of high-temperature silicate melts [8]. The flow of plasma-supporting gas in this plasma apparatus serves as a heat-transfer agent generated from a plasma nozzle. The LTP jet enters the melting space of the water-cooled furnace to produce melting of refractory silicate raw material. The batcher and worm feeder designed for the plasma apparatus allow feeding the raw material in the plasma jet area. The powder raw material is introduced in the depth of the previously formed silicate melt and melted over the whole melting space of the furnace owing to Joule heating.

The plasma apparatus specifications [9] allow using it for quartz melting.

2.1. Task description

It is advisable to divide the whole period of the melting process into time intervals of $0 < t \leq t_1$ and $t_1 < t \leq t_2$, where t_1 is the time of quartz particle heating up to melting temperature T_m; t_2 is the time of completion of the melting process; t_3 is the time of stabilization of the equilibrium state (the temperature of particles equals to that of silicate melt).

The particle energy at the stage of heating up to melting temperature can be obtained from

$$c_p \rho_p V_p \frac{dT_p}{dt} = \alpha S_p (T_{am} - T_p) + Q_{el} . \tag{1}$$

$$T_p(0) = T_i . \tag{2}$$

The following notations are suggested for equations (1) – (2):
t is the time; T is the temperature; c is the specific heat capacity; ρ is the density; α is the heat exchange rate; $S_{p,0} = 4\pi r_{p,0}^2$ is the surface area of quartz particle at its initial radius $r_{p,0}$; $V_{p,0} = \frac{4}{3}\pi r_{p,0}^3$ is the quartz particle volume at its initial radius $r_{p,0}$; Q_{el} is the heat generated by electric heating.

Indices p, am and i denote parameters relating to the quartz particle, ambient air, and initial state, respectively.

The heat exchange rate can be obtained using the Nusselt number:

$$\alpha = \frac{\lambda_{am} Nu}{r_p} .$$

Let us assume that upon the achievement of the melting temperature by quartz particle, the amount of heat transferred to it due to heat exchange and Joule heating, is spent for its melting.

Taking the above assumption into account, the volume of the solid (crystalline) portion of quartz particle is changed during the melting stage:

$$\rho_p q_m \frac{dV_p}{dt} = \alpha S_p (T_m - T_{am}) - Q_{el} , \tag{3}$$

$$V_p(0) = V_{p,0}, \; S_p(0) = S_{p,0} , \tag{4}$$

where q_m is the specific heat of melting; $m_p = \rho_p V_p$ is the mass of quartz particle.

3. Results and discussion

Equations (1), (3) consider the effect from the electric heating on the whole process. In order to determine the heat generation produced by the electric source, the Joule-Lenz's law should be used. According to this law, the amount of heat released is proportional to the square of the current such that

$$Q = I^2 R ,$$

where I is the current in the unit.

Quartz particle releases the amount of heat equaling to

$$Q_{el} = i^2 r ,$$

where i is the current passing through quartz particle; $r = \rho_{sp} \frac{l_q}{S_q}$ is the resistance of quartz particle; ρ_{sp} is the specific resistance of quartz particle.

Let quartz particle be a cylinder with the diameter equaling to its height, i.e. $d_{cyl} = l_{cyl}$. At the same time, the diameter of cylinder is selected such that the volume of a cylindrical particle equals to the volume of a spherical particle, i.e. $\frac{4}{3}\pi r_p^3 = \frac{\pi d_{cyl}^3}{4}$.

The specific resistance ρ_{sp} ranges from 10^{14} Ohm·m at 20°C to 0.9 Ohm·m at 1600°C.

Let us analyze the intensity of the current passing through quartz particle at the stage of its heating. Quartz particle is in the silicate melt the specific resistance of which is low and in the order of ρ_{sp} = 1 Ohm·m. Quartz particle has 10^{14} Ohm·m specific resistance. Let us assume that the volume fraction ratio of the silicate melt and that of unmelted quartz particles is 10:1 (Fig. 1).

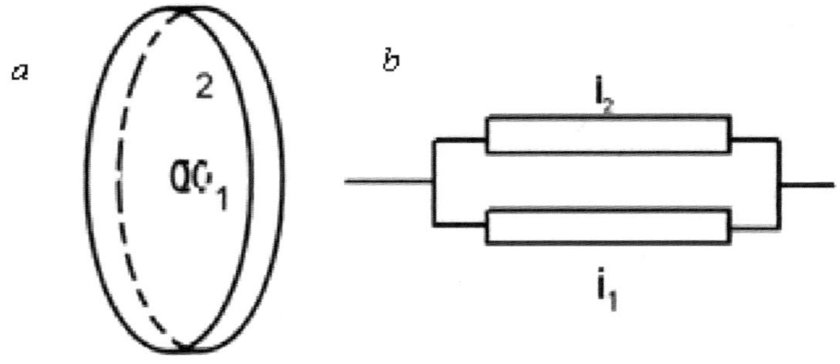

Figure 1. Schematic view of the circuit section: *1* – quartz particle; *2* – silicate melt; i_1 – current in quartz particle; i_2 – current in silicate melt.

The section of the circuit shown in Figure 1*a* is presented by a parallel combination of two conductors shown in Figure 1*b*. The electric current in the section (Fig. 1*a*) can be obtained from

$$i = \frac{I(s_q + s_{mt})}{S_f}, \tag{5}$$

where s_q, s_{mt} and S_f are the areas of quartz particle, silicate melt, and furnace, respectively.

For the parallel combination the following condition is met:

$$i = i_1 + i_2, \tag{6}$$

where i_1 is the current in quartz particle; i_2 is the current in the silicate melt. At the same time, by Ohm's law:

$$u = ir = i_1 r_1 = i_2 r_2, \tag{7}$$

$$r = \frac{r_1 r_2}{r_1 + r_2},$$

where r is the total resistance; r_1 and r_2 are resistances of quartz particle and silicate melt, respectively.

From (5), the values of i_1 and i_2 can be obtained:

$$i_1 = \frac{i\,r}{r_1},\quad i_2 = \frac{i\,r}{r_2}.$$

Resistances r_1 and r_2 can be obtained from:

$$r_1 = \rho_{sp1}\frac{l_q}{s_q},\quad r_2 = \rho_{sp2}\frac{l_{mt}}{s_{mt}},$$

where $\rho_{sp1} = 10^{14}$ Ohm·m; $\rho_{sp2} = 1$ Ohm·m; $l_q = 1$ mm; $l_{mt} = 1$ mm; $s_q = \frac{\pi l_q^2}{4}$, $s_{mt} = 10 s_q$.

Then $r_1 = \frac{4\cdot 10^{17}}{\pi}$, $r_2 = \frac{10^2}{\pi}$, $r = \frac{r_1 r_2}{r_1 + r_2}$.

The ratio between current values i_1 and i_2 can be found as follows:

$$\frac{i_1}{i_2} = \frac{r_2}{r_1} = \frac{1}{4}10^{-15}. \tag{8}$$

Suppose that the current in the unit is 400 A; then equation (5) can be used to determine $i = 0.216$ A. And equations (6) and (8) determine $i_1 = \frac{0.216}{1 + 4\cdot 10^{15}} \approx 5.4\cdot 10^{-17} A$.

Substituting this value in $Q_{el} = i^2 r$ we get $Q_{el} = 37.1\cdot 10^{-17}$ J/s.

However, the specific resistance of quartz particle decreases with the increase of temperature. Therefore, the model simulation is used such that $Q_{el} = 2\cdot 10^{-2}$ J/s. Calculations are made for the case of the quartz particle heating and melting in the silicate melt accounting for the heat exchange rate that depends on the radius of the solid (crystalline) quartz particle portion at its melting stage. Phase transformations are neglected.

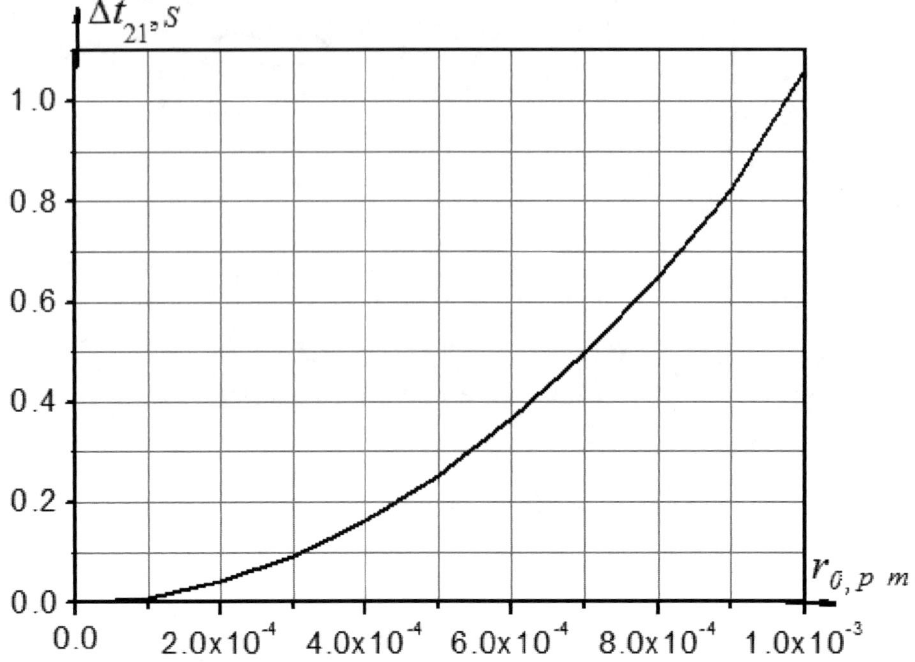

Figure 2. Dependence Δt_{21} and the particle initial radius $r_{0,p}$

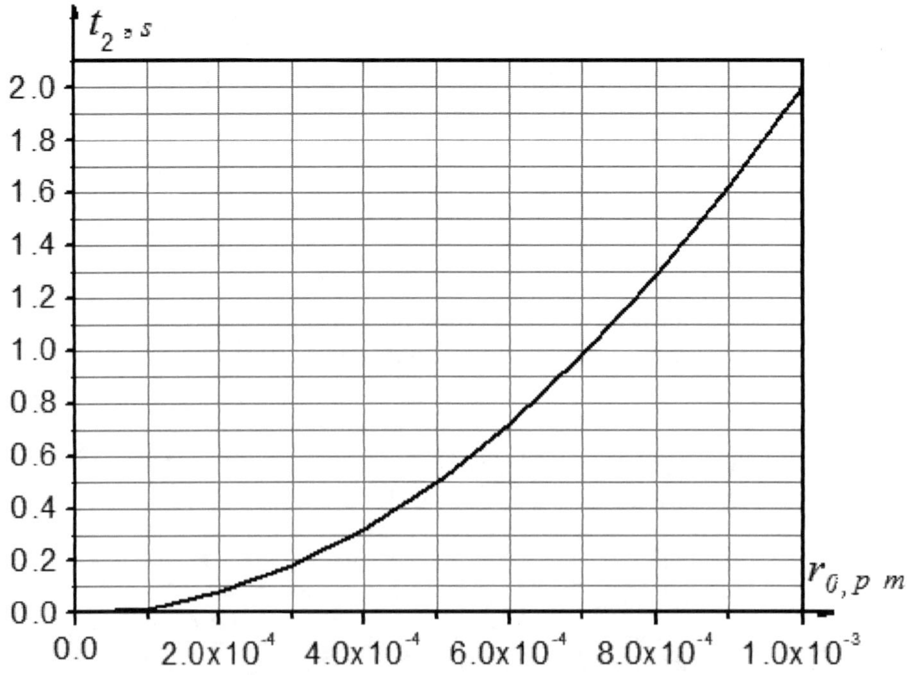

Figure 3. Dependence t_2 and the particle initial radius $r_{0,p}$

Figure 2 contains the plot of the dependence between the melting time $\Delta t_{21} = t_2 - t_1$ and the quartz particle radius accounting for Joule heating and the dependence between the heat exchange rate and the radius of the solid (crystalline) quartz particle portion at its melting stage. Figure 3 contains the

IOP Conf. Series: Materials Science and Engineering **93** (2015) 012071 doi:10.1088/1757-899X/93/1/012071

plot of the dependence between time t_2 and the quartz particle radius accounting for Joule heating and the dependence between the heat exchange rate and the radius of the solid (crystalline) quartz particle portion at its melting stage. Results of calculation of Δt_{21} and t_2 are also given in the Table below.

Table. Dependence of Δt_{12} and t_2 on the quartz particle radius.

$r_{0,p}$ [m]	10^{-6}	$2\cdot10^{-6}$	$5\cdot10^{-6}$	10^{-5}	$2\cdot10^{-5}$	$5\cdot10^{-5}$	10^{-4}	$2\cdot10^{-4}$	$5\cdot10^{-4}$	10^{-3}
Δt_{21} [s]	10^{-6}	$8\cdot10^{-6}$	$2.5\cdot10^{-5}$	10^{-4}	$8\cdot10^{-4}$	$2.5\cdot10^{-3}$	10^{-2}	$8\cdot10^{-2}$	0.25	1.02
t_2, [s]	$2\cdot10^{-4}$	$4\cdot10^{-6}$	$5\cdot10^{-5}$	$2\cdot10^{-4}$	$4\cdot10^{-4}$	$5\cdot10^{-3}$	$2\cdot10^{-2}$	$4\cdot10^{-2}$	0.5	2.0

4. Conclusion

The results showed that electric heating had no a significant effect on the time period of quartz particle heating and melting. After the heating, the melted layer around the crystalline particle was rapidly heated up to the melting temperature due to the electric current that accelerated the melting process.
The research was financed by the Ministry of Education and Science of the Russian Federation (State Order N 11.351.2014/K).

References

[1] Shchedrovitskii Ya.S. Vysokokremnistye ferrosplavy [High silica ferroalloys]. Sverdlovsk: Sverdlovsk, GNTI, 1961.

[2] Pivinskii Yu. E., Romashin A.G. Kvartsevaya keramika [Quartz ceramics], Moscow, Metallurgiya, 1974.

[3] Ailer R. Khimiya kremnezema [Chemistry of silica].Moscow, Mir, 1982.

[4] Volokitin O.G., Skripnikova N.K., Shekhovcov V. V. Technology for plasma treatment of molybdenum ore tailings. Applied Mechanics and Materials. 2014. V. 682, Pp. 515-518.

[5] Bondareva N.S., Volokitin O.G., Morozova O.O., Sheremet M.A. Unsteady regimes of hydrodynamics and heat transfer at production of high-temperature silicate melts. Thermophysics and Aeromechanics. 2013. V. 20, N5, Pp 621-629.

[6] Volokitin O.G., Volokitin G.G., Skripnikova N.K., Shekhovtsov V.V. Definition of nanostructural characteristics melting product of quartz sand. International conference on physical mesomechanics of multilevel systems 2014. AIP Publishing LLC. V. 1623, Pp 655-658.

[7] Volokitin O.G., Shekhovcov V.V., Maslov E.A. Plasma treatment technology for silicate melt used in mineral fiber production. Advanced Materials Research. 2014. V.880, Pp 233-236.

[8] Volokitin O.G. Timonov E.V., Volokitin G.G., Nikiforov A.A., Chibirkov V.K. Plazmennaya ustanovka dlya polucheniya tugoplavkogo silikatnogo rasplava [Plasma apparatus for the production of refractory silicate melt]. RF Patent 2503628. 2014.

[9] Nikiforov A.A., Maslov E.A., Skripnikova N.K., Volokitin O.G. Study of plasma technology for silicate refractory melts production. Thermophysics and Aeromechanics. 2009. V. 16, N 1, Pp 149-153.

Functional and aesthetic approach to design of bird feeders

A Kukhta [1,a] , M Kukhta [2,b]

[1] *Tomsk State University, Tomsk, Russia*

[2] *Tomsk Polytechnic University, Tomsk, Russia*

E-mail: [a] artkuh@mail.tomsknet.ru, [b]eukuh@mail.tomsknet.ru

Abstract. Anthropogenic objects which load the urban environment negatively affects the human psyche. The alternative is attracting elements of the natural environment into urban environment, of which some of the most frequently identified are birds. Attracting birds in the city is possible by means of feeders and artificial nests, however, both must be harmonious. The aim of this study is to analyze the essential functions of the feeders, and their integration into the environmental design and development of the city. On this basis an original feeder which is convenient for use by birds and attracts people's attention is developed. In this paper we apply comparative analysis of different types of feeders encountered in Tomsk, bird watching, and evaluate usability of different types of feeders from the position of their convenience both for birds and human beings. Historical-cultural analysis for determining features of the architectural and environmental design of Tomsk is carried out, the method allows us to solve engineering problems. In this study the feeder convenient for bird use is designed which blends harmoniously with the architectural design of Tomsk.

Keywords: urban environment, bird feeder, wooden architecture, birds usability.

1. Introduction

Technogenic progress is deeply embedded in the life of modern man and in the space around leaving fewer and fewer places to nature. The evolution peak of the urban environment is perhaps the urban areas where the natural environment is absent or is under the control of the human being. On the other hand the human being is also a part of the nature and anyway needs communication experiences with it.

In particular throughout the world there is a tendency to attract animals which are usually birds as the most prominent representatives of the animal world. They always attract attention and arouse interest among residents of the cities [1]. Besides the creation of parks, recreational areas, there are other ways to attract wildlife particularly bird feeders are probably an integral part of modern urban environment[2]. Animals` attraction is possible when offer them resources to meet the vital needs, particularly birds feeding in winter is an important necessity in the life of the bird on the one hand, and a big step for rapprochement between man and nature on the other. Feeding allows birds survive in unfavorable period of the year by means of the availability of food resources and hiding place. Popularity of the alone "window" feeders which are hung on the windows in the winter is an average of 1 pc. / 110 windows (Tomsk). At the same time they are used by 14 species of birds besides the typical Sinanthropus (dove, house sparrow, tree sparrow) includes species such as: great tit, bullfinch, waxwing, coal tit, willow tit, great spotted woodpeckers and gray-headed woodpeckers. In addition feeders indirectly attract birds of prey, sparrow hawk and black kite purposefully appear around them hunting for feeding birds. The ease of fabrication of the feeders and the ability to use the most common materials [3] contribute to their wide variety, in fact it`s difficult to find two of the same through. However under the conditions of the urban environment the feeder besides its direct function of attracting birds performs an equally important role in the provision of aesthetic perception of the environment.

The purpose of research is to develop and offer a bird feeder that would be useful not only for birds but also would be easy to use by human being, and perfectly blend in urban environment of Tomsk.

2. Research

Different types of original feeders are used in Tomsk in winter for the birds feeding among the variety of which there are several major types:

1. Window feeders for frequent use which hang out of the window openings and destined for birds feeding from the apartments and premises. These feeders are usually made from scrap materials and compact, adapted for mounting in the window opening.
2. Street feeders for general use, designed for use by a wide range of people. Such feeders are usually arranged in trees or on man-made elements and are freely available to people (Fig. 3, 4). This is the largest group of feeders, which includes a wide variety of shapes and sizes of containers for food. Usually they are also made from scrap materials.
3. Design feeders. The smallest group of feeders intended for use as a window opening and on the street. These feeders are characterized by original design solutions, the use of hard-treated materials. In general, the production of this type of feeders takes significantly more time and effort than first two types figure 1 and figure 2.

In the whole the analysis has showed that the most frequently used material for the manufacture of feeders is plastic containers (about 83%). In recent years cardboard feeders become to give way significantly to plastic (14%) because of its fragility. Least of all the feeders made of wood and other "designer" materials can be meet (3%) [4]. At the same time these are the feeders that attract people more from an aesthetic point of view and here there is a wide field for creativity from the perspective of material selection, formation, function and so on. Due to the popularity of bird feeders in the – just because of design decisions in designing and manufacturing there is an opportunity to improve visually the aesthetic appearance of the city and not only attract birds but also fit harmoniously into the overall appearance of the feeders.

Considering the specificity of Tomsk architecture it is advisable to offer design feeders associated with motifs of wooden ornaments and because the city is known primarily for its old wooden architectural structures. The small area of the city as a whole (300m^2) and compact construction are preserved wooden houses of the old buildings of the XIX century. Some of them are residential and some restored and compose the part of the museum complex - the monuments of wooden architecture. In the city there are a large number of old stone buildings built in XIX-XX centuries and made in the style of the Siberian baroque and classicism. For the moment the city has more than 700 buildings which are architectural monuments and the "hallmark" of the city [5,6]. Using the tree in the structural elements of the feeder it emphasizes the stylistic specificity of Tomsk integrating wooden ornaments in urban design elements feeders. The most typical of them and the possibility of their use in functional design of feeders:

1. Wooden openwork ornament is one of the most visible and attractive element of carved wooden houses. In the design of the feeder the openwork lattice fence helps to protect the food from the blowing wind and in cold period such additional feeder becomes a shelter for birds. Openwork lattice prevents pigeon penetration into the feeder that do not provide food for smaller birds, openwork structures with different holes - cannot let in house sparrow which takes off other birds [7]. Openwork lattice most needful for use in outdoor rigidly fixed feeder. Openwork elements also look aesthetically as props in the construction of window feeders.

2. Bay windows and balconies bay type are quite spacious rooms that extend beyond the walls of the main building and balconies bay type, tend to be based on the porch and rely on massive props and have a large glass area. Such design solution is important for large rigidly fixed feeders placed on the balcony or on the trunk. In this case carved props or props with banisters provide a reliable basis for additional binding and several bay windows let birds feeding without interfering with each other.

3. The shape of the roof. The climate in Tomsk region is characterized by high humidity because fallout is significantly higher than the evaporation [6]. In this regard rational form of the roof is sloping roof with abrupt slope, so that in winter the snow did not stay on it. Abrupt slopes are reflected in the decorative turrets that adorn many homes of the city. Abrupt slopes of the roof will play the same role in design of the feeder as for real home and won`t let snow lay over, turret will serve as a convenient "handle", which is used to remove the entire roof or for fixing the feeder to a branch. The manufacture of such a roof would be done with the expectation that the snow is not rolled down on bird table which may hinder birds feeding or even make it impossible.

4. Protruding elements of the wooden ornament, spiers, jesting beams. Protruding elements are common for wooden architecture in general but if the wooden spiers can be seen in many houses then, for example, the stylized "wooden horse" has become the landmark of Tomsk wooden architecture. These elements in stylized form are relevant to all types of feeders, especially on the window where the birds can use them as perches and a longer time to linger near the window.

5. Openwork window frames are typical for wooden architecture in general. As for the design of the feeders openwork frame entrance can accentuate the aesthetic perception of the structure as a whole. It is important that the entrance in this case should be large enough and the ornament must not have sharp edges that could interfere bird feeding.

6. Log and cobbled walls. The walls of the feeder, neatly made of tiny bars or logs has already attracted the attention of the person and associates with wooden buildings. Using bars of different lengths you can create a comfortable environment for the birds` perch [8, 9].

Wooden feeders with elements of decorative carving are quite time-consuming to produce in comparison, for example, with simple feeders from plastic bottles or boxes, but their design and manufacture may be no less interesting creative process, which contributes to the development of children's sense of care and responsibility towards the environment in family or at school lesson labor [4].

Figure 1. Designer window-type feeder (Design Kukhta A)

Figure 2. Designer window-type feeder (Design Toichkin A)

According to the principles of aesthetic perception, and considering feeder from the position of birds usability we have proposed a number of parameters which satisfy the requirements.

1. The feeder should own a large margin of safety. With daily use, harsh weather conditions the feeder should retain sufficient strength.
2. The feeder should consist of durable materials that need a long time to withstand adverse weather conditions without loss of functional and aesthetic qualities.
3. The feeder should be suitable for use by birds so that it should provide a sufficient number of perches, hold a lot of food and thus hide it from adverse natural phenomena (wind, precipitation).
4. The feeder should be easy to use. Its design should be very simple, provides the convenience of adding food and at the same time it should be aesthetically good-looking not only for birds but also for the people. [3, 8].

3. Results and discussion

On the basis of the foregoing parameters the feeder was designed and constructed which satisfies necessary criteria. Figure 3 is a model made in the program SolidWORKS based on wooden architecture of Tomsk.

Figure 3. Feeder made after wooden architecture of Tomsk. (Design Fedorenko E)

The analysis has showed [10, 11] that the most effective material for the feeder is wood and plywood. This feeder has the following design solutions that provide the required functions of the feeders: bird watching, additional food for birds, operational and aesthetic functions are given in table 1.

Table 1. Design solutions that provide the required functions

function	construction
1. Observation function	
convenience of counting birds	There is one main approach to the feeders through the mainstream rectangular opening.
the possibility of observing birds from different angles	high perches for birds all the way around the feeder, viewed through transparent ornamental panels.
2. Functions of additional food:	
feed protection against weather elements	large gable roof with an angle of 120 °.
protect food from the blowing wind	skirting around 30mm high and ornamental panels on three sides.
eliminating undesirable use of feeders by synanthropic species of birds primarily rock pigeon	Box in the ornamental panels measuring 20x20 mm and smaller, high and thin perches, feeder location at a height from the ground which hinders its use by pigeons.
attract various species of birds, the use of different types of feed	a separate device - wooden pins 100mm in length pushed on the possibility of fat and other volume food items that can further attract woodpeckers and different kinds of tits.

3. Exploitation functions:	
providing servicing ease by means of removable fronted welt on the screws for adding feed and cleaning	Adding feed can be also made directly to the central box which allow completely pass the hand into the feeder.
4. Aesthetic functions:	
providing visual aesthetics associated with peculiarity of wooden architecture of Tomsk	Decorative side panels, perches, skating on the roof are made in the style of wooden ornament of Tomsk.

In such a way the proposed approach to the design of the feeder satisfies the required functions and is the subject of a harmonious urban environment corresponding to all the requirements of visual aesthetics associated with wooden architecture which became the architectural landmark of Tomsk. Recommendations and methods proposed in this study can serve as a basis for creating objects of environmental design for various purposes.

References

[1] Blagosklonov K 1991 Nesting and birds`attraction in gardens and parks. *Moscow* p 251
[2] Reimers N 1990 Nature. Dictionary guidebook. *Moscow* p 637
[3] Sokolov A, Kukhta M, Pelevin E 2014 Modern technologies of decorative surface treatment *Mechanical Engineering, Automation and Control Systems : Proceedings of International Conference* pp 1-4
[4] Kukhta A Moskvitin S 2011 Aesthetic and functional peculiarities of the bird feeders design. *Modern techniques and technologies: proceedings of the XVIII International scientific-practical conference of students, graduate students and young scientists: Tomsk vol 3* pp 334-336
[5] Zaitsev Z 2004 Wooden architecture of Tomsk. *Tomsk* p 370
[6] Kosova L 1999 The nature of the city of Tomsk. *Tomsk* p 115
[7] Kukhta A Moskvitin S 2014 Using technological elements of agricultural complexes by birds in the vicinity of Tomsk. *Vestnik TuvSU, natural and agricultural sciences in 2014, - vol 2.* pp 30-37.
[8] Kukhta A 2014 Design peculiarities of structural parts of the bird feeders in the city of Tomsk. *Modern Equipment and Technology: Proceedings of XX Intern. Scient. Conf. students, graduate students and young scientists. - Tomsk Vol 3* pp 251-253
[9] Krauinsh D, Zhukova N, Volkova M, Kukhta A 2014 Alternative technologies of manufacture and decoration of wood balusters Welding technologies in art processing of metal. *IOP Conf. Series: Materials Science and Engineering Vol. 66 - №. 1, (art. 012043)* pp 1-5
[10] Kukhta M, Kazmina O, Sokolov A, Arventjeva N, Soroka A, Homushku O, Zaitseva S, Sergeyeva M 2014 The influence of glass and metal properties on the peculiarities of an item of art's shaping in ethnostyle. *IOP Conf. Series: Materials Science and Engineering Vol. 66 №. 1, (art.012046)* pp 1-6.
[11] Kukhta M, Sokolov A, Pelevin E. 2014 Welding technologies in art processing of metal. *IOP Conf. Series: Materials Science and Engineering. Vol. 66 - №. 1, (art.012044)* pp 1-5.

Category's analysis and operational project capacity method of transformation in design

S V Obednina[1] **and T Y Bystrova**[2]

Department of Cultural Studies and Design, Institute of Arts and Humanities, Ural Federal University named after the first President of Russia B.N. Yeltsin, 19, Mira street, 620002, Ekaterinburg, Russia

E-mail: svetaobednina@gmail.com[1], taby27@yandex.ru[2]

Abstract. The method of transformation is attracting widespread interest in fields such contemporary design. However, in theory of design little attention has been paid to a categorical status of the term "transformation". This paper presents the conceptual analysis of transformation based on the theory of form employed in the influential essays by Aristotle and Thomas Aquinas. In the present work the transformation as a method of shaping design has been explored as well as potential application of this term in design has been demonstrated.
Keywords: design, form, theory of form, method of shaping design, transformation, criteria of transformation, shaping design, graphic design

Problem

The human spatial area has been transformed a lot due to the tremendous development of modern life. Nowadays the demand for flexible, effective, and alternative solutions concerning the elements of interactivity and/ or games is of great importance. That is why the transformation is thought to be a relevant method in design projects. The commitment of designers in design shaping to human concerns reduces a psychological distance between the consumer and the design product, and thus might contribute to the commercial success of the product. The designer needs special tools for the achieving these goals. However, the transformation in design as a research method remains largely unstudied; there are many vague interpretations of this term in relation to design project (transformation method) as well as design product ("transformers", transformable object, etc.).

Aim

This work aimed to identification of the specifics of the transformation method`s implementation in design products based on preliminary categorical analysis of the "transformation design" concept.

Methodology

Much research in recent years has focused on the specifics of shaping design (see works of D. A. Norman, V. Papanek, V. L. Glazychev, N. A. Koveshnikova, M. S. Kukhta, K. Cantor, S. O. Khan-Magomedov, E. Crenshaw, V. F. Sidorenko, Yu. Bystrov, N. V. Voronov, S. M. Mikhailov, N. P. Garin, V. F. Runge, L. M. Kolmannskij, A. S. Shchipanova, and others).

The rationalist interpretation of the form as an integral entity determined by the specific goal has also been addressed by many works of Aristotle, Thomas Aquinas, J. W. Goethe, and I. Kant [1; 128–131]. In 1964 T. Maldonado developed the classical definition of form further with the introduction of design.

A few works were devoted to the issues of transformation in the design (Z. T. Akilov, G. I. Petushkova, A. A. Paceviciute, L. S. Shamuhitdinova, S. A. Goncharov, C. B., Zvyagintsev, V. Semkin and V. F. Sidorenko). The term "transformation" first appeared in the work of the Russian physicist and meteorologist A. D. Michelson in 1865: transformation (trans – through, and formatio – formation of the species) is understood as a change of form, metamorphosis, transformation.

Discussion. Transformation as a method of design project and as a characteristic of the design product. In english the word "design" is used extensively, the meaning is strachted, it's too general. Sometimes it results in misunderstanding the the abilities of the profession. We should realise the differance between the designer and others who use projecting as a part of their professional activities. But we also have theoretical definition. Let's remember the defenition given by Tomas Maldonado in 1964. "Industrial design is a creative activity that aims to determine the formal qualities of objects produced by industry. These formal qualities are not only the external features but are principally those structural and functional relationships which convert a system to a coherent unity both from the point of view of the producer and the user" [10]. This definition is really important for all the design types because it defenitly figures out the specific features of design.Well:

1. A designer makes a form, creates the form, using the qoutation of russian philisofer Nikolay Berdyaev "creates something new brings into being something that has never existed before".
2. Designing is the activity that is directed to some certain purpose. Such activity isn't regarded just as creating it has a an aim. The aims are mainly related to the customer's needs [10].
3. A designer creates the form of the industrial product. I mean the product that is worth to be fully or partly manufactured industrially using the knowledge of specific technologies. Therefore as a result it allows to speed up the prosess and cut the cost. In the middle of the 20th century Tomas Maldonado suggests a concept of democratic design which makes the products available for massmarket customers. These days his idea is becoming much more popular.
4. Meeting the goal but not coming across the lack of quality. How?
   - visual characteristics;
   - structural bonds;
   - functional bonds.

Transformation is one of the methods of work of the designer with the structural and functional connections.

The term "transformation" had been introduced in a variety of disciplines such as biology, physics, engineering, architecture, linguistics and others. We argue that in design sphere it is characterized by the specific features due to (a) project nature of design activity [1; 8; 15]; b) inclusion in shaping design as the main criterion and the result of design ("shape grammar", Lawson, 2005). The transformation of the object can be designed, modeled, calculated by the designer. The resulted form is supposed to be flexible, variable in terms of functional purpose and, at the same time, being unchanged in their basic characteristics [2; 3; 6; 7]. This aspect is crucial, because this term can also be used in our daily life to refer to the transformed elements in different constructions (for example, furniture). Such products typically associated with modules and can be defined as *constructors*, not the *transformers*. They are differentiated form each other due to the fact that every element of the constructor can be easily removed, whereas the elements of transformers are expected to be undivided.

Transformability as a *characteristic* of design product is manifested in the process of its existence and using, i.e. being asserted a fact concerning external factors and environmental changes. It is important that the movement is a factor of form creating [7]. Object of design is thought to have a variety of functions and could be developed in time and space, due to the activity of the consumer.

One can say about the transformation as a *method* of design project. This means that the designer creates a project focused on the presence of several variants of implementation of the form ("transform"). The parameters of the transform are determined by the specific functional goals. Perception of the form as an integrity supports the project as a whole and saves its compositional and aesthetic characteristics [3; 5; 14].

IOP Conf. Series: Materials Science and Engineering **93** (2015) 012073 doi:10.1088/1757-899X/93/1/012073

The advantages and limitations of the products made with the use of method of transformation

1. ***Multiple functions.*** This quality presupposes the use of the same form in accordance with the different objectives, i.e. the replacement of several products by one`s [14]. It can be straightforward successful only in case, when the functions do not contradict to each other and the resulted form is not destroyed by the additional elements. "Improved versatility for transformable tops is one of the core variables crucial to increasing sustainability in the clothing's lifecycle. Transformable tops are expected to replace many tops with their changeable design options. Thus, wearers may not need to purchase five different black T-shirts with different necklines but only purchase a single black T-shirt with five transformable necklines" [11, 115-116]. All elements are used on regular basis and are supposed to be balanced and integrated. At the same time, within such products are not allowed to use more than one function simultaneously.
2. ***Flexibility.*** Transformable design product refers to the needs of the customer in a specific moment of time. "Transformation indicates in general a change of shape, form, or structure without loss of substance. The most important method of transformation is variation" [5, 403].
3. ***Economical efficiency*** of transformable design product is associated with increasing of its working lifespan and maintaining of its high consumer qualities. This aspect is supposed to be particularly relevant for the design products for kids; these products must be optimized in accordance with the necessity to grow up in children case. "It is shown that higher upfront investment leading to reduced cost for future change is economically justified in certain scenarios. The value of flexibility increases with increased time horizon and increased uncertainty in the market price…" [4].

The use of the method of transformation in design is thought to have several positive consequences. The number of required design products in environmental milieu decreases due to the multifunctional character of the product, which assert a fact of reduction the need for recycling and increases environmental credentials. This method of shaping design might increase the working lifespan of design product and thus might be deal with the improving of its functionality. This shaping design method saves resources and space. "Transformable garments have the potential to lead consumers' natural engagement with sustainable acts by satisfying their various needs and wants… behavior without their awareness" [10].

At the same time the advantages and disadvantages of this approach need to be addressed to. It is thought to be a complexity of the construction as well as redundancy of form; economic failure of the decision making; the impossibility of simultaneous use of multiple functions or conditions of the transformation. "Thus, having many options is not the best solution for transformable garment designs; and when developing transformable tops, designers need to consider how to simplify consumers' decision making process by understanding how the garments will be used" [11, 121].

Transformation is one of the methods of work of the designer with the structural and functional connections. The idea of "liquid modernity" was introduced in the context of the turn of the 19th-20th centuries by American sociologist Zygmunt Bauman. As he noticed significant social changes he offered the metaphor of civilization conversion from solid to liquid state. To describe modern society he draws a parallel with physical substances: solid subjects (analogy to social patterns and durable social relations) can not only be destroyed but are also able to change their volume; liquids keep volume and thus are easily transformed. That "solidity" of western society traces the roots back to the XVIII century with its rationalistic philosophy and science striving to establish the rational regime of life.

Evolving Bauman's ideas in the context of fast sociocultural changes, we face at the beginning of the XXI century, it's reasonable to describe "solid" thinking people have in the Era of modernism as linear, flat and since a certain moment schematic. The revival of positivistic ideas which is also noticed in liberal education implies mathematization (even arithmetization) of thinking therefore fictitious simplicity of following the algorithms becomes attractive for the majority of people.

Conclusion

In this study transformation in design is considered to be a variant of the designer's work with the integral form as a closed, but flexible system. On the one hand, it implies a stable number of the elements of form; on the other, the remarkable variety of the structural connections between them.

The application of the method of transformation in design requires an understanding of the feasibility of combining functions. It provides a significant effect in case of the development of only one function oriented to necessity to grow up in case of children. In this point of view, you must choose the materials which can be used to reduce the risk of physical depreciation of the objects. The method of transformation is relevant for the objects used in small spaces where a single design product can be multifunctional. The number of the options for transformation is thought to be efficiently excellent and sufficient. The algorithm of use of this design product might integrate this number of options in an optimally-effective design shape.

Acknowledgments
The paper had been written with the financial support of the research group "Open City" at Ural Federal University.

References
[1] Bystrova T. Y. Object, Form, Style: An introduction to Design Philosophy. Ekaterinburg: Ural State University, 2001. 223 p.
[2] Crenshaw E. E. Transforming together: Reconsidering adaptive reuse: diss. … degree of Master of Architecture / E. E. Crenshaw ; Univ. of Maryland, College Park, 2013. 126 p.
[3] Goncharova S. A. The development of methods of morphological transformation in the design of clothing for children. Author's abstracts, Shachty: South-Russian State University of Economics and Service, 2001.
[4] Greden L. V. Flexibility in building design: A real options approach and valuation methodology to address risk: diss. … Dr. of philosophy. Author`s Abstract / L. V. Greden ; Massachusetts Institute of Technology, 2005.
[5] Design. Illustrated Thesaurus / eds. G. B. Minervin, V. T. Shimko, etc/. Moscow: Architecture, 2004. 288 p.
[6] Design Dictionary. Perspectives on Design Terminology / Erlhoff M., Marshall T. (Eds.). Birkhauser Verlag AG, 2008.
[7] Dynamic and kinetic form in Design: Proceedings. M.: VNIITE, 1989. 80 p.
[8] How Designers Think. The design process demystified/ Lawson B. Fourth Edition, 2005.
[9] Head First Design Patterns / Freeman E., Freeman El. O`Reilly Publ., 2004.
[10] ICSID. URL: http://www.icsid.org/about/about/articles33.htm
[11] Koo H. Design Functions in Transformable Garments for Sustainability: diss. … dr of philosophy. Author`s Abstract / H. Koo; Univ. of Minnesota, 2012. - 185p.
[12] Norman, D. The design of everyday things. – MIT Press, 1998.
[13] Real Design. Proceedings of the IV International Festival of Design. Yekaterinburg – Omsk – Moscow – New York – Tyumen – Einhoven / Comp. G. V. Vershinin. – Tyumen: Tyumen design center of the Union of Designers of Russia, 2001. 128 p.
[14] Zvyagintsev S. V. The development of methods of complex design projecting in the sphere of articles of clothing. Author`s Abstract. M: MGAP, 1999. URL: http://www.dissercat.com/content/razrabotka-metodov-kompleksnogo-proektirovaniya-komplektov-vzaimozamenyaemykh-i-transformiru.
[15] Singh V., Walther B., Krager J., Putnam N., Koraishy B., Wood K. L., Jensen D. Design for Transformation: Theory, Method and Application // Proceedings of IDETC/CIE 2007. ASME 2007 International Design Engineering Technical Conferences & Computers and Information in Engineering. Conference. Las Vegas, Nevada, September 4–7, 2007 URL: http://www.sutd.edu.sg/cmsresource/idc/papers/2007-_Design_for_transformation-_theory_method_and_application-_DETC2007-34876.pdf

Semantics and technologies in modern design of interior stairs

M Kukhta [1,a], A Sokolov [1,b], E Pelevin [1,c]

[1]*Tomsk Polytechnic University, Tomsk, Russia*

E-mail: [a]eukuh@mail.tomsknet.ru, [b]iscanderaga@rambler.ru, [c]eapelevin@gmail.com,

Abstract. Use of metal in the design of interior stairs presents new features for shaping, and can be implemented using different technologies. The article discusses the features of design and production technologies of forged metal spiral staircase considering the image semantics based on the historical and cultural heritage. To achieve the objective was applied structural-semantic method (to identify the organization of structure and semantic features of the artistic image), engineering methods (to justify the construction of the object), anthropometry method and ergonomics (to provide usability), methods of comparative analysis (to reveale the features of the way the ladder in different periods of culture). According to the research results are as follows. Was revealed the semantics influence on the design of interior staircase that is based on the World Tree image. Also was suggested rational calculation of steps to ensure the required strength. And finally was presented technology, providing the realization of the artistic image. In the practical part of the work is presented version of forged staircase.

1. Introduction

In today's world stairs becomes dominant in the interior of apartments and houses. Development of new technologies and the variety of materials (stone, concrete, metal, glass) allow to constantly improve its appearance and decoration. However in addition to functional, aesthetic and artistic value staircase has a symbolic meaning, which manifests itself in many aspects of our lives.

The aim of the article is to study the symbolism of staircase image and to identify interferences of symbol, style and production technology of interior spiral staircase made of metal.

The main objectives are:

- to identify the symbolic aspects of staircase image;
- to consider the theme of staircases in the history of cultures;
- to submit relationship of the symbolic image and stylistic solutions of staircase;
- to propose and justify a method for manufacturing the stairs.

2. Research

The symbolism of staircase is connected with possibility of people, gods, inhabitants of the underworld to move. Thus, the staircase in the myths reflects the hierarchical levels of cosmic zones. Image of Osiris on the top the stairs symbolized his resurrection from the dead. In Buddhist mythology Shakyamuni Buddha descended to earth via a staircase, broughted to him by Indra. The tribes of Siberia perceived stairs as an image of the World Tree, by which the supreme shamans got in a top world. In Christianity is popular the image of Jacob's ladder, which the biblical patriarch saw in a dream. Angeles were going up and down the Jacob's ladder. The Virgin in the iconography depicted sitting on a throne with a ladder in the hands [1].

Going upstairs symbolizes good luck, knowledge, holiness, perfection; downstairs – degradation, misery. In Norse saga hero King Olaf dreamed that he climbed the stairs to the high heavens; this dream was interpreted as a sign of impending death and ascension to heaven as a saint. In Islamic tradition Muhammad describes the ladder by which believers climbed to reach God.

Stairs as an attribute of ascension stands as analogue of column, tree or mountain (also symbolizing the world axis) [2]. Stairs also was considered as an attribute of the descent, in which the hero falls into the underworld.

The semantics of stairs image is multivalued and the main thing in it is the vertical dimension, where the rise signifies the triumph of overcoming the Earth's gravity, as well as the triumph of the spirit over the flesh. Thus, we can identify the following symbolic aspects of stairs image:

- communication between the different levels of the universe and the transition from one to another;
- ascension and evolution, presented as a sequence, gradual, stepwise process.

These meanings of ancient interpretation of stairs appear so far - staircase connects the floors of buildings, creating a connection between them, help the movement of people, contribute communications.

In today's world the presence of stairs in the house is a testament of prosperity. However was not always so, for example, theorists of Quattrocento (early Renaissance) considered the ladder as an annoying functional necessity of city construction. In his treatise "On the Art of Building" Alberti recommends hiding the stairs at the buildings' corners or behind columns in order not to "spoil" the interior space. All these difficulties are constructional, they are experienced by the one who builds stairs, but not by the one who has to climb them; it is not the difficulty of use. The task the architect is to make stairs comfortable and safe as much as possible [3].

Thus, Alberti gives inversed interpretation of stairs symbol (by which raises the righteous and overthrown sinners) - way up is extremely relieved, and accidental drops is safe. In the painting of the Renaissance stairs theme was also secularized. Staircase of Ghirlandaio devoid of semantic meaning and does not lead to the entrance of the temple. As Alberti advised, it only leads to an intermediate platform for a relaxing.

In the second half of the 16th century, at the end of his career Michelangelo returned to the theme of stairs. Laurentian stairway roughly rushes down and it seems impossible to climb over it. Deprived of railing, it is like a marble waterfall catastrophically clamped between too narrow walls. This ladder does not lead to the exit; it almost rests against the blank wall of the Library's lobby. This stairway is "hopeless". [4]

In 17th, 18th century stairs image gradually loses its sacred symbols. Main staircase is a way of significance, ceremonial ascent and descent.

At the end of the 19th century the development of the Art Nouveau style has shown a new attitude to stairs, with the special attention to decor of railings. Modernity is characterized by lightness and elegance of lines, graceful asymmetry of floral motives [5]. Staircase becomes "accent" of interior around which the script is tied. In the author's design project stairs "remembers" its symbolic meaning. Like the Tree it grows and unites the upper and lower space of the house, spiraling around the support pillars (pipes, transformed into a spiral). Decor of handrails emphasizes the symbolism of Tree-Stairs with fanciful twists of branches, dynamics of the rhythm which is directed upward (Fig. 1). From a technical point of view, this stairs refers to the spiral staircase [6].

Figure 1. Decor of forged railings. Design project by A. Sokolov.

In the design of stairs it is necessary to provide its maximum flatness. Thus the length of walkline increases. When designing a spiral staircase walkline is carried out not in the middle of step, but at a distance of 2/3 of the step, starting with the pillars. Therefore the spiral staircases are more compact than stairways.

To calculate the ideal step size of the stairs is recommended method of French mathematician, engineer and architect François Blondel:

$2s + a = 63$ cm,
with **s** the step height, and **a** the step width [7].

However, anthropometric and ergonomic parameters, reflected in the building regulations as well as state standards [8] for stairs allowed to specify the above formula:

$s = 20$ cm – the maximum step height;
$a/s = 1.5$ – the recommended ratio of the width and height of the steps;
$a = 25$ cm – minimum width of the steps.

The ratio ($a/s = 1.5$) provides comfortable movement along the stairs. Thus, was suggested the ideal measurement for the stairs: step height - 19 cm; 28 cm - width of the steps, calculated by walkline. It is noteworthy that for spiral staircase the walkline can be varied according to one's individual anthropometric data to make climbing as comfortable as possible [9,10].

3. Results and discussion

The technical project specifies the size of stairs and the length of walkline. It was necessary to take into account the following requirements: small size of top platform and access to the window.

In drafting was decided to abandon the wooden bearing structure and to make it out of steel, while steps' cover of wood. Modern technologies and automated tools simplify the production of stairs [11,12]. The complex shape of fencing is carried out using the methods of artistic forging [13].

Figure 2, 3 shows the construction of stairs, which can be divided into three parts: pillar, steps, railing. For pillar was used 325 mm pipe with 6 mm wall thickness and a length of 2.4 meters. In order to lighten the stairs, as well as saving material, a section of pipe was cut along a helical line (double-threaded spiral). As a result we got two parts that was welded together to form a pillar.

Steps were also performed with welding [14]. Each of the steps consists of a plate and the reinforcement rib. For the ribs was used 89 mm pipe. It was sawn along the cutting plane and welded to plate. The resulting billet was machined on the edges to get surface close to column's surface. These steps were welded to pillar to provide a predetermined length of walkline, building codes and standards of ergonomics. During the tests of produced staircase steps elastically deformed when walking.

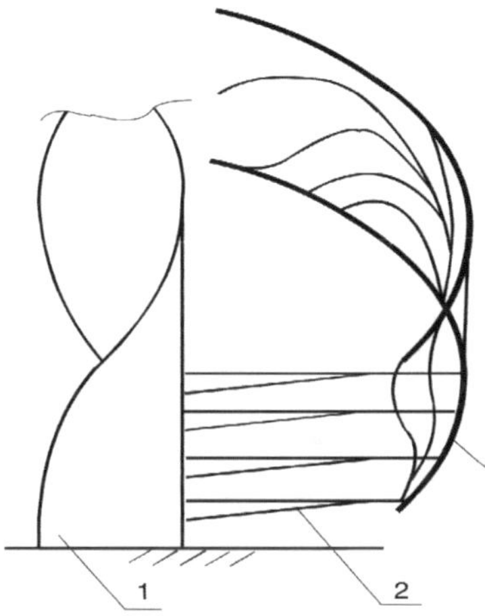

Figure 2. Construction of stairs:
1 - pillar; 2 - step; 3 - fencing

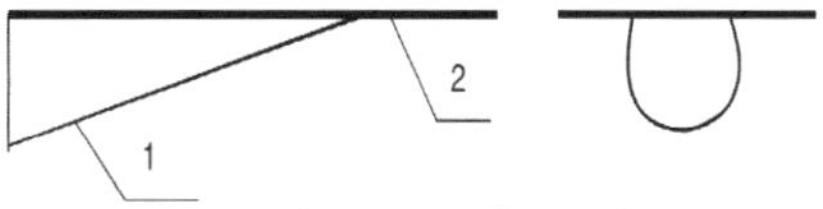

Figure 3. Construction of steps:
1 - reinforcement rib; 2 – plate

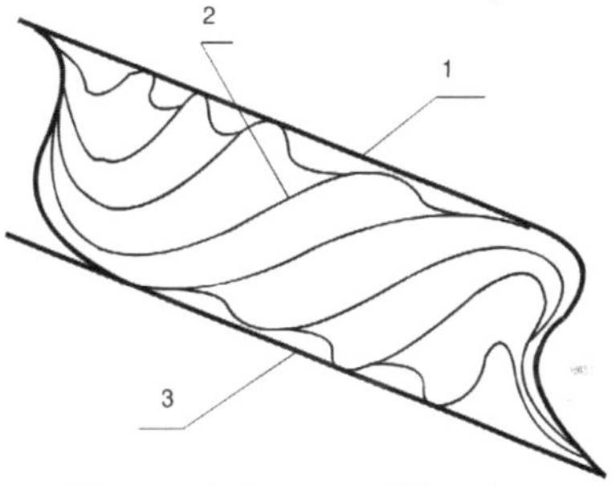

Figure 4. Sweep of fencing:
1 - rails, 2 - grille, 3 - stringer

To increase the rigidity was strengthen links between steps and fencing. The bases of fencing are stringer and railing, which are made of 26 mm pipe. Stringer and railing are connected by 14 mm steel rods curved to match the style of Art Nouveau (Fig. 4). The result is a rigid railing, which increased hardness of steps to the required level.

For convenience of manufacture and installation the fencing was divided into three sections of varying lengths. The middle section can be pulled off to provide access to the window. In the design of fencing also is fully used dignity of Art Nouveau, which allowed performing the middle section, on the basis of functional requirements and without creating aesthetic "tension" of joining elements.

Fencing was performed by the following technology. After joining steps to a pillar at each step were installed removable vertical racks to which was attached a sheet of hardboard. On this sheet with scaling method was drawn the reamer of all sections of the fence. By plastic deformation were prepared sections of stringer and railings. Details of stringer were joined by bolts to the steps; railing sections were fixed on the racks, and then all parts of grilles were connected by tack welds [13]. As a result of these operations have been received three sections of the fencing. After disconnecting them from steps further surface processing was carried out outside.

Thus, in the space of modern interior recreate the image of the stairs, the semantics of which is associated with the World Tree. Organicity of Art Nouveau and ancient symbolism embodied in the design project, because of the opportunities of art metal treatment technologies. The performance

IOP Conf. Series: Materials Science and Engineering **93** (2015) 012074 doi:10.1088/1757-899X/93/1/012074

characteristics are provided due to correct choice of sizes that correspond to the anthropometric and ergonomic parameters.

References

[1] Toporov V 1988 Stairs Myths of nations of the world *Soviet Encyclopedia* pp 50-51

[2] Kuhta M 2013 Broadcast of the arbor mundi archetype in the design of decorative panels *Proceedings of the Academy of technical aesthetics and design* №**2** pp 27-30

[3] Leon Battista Alberti 1937 Ten Books on Architecture *Moscow* pp 38-39 (in Russian)

[4] Kuhta M and Sokolov A 2011 Interrelation of symbol, style and technology in the design of interior stairs *Design. Materials. Technology №* **1** (16) pp 28-31

[5] Kukhta M, Kazmina O, Sokolov A, Arventjeva N, Soroka A, Homushku O, Zaitseva S, Sergeyeva M 2014 The influence of glass and metal properties on the peculiarities of an item of art's shaping in ethnostyle *IOP Conference Series: Materials Science and Engineering vol* ***66*** *(art 012046) pp* 1-6

[6] Alessandro B, Ileana C 2013 Equilibrium models for helicoidal laterally supported staircases *Computers and Structures* ***124*** pp 21–28

[7] Blondel J 2002 Cours d'architecture *Monum, Éd. Du Patrimoine*

[8] GOST 23120-78 Stairs march, platforms and fences of steel

[9] Richard F, John H, John B, David E 2014 Safety on stairs: Influence of a tread edge highlighter and its position *Experimental Gerontology vol* ***55*** pp 152–158

[10] Max P, Songning Z, Clare M, Jeffrey F, Jeffrey R 2014 Effects of increased step width on frontal plane knee biomechanics in healthy older adults during stair descent *The Knee vol* **21** pp 821–826

[11] Kukhta M, Sokolov A, Pelevin E 2014 Welding technologies in art processing of metal *IOP Conference Series: Materials Science and Engineering vol* ***66*** *№ 1 (art 012044)* pp 1-5

[12] Sokolov A, Kukhta M, Pelevin E 2014 Modern technologies of decorative surface treatment *Mechanical Engineering, Automation and Control Systems : Proceedings of International Conference* pp 1-4

[13] Ledzinsky I 1994 Modern artistic forging *Metallurgy* p 480

[14] Glizmanenko D 1974 Welding and cutting metals *Higher School* p 479

IOP Conf. Series: Materials Science and Engineering **93** (2015) 012075 doi:10.1088/1757-899X/93/1/012075

Technology and methods of making decorated balusters

D Krauinsh[1] and A Selivanova[1]
1 Tomsk Polytechnic University, Tomsk, Russia

E-mail: dima_krauinjsh@tpu.ru

Abstract. Woodworking companies looking for a simple solution to simplify and speed up the whole process of making a carved balusters. Tech solution is the replacement of threaded elements solid wood balusters on plastic threaded overhead items secured by adhesive to the surface of the workpiece chiseled. This solution allows to reduce the cost of manufacturing carved balusters, without resorting to a woodcarver and using expensive equipment. Through further improvements, such as painting, patina, it turns out a product that is difficult to distinguish from solid carved balusters.

1. Introduction

The starting point was the investigation of the claimed idea of developing designs and technologies to produce a "massacre" balusters of composite materials that reduce the complexity of manufacturing and accelerate the production process. Such technology could be the example of the high resource efficiency and is a useful model that has the potential to be implemented in the near future.

With the development of low-rise construction in Russia (cottages, townhouses) is a growing demand for the manufacture of wooden stairs, invitations pillars, columns, balusters, making it necessary to find new manufacturing techniques [1, 2].

Baluster - support column railing, made in the form of a columnar shape. Traditional wooden baluster is manufactured by turning on the universal lathe wood lathe or copying machines for wood with manual control (Figure 1). This profile chiseled surface can be quite varied, and the actual process of turning a relatively cheap.

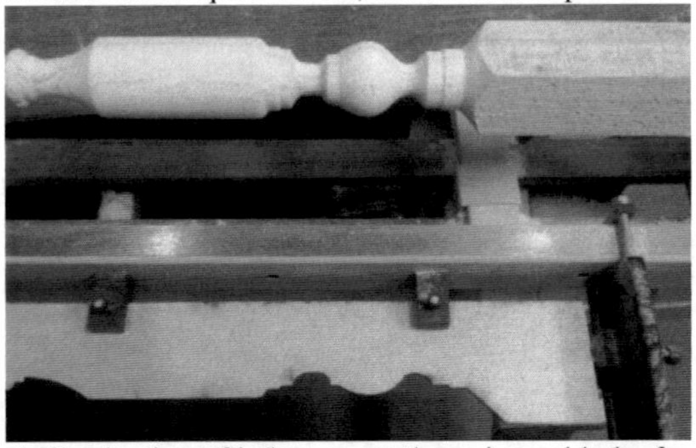

Figure 1. Production of balusters on the universal lathe for wood

Carved wooden balusters performed woodcarver for milling and copying machine with manual control (single or multi-position) or lathes with computer numerical control (CNC) (Figure 2). The cost of manufacturing such balusters becomes very high.

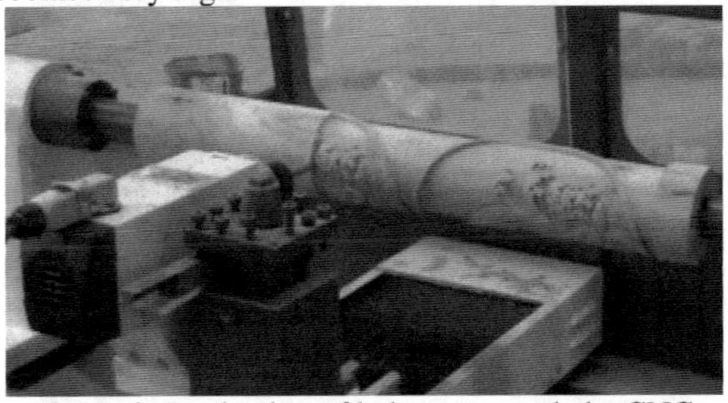

Figure 2. Production of balusters on a lathe CNC

Besides the basic functions - security stairs, balusters perform decorative function [1, 2]. To create a complex carved balusters necessary work of the artist-technologist (Figure 2). A large number of customers prefer painted or patinated stairs and balusters, hide the texture of wood. At the same time the customer wants to see the unique carved balusters, which no one has. The proposed technology allows the manufacturing process to make a "massacre" balusters relatively cheap and fast enough.

The main technological method of decorating the baluster is the imposition of the relief decoration of plastic or sealant to the prepared surface chiseled [3]. The curvature of the surface is different, so it is very important decorative element plasticity. Plastic tends to remain pliable, with a special component ratio. In addition, by adding colorants in a plastic mass can be obtained deco coinciding in color with chiseled workpiece. Considering the aforementioned advantages the plastic can be concluded that it is a suitable material for forming a decor. The article described in detail the technology of creating a plastic decoration and carved balusters in general.

2. Technology

The process of manufacturing plastic decorative elements:

1) Creating a model element in the (scanning);

2) Construction of an electronic model, with a software module Artcam or any other 3D-design program to produce the output file with the stl, which can be opened in Artcam;

3) Production of wooden models on the three axis machines with numerical control (CNC);

4) Further development of the model by hand engraver, if you want to identify the shadow surface of the model, which cannot be obtained in a simple three-axis CNC machines;

5) Production of matrix. It will be appreciated that the matrix materials, for the manufacture of plastic decoration or decoration on the basis of the sealant will be different. Silicone and gypsum, respectively. In this case, we used a two-component silicone brand Compound "Super silicone" Blue:

- Place a wooden model in a box;
- Measure the necessary amount of base and catalyst in a ratio of 100: 5. Stir stick until smooth;
- Slowly pour the material in the box to the mixture evenly spread over the model, which will minimize the amount of air bubbles. Additionally, you can apply this process some vibration, which will provide more intense output of air bubbles;
- Allow material to harden at room temperature for 2-3 hours;
- Remove the model from the cured silicone matrix [4].

6) Production of plastic element. To work was selected two-white liquid plastic polyurethane brand Smooth-Cast 300:

- Equal volumes of components A and B are thoroughly mixed in a container (metal, glass or plastic);
- Slowly pour the plastic mass in the box to the mixture evenly spread over the model that will minimize the risk of air bubbles (Figure 2). Additionally, you can apply this process some vibration, which will provide more intense output of air bubbles;

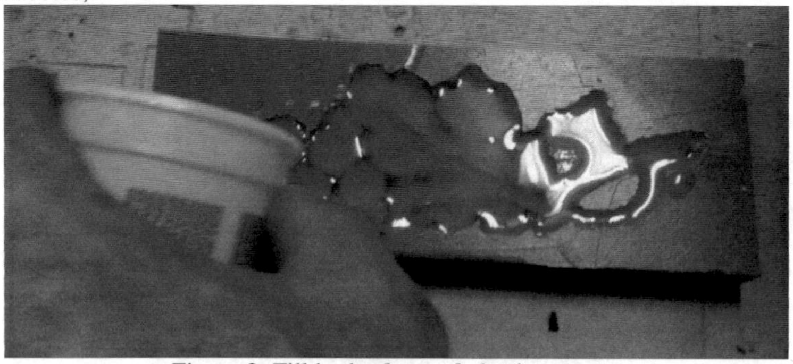

Figure 2. Fill in the form of plastic masses

Remove the product from the matrix after a time calculated according to the formula, to ensure its plasticity (Figure 3):

$t_2 = t_1 \cdot {}^3/_4$,

where t_2-time extraction,

t_1-curing time.

For the two-component liquid plastic grade white Smooth-Cast 300:

$t_1 = 10$ minutes;

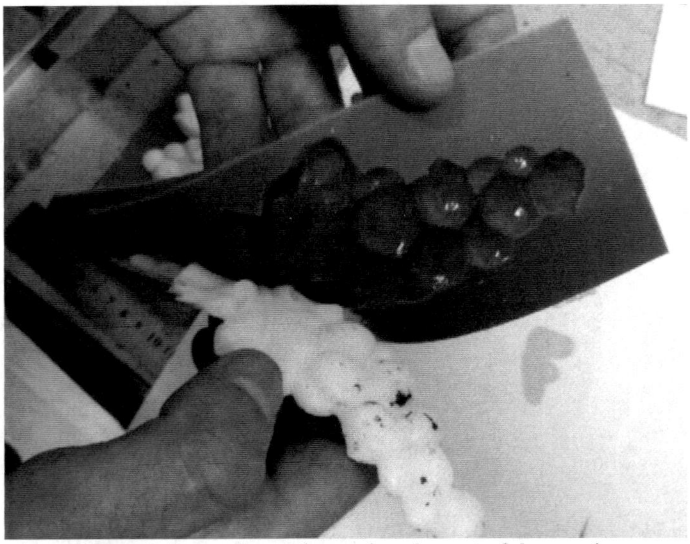

Figure 3. Removing a decorative element of the matrix

• Apply the product to the desired surface carved balusters, neatly fixing stretch film, or other alternative materials. After curing separate element (Figure 4). The hardened elements do not lose their shape;

Figure 4. Ready plastic decor

7) Make any other necessary accessories and decorative objects, repeat the steps above;

8) Decorate chiselled baluster prefabricated elements, pasting them, for example, the adhesive COSMOFEN CA 12 (Figure 5). Elements glued to the place where they are fixed for their final solidification in the manufacturing process;

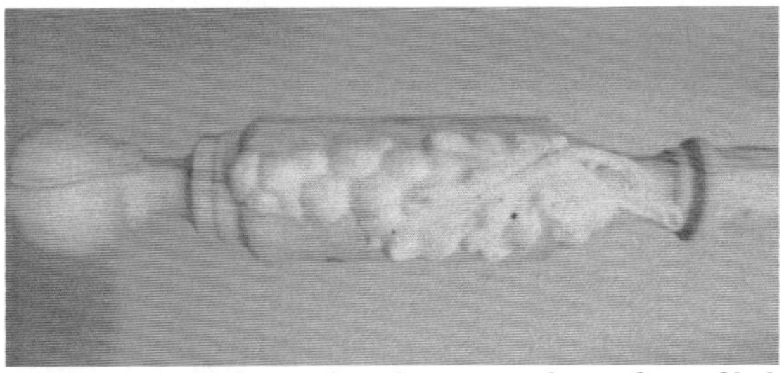

Figure 5.Zakreplenie decorative element on the surface of balusters

9) Paint the baluster to the desired color (Figure 6).

Figure 6. Ready balusters and possible overlay on the flat surface of the product

The required set of elements for decorating balusters allows different options to combine them and arrangement relative to each other and to get an excellent alternative to a decorative thread [5].

3. Conclusion

The study has been developed and investigated a new technology decorating balusters, based on the fact that plastic and plaster are the good alternative to thread, excluding the work of the engraver and turning and CNC milling machines. The advantage of this technology - reducing time spent on production of carved balusters and correspondingly low cost of finished balusters, against the carved balusters, obtained by milling on turning and milling machines. Monochrome painting and patina for easily hides the difference between textures of wood and plastics. The described technology allows increasing the diversity of the range of products and their aesthetic expression while minimizing manufacturing costs. The equipment allows you to place production on a smaller area, resulting in reduced costs. Documented in the technology can be applied to the decoration of wooden stairs, columns, pillars invitation ladders, wooden furniture.

References

[1] Krauinsh D et al 2014 Alternative technologies of manufacture and decoration of wood balusters. IOP Conf. Series: Materials Science and Engineering 66 (2014) 012043

[2] Krauinsh D et al 2014 Transformation of form as alternative manufacturing techniques of wood balusters Design. Theory and practice 18 pp. 30-37 [in Russian]

[3] Krauinsh D 2013 A method for producing a decorative stucco in product design with a curved surface *Proceedings of the Academy of Industrial art and design* **2** pp. 5-7 [in Russian]

[4] Club soap-boiler [electronic resource]. - Access: URL: http://varimylo.ru/product/4150/.html (date of treatment: 3/15/15)

[5] Chernich M 2011 Classification of wood art. Features engraved ornament decoration Design. Materials. Technology. 4 (19) pp.30-35 [in Russian]

IOP Conf. Series: Materials Science and Engineering **93** (2015) 012076 doi:10.1088/1757-899X/93/1/012076

The design features of the body of the portable electrocardiograph «ECG-EXPRESS»

V Seryakov[1], **Yu Khmelevsky**[1], **G Mamontov**[1]
[1] Tomsk Polytechnic University, Tomsk, Russia

E-mail: seryakov@tpu.ru

Abstract. The sequence of the industrial product shaping is specified, the main modeling characteristics of the configuration and some of its parts are designated subject to the electrocardiograph design and the technologies used for the production of its body.

1. Introduction

Currently the industrial design is a fundamental tool for the development of the competitive product, while the manufacturers experience much difficulties to stand among the competitors and amaze the customers with the modern and esthetically attractive design of the industrial products.

The relevance and potential of this trend is confirmed by a variety of the industrial products configuration introduced to the market. The quality, usability and ergonomics of the designed product is largely caused by the consistent regular shaping, which is dependent on the advanced materials and technologies, relevant configuration, stylistic and color solutions, that in general excludes the design errors and defects at the various stages of design [1].

2. The process of the product development and use

The main process stages of the product development and use are presented in Figure 1, including the following product design stages:

1. The preparatory stage
2. The draft design of the new product
3. The technical specifications
4. The engineering proposal
5. The basic design
6. The volume visualization
7. The engineering design (modeling, design, prototyping)
8. The working design documentation
9. Support [2]

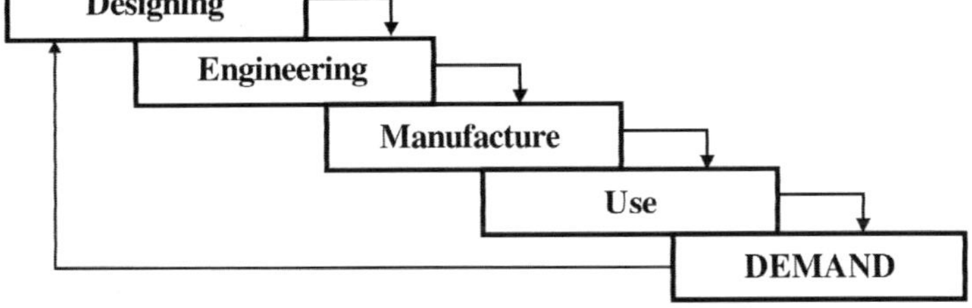

Figure1. The process scheme of the product development and use

In the configuration development of the industrial medical device, in particular electrocardiograph, it is important to know its working principle and the scope of application: the self-control of the heart rhythm disorder and the ECG data storage for the subsequent analysis and study. This device is operated by sensors, which receive and convert the incoming signals, and then register them as a curved line on the paper.

The portable electrocardiographs are the terminated devices being often equipped with the screen which allow making a rough analysis immediately after the measurement, as well as, if needed, sending the recording for the detailed interpretation. These devices are wireless, and the measurement is made using the body built-in electrodes. The standard application scheme requires putting both thumbs to the electrodes on the surface of the device and wait for 30 seconds.

3. The types of the portable electrocardiograph

A key advantage of the hand-held electrocardigraphs is their portability and the diagnosis ability wherever and whenever possible, that is particularly important for the people suffering from the heart disease. The hand-held electrocardigraphs became available only 2-3 years ago, and there are only a few analogues on the market. Their comparative analysis is given in Table 1.

Of these devices currently the cardiograph Armed only is certified and sold in Russia, while being a Chinese development of Easy ECG, certified and imported in Russia by Armed. There are also attempts to adapt HeartCheck PEN to Russia. The strengths of these analogues, in particular, of the device Armed are their low price, achieved due to a significant reduction in the technical characteristics, and the huge advertising campaign.

As compared to the analogues, at the current reasonable price the electrocardiograph ECG-Express offers a wider range of functions (not only the diagnosis of the heart rhythm disorder, but for other diseases as well). And its innovative features will allow maintaining its competitiveness and attractiveness for the customer during several years, required to enter the market (figure 2).

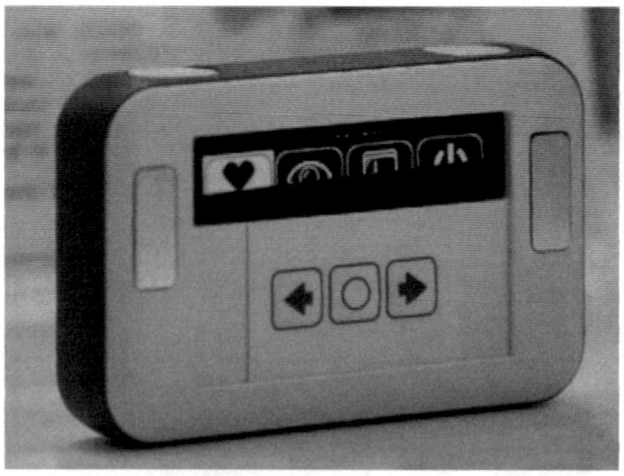

Figure 2. The portable electrocardiograph «ECG-EXPRESS»

IOP Conf. Series: Materials Science and Engineering **93** (2015) 012076 doi:10.1088/1757-899X/93/1/012076

Table 1. The types of the portable electrocardiograph

Name	Description
Armed PC-80A   Armed PC-80B 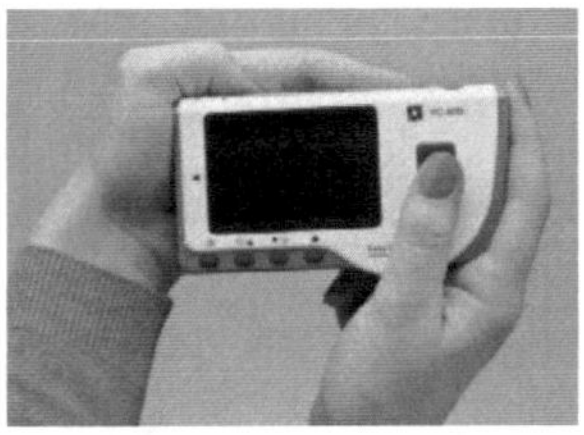	Manufacturer «Armed» (there are wire-based and built-in electrodes) - the light cardiograph, weight 78 g; - the compact size; - the handy packaging; - the speed rate of the results generation; - ease of use; - the possible data transfer into PC; - the possible record onto a memory card; - the recording memory of 24 and 300 measurements; - the display backlight; - the automatic shutdown; *Design*: - the simple design; - the asymmetrical configuration; - the blank shaping; - the dull color scheme *The target audience:* - women and men (from 30 to 50 years); - the seniors (55 and up)
2. HeartCheck PEN 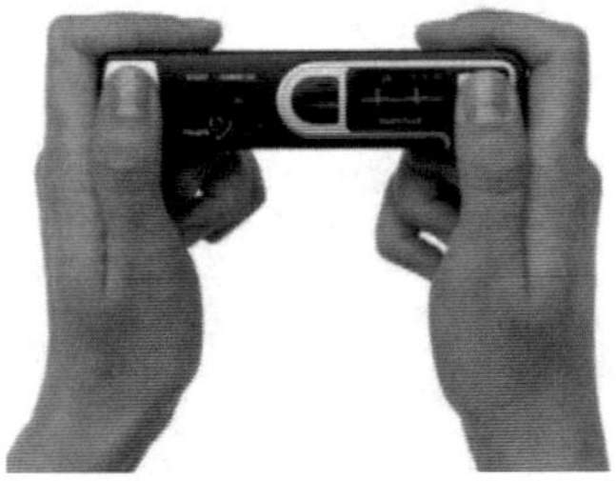 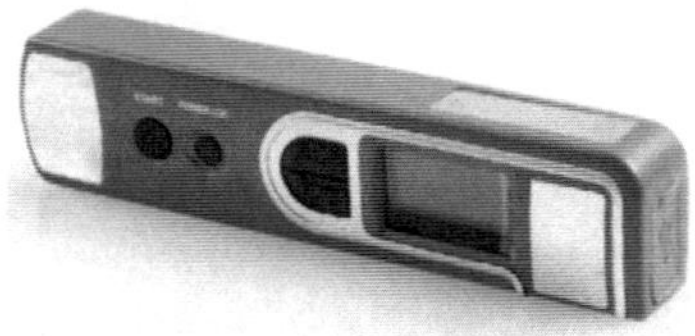	Manufacturer «HeartCheck» (2 built-in available electrodes) - the light cardiograph; - the compact size; - the speed rate of the results generation; - ease of use; - the possible data transfer into PC; - the display backlight *Design:* - originality and usability of the configuration; - the limited color scheme; *The target audience:* - the youth (from 18 to 30 years); - women and men (from 30 to 50 years); - the seniors (55 and up)

IOP Conf. Series: Materials Science and Engineering 93 (2015) 012076 doi:10.1088/1757-899X/93/1/012076

<table>
<tr>
<td>3. ECG Monitor MD100B
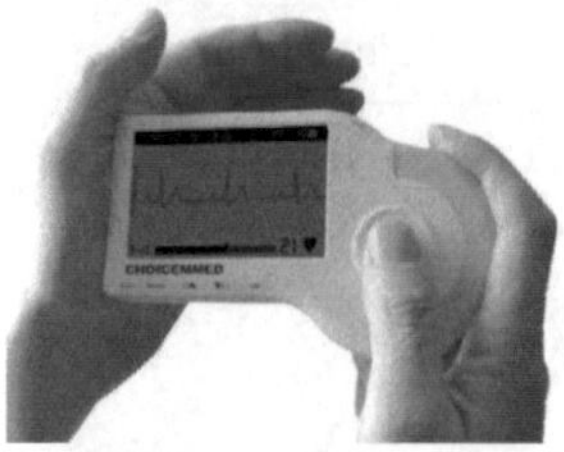

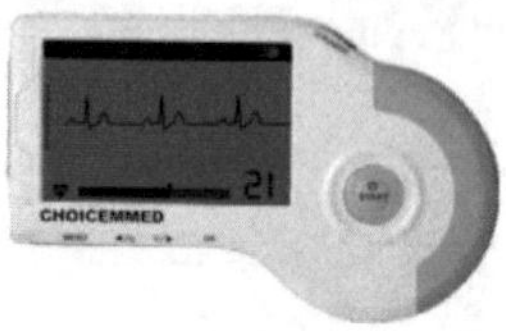
</td>
<td>- the light cardiograph;
- the speed rate of the results generation;
- ease of use;
- the possible data transfer into PC;
- the display backlight
Design:
- the configuration usability;
- the incompact size;
- no color options
The target audience:
- women and men (from 30 to 50 years);
- the seniors (55 and up)</td>
</tr>
<tr>
<td>4. ECG Check – the peripheral equipment for iPhone with the built-in electrodes
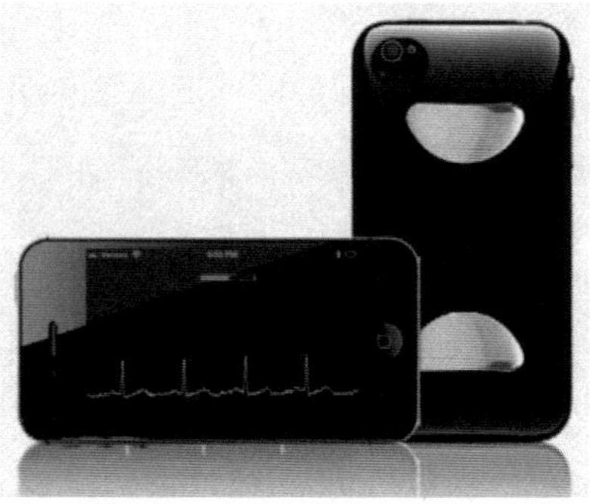</td>
<td>- ease of use
- the automatic analysis in seconds
- the local storage of ECG data, reports and analysis
- the compatibility with the web portal for the long-term storage, analysis, communication and reporting
- the heart data accessibility at any time from anywhere globally
Design:
- the configuration usability;
- the iPhone style;
- the black and white cover piece for the body
The target audience:
- the youth (from 18 to 30 years);
- women and men (from 30 to 50 years)</td>
</tr>
<tr>
<td>5. AliveCor Heart Monitor - the peripheral equipment for iPhone with the built-in electrodes

</td>
<td>- ease of use
- the local storage of ECG data, reports and analysis
- the compatibility with the web portal for the long-term storage, analysis, communication and reporting
- the data accessibility at any time
Design:
- the configuration usability;
- the iPhone style;
- the black cover piece for the body
The target audience:
- the youth (from 18 to 30 years);
- women and men (from 30 to 50 years)</td>
</tr>
</table>

The analysis of the analogues configuration allowed identification of the most ergonomic configuration of the cardiograph ECG-EXPRESS with the choice of the color scheme, thereby not only expanding the target audience, but also detecting the only competitive analogue not presented on the Russian market, i.e. ECG HeartCheck PEN.

4. The working principle of the electrocardiograph

The basic device with a fundamentally similar working principle is the joypad for the games console, since the position of the hands when playing is the most practical from an ergonomic point of view. The position of the hands and fingers has determined the location of the main electrodes on the body panel of the product, that meets the functionally ergonomic criteria (figure 3).

Figure 3. The working principle of the electrocardiograph

The weight of this device is about 108 grams, the length is 145 mm, the height is70 mm and the width is 20 mm (figure 4). The cardiograph takes a little space in the bag and in spite of its small size the cardiograph is handily placed in the hand and does not slip. The required minimum number of buttons is located on the central part under the monitor, the plugs are on the lateral side. The electrode-buttons used to record the measurements are located on the front and lateral sides. The slot for convenient removal of the back cover, which is very tight to the body, is located on the rear side. The mini-USB plug, which is designed to charge the cardiograph or to PC connect, is located on the bottom. The extended round configuration of the device is caused by the ergonomics criteria, the working principle and the configuraton of the modern phones and gadgets used every day [4, 5].

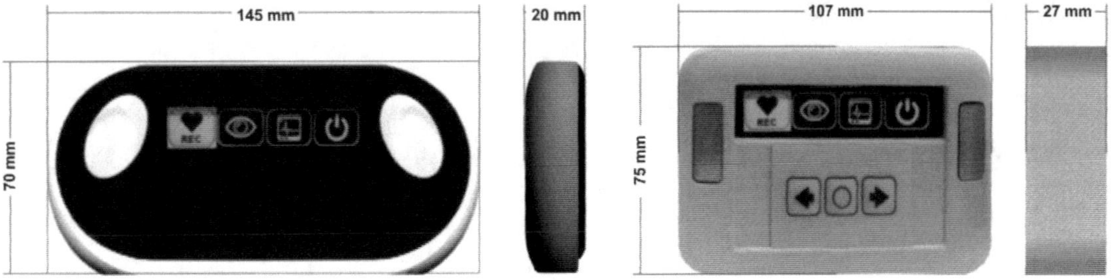

Figure 4. The modified configuration of the portable hand-held electrocardiograph «ECG-EXPRESS»

The available color scheme for the cardiograph body is offered subject to the common gadgets and their colors (figure 5). The significance and originality of the cardiograph configuration is determined by the balanced ratio of brightness, color, size and location of the various components. The main categories of the composition are the tectonics and the volume-spatial structure. The cardiograph configuration reflects the characteristics of its construction, the force distribution, the material behavior, that is characterized by the phenomenon of tectonics. The cardiograph shape also communicates with the environment, in this case it can be referred to the one-piece structure with the built-in hidden mechanism [6].

The color scheme is determined based on the configuration composition of the industrial product and the product color matrix, established in 1994 by Weinberger, Campbell and Brodie, according to which the selected colors are referred to the low-risk group in connection with their dyeing. The main characteristics of the visual perception are given in Table 2 [7]. By physiological and psycho-physiological effects on the humans the yellow color initiates the brain function, green and blue cause a sense of freshness, remove exitation, calm, reduce the blood pressure, slow down the pulsation. The green colour is disciplinary making people self-control. Blue, green, yellow have an overall positive psychoactivity effect, improving the working ability [5].

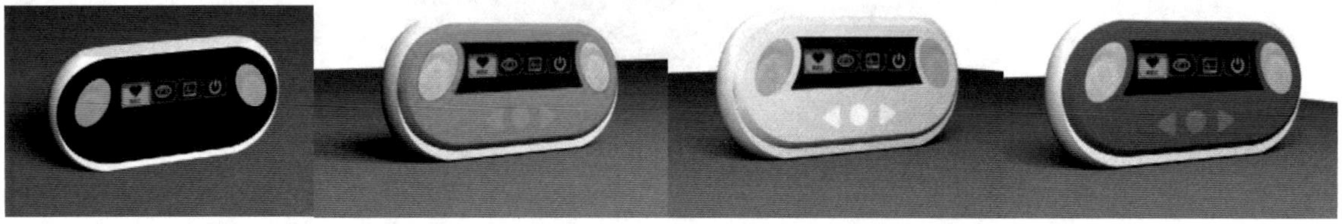

Figure 5. The options of the color scheme of the electrocardiograph body

Different plastics are used for the manufacture of the body frame, i.e. ABS, HDPE, LDPE, polypropylene, polyacetal, polycarbonate and polyamide. Most often ABS plastic (acrylonitrile butadiene styrene) is used in the production. It is a good material to manufacture the body frame of the electronic systems, which is used under low temperatures and if the high ductility properties are required. Most often, this material is used in the manufacture of the body frame components [8].

Table 2. The psychology of the visual perception

Color	The visual perception of			
	the distance	the temperature	the mental feeling	the hygienical effect
green	neutral	very cold	too mild	fresh
yellow	close	very warm	fascinative	-
blue	far	cold	mild	clear

The body frame is made of the ABS plastic or two-component polyurethane plastic by casting method in a silicone mold under the ambient pressure (the autoclave molding), or by vacuum casting method, depending on the configuration, the dimensions of the component and the set task. The two-component plastics are not "standard" plastics, which are casted under pressure into the mold, and there are some restrictions on use, such as the wall thickness (the wall thickness of the piece can not be less than 1 mm), it is prohibitted to use the polyurethane plastic under temperatures above 100-120°C. Other characteristics are similar to the common thermoplastics.

All materials used for the manufacture of the products from plastic, are painted in a variety of colors and combinations by adding the dye (pigment). The technology of the plastics production with different colors has a small weak point, that is the colouring of the plastics requires the specific equipment, which only a few manufacturers possess, thus very often it turns out that the color shades

IOP Conf. Series: Materials Science and Engineering **93** (2015) 012076 doi:10.1088/1757-899X/93/1/012076

of the plastics depend on the suppliers. The highly concentrated plastic granules are used to facilitate the technology and avoid the dependence on the pigments' suppliers. About 2% concentrated plastic granules are added in the uncolored plastic to produce the desired color. The choice of the available colors of the super concentrate is highly diverse, but still limited, that should be taken into account for the plastics molding.

The superficial coloring is used to make the plastic parts of the body frame in the silver or metallic colors. It is the manual painting of the details with the standard spraygun followed by drying process in a special chamber. A layer of the protective varnish is applied on top of the paint, which increases the coating resistance and assures the extra shine of the surface. The major processing complexity of the superficial coloring is the requirement to work in the clean rooms with the minimum dust.

5. Conclusion

Thus, the modern portable devices for the diagnosis of the cardiac rhythm have become the most relevant means of the cardiovascular disease prevention and entered in our life, that generally affects their configuration and appearance. The ease of use, ergonomics, functionality and appearance of the cardiograph prevail in the development of its configuration.

References

[1] Kuchta M, Seryakov V, Sokolov A 2012 Theory and practical design of the exhibit and show-case. *Monograph* 168
[2] Lotsmanenko V, Kochegarov B 2004 Design and engineering (fundamentals). *The teaching guide* 96
[3] Kuchta M, Kumanin V, Sokolova M, Goldschmidt M 2013 Industrial design. *A textbook: Publishing house TPU* 310
[4] Gregory R 1970 The eye and the brain: the psychology of the visual perception. *The teaching guide* 279
[5] Runge V, Manusevich Yu 2005 Ergonomics in the environment design. *The teaching guide* 328
[6] Kochegarov B 2006 Industrial design. *The teaching book* 297
[7] Wolfgang H 2006 Anthropometry, ergonomics, the color in industrial design. *The technical aesthetics and industrial design* pp 9-11
[8] Kukhta M, Seryakov V 2010 The Specifics of Industrial Display Design's Forming and Structuring. *Design. Materials. Technologies (DMT)* **1** pp 18-24

The technology for creating of decorative plywood with low formaldehyde emission

R R Safin, R R Khasanshin, A R Shaikhutdinova and R R Ziatdinov

Kazan National Research Technological University, Kazan, Russia

E-mail: cfaby@mail.ru

Abstract
The development of new technologies of processing of industrial art products to improve their functional, ergonomic and aesthetic properties is one of the important directions of improvement of product design. The article presents the technology of preliminary contact thermal modification of sheets of veneer in the production of low-toxic decorative plywood and laminated products, which lets significantly improve their water resistance. It has been established that thermal modification of wood material causes a decrease in density with increasing temperature and duration of treatment. A mathematical model describing the process of heat treatment of wood veneer and allowing predicting the degree of thermal modification of wood material depending on the modes of modification has been developed. Several studies to determine the basic properties of plywood made on the basis of heat-treated veneer were conducted in the work. Studies have shown that the thermal effect on veneer in the manufacture of plywood provides improves water-repellent properties of it, while not increasing its toxicity. It has also been found that the alternating of layers of wood with varying degrees of heat treatment when creating laminated products greatly enhances the decorative features of the products.

1. Introduction

The main challenges for improving the technology of plywood production nowadays are aimed at improving the environmental sustainability of the processes, and to receive the finished product with low toxicity, enhanced durability and minimal shape variability in conditions of humidity and temperature influences [1].

Plywood is hygroscopic as well as the solid wood. Its operation in a humid environment leads to the loss of strength, to shape variability and biological damage. Without special treatment plywood production has a limited lifespan.

The resin, which consists of toxic formaldehyde, the selection of which from the material is a serious danger to human health is applied in Russia for the production of plywood of increased water resistance. In this regard there is a prohibition on the use of water-resistant plywood in the conditions of the premises. According to the Russian standard the waterproof plywood can be only used in external conditions.

At the same time it is known that thermal modification – the technology of high-temperature treatment of wood without oxygen access air allows reducing the water absorption and enhancing the resistance to decay.

There are various methods of thermal modification of wood nowadays: in hydrophobic liquids, in the environment of superheated steam, in the environment of flue gases, vacuum contact method [2-5]. Heat treatment of wood significantly reduces swelling, shrinkage, improves the heat insulation qualities. Improvement of these properties of wood raw material is determined by the influence of high temperature on its chemical structure - mainly due to the thermal degradation of hemicelluloses. In addition, the results of the tests have shown [6, 7], that after heat treatment the ordinary tree species acquire the characteristics approaching to the rare and exotic breeds. Thermo wood acquires the intense colour throughout the thickness, more smooth and dense texture.

The researches of the influence of preliminary heat treatment of wood filler on the properties of composites have been conducted by the scientists worldwide in the last 6-7 years.

The technology of thermal pre-treatment of eucalyptus wood using steam at a temperature of 453 K with subsequent production of wood polymer composite is studied in the work [8]. Studies have shown that the thermal effect on wood raw material in the manufacture of WPC reduces the swelling and water absorption of the composite with 80 % and 70 % respectively after 28 days of immersion. In addition, the modulus of elasticity in bending has been reduced to 22 %.

In the work of Matjushenkova the method of oil heat treatment of plywood larch wood for increasing the water resistance is presented. The process of getting this plywood consists of impregnation of veneer by specially prepared tall oil from hardwood with subsequent heat treatment. The water and moisture resistant non-toxic building product is made in this case. But the proposed method of oil heat treatment of plywood is time-consuming on an industrial scale [9].

The work [10] presents the results of investigations of the basic properties of a composite material created on the basis of thermally modified wood and Portland cement grade 400 as a mineral binder. It has been established, that the thermal impact on the filler reduces the water-cement ratio that can be explained by a significant reduction of water absorption of thermally modified wood particles compared to untreated. Studies of the frost resistance of these types of samples allowed establishing the increase of this parameter with increasing of treatment temperature.

The results of the presented studies indicate an increased interest in issues of pre-heat treatment of wood in the manufacture of wood-based composites because of the advantages of using a less hydrophilic wood [2-11]. Despite the fact, that for some types of composites decreased mechanical properties were found, preliminary thermal modification of wood particles allows significantly increasing of the performance characteristics of the composite and, consequently, expanding the sphere of its possible use. Analysis of literature has shown that there are the active researches in the field of composite materials based on thermally modified crushed wood in recent years. While researches, aimed at improving of operational characteristics of the laminated material by heat-treatment of wood veneer has not been carried out.

2. Materials and methods

In this regard the studies on modifying the properties and characteristics of the plywood material, created on the basis of thermally modified veneers and heat-cured glue based on urea-formaldehyde resin have been conducted.

During the experiments birch peeled veneer sheets were used as wood filler. They were thermally treated by the contact method in the temperature range 180-260 °C. For thermal modification the installation, diagram of which is shown in Figure 1 was established. The unit is a sealed heat-insulated heat chamber 1 with the insulated lid 2. The internal space of the chamber 2 through a valve 3 communicates with the pumping system, including the circulator 4, a capacitor 5 and the condensate collector 6. Thermal energy supplies of to the veneer 7 by a contact method using heating plates 8.

To determine the hygroscopic properties of the thermally modified veneers the method of treatment of wood in desiccators with a saturated solution of soda for 30 days was used. The degree of saturation of the medium was 0.75. To determine the water absorption, the samples were kept in distilled water for 50 days. The moisture content of samples was determined by the gravimetric method: by measuring the current mass, followed by final drying to absolutely dry state at a temperature of 105 ^{0}C.

To determine the water absorption of plywood the method, according to which the samples were immersed in desiccators containing distilled water and were kept for five days was used. The samples are periodically weighed and measured, and the first weighing and measuring is in a day, counted from the moment of immersion of the samples in water, then through two, three and five days.

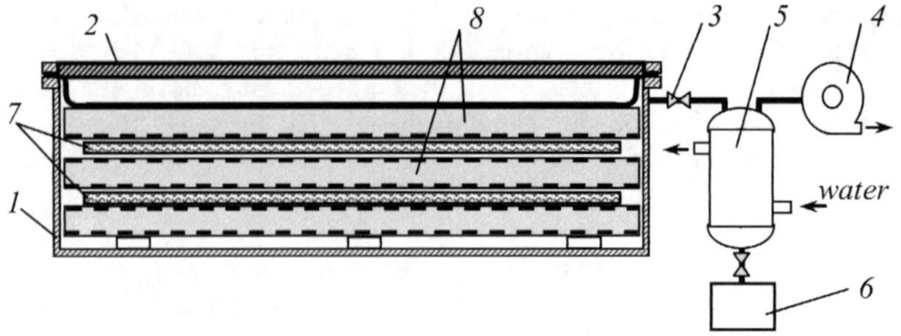

Figure 1. The scheme of installation of the vacuum-contact thermal modification of veneer

3. Results

In the process of heat treatment of the veneer on the unit, shown in Fig. 1, the kinetic changes of the density of the veneer sheets is founded (Figure 2).

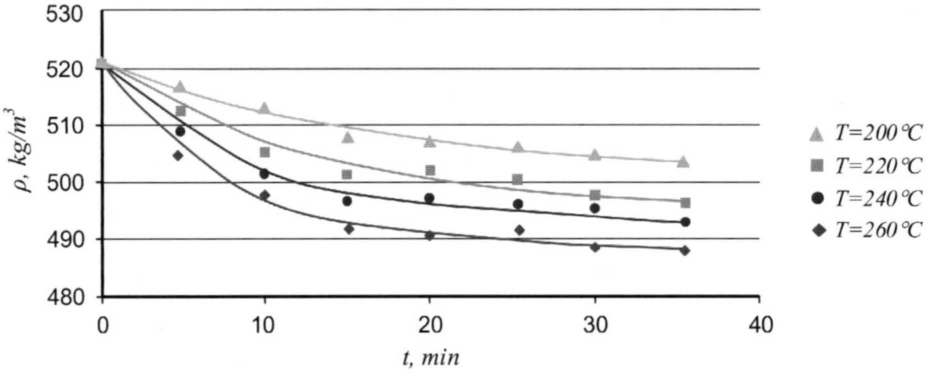

Figure 2. Kinetic curves of the change in the density of veneer thickness of 1.5 mm in the process of thermal modification

In this regard the relative value is offered to determine the degree of thermal modification (1)

$$Q=(\rho_s - \rho_i)/(\rho_s - \rho_f), \tag{1}$$

where: ρ_s – the initial density of the sample, (kg/m^3); ρ_i – is the density of the sample after heat treatment, (kg/m^3); ρ_f – density for a given wood sample, achieved at the maximum temperature for the process of thermal modification – 260 °C (kg/m^3).

The proposed method of classification of thermally modified veneers through a change in density allows quantifying the degree of thermal modification.

Using the kinetic curves of the change of density of the veneer in the process of contact thermal treatment (Figure 2) the graphic dependences of the duration of the stage of heat treatment to the required degree of thermal treatment which are presented in Figure 3 were obtained. Received chart allows us to select the mode of the process of thermal treatment by reducing unsustainable options that require significant duration of the process.

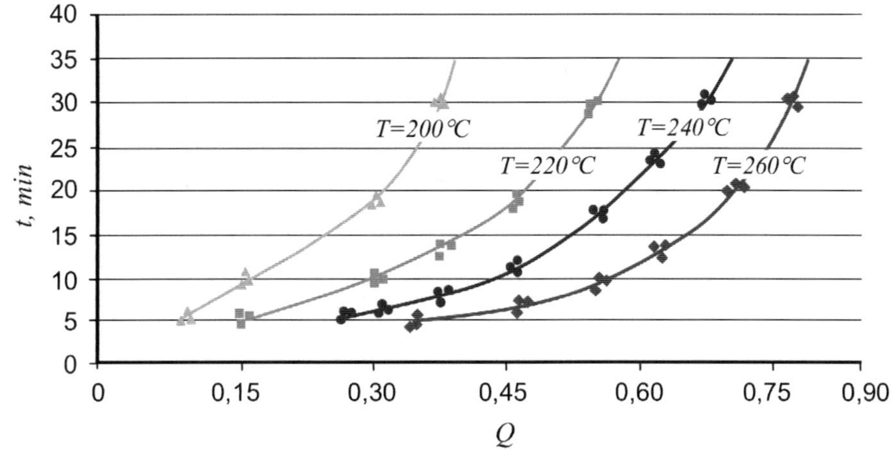

Figure 3. The degree of thermal treatment of birch veneer at thickness 1.5 mm

On the basis of experimental design and results of experimental studies of thermal modification of wood sheet material a mathematical model (2) allowing predicting the process of thermal modification of veneer depending on time (t), temperature (T) and the thickness of the veneer (δ) was developed.

$$Q = -1{,}35 + 0{,}0803t + 0{,}0056T - 0{,}54\delta + 0{,}00183T\cdot\delta - 0{,}00169t^{2}. \quad (2)$$

As a result of the conducted research the change of colors of thermal veneer depending on the degree of thermal modification has been found. The results are presented in the table.

Table. The color scheme and the degree of thermal modification for veneer from birch

Temperature processing	The thickness of the veneer					
	δ = 1,5 mm			δ = 2 mm		
	Processing time 10 min	Processing time 20 min	Processing time 30 min	Processing time 10 min	Processing time 20 min	Processing time 30 min
< 210°C	0,33	0,48	0,57	0,27	0,44	0,51
> 230°C	0,44	0,65	0,76	0,41	0,60	0,70
> 260 °C	0,75	0,97	0,99	0,71	0,90	0,98

Further the thermal veneer sheets were sorted by texture and color and were bonding. As a result of expert analysis of texture and color when creating the decorative plywood the degree of thermal modification of outer layers in the range of 0.5-0.75 was recommended. In addition, to create the products with high decorative characteristics the alternation of veneer sheets with different degrees of heat treatment that allows you to create curved supporting structures with interesting cross texture and to get Souvenirs by the method of milling was proposed (Figure 4).

Figure 4. Samples of plywood and Souvenirs made of birch veneer with varying degrees of thermal modification

Further samples of plywood created from veneer sheets with the same degree of thermal modification were tested for water absorption.

Figure 5 shows the kinetic curves of the change in the humidity of the samples of plywood in the process of treatment in distilled water. Chart analysis shows that the increase in the degree of thermal processing of raw wood provides a lower water-absorbing ability.

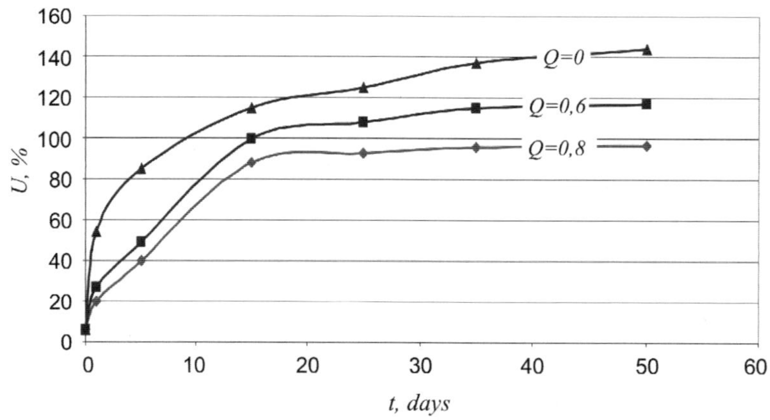

Figure 5. Water absorption of samples of plywood from thermally modified veneers

Analysis of the results of the comparative evaluation of shear strength of plywood in the dry state and after boiling in water for 3 hours (Figure 6) indicates the lower limit of the compressive strength of dry plywood with a higher degree of heat treatment of veneer. Thus, a significant reduction in strength is observed when the veneer with the degree of thermal modification is higher than 0.75. At the same time, for samples of plywood veneer with degree of thermal modification in the range 0.45-0.55 boiling does not almost cause reduction of the strength characteristics.

As the result of studies of the mechanical characteristics of plywood from thermally modified veneers rational degree of heat treatment of wood material in the range of 0.45 to 0.6 for plywood of general purposes, providing high aesthetic characteristics with high moisture and water resistance and maintaining the basic strength parameters is determined.

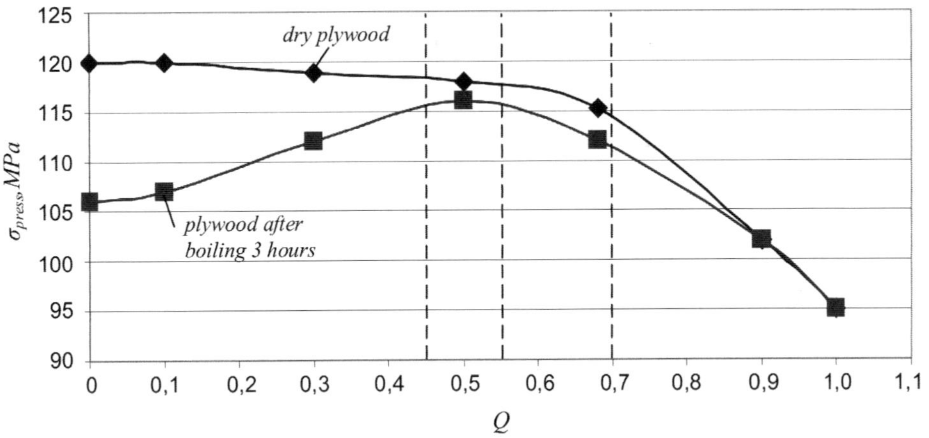

Figure 6 The change in compressive strength of samples of plywood depending on the degree of thermally modified veneer

4. Conclusions

It has been revealed on the basis of an expert analysis that the thermal modification of veneer sheets in the manufacturing process of plywood allows us to create the products with high decorative features. In addition it has been found that the thermal modification of sheets of veneer can improve the water resistance of plywood without reducing its environmental properties.

References

[1] Gangi M, Tabarsa T, Sepahvand S and Asghari J 2013 Reduction of formaldehyde emission from plywood *Adhesion Science and Technology* **27** pp 1407–1417
[2] Safin R, Khasnshin R, Shaikhutdinova A and Safina A 2014 Research of heating rate while thermo modification of wood *World Applied Sciences Journal* **30** pp 1618–1621
[3] Kariz M, Kuzman M-K and Sernek M 2013 The effect of the heat treatment of spruce wood on the curing of melamine–urea–formaldehyde and polyurethane adhesives *Adhesion Science and Technology* **27** pp 1911–1920
[4] Khasanshin R, Safin R and Razumov E Industrial plant for thermo modified timber the average flue gas *Bulletin of Kazan National Research Technological University* **16** pp 122-124
[5] Bal B and Bektas I 2013 The effects of heat treatment on some mechanical properties of juvenile wood and mature wood of eucalyptus grandis *Drying Technology* **31** pp 479–485
[6] Ayrilmis N, Jarusombuti S, Fueangvivat V and Bauchongkol P 2011 Effects of thermal treatment of rubberwood fibres on physical and mechanical properties of medium density fibreboard *Tropical Forest Science* **23** pp 10–16
[7] Saputra H, Simonsen J and Li K 2004 Effect of extractives on the flexural properties of woodplastic composites *Composite Interfaces* **11** pp 515–524
[8] Ayrilmis N, Jarusombuti S and Bauchongkol P 2011 Effect of thermal-treatment of wood fibres on properties of flat-pressed wood plastic composites *Polym Degrad Stabil* **96** pp 818–822
[9] Matjushenkova E, Chubov A and Tsarev G 2011 Rationale modes of oil heat treatment of plywood *News of the St. Petersburg Forestry Academy* **194** pp 129–137
[10] Razumov E, Safin R, Khasanhin R, Barcik S and Kvietkova M 2013 Studies on mechanical properties of composite materials based on thermo modified timber *Drvna Industrija* **64** pp 3–8
[11] Shahzad A 2011 Effects of fibre surfa treatments on mechanical properties of hemp fibre composites *Composite Interfaces* **18** pp 737–754

FORMING ENGINEERS' SOCIOCULTURAL COMPETENCE: ENGINEERING ETHICS AT TOMSK POLYTECHNIC UNIVERSITY

E Galanina[1], A Dulzon[1], A Schwab[2]
[1]National Research Tomsk Polytechnic University, Tomsk, Russia
[2]Karlsruhe Institute of Technology, Karlsruhe, Germany

Email: galanina@tpu.ru

Abstract. The aim of the present research is to discuss Tomsk Polytechnic University in respect of forming engineers' sociocultural competence and teaching engineering ethics. Today international standards of training engineers cover efficient communication skills, ability to understand societal and environment context, professional and ethical responsibility. This article deals with the problem of contradiction between the need to form engineers' sociocultural competence in Russian higher education institutions in order to meet the requirements of international accreditation organizations and the real capabilities of existing engineering curricula. We have described ethics teaching experience of TPU, studied the engineering master programs of TPU to see how the planned results are achieved. We have also given our recommendations to alter the structure of TPU educational curricula, which can also be applied in other higher education institutions.

1. Introduction

Science and engineering are of crucial importance these days. The époque we're living in is the time of rapid hi-tech development, science driven production and innovation projects. The development of nuclear power industry, information and telecommunication technology, genetics, nanotechnology, chemical technology and new materials raises the issue of responsibility when it comes to the future generations along with the environment and society.

Moral and value-related neutrality according to which an engineer should focus exclusively on scientific and technical part of a project disregarding moral evaluation proved to be absolutely inconsistent in the 20th century [1].

During the period of classical science development engineering stuck to the ideals of objective truth, discovery for the sake of discovery not related to morals in any way. Today scientific discovery's social orientation is becoming extremely important. As it is noted in "Mount Carmel Declaration on Technology and Moral Responsibility" (1974), no aspect of modern engineering can be seen as neutral in terms of morality [2].

Using leading-edge technology for military purposes, negative consequences of technogenic impact on the environment and society throughout the 20th century raised the issue of engineers' moral responsibility. Investigating numerous technogenic disasters, e.g. the Chernobyl NPP accident in 1986, the Challenger shuttle disaster in 1986, oil production platform explosion in the Gulf of Mexico in 2010, has shown the key role of engineers' moral responsibility in the course of events. Today, we cannot afford to ignore or neglect the pressing ethical problems of technology and applied science [3]. The pace and the scale of sociocultural changes led to by the technological advancement make the issues of urgent technological innovations very important. The problems of social control in science and in the sphere of high technology, social estimation and hi-tech projects expertise are becoming really topical.

However, despite the trend mentioned not all engineers realize the ethical component of what they create. This is why we chose to highlight the magnitude of forming engineers' sociocultural competence, especially by teaching the course of "engineering ethics" in universities.

2. Forming engineers' sociocultural competence

In the 21st century the new paradigm of technical education is being formed, engineering training standards undergo vast changes. These changes generally have to do with involving people with their system of values, convictions and ethical assumptions into the structure of scientific research and engineering solutions. This is why we see gradual rejection of the narrow engineer training based solely on specific technical subjects.

Sociocultural competence of engineers is the present day need which will help them self-actualize in rapidly changing professional engagement. This kind of competence includes the ability and readiness to communicate efficiently, to function in multicultural teams, to understand professional and ethical responsibility, to assess engineering prospects and consequences of operation [4].

Sociocultural competence in place including following the engineers' ethics code determines the efficiency of professional practice in today's sociocultural environment. It indicates the ability to achieve results that would satisfy the needs of society and business.

According to "The CDIO Syllabus v2.0", sociocultural competence means professional ethics, ability to think creatively, critically and systemically, to work in groups and communicate effectively, to understand societal and environmental context. Engineers must possess a wide array of personal, interpersonal, and system building knowledge and skills that will allow them to function in real engineering teams and to produce real products and systems, meeting industrial and societal needs [5].

Along with professional competences Accreditation Board for Engineering and Technology (ABET) states the need in the following Student Outcomes: (a) ability to design a system, component, or process to meet desired needs within realistic constraints, such as economic, environmental, social, political, ethical, health and safety, manufacturability, and sustainability (b) ability to function in multidisciplinary teams; (c) understanding professional and ethic responsibility; (d) ability to communicate effectively; (e) broad education necessary to understand the impact of engineering solutions in global economic, environmental, and societal context [6].

Thus, the international standards are designed to train engineers who would be able to assess the results of engineering solutions, communicate professionally and efficiently, bear responsibility for result within the whole engineering spectrum.

However, unfortunately, a great deal of curricula in Russian universities does not correspond to the international criteria mentioned above. Designing engineering curricula most frequently means using the traditional method: a limited number of social and humanitarian subjects are included into the training process, but those subjects are not integrated with the major course. This method does not prove to be effective. Students don't see the opportunity of applying the gained knowledge in their work. On the other hand, they experience the need in understanding sociocultural engineering context and the transformations that take place in modern society.

And here is the problem – how do we neutralize the contradiction between the growing need in sociocultural training of engineers and the inefficiency of its practical realization in Russian higher education institutions?

At this point the solution can be found in starting supplementary curricula designed to form sociocultural competence along with a number of business-related and managerial competences for engineers. The curriculum of "The elite engineering education" that has been practiced in TPU since 2004 can serve as an example. It is destined to prepare the engineering leaders ready to foresee technological development situation and to communicate efficiently in the course of their work out of the graduates able to work in a team.

Another one could be applying modules (e.g. business communication, intercultural communication, engineering ethics, applied ethics, engineering leadership and entrepreneurship) integrated with majors in engineering curricula.

3. Engineering ethics

Most certification and accreditation organizations and professional engineering associations (ABET, FEANI, WFEO, NSPE, ASCE, IEEE, ASME, BMES, IPENZ) require ethics observance and understanding of professional and ethical responsibility.

For instance, IEEE Code of Ethics starts with the words: "We, the members of the IEEE, in recognition of the importance of our technologies in affecting the quality of life throughout the world, and in accepting a personal obligation to our profession, its members and the communities we serve, do hereby commit ourselves to the highest ethical and professional conduct…" [7]. Society of Petroleum Engineers Strategic Plan (2013-2017) also pays attention to the growing interest in ethics and the need for ethical education [8].

We see engineering ethics as the technology of finding efficient solutions when dealing with complex engineering problems basing on moral assumptions and understanding engineers' responsibility for short- and long-term consequences of the decisions made. Today engineering ethics is a branch of scientific studies and a subject in universities in the framework of which ethical aspects of engineers' work are reviewed.

Engineering ethics studies the problems of engineers' moral responsibility, commitment to colleagues, customers, employers, society in general, safety issues, cooperation with other decision makers, objectivity of professional analysis, privacy issues, copyright and commercial secrets.

Ethics codes, e.g. NSPE Code of Ethics for Engineers, ASCE Code of Ethics, IEEE Code of Ethics, represent series of moral principles setting standards and restrictions regulating engineers' actions. Those principles are shared and accepted by national professional communities. Such codes also exist in Russia: Code of APEC Engineer Professional Ethics, Science and Engineering Code of Ethics worked out by Russian Union of Scientific and Engineering Public Organizations [9].

The main thesis of ethics codes is that engineers shall recognize that the lives, safety, health and welfare of the general public are dependent upon engineering judgments, decisions and practices incorporated into structures, machines, products, processes and devices [10]. Modern engineers should understand professional and ethical responsibility at all levels of their practice from designing to realization. Mistakes in construction and technology made by engineers may lead to catastrophe for society. That is why ethical regulation of engineering in higher education institutions is a necessity today.

The requirements for the competence formation which is an important factor of a specialist's readiness for work are listed in the criteria of Association for Engineering Education of Russia [11], International Engineering Alliance (IEA) [12], Accreditation Board for Engineering and Technology ABET [13], European Federation of National Engineering Associations (FEANI) [14], World Federation of Engineering Organizations (WFEO) [15], etc. Experts of accreditation agencies and professional engineering associations assess an engineer's aptitude in terms of following the ethics code.

The foundation of engineers' professional culture lies in university. In the course of study it is possible to form students' adherence to norms of professional ethics, understanding of moral responsibility, ability to foresee and estimate consequences of operations and results of applying innovations. Today many academic institutions are recognizing the importance of these societal and ethical responsibilities and have taken initiatives to incorporate them into engineering curriculum along with traditional technical materials [16].

The course on engineering ethics is read in all leading universities of The United States and Europe, e.g. Massachusetts Institute of Technology, Stanford University, California Institute of Technology , Karlsruhe Institute of Technology and Texas A&M University where the course in engineering ethics was established due to the rising interest in the subject [17]. A lot of scientific projects devoted to ethical issues are realized in Karlsruhe Institute of Technology, e.g. *Futures, Visions and Responsibility – The Ethics of Innovation, Global Ethics in Science and Technology, Synth-Ethics* [18].

This experience shows the overall success and necessity of the course. It may be an independent engineering ethics course or modules on ethics included into the structure of engineering subjects. The systematic approach to engineering ethics education considers the context-specific needs of different engineering disciplines in ethics education and leverages the collaboration of engineering professors, practicing engineers, graduate students, ethics scholars, and instructional design experts [19].

Most frequently studying professional ethics is based on the "case method" approach that provides an opportunity to assess knowledge and understanding of the engineering ethics codes by students as well as the ability to take ethical responsibility while making decisions. Today the case method is one of the most popular methods for engineering ethics instruction in The United States [20]. The US teaching of ethics relies upon the use of hypothetical and "real" scenarios along with open discussions framed by the codes [21].

According to M S Pritchard, ethics lessons are often learned only after something has been overlooked or has gone wrong. There is no wholly adequate substitute for actual engineering experience. However, having students reflect on realistic case studies can provide some helpful preparation for dealing with ethical issues they are likely to face once they start their engineering practice [22]. Engineering education helps students understand and deeply internalize the core values of safety and environmental protection [23].

Along with the case studies there are other methods of integrating ethics into the engineering curriculum. They include external course work (e.g. philosophy classes), service-learning projects, team-based senior design course work, and the cross curricular approach (integration of ethics in multiple courses throughout the academic career) [24]. Workshops are also an efficient method to improve engineering students' awareness on ethics; it gives an opportunity to meet engineering leaders including those who have chosen the ethics-related career [16].

When it comes to Russian universities, ethical issues are generally covered by philosophy classes, which is not effective enough for the course material is destined to teach metaphysical aspects of science and engineering. Studying engineering ethics should be based on empiric practical material.

4. The Experience of Teaching Ethics at Tomsk Polytechnic University

Teaching ethics in TPU started in 1990s. The course "Applied ethics" and the student's book [25] were written with the assistance of German professors A Schwab and H Lenk.

The idea that ethics should be seen as a "moral guardian discipline" for a professional is Lenk's basic principle [3]. Despite its limitations applied ethics is a necessary step on the way of stimulating and supporting ethical behavior. Everyday life is a continuous ethical challenge. Applied ethics brings ethics back to where we live and work [26].

The experience of teaching applied ethics in TPU turned out to be quite contradictory. The University's Administration, students and attendees of postgraduate training programs met the new course quite tepidly. Many students reacted to the discussed ethical dilemmas rather cynically. Probably that was the result of the former system of values breakdown that took place in 1990s when the USSR collapsed.

Nowadays students start showing interest in ethical problems. The module "engineering ethics" is read quite successfully to the students of the "Elite engineering education" course. Russian society of today tends to return to traditional norms of morality. And that brings up the issue of developing a new course on engineering ethics in TPU again.

5. Materials and Methods

Research goal is to study the correspondence of the planned teaching results (in the field of engineering ethics) in the framework of TPU master programs with the actual ones.

Object of research is TPU engineering master programs of 2014. We have analyzed 27 master programs which constitute 80% of all master programs in TPU (2014). Each program covered 120 ECTS credits.

At the first stage we analyzed the planned study results in accordance with the following theses: "to follow the norms of professional ethics", "to follow the ethics code", "to demonstrate responsibility for the work results". We have also estimated their credit value.

At the second stage we analyzed curriculum contents in every subject of every master program through the perspective of forming the required study results. We have also estimated their credit value.

Research methods: analysis, comparison.

Table 1. Master programs of TPU

№	Program title	Study result code	Study result credit value
1	Applied mathematics and IT	R 4	15
2	Physics	R 5	15
3	Geology	R 7	13
4	IT and computing	R 10	6
5	Information systems and technology	R 11	7
6	Applied IT	R 8, P9	14
7	Electronics and nanoelectronics	R 8	5
8	Instrument engineering	R 3	10
9	Optical engineering	R 7	6
10	Bioengineering technology and systems	R 10	3
11	Thermal engineering and thermotechnics	R 3, P4	7
12	Electric power engineering and electrotechnics	R 3	7
13	Power engineering	R 2	5
14	Design engineering for mechanic production	R 9, P10	9
15	Mechatronics and robotics	R 10, P11	10
16	Applied physics	R 12	15
17	Chemical engineering	R 9, P10	10
18	Energy and resource efficiency processes in chemical engineering, petrochemical industry and bioengineering	R 6, P9	15
19	Technoshere safety	R 7, P9, P10	15
20	Environmental engineering and water resources management	R 7, P9	8
21	Land registry and utilization	R 4, P6	14
22	Material engineering and technology	R 8	7
23	Standardization and metrology	R 7, P10	10
24	Quality control	R 7, P9	6
25	Administration in engineering systems	R 8, P9	11
26	Innovation studies	R 12	8
27	Design	R 8, P9	21

6. Result and Discussion

Figure 1. TPU master programs research results

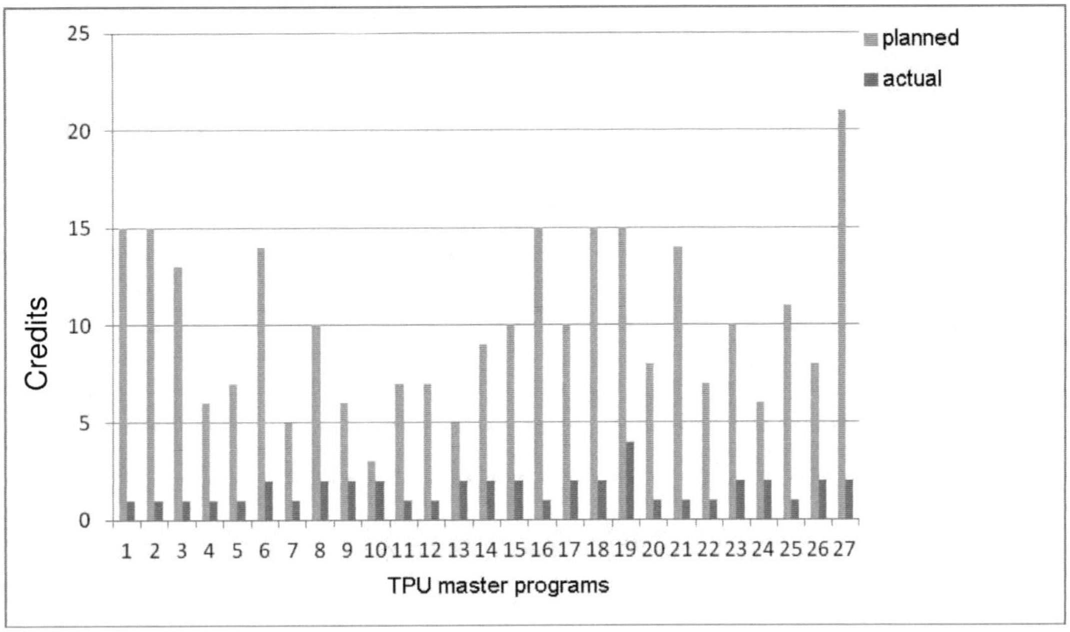

The research results have shown that, on one hand, the planned results correspond to common cultural and professional competences indicated in Federal State Educational Standards and correlate with criterion 5 of Association for Engineering Education of Russia, international standards EUR-ACE and FEANI. Credit value of the results varies from 3 to 21, on the other hand, the curricula structure makes it impossible to achieve those results for the absence of an independent engineering ethics course. There are minor modules on ethical issues within the framework of other disciplines, but their credit value does not exceed 1-2 credits on average.

Basing on the results of the research we can give the following recommendations.

Firstly, all master programs of TPU need to include the engineering ethics course. This suggestion is based on the fact that ABET considers that course to be one of the conditions for international programs accreditation. Chairs in social and human sciences are to solve the problem. However, the experience of teaching ethics in TPU has shown that a separate course is not efficient enough in terms of project-oriented paradigm.

Secondly, we need to work out a unified and integrated system of ethics study within the framework of most subjects, the system that would work throughout the whole period of study at university. Ethics should be embedded even in "hard" engineering modules and not only in ones of professional development [27]. To do so we need to consider ethical problems of concrete disciplines, to upgrade the university stuff qualification in terms of teaching ethics and to formulate the learning and teaching aids including case study tasks based on Russian engineering practice.

The "project management" course read by Chair of Engineering in Business of TPU could be an example. This course implies starting with the Ethics code of project managers and moving on to project stakeholders, analysis of projects' influence on society, environment and further generations.

The main problem of estimating students' ethical competence is that checking the ability to follow the ethics code is possible only in the process of supervising students during working on real engineering projects. That is why the best way to create proper conditions for mastering engineering ethical principles by students is project work in teams. Problem- and project-oriented teaching during working on real projects in teams helps students grasp the analysis of ethical projects and apply it in their future work realizing the constructive role of ethics in real life.

7. Conclusions

Today it is extremely important for Russian higher education institutions to pay more attention to teaching ethics. Engineers need to acquire moral principles and standards, give the graduates proper knowledge in ethics for solving moral dilemmas. Following engineering ethics codes while doing their job they will be able to assure decent quality and high standards of living for the future generations.

References

[1] Cooley M 1995 The myth of the moral neutrality of technology *AI & SOCIETY* **9(1),** pp 10–17
[2] The Mount Carmel Declaration on Technology and Moral Responsibility 1974
[3] Lenk H 1983 Notes on Extended Responsibility and Increase Technological Power *Philosophy and Technology* **80** pp 195–210
[4] Galanina E, Bikineeva A and Gulyaeva K 2015 Sociocultural Competence Training in Higher Engineering Education: The Role of Gaming Simulation *Procedia – Social and Behavioral Sciences*, **166** pp 339–343
[5] The CDIO Syllabus v2.0. An Updated Statement of Goals for Engineering Education 2011 *Proceedings of the 7th International CDIO Conference*. Retrieved from http://www.cdio.org/files/project/file/cdio_syllabus_v2.pdf
[6] Criteria for Accreditation Engineering Programs, ABET, 2013-2014. Retrieved from http://www.abet.org/wp-content/uploads/2015/04/eac-criteria-2013-2014.pdf
[7] IEEE Code of Ethics. Retrieved from http://www.ieee.org/about/corporate/governance/p7-8.html
[8] SPE Strategic Plan, 2013-2017. Retrieved from http://www.spe.org/about/docs/strategicplan.pdf
[9] APEC Code of Ethics. Retrieved from http://portal.tpu.ru/apec_eng/certification/Requirements%20to%20be%20registered/Ethics
[10] ASCE Code of Ethics. Retrieved from http://www.asce.org/code-of-ethics/
[11] Association for Engineering Education of Russia. Retrieved from http://aeer.ru/en/index.htm
[12] International Engineering Alliance. Graduate Attributes and Professional Competencies Paper. 2013. Retrieved from http://www.ieagreements.org/GradProfiles.cfm
[13] Accreditation Board for Engineering and Technology. Retrieved from http://www.abet.org/
[14] European Federation of National Engineering Associations. Retrieved from http://www.feani.org/
[15] World Federation of Engineering Organizations. Retrieved from http://www.wfeo.org/
[16] Ooil P C and Tan M 2015 Effectiveness of workshop to improve engineering students' awareness on engineering ethics *Procedia - Social and Behavioral Sciences* **174** pp 2343–2348
[17] Introducing Ethics Case Studies Into Required Undergraduate Engineering Courses. Retrieved from http://ethics.tamu.edu/
[18] Project Overview of ITAS, Karlsruhe Institute of Technology. Retrieved from http://www.itas.kit.edu/english/projects.php
[19] Li J and Fu S 2012 A Systematic Approach to Engineering Ethics Education *Science and Engineering Ethics* **18(2),** pp 339–349
[20] Herkert J R 2000 Engineering ethics education in the USA: content, pedagogy and curriculum *European Journal of Engineering Education* **25(4)** pp 303–313
[21] Bucciarelli L L 2008 Ethics and engineering education *European Journal of Engineering Education* **33(2),** pp 141–149

[22] Pritchard M S Teaching Engineering Ethics: A Case Study Approach. Retrieved from http://courses.washington.edu/cee440/ETHICS1A.htm
[23] Colby A and Sullivan W M 2008 Ethics Teaching in Undergraduate Engineering Education *Journal of Engineering* Education **97(3),** pp 327–338
[24] Yadav A and Barry B E 2009 Using case-based instruction to increase ethical understanding in engineering: What do we know? What do we need? *International Journal of Engineering Education* **25(1)** pp 138-143
[25] Dulzon A and Vasilyeva O 2004 *Applied ethics* 252 p.
[26] Schwab A 2000 Applied Ethics: a third-millennium approach *IEEE Spectrum* Nov, pp 22-25
[27] Ocone R 2013 Engineering ethics and accreditation Education for Chemical Engineers **8(3)** pp e113–e118

How to attract young people to science? (based on materials of sociological research)

D A Karpov[1,2], A Yu Karpova[2], Yu Yu Kryuchkov[2]

[1] New Mexico State University, Las Cruces, USA
[2] National Research Tomsk Polytechnic University, Tomsk, Russia

E-mail: dkarpov@nmsu.edu

Abstract. Involvement of young people into science at the present time is relevant not only in Russia but as well in countries with big experience in this process. The article states that profession of scientist is considered prestigious in the United States and positioned at 4th place in the rating, wheras in Russia it is only at 19th place in the similar rating. The conclusion is based on the sociological studies conducted in the United States and Russia. The authors speculate that changes in public policy in Russia, aimed at recovering of scientific potential, had an impact on young people's ideas and motivation for scientific work. The article provides an analysis of the sociological study conducted by the National Research Tomsk Polytechnic University (TPU), which aims to determine the willingness and possibility to engage in scientific work. The authors note that TPU entered the federal program "5-100-2020" at 4th place in the ranking of the best universities in Russia and has extensive experience, research base, international training programs, exchanges, and internships with best universities in the United States and Europe. The main conclusions of the study is that master students are ready to engage in scientific work; conditions created at the university are the backbone for the development of scientific career of the students; the highest motivation for students is the satisfaction in their scientific advisors.

1. Introduction

Almost 20 years ago, a prominent populariser of science, astrophysics, astronomer, famous scientist Carl Sagan wrote: "In opinion poll after opinion poll science is rated among the most admired and trusted occupations, despite the misgivings" [1]. Obviously, the question of how to attract students to science remains relevant today. It is relevant not only in Russia but also in countries with extensive experience in the involvement of young people in science. We can see it in G8 countries: UK, Germany, Italy, Canada, USA, France, Japan. Despite the fact that these countries have a strong scientific and industrial base, science parks, exchange programs, support programs for talented youth, development of government-supported private partnership of major companies with universities, the search for new ways to involve young talented researchers in science not only continues, but is also "gaining momentum". It is worth emphasizing that in Germany, France, Belgium and most of European countries studying in undergraduate, graduate, and postgraduate programs is free and the base for fundamental scientific research is strong. On the other hand studying in the United States requires payments while granting access to high quality of educations and tools. Still the school system of the United States remains behind the University standards making the States an attractive place for talented youth from all over the world. The basis for fundamental and applied research in the US is as strong as in Europe. By the number of research centres the United States overtake any of the European countries. The climate for the development of scientific and innovative ideas in the United States is maybe the most favourable in the world. The active work for integration of young people in different research projects in the United States comes through:

1. Governmental organizations such as: USAID (United States Agency for International Development);
2. Non-profit, non-governmental organizations such as ACIE (American Councils for International Education), CIEE (Council on International Educational Exchange).

The main objective of these organizations is educational and research programs of academic exchange, programs on international scientific collaboration, programs for research projects support, investments search supports and consulting. But what should be particularly emphasized is that these programs are focused to attract foreign research staff as one of the most important directions of their activities. Here is the list of the most important programs of that kind:

- Hubert H. Humphrey [2];
- Fulbright Program [3];
- Global UGRAD Program (The Global Undergraduate Exchange Program of the Department of State) [4];
- Edmund S. Muskie Graduate Fellowship [5].

2. How sociologist sees the profession of scientist

A vivid picture of the prestige/ non-prestige of the profession of a scientist can be drawn on the basis of the sociological data. For example, according to the US research firm Harris Interactive for 2014, in the top ten most prestigious professions in the US the profession of scientist ranks fourth (76%). Moreover, in the age group 18-37 years, this figure is even slightly higher - 78%. Additionally, in the list of 23 occupations that parents would choose in the future for their child, the profession of scientist holds the third place - 91%, after the engineering profession - 93%, and the medical doctor - 91% [6]. As of 2002, the profession of the scientist took first place in the ranking of the most prestigious professions, 51% of respondents put it in the first place. The medical profession was the second place - 50%, and the profession of engineer - seventh place (34%). There is no doubt that the position occupied by the profession of the scientist in the today's US is much higher than in Russia. For example, according to the Russian Public Opinion Research Center, in 2012 the profession of scientist was ranked at 19 position (2%) among the most prestigious professions [7]. Center for Social Forecasting in Russia in 2006 published a digest "Science in Russia: the sociological analysis". The authors of the evaluation of the students at Russian university concluded that "the academic profession is of public importance right after the medical profession. However, it is only the 8th among the prestigious professions, and profitability-wise shares 7-8th places with the military" [8]. In the same book the indicators of potential of scientific specialists among university students shows that 60% of students do not show interest in scientific work and only 6-8% have a tendency towards scientific work, 30.8% participate in scientific research (university/ faculty) and only 14% plan to apply to graduate school. In addition, the utility as a motivator to participate in the scientific work is evaluated by the students of Russian universities in the following way: helps to expand professional knowledge - 18.8%; develops research skills - 11.1%; provides professional practice and will help to quicker adapt to the workplace - 10.0%; helps to understand whether it makes sense to do science in the course of the working live or not - 7.3% [8].

IOP Conf. Series: Materials Science and Engineering **93** (2015) 012079 doi:10.1088/1757-899X/93/1/012079

Referring to the Russian reality, the problem of aging of scientific specialists in Russia had started to be pronounced since the early 90s, after the liquidation of the Soviet Union. Whole generation of young talented researchers "dropped out" from science. By the beginning of "zeros" Russian science has been recognized as the oldest in the world, 40% of graduate students due to a low income had left science. "The average age of the candidate of sciences has reached 50 years, every sixth scientists is a pensioner, half of the researchers are older then 50. Only 15% of the college students continue into science after graduation. Ratio of scientists younger then 40 decreased to 12%" [9].

According to sociologists, the reasons for the reluctance of students to engage in scientific work in Russia are due to conditions that dominated in the early 90s and until the mid-2000s: low wages; the absence of a material base for research and development; bureaucratization of the organization of scientific work; low prestige of scientific work [8]. And speaking about the interest in scientific work, it is first necessary to understand that in any creative profession, "and science is a creative profession, and forced creativity is impossible. - As emphasized by sociologists - one in ten students [in Russia] is involved in research work in 'voluntarily-forced' manner" [8].

Since the mid-2000s, the restoration of Russian scientific specialists was among the government priorities. Today, young scientists receive financial support through both domestic and foreign programs through grants, foundations, and governmental organizations.

At the state level the target program "Scientific and scientific-pedagogical specialists of innovative Russia" was designed 2014 - 2020 years and is currently operational. The purpose of the program is the "development of an effective system of reproducibility of highly potential specialists of scientific and educational spheres and increase of their competitiveness on a global scale" [10].

Among the non-state funds to support research the best known are: Alferov Foundation, Vladimir Potanin Charity Foundation, Bortnik Foundation, Prokhorov Foundation, Non-Governmental Ecological Foundation of V. I. Vernadsky, National Specialists Training Foundation, Charitable Foundation to promote domestic science etc.

The Russian Bureau of International Educational Programs "Straight talk" regularly presents new services in the field of education, training, exchange, *etc*, which are provided by the most well-known universities, primarily from the countries of G8. In addition, scholarship programs developed by the President and the Ministry of Education for the period 2014-2017 provide grants of € 32 000 for tuition fees (master or doctoral programs in the best universities of the world). "One of the main conditions of the grant will be signing by the involved student an agreement that he will return to Russia at the end of the program and will work in a state-owned company or organization for 3 years" [11].

3. Results of sociological research at TPU

An example of the effective involvement of young talented youth in science is the Tomsk Polytechnic University, who entered the program '5-100' - a project of Ministry of Education aimed to improve the competitiveness of the leading Russian universities among the world's leading research and education centres. Tomsk Polytechnic University has great experience, research base, international training programs, exchanges, internships with the best universities in the US and Europe. In 2014, according to Vladimir Potanin Foundation, TPU received 4th place in the ranking of Russian universities. Among the 14 universities of Russia in the ranking of QS (for 2014) TPU occupies the 10th position and 45th place in the top 50 (QS) rankings of universities in developing countries of Europe and Central Asia, showing the best result among Russian technical universities located outside Moscow and St. Petersburg.

In 2014 TPU carried out a sociological study to examine the readiness of students and the available opportunities for the students to engage in scientific work. The main method of obtaining empirical results was conducting a survey in the form of a questionnaire. The selected for questionnaire were the students of 1st and 2nd year in master program. The number of graduate students at the time of survey was 1 686 people. The sampling quota was managed according to the profile of training in the master program. 80 students were surveyed. The questionnaire included 100 items, grouped in the following blocks: personal information; objectives, results and plans; daily routine, leisure time; scientific work.

The questions were of different types: multiple choice (no more than 3 options), closed questions, the assessment on a scale from 1 to 5. For further analysis, we can identify the following types of indicators:

- Quantitative. For example: "How much of free time do you have per day?";
- Ordinal. For example: "How do you assess, on a scale of 5, the most important results that you expect from your master degree?";
- Nominal. For example: "Do you consider as your possible future work in the next 5 years, the following options?", with the options indicated.

The hypothesis of the study: TPU has highly professional scientific atmosphere, there are the necessary resources to carry out research (laboratory space, office space, equipment, material resources and qualified supervisors, opportunity to participate in scientific conferences and other events held in the TPU, Russia, and abroad), availability/ opportunity to engage in scientific work depends only on the level of personal self-organization of the student.

Analyzing the data obtained during the study, we concluded that the majority of students (75%) at master program are former undergraduate students from TPU, it indicates that the outflow of students to other schools is minimal and the university is actively informing, advising, carrying agitation work to attract students in the master's programs. But at the same time, a fairly low rate of participation are in research projects of TPU, only 25%; the number of students who combine the work in the TPU projects and according to their specialization is only 8%, while almost half of the students combine school and work (48%). Consequently, the last category of the students has the time to participate in research projects. Distribution of answers to questions by category is shown in Figure 1.

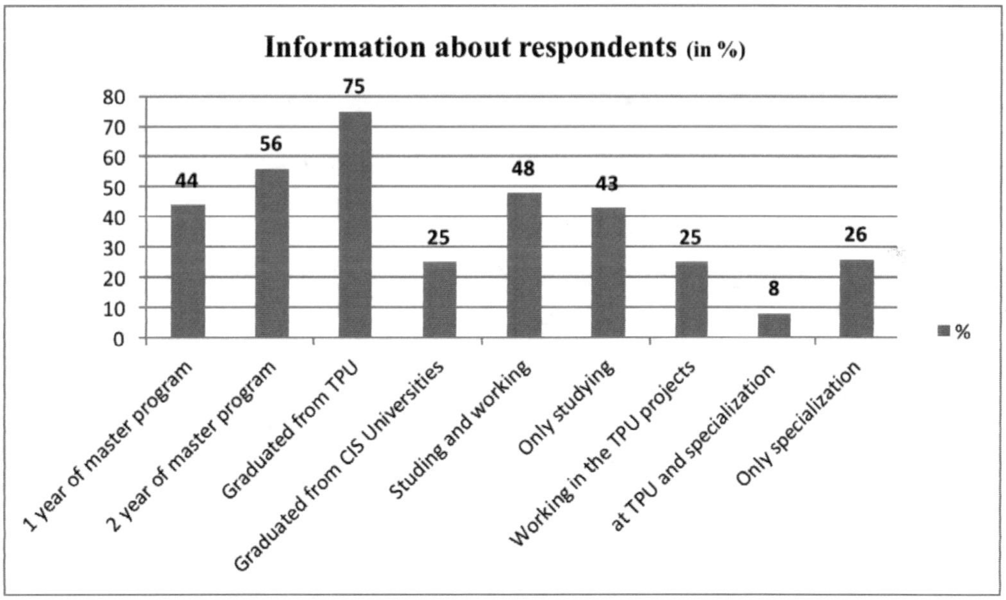

Fig. 1. Diagram of distribution of responses on the block "information about respondents".

It may be noted that the application of students for the master program is a sensible decision. Perhaps this decision was done at school or while studying for bachelor degree, as "acquiring of knowledge and a diploma" among the possible answers (no more than 3) is the choice of 74% of respondents. Relatively fewer responses selected from the category of "on demand" and "no alternative", which presents options haven't found a job - 8%, forced necessity (Army, influenced by parents) - 18% (Figure 2).

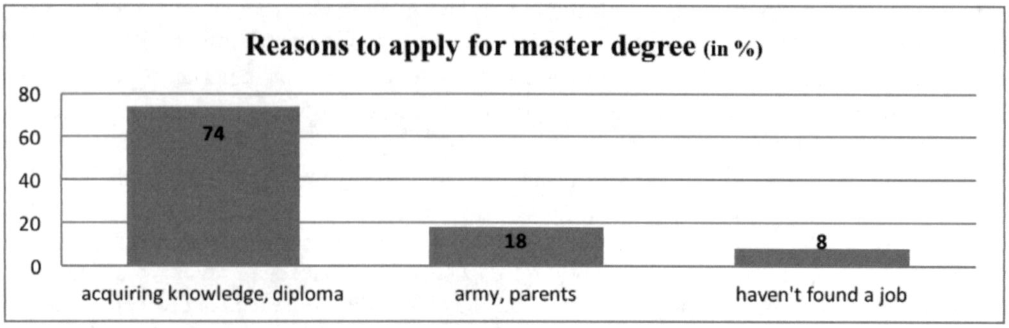

Fig. 2. Diagram of distribution of responses on the block "reasons".

As results of the acquiring master degree students have chosen the answers that can be attributed to the group of non-material incentives. This is primarily a prestige (63%). To get practical skills, a high level of knowledge and interesting work is the aim of more than a half of the respondents, whereas a financial incentive, such as a high-paid job, and a diploma, is present in one third of all respondents. Quite interesting is that (although a small number of 19%) respondents have chosen as the end result of learning "the ability to acquire knowledge" (Figure 3).

This allows us to hope that in the future these students plan to relate their further education to the TPU's graduate school. Perhaps this group is the future scientific potential. And how well suited the university teachers to motivate this group depends on how many of them will be not just staying in Russia, but in science in general. At least for the moment, plan to continue the education in graduate school in Russia is the choice of 39% of respondents, stay and work in Russia plans 76%. The number of students who are planning to work abroad is only 35%, going to graduate school abroad is the choice of an even smaller number of respondents, only 20% (Figure 4). We can assume that this is due to "language barrier". We believe that in the near future this figure will change for the better, because TPU deployed multiple language training programs for different levels of language proficiency.

A special place in the study was given to the study of the student's self-organization and the allocation of their time between studying, scientific work and the leisure time during a week. About a third of the students spend on their studies and research work only 8-10 hours a week, while on entertainment, sports and hobbies they spend 16-28 hours a week. To make time for studies of 10 and more hours is the answer of 40% of students, 5-10 hours and more for the scientific work is the answer of 1% to 10% of the master students. To monitor personal free time and to estimate that it is of enough quality is the skill and answer of less than half (42%), while 58% do not know how to dispose their free time and are feeling constant shortage of free time. Moreover, 42% are spending 6-7 hours on sleep, the remaining 58% are spending 8-12 hours of sleep. 60% of respondents try to stick to the agenda, for 40% of the others there is no agenda.

Interest in scientific work is expressed by more than half of master students (78%) and 89% believe that in the TPU there is highly professional scientific atmosphere. According to respondents, TPU has the necessary resources for the organization of scientific work (from 74% to 89% of the students choose this answer). Here the resources considered are: financial and material. 66% of the students participate in scientific events at TPU, 80% in Russia, and 35% abroad.

Among the key problems hindering scientific work are the following: laziness prevents - 75%, the lack of free time is experienced by 78%, family obligations prevent 60% to engage in scientific activities, and the need for jobs is the reason of 54% of the students. We can say that in general, students do not know how to build a schedule, to control the distribution of their free time, to overcome laziness. But you should pay attention to the fact that the involvement of students in research work that pays well, will reduce the number of those who are forced to look for jobs on the side to provide themselves and their families with the necessary level of prosperity.

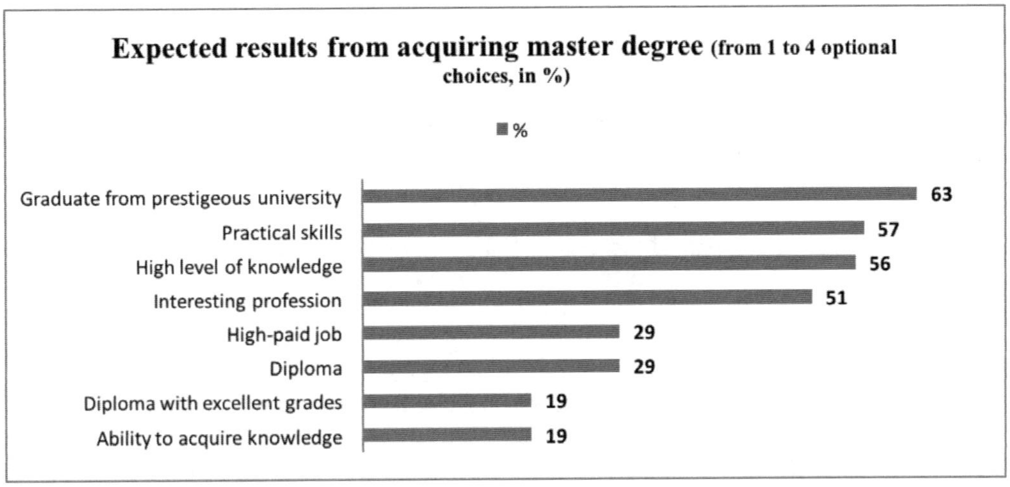

Fig. 3. Diagram of distribution of responses on the block "Expected results".

According to the analysis the satisfaction with scientific supervisor is among the essential incentives of students to involve in research. Most students agree with these statements: supervisor has good leadership skills - 96%; supervisor is interested in my participation in scientific activity - 94%; supervisor shows me respect - 92%; supervisor gives me time to such an extent that I need - 94%; I'm interested in the proposed by supervisor themes - 84%. It should be emphasized that the majority of students is ready to work in a team (71%).

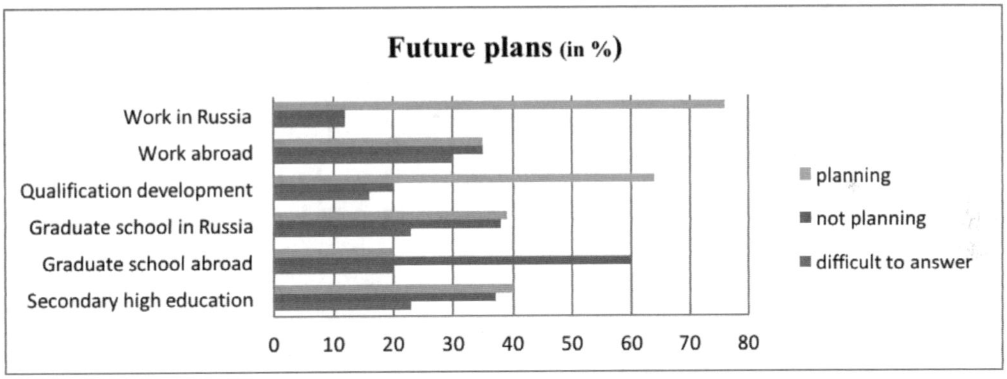

Fig. 4. Diagram of distribution of responses on the block "Future plans".

4. Conclusions

The conducted study shows that the master students of Tomsk Polytechnic University are willing to do research. In general, the study confirmed the hypothesis. Non-material incentives such as personal interest, prestige, knowledge prevail over material incentives. The biggest problem hindering the organization of scientific work of the students is the lack of skills and the ability to self-organization: the daily schedule, the self-control over personal free time, laziness. The study has revealed a problem that requires deeper analysis and attention. This is a lack of awareness of students about opportunities to participate in research projects, grants, conferences, internships, and other events.

Available at TPU base for research activities, including financial, material resources, highly qualified scientific executives allows to make optimistic forecasts. First of all, about the fact that 'voluntarily-forced' form of organization of scientific work in the TPU is not the case. The conditions created by the leadership of the university, are the backbone for the development of scientific career of students. Secondly, the highest incentive motivation is the satisfaction of students with their scientific supervisor.

IOP Conf. Series: Materials Science and Engineering **93** (2015) 012079 doi:10.1088/1757-899X/93/1/012079

In our opinion, the scientific career of famous scientists always shows an example how a good scientific supervisor pushed young scientists towards their future careers. In his memoirs Carl Sagan writes on his experience at the University of Chicago: "The status of the teachers in the Hutchins curriculum had almost nothing to do with their research; perversely – unlike the American university standard of today – teachers were valued for their teaching, their ability to inform and inspire the next generation" [1].

Acknowledgement
This research was completed as part of the project "The Youth's Portrait" of the Future: Methodology of Investigating Representations" funded by the Russian Humanitarian Scientific Fund, Grant Number 15-03-00812a.

References
[1] Sagan C 1997 The demon-haunted world: Science as a candle in the dark (New York: Ballantine Books)
[2] Hubert H. Humphrey Program, URL: https://www.humphreyfellowship.org/ (accessed 7/12/2014)
[3] Fulbright Program, URL: http://eca.state.gov/about-bureau (accessed 7/12/2014)
[4] Global UGRAD, URL: http://www.irex.ru/programs/ugrad/about (accessed 7/12/2014)
[5] Edmund S. Muskie Graduate Fellowship Program, URL: http://www.irex.org/project/edmund-s-muskie-graduate-fellowship-program (accessed 7/12/2014)
[6] The Harris Poll of 2,537 adults surveyed online between August 13 and 18, 2014. URL: http://www.harrisinteractive.com/vault/Harris%20Poll%2085%20-%20Prestigious%20Occupations_9.10.2014.pdf (accessed 06/12/2014)
[7] Russian Public Opinion Research Center, Survey results, URL: http://wciom.ru/zh/print_q.php?s_id=868&q_id=60673&date=16.09.2012 (accessed 6.12.2014)
[8] Sheregi F and Strihanov M 2006 Science in Russia: The sociological analysis (Moscow: CSP)
[9] Why Newton hasn't eaten the apple, URL: http://izvestia.ru/news/254543 (accessed 5.12.2014)
[10] Resolution of the Government of the Russian Federation "On the federal target program" Scientific and Scientific-Pedagogical specialists of innovative Russia" for 2014-2020", URL: http://xn--80abucjiibhv9a.xn--p1ai/%D0%B4%D0%BE%D0%BA%D1%83%D0%BC%D0%B5%D0%BD%D1%82%D1%8B/3332 (accessed 07.12.2014)
[11] Russia in the field of international educational programs, new scholarships for Russian students to study abroad, and new norms for foreign students studying in Russian universities, URL: http://www.directtalk.ru/article/rossiya-v-sfere-mezhdunarodnyh-obrazovatelnyh-programm (accessed 07.12.2014)

Project management practices in engineering university

Y Sirazitdinova[1,a], A Dulzon[1,b], B Mueller[2,c]
[1]*National Research Tomsk Polytechnic University, Russia*
[2]*University of Applied Sciences Aachen, Germany*

[a]sirazitdinova@tpu.ru, [b]vizepres@tpu.ru, [c]burghard.mueller@fh-aachen.de

Abstract. The article presents the analysis of usage of project management methodology in Tomsk Polytechnic University, in particular the experience with the course Project management which started 15 years ago. The article presents the discussion around advantages of project management methodology for engineering education and administration of the university in general and the problems impeding extensive implementation of this methodology in teaching, research and management in the university.

1. Introduction

In the last century a project was defined mainly as "a developed plan of buildings, construction, manufacturing or reconstruction of something" [1]. Today, our country and the world are literally bitten with "project mania", and the concept of a project includes almost any type of activity. In the process of implementation of major military projects and aggregation of lessons learned there was developed and introduced in Tomsk Polytechnic University (TPU) a special discipline "Project Management". The analysis of this experience has shown that the application of project management methodology may result in savings up to 20% of the funds on average, and reducing the time of projects implementation by 20% on average. There is another weighty argument in favor of the use of modern project management methodology (PM). To justify the approach, one can use the classical Marxism idea of K. Marx which states that the worst architect differs from the best bee from the outset because before building something, he create an ideal image of a future construction in mind. The processes of initiation, preparation and planning a project together with relevant stakeholders and the execution of a risks analysis enable dramatically reduce of errors, as well as the likelihood and consequences of unexpected events. This, in turn, reduces the overall stress level in the process of project implementation. When people know that during the work execution an accident, explosion, etc. might happen, they are mentally prepared for the occurrence of such events, do not lose capacity, and can immediately initiate the necessary actions.

2. Project management evolution

Following the publication of the first standard of project management (PMBOK) in 1987 a rapid spread of project management activities began across countries in different forms of appearances, including unconventional ones. If at the initial stage of PM methodology development it was regarded only in relation to large and complex defense projects, then later it was realized that many of its elements can be effective for the implementation of medium and small-scale projects in the private sector in any form of activity. Project management has become widely used in the social and economic projects, assistance programs, including personal projects (organizing holidays, anniversaries, weddings, etc.). And such projects are performed and accounted for many millions yearly. Many of them are successful, although the application does not need special knowledge in project management techniques.

There is no doubt that the methodology of Project Management makes available a range of useful tools for project performers. It may improve significantly efficiency and effectiveness of any kind of venture.

However, even in major universities, large companies, and in the government, where every year hundreds of projects are carried out, modern PM methodology is hardly used. When it comes to the research there is strong evidence supporting the success of project management methodology use in construction, IT and healthcare industries while there is definitely a lack of research that would address specifically the state of PM implementation in higher educational institutions [2]. A legitimate question arises: having such benefits, why is the project management methodology not applied in many more cases?

There are several reasons for this situation. Firstly, many experts and organizations are not aware of the existence of the Project Management methodology, although the information about this is gradually spreading across Russia, which in recent years contributes to vigorous activity of the Russian Project Management Association "Sovnet". Secondly, in most education programs such discipline is absent. In TPU this discipline is included only in two educational programs of the Institute of social and humanitarian technologies: “Management” and “Innovation studies”. There is an attempt to introduce all the students of TPU to project management methodology through the course of “Creative projects”, which is an obligatory course for the first and second year students. It lasts three semesters, and the students are supposed to work on creative projects at any department of their Institute, so that by the end of the second year they can decide on the bachelor program specialization for the third and fourth years of studies. The particularity of this course is that it is not supervised by a teacher (the teacher gives only two hours of introduction into the course), and the students run projects by themselves without methodological support. Third, many managers prefer to carry out projects in the old way, because they do not know or do not believe in the benefits of modern PM technology. In support of the latter there is some evidence, because as notices P. Morris [3] the cause-and-effect relationship between the use of formal project management and project results is not experimentally proven. Finally, fourthly, in many cases leaders of different ranks resist to the introduction of the methodology exactly because of its advantages: PM methodology ensures the transparency of the project, exposing the absurdity of some ideas, alternatives to the proposed solutions, projects inefficiency, disclosure of hidden implicit goals and corruption schemes.

3. Project management training

Problem-oriented and project-organized education, which is considered to be an effective means of improving the quality of specialists training, can be effectively used to improve the quality of training. The importance of this approach for the training of engineers is difficult to overestimate. The table 1 presents advantages and problems of project-organized education, the latter helps to understand better the reasons for low level of project management methodology implementation in TPU.

Table 1. Advantages and problems of project-organized education.

Advantages	Problems
Students have a vast interest in the subject	It is difficult to assess the complexity of the overall project for a team of students, since it depends not only on the content of the project, but also on the degree of preparedness of students
Verification of residual knowledge in several months confirms the strength of the information mastering	If in teams among students there are dishonest ones, the workload can grow significantly for the rest of the students; as a result there might be the situation when the project is not carried out which places the teacher and conscientious students in a "stalemate" situation
Students acquire skills of independent work planning	Since a teacher cannot assess the contribution of each student in the project, and the distribution of points for project work among team members is assigned to the team itself, with a high degree of conformity that does not eliminate conflicts and does not always provide fair grades
Students learn the basics of system analysis	Project design of teaching is more difficult for the teacher than classical design with lectures and more time-consuming than seminars and lectures
Students develop real	If working hard, the team will soon gain a lot of information that may go

teamwork skills	beyond the knowledge of the teacher. Therefore, substantial part of the projects should be sufficiently well known for the teacher. Besides the teacher requires sufficient psychological stability. The questions during the consultations may bring the teacher to a nonplus
Students gain the skills of interaction with external structures in solving problems	Project design of teaching is not provided by local University regulatory documents, which creates difficulties in determining the workload of the teacher Teacher workload in project design of classes compared to conventional lecture course increases by a multiply due to: • preparation of project subjects; • initial data preparation; • a big number of consultations (both for the course and for the project); • constant support of teams.

There is quite a large number of PM Training courses offered by well known and unknown providers on the market. The acquaintance with the proposals for project management training (primarily commercial) by different organizations, as well as participation in a number of short courses and workshops allow marking two most common drawbacks.

The first one relates to the fact that the main and often exaggerated attention is devoted to PM processes and technology, international and national standards, specialists' certification, software products. Without denying the relevance and usefulness of learning process, it is necessary to pay attention to the fact that they do not provide the mastery of PM discipline. Sometimes it comes even to identifying PM with software products, in particular, with Microsoft Project.

The second drawback in PM training is related to the insufficient attention to the deeper foundations of effective PM: methods of analysis and problem solving, correct formulation of goals, development of on time and resource base action plan, budget breakdown, technical requirements and objectives, choice of an alternative to the proposed problem solution, rational improvement of decisions.

4. TPU experience in teaching project management

In Tomsk Polytechnic University the discipline "Project Management" has been included in the curricula of management educational program in 1999. The course of PM for students, including MBA program students, students of presidential management training program for the national economy as well as in various advanced training programs, assumed ideally the implementation of real projects, usually in teams of 6-10 people. This allowed taking on projects of 1000 man-hours complexity and more. Accordingly there were prepared textbook and teaching aids for project management course [4-5]. The experience in project management training in TPU is discussed in detail in the article [6]. A team work design in teaching the discipline "Project Management" has been investigated and supported as a technique enabling students to go beyond their zone of comfort and deal with the challenges they will face with in professional life [7]. The team work is one of the best instruments to develop craft knowledge or "soft skills" that are considered to be of the same importance as codified knowledge or "hard skills" in project management, and these two should be balanced [8].

A complex and often conflicting task for the teacher is to determine the individual grade for all members of the team, especially since the vast majority of the work is done outside of the classroom. The teacher evaluates the project as a whole. The team is provided with a detailed review stating the number of points (total and by project sections) and explaining the reasons for points withdrawal. In order to ensure as far as possible a fair and proper individual assessment the definition of labor participation rate (LPR) of each team member is calculated. Although it is time consuming, the procedure seems to be appropriate for two reasons: firstly, the definition of the LPR is one of the common tasks of any manager; secondly, the experience has shown that students are often unable to allocate points without conflict. To determine the LPR the team at the meeting sets out the basic assessment criteria (preferably no more than three to five), for example, the number of working hours spent for the project, the difficulty of work performed, interaction with colleagues, creativity, etc. To calculate the LPR one can use the expert method of paired comparisons, in which each team member fills in a square matrix where in rows and columns all team members' names are written. Then the relative eigenvector of summary matrix is computed to reflect the contribution LPR_i of each team member. After that the individual grade k_i of each team member can be calculated:

$$k_i = \frac{5\, LPR_i * S * N}{B},$$

B – maximal number of points for the whole project,
S – the number of points received after evaluation,
N – the number of team members.

5. Discussion

The lack of culture of project management approach to projects in the training of engineers is a *considerable deficit*. However, the inclusion of PM course in educational programs will not solve the problem. It is required to train all faculty staff of the university in this area so that the project management methodology is being applied routinely in research activities of departments, including research work of students. But this is not enough. It is necessary to create a culture of project management at the university as a whole, including the administrative and economic departments. Without any doubt, the shift towards formal PM methodology usage in university research should be balanced with regard to values, norms and principles of open-ended research, ensuring that "backstage" research and "front-stage" administration realize and share their roles and responsibilities when dealing with a research project [9].

Large universities annually organize hundreds of conferences, seminars and meetings of different level and scale. One can add here the execution of contractual and state budgetary research works, carrying out a variety of sports activities, development of regulatory documents, not to mention the construction and repairing works. In universities it is relatively easy to evaluate the projects results concerned with the repairing and reconstruction of facilities. In conjunction with project management methodology the cost savings can reach 10-20% of the amount of work, which may reach millions of rubles for a large university. The result may be expressed not only in the money-saving expenditure of funds, but also in reduction of possibility of corruption development. It is more difficult to assess the results of complex multi-dimensional projects of the university, such as the language training program, establishment of the Institute of International Education, establishment of the Center for academic mobility and others. There can be a number of difficultly appreciable side effects. For example, the creation of the Institute of International Education is not only related to earning money from teaching international students, but also to teacher training programs, modernization of curricula, creation of modern teaching aids, etc.

TPU performs each year hundreds of different projects, but the project management methodology is not used. There is a number of reasons explaining this situation that were mentioned above. But one aspect should be noted. There has almost never been carried out the balance of the project costs (money, material and human resources, time, motivation) and results – positive (increase in the labor efficiency, quality of specialists, results of research and development, profits, etc.) and negative ones (loss of time, inefficient material resources usage, lack of personnel motivation, etc.). At the same time, one can give many examples of costly projects that have significant adverse impacts and questionable benefits. The widespread use of project management methodology would not only substantially enhance the effectiveness of specific projects implementation but also create a modern culture of project activity at the university.

References

[1] Russian language dictionary in four volumes, Vol.**3** // M., 1983.
[2] Austin Ch. et al. Application of Project Management in Higher Education // *Journal of Economic Development, Management, IT, Finance and Marketing*. – 2013. – Vol. **5**(2). – P. 75-99.
[3] Morris P. W. G. Science, objective knowledge, and the theory of project management // *Civil Eng.: proc. Inst. of Civil Engineers*. – 2002. – Vol. **150** (2). – P. 82–89.

[4]. Dulzon, Alfred Andreevich. Upravlenie proektami: uchebnik / A. A. Dulzon; *Nacional'nyj issledovatel'skij Tomskij politehnicheskij universitet* (TPU). — Tomsk: Izd-vo TPU, 2013. — 334 p.
[5]. Dulzon A. A. Upravlenie proektami: ucheb.-metod. posobie po kursovomu proektu / A. A. Dulzon; Tom. politehn. un-t. – Tomsk: Izd-vo Tom. politehn. un-ta, 2013. – 112 p.
[6]. Dulzon A. A. Opyt obuchenija upravleniju proektami // *Vysshee obrazovanie v Rossii,* 2013. — № 10. - P. 83-90.
[7] Córdoba J-R., Piki A. Facilitating project management education through groups as systems // *International Journal of Project Management*. – 2012. – Vol. **30**. – P. 83–93.
[8] Louw T. Project Management Training Curricula at South African Public Universities: Is the Balanced Demand of the Profession Sufficiently Accommodated? // *Project Management Journal*. – 2012. – P. 70-80.
[9] Fowler N., Lindahl M., Sköld D. The projectification of university research // *International Journal of Managing Projects in Business*. – 2015. – Vol. **8** (1). – P. 9-32.